Primate Conservation

Contributors

S. K. BEARDER

JOHN CASSIDY

DAVID J. CHIVERS

ADELMAR F. COIMBRA-FILHO

JOHN M. DEAG

WOLFGANG P. J. DITTUS

G. A. DOYLE

FRANK V. DUMOND

R. I. M. DUNBAR

ROY FONTAINE

ALAN G. GOODALL

STEVEN GREEN

COLIN P. GROVES

S. S. KALTER

LOIS K. LIPPOLD

B. ANTHONY LUSCOMBE

HERNANDO DE MACEDO-RUIZ

KAREN MINKOWSKI

RUSSELL A. MITTERMEIER

J. F. OATES

JOHN R. OPPENHEIMER

JEAN-JACQUES PETTER

M. FAROOQ SIDDIQI

CHARLES H. SOUTHWICK

Primate Conservation

Edited by

His Serene Highness Prince Rainier III of Monaco
Centre d'Acclimatation Zoologique
Monaco

Geoffrey H. Bourne
Yerkes Regional Primate Research Center
Emory University
Atlanta, Georgia

ACADEMIC PRESS   New York   San Francisco   London   1977

A Subsidiary of Harcourt Brace Jovanovich, Publishers

ACADEMIC PRESS, INC.
111 Fifth Avenue, New York, New York 10003

United Kingdom Edition published by
ACADEMIC PRESS, INC. (LONDON) LTD.
24/28 Oval Road, London NW1

Library of Congress Cataloging in Publication Data

Main Entry under title:

Primate conservation.

    Includes bibliographies and index.
    CONTENTS: Petter, J.J. The aye aye.–
Doyle, G. A. and Bearder, S. K. The galagines of
South Africa.–Mittermeier, R. A. and Coimbra-
Filho, A. F.   Conservation of the Brazilian lion
tamarins. [etc.]
    1.   Primates.   2.   Wildlife conservation.
I. H.S.H. Prince Rainier III of Monaco Date
II.   Bourne, Geoffrey Howard,   Date
QL737.P9P6723   599'.8   76-42978
ISBN 0–12–576150–3

Contents

Contents

8 The Status of the Barbary Macaque *Macaca sylvanus* in Captivity and Factors Influencing Its Distribution in the Wild

JOHN M. DEAG

9 The Lion-Tailed Monkey and Its South Indian Rain Forest Habitat

STEVEN GREEN AND KAREN MINKOWSKI

Contents   ix

14 *Presbytis entellus*, The Hanuman Langur

JOHN R. OPPENHEIMER

15 The Douc Langur: A Time for Conservation

LOIS K. LIPPOLD

Contents

List of Contributors

Numbers in parentheses indicate the pages on which the authors' contributions begin.

S. K. BEARDER (1), Primate Behavior Research Group, University of the Witwatersrand, Johannesburg, South Africa

JOHN CASSIDY (95), Lagothrix Program, Bogotá, Colombia

DAVID J. CHIVERS (539), Sub-Department of Veterinary Anatomy, University of Cambridge, Cambridge, England

ADELMAR F. COIMBRA-FILHO (59, 117), Department of Environmental Conservation, FEEMA, Rio de Janeiro, Brazil

JOHN M. DEAG (267), Department of Zoology, University of Edinburgh, Edinburgh, Scotland, and Department of Psychology, University of Bristol, Bristol, England

WOLFGANG P. J. DITTUS (237), Office of Zoological Research, National Zoological Park, Smithsonian Institution, Washington, D. C.

G. A. DOYLE (1), Primate Behavior Research Group, University of the Witwatersrand, Johannesburg, South Africa

FRANK V. DUMOND (167), Monkey Jungle, Inc., Goulds, Florida

R. I. M. DUNBAR (363), Department of Psychology, University of Bristol, Bristol, England

ROY FONTAINE* (167), Department of Psychology, Bucknell University, Lewisburg, Pennsylvania

ALAN G. GOODALL (599), Biology Department, Paisley College of Technology, Paisley, Strathclyde, Scotland

STEVEN GREEN (289), Field Research Center for Ecology and Ethology, The Rockefeller University, Millbrook, New York

COLIN P. GROVES (599), Department of Prehistory and Anthropology, School of General Studies, Australian National University, Canberra, Australia

* Present address: Field Research Center for Ecology and Ethology, The Rockefeller University, Millbrook, New York.

S. S. KALTER (385), Southwest Foundation for Research and Education, San Antonio, Texas

LOIS K. LIPPOLD (513), Department of Anthropology, San Diego State University, San Diego, California

B. ANTHONY LUSCOMBE (95), Miraflores, Lima, Peru

HERNANDO DE MACEDO-RUIZ (95), Museo de Historia Natural "Javier Prado," Universidad Nacional Mayor de San Marcos, Lima, Peru

KAREN MINKOWSKI (289), Field Research Center, The Rockefeller University, Millbrook, New York

RUSSELL A. MITTERMEIER (59, 95, 117), Department of Anthropology and Museum of Comparative Zoology, Harvard University, Cambridge, Massachusetts

J. F. OATES* (419), New York Zoological Society, New York, New York

JOHN R. OPPENHEIMER (469), International Center for Medical Research and Department of Pathobiology, The Johns Hopkins University, Baltimore, Maryland

JEAN-JACQUES PETTER (37), Muséum National d'Histoire Naturelle, Equipe de Recherche sur les Prosimiens, Laboratoire d'Ecologie Générale, Brunoy, France

M. FAROOQ SIDDIQI (339), Department of Geography, Aligarh Muslim University, India

CHARLES H. SOUTHWICK (339), Department of Pathobiology, The Johns Hopkins University, Baltimore, Maryland

* Present address: Smithsonian Tropical Research Institute, Balboa, Canal Zone.

Preface

From 1950 to 1970, "medical research" consumed large numbers of nonhuman primates. The term "medical research" included, however, monkeys that were used for vaccine development and testing; at the time polio vaccine was being developed very large numbers of monkeys were used. The pet trade has also exerted a sizable drain on the monkey population over the years; the requirements of zoos have been modest by comparison. The number actually used in research, pet trade, etc., is misleading in the sense that it does not include the losses that occur during the capturing, holding, and transport periods. These losses may be as high as 85%; for example, if 35,000 live monkeys arrive in the United States from Iquitos in Peru, it means that about 180,000 monkeys were originally taken from their native environment to permit the 35,000 to arrive live in the United States. This loss has been so great over the years that in many areas in South America it is difficult to find monkeys near rivers and highways, and each year hunters and trappers have to go deeper into the forest to find an adequate supply for their purposes.

Another significant loss of nonhuman primates in the wild is due to their consumption as food. An official of the flora and fauna section of the Agricultural Ministry of Peru has estimated that 7.5 million monkeys were eaten from 1964 to 1974 in Peru compared with 1.5 million killed or captured for export. We know that monkeys and apes are also used for food by the indigenous population in many parts of Africa. The forest people of the African Coastal rain forest prefer monkeys as food to any other game. The increase in population in these areas with their growing demands for protein makes the future of primates in many areas uncertain, apart from their use in the Western world for medical research and for other purposes.

How serious is the situation among nonhuman primates? It varies for different species. Among the apes the condition of wild orang stocks rang the alarm through the efforts of Dr. Barbara Harrisson some years ago. In the early 1960's she estimated that only 3000–5000 orangs were left alive in the wild. This was undoubtedly an understatement, but it led to the establishment of many conservation measures which greatly reduced the drain on the wild population of this species. H. D. Rijksen, working in Sumatra, recently estimated that there are 15,000 orangs left in Sumatra, and since there are probably about 5000 left in Borneo an estimate of 20,000 for the world population of orangs is a reasonable one. However, Dr. Rijksen pointed out that the habitat for half the Sumatran orangs was in the process of being eliminated. The orang reservation at Sabah in North Borneo will remain intact so long as the Government continues to resist the pressures of lumber companies that would like that choice piece of tropical forest located only 15 miles from a timber port. The tropical rain forests throughout the world are, in fact, being destroyed at a rate without precedent in human history. According to Thomas Lovejoy (1976), Program Director of the World Wildlife Fund (United States), two-thirds of southeast Asian rain forests and half the African rain forests have already been eliminated, and even in virgin forests of the Amazon, of which one-third was covered almost completely by forest, only 17% of the area is now forested.

The eastern gorilla is suffering from the intense pressures for land which accompany population growth; the western gorilla and the chimpanzee are subjected to similar pressures as are many species of monkeys. Most of us have thought of the rhesus monkey as being abundant, and they have certainly been used in a prodigal fashion as if they were. A million rhesus monkeys were trapped and removed from India in the last twenty years. In the late 1950's, 100,000 rhesus monkeys were exported a year; by the mid-1960's this had dropped to 50,000 per year, and during the 1970's the number was reduced to between 30,000 and 40,000 a year. Now the number available has been drastically reduced by the restrictions imposed by the Indian Government. In this volume Southwick and Siddiqi indicate that unless there is greater protection of rhesus monkeys by local people "they will be eliminated from most of the agricultural areas of India within the next 25 years." At least the rhesus monkey is protected from being eaten by the Indian population because of the Hindu religion; had they not been protected in this way there would almost certainly not have been any left today.

Most conservationists advocate the establishment of primate reservations. This suggestion is well meaning, but is not entirely practicable since it is impossible to prevent the indigenous populations from hunting and killing the primates on the reservation for their own purposes. The establishment of a reservation should not be the only method of conservation. No country can afford the money and manpower for the policing necessary to protect these areas. There has to be a multipronged approach to conservation. Endangered animals can be brought into areas in which they can be more easily supervised, and this need not necessarily be in their country of origin. A good example is the proliferation of vervet monkeys in the Virgin Islands and the successful transplantation of rhesus monkeys to Cayo Santiago. In some cases it may be easier to preserve a species in a part of the world not normally inhabited by them. It may even be necessary to preserve some endangered species in zoos or laboratories where they can be given top veterinary care. A good example of this is the Yerkes Center's colony of orangutans. Thirty-five orangs already living in the United States were bought in the early 1960's, most of them around 1962. They were all young animals at that time. They have now grown to maturity, and with skilled veterinary care and excellent nutrition have produced thirty-five live offspring. This colony provides a unique opportunity for the scientific study of the orang which is unequaled in the world, and which will never again be possible to duplicate. The Yerkes Center has bred 299 chimpanzees, orangutans, and gorillas since 1930, an average of nearly 7 apes a year for 46 years.

There is no doubt that the ideal method of conservation is to establish a wild reservation in the country of origin of the animals. The reality is, however, that these reservations are unlikely to continue indefinitely in the face of increasing world population and hunger, especially in the developing countries in which the world's nonhuman primates are located. As an alternative, controlled breeding groups of animals in natural surroundings, such as the Government of Zaire is planning to do with its pygmy chimpanzees, has much in its favor.

Since there is a good chance that despite all our attempts the preservation of endangered species of primates by the establishment of reservations in the wild will be unsuccessful, a few high quality zoos throughout the world could be selected to receive and breed endangered species. The Yerkes Center could participate in this program and could at the same time make a scientific study of these animals. The demand on wild populations of nonhuman primates for pets has now been eliminated in the United States, and hopefully other countries will follow suit. For vaccine production and medical research the answer is to set up enough breeding stations by the user countries to meet their requirements and eventually to eliminate completely their demands on the wild animals.

The International Union for the Conservation of Nature is attempting to establish a set of guidelines on the use of nonhuman primates in medical research. The list drawn up by the Specialist Group, I.U.C.N., and the International Primatological Society is an excellent document, and should be supported both by the conservationist and the medical scientist.

The problems of conservation of many species of nonhuman primates are discussed in this book by distinguished scientists who are experts in their knowledge of the animals they write about and who have firsthand knowledge of the problems of conserving them. We cannot deal in one book with all the endangered nonhuman primates, but we have selected animals ranging from *Galago* to the *Gorilla* to serve as examples of the types of problems that the conservationist faces.

All authors of this book agreed that the royalties earned should be used to further primate conservation.

His Serene Highness
Prince Rainier III of Monaco

Geoffrey H. Bourne

1

The Galagines of South Africa

G. A. DOYLE and S. K. BEARDER

I. Introduction

Of 33 known extant prosimian primates, eight are indigenous to the African continent and two, *Galago crassicaudatus* and *Galago senegalensis*, the thick-tailed bushbaby and the lesser bushbaby, respectively, are indigenous to South Africa. As far as is known neither of these two species is threatened with extinction but then, not many years ago, the same could have been said of all those species of primate whose existence is today so threatened that sadly some will be extinct in our lifetime.

Man is the chief threat to the existence of his fellow primates and Africa, with its ever-growing and largely hungry population, is witnessing a continual invasion of those areas in which many primate species live in harmonious adaptation with varieties of other species. The continued survival of these species of primate will depend both on the foresight and hard work of the conservationists and on the scientists whose studies of these primates both in their natural habitats and in the laboratory are adding to our understanding of their biology and without which conservation programs cannot succeed.

This chapter is concerned with the ecology of two species of galagine in South Africa in the hopes that the information provided will assist conservationists in steps that they will quite likely have to take in the foreseeable future to protect them from possible extinction.

II. Distribution

Galago senegalensis is divided into nine subspecies and *G. crassicaudatus* into eleven subspecies (Hill, 1953; Hill and Meester, 1969) on the basis of both anatomical and morphological variations in relation to geographical range. Not all authors agree with this subspecific classification (Buettner-Janusch, 1963, 1964) and recent karyotype studies on *G. crassicaudatus*, for instance, have further clouded the issue (De Boer, 1973). Whatever the outcome, the division of a species into a subspecies is meaningful if it defines breeding populations whose members have characteristics in common and which are separated from one another by effective physical barriers. In bushbabies the tendency toward isolation of breeding populations is very great because of the nature of the habitat to which they have adapted. Patches of forest may be cut off from one another by tracts of unsuitable bush or open grassland which the animals are unlikely to cross. While individuals of one population may show a considerable range of characteristics which they share with those of another population they will nevertheless tend to conform to one particular type and this may partially explain the large variety of subspecific characters which have been described.

A. *Galago crassicaudatus*

The thick-tailed bushbaby is one of the most widely distributed of the African prosimians. Its northern-most limits are in the east just north of the equator in southern Somalia, in Kenya, and in Uganda in the region of Lake Victoria. It extends southward in the east through Tanzania, Mozambique, the eastern Transvaal, Swaziland, and as far south as Kwazulu (Hill, 1953) and the southern part of Natal (Pringle, 1974). Further west it is reported to be widespread in Malawi, central and eastern parts of Rhodesia, and western parts of Zambia but it is sparse in western Rhodesia (Smithers, 1968). Hill (1953) suggests that it may also be found in the eastern and southern parts of Zaire extending westward to the coast of Angola south of Luanda. It is absent from Botswana and from most, or all, of Southwest Africa, while in Central and West Africa it is replaced by a number of other species of lorisid (Hill, 1953).

The approximate known southerly distributions of *Galago crassicaudatus* and *G. senegalensis* are shown in Fig. 1. Within the overall limits of distribution there are many discontinuities between suitable habitats in which the species are found.

The two South African forms are found at altitudes from sea level to 1800 m. *Galago garnettii* is found in the Natal coastal regions extending westward into Kwazulu, and *Galago c. umbrosus* is found in the northeast Transvaal (Roberts, 1951; Hill, 1953; Hill and Meester, 1969). Reference will also be made to the Rhodesian form, *Galago c. lönnbergi*, found in eastern Rhodesia (Hill, 1953; Hill and Meester, 1969).

B. *Galago senegalensis*

The lesser bushbaby is even more widely distributed than *G. crassicaudatus*. South of the Sahara it is found throughout most of Africa from Senegal in the

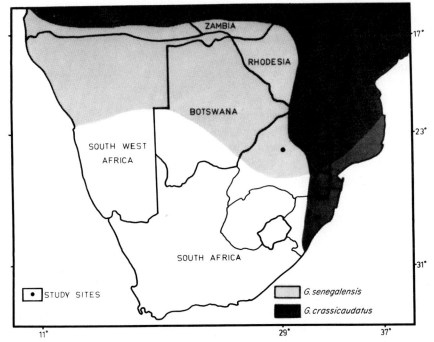

Fig. 1. Map of southern Africa showing the approximate distribution of *Galago crassicaudatus* and *Galago senegalensis* (modified after Hill, 1953, and Bearder, 1975).

west extending directly east as far as Somalia and south through East Africa as far as southern Mozambique and west through the Transvaal, Rhodesia, Botswana to the northern parts of Southwest Africa. Within these northern and southern boundaries it is excluded only from central and Central West Africa (Hill, 1953; Hill and Meester, 1969).

As far as is known only one form is found in South Africa, *Galago s. moholi*, which thrives in the bushveld area of northern and central Transvaal west of the escarpment from about 1000 to 2000 m (Shortridge, 1934; Roberts, 1951; Hill, 1953; Hill and Meester, 1969).

III. Description

A. *Galago crassicaudatus*

1. *Galago c. umbrosus*

This subspecies is distinguished from others by its smoky-grayish-brown color on the upper parts and tail with the underparts smoky-grayish. The brown coloration is particularly marked on the limbs and dorsal surfaces of hands and feet (Hill, 1953; Bearder and Doyle, 1974). Approximately 8% of the population had a noticeably darker overall hue and dark tips to the tail. Bearder (1975) reports weights in two adult males as 1250 and 1820 gm, and in three adult females as between 910 and 1320 gm.

2. *Galago c. garnettii*

This subspecies is similar, if not identical, in size to *G. c. umbrosus*. It is overall brown to buff-brown on the upper parts and tail with the underparts much lighter tending to whitish. The digits tend to be blackish and approximately 80% of the study population had dark tips to the tails (Hill, 1953; Bearder and Doyle, 1974).

B. *Galago senegalensis moholi*

This subspecies is a large-eared race predominantly gray in color on the upper surfaces with the back washed in faint otter-brown and the hindlimbs yellowish. The underparts are whitish washed with creamy buff. The face has a whitish nose stripe and distinctly black circumocular rings. Unlike *G. crassicaudatus*, the tail shows no tendency to bushiness (Hill, 1953; Bearder and Doyle, 1974). *Galago s. moholi* is a small race and Doyle (1977) reports a range in weight of 150 to 252 gm in 10 females and 188 to 301 gm in 7 males.

IV.   Habitat

A. *Galago crassicaudatus*

In South Africa the thick-tailed bushbaby is commonly found in dense evergreen indigenous forests and tracts of evergreen bush and occasionally in open wood- and parkland. They may also be found in stands of commercial timber trees but only where these adjoin natural bush. Such suitable vegetation occurs in the well-watered coastal regions or inland as montane forest in steep-sided valleys or as riparian bush flanking the beds of mountain streams and rivers. Much of this habitat has been exploited by man in the past and replaced by cropland or relatively sterile timber plantations. Despite this, a number of government-protected forest areas still remain where bushbabies thrive.

The discontinuous nature of suitable vegetation in the Transvaal and Natal is generally reflected by the patchy distribution of *G. crassicaudatus*. Yet some isolated forests, which provide suitable habitats, do not contain bushbabies suggesting that the forest belt was never continuous or that forests became separated at an early age before the spread of bushbabies.

Galago crassicaudatus tolerate marked seasonal and diurnal variations in temperature, from 28° to an extreme maximum of 41.9°C. Mean maximum temperature per month varies from 22.0° to 29.1°C and mean minimum temperature from 5.0° to 18.1°C in the study areas. The winter months from June to September are usually dry and sunny with a large daily temperature range (up to 17°C). There is a considerable drying back of vegetation and leaf growth begins before the first good rains, with severe consequences if they are late. The hot summer rainy season occurs between September and April but rain may be recorded in any month. Thunderstorms contribute a large proportion of the total precipitation and long dry spells may occur between them, while periods of continuous rain may also occur, following which the vegetation becomes lush and almost impenetrable. The distribution and annual total rainfall is highly variable from year to year. In one season the annual rainfall in the Transvaal study area was 1768 mm with mean relative humidity per month from 73 to 88. In the Natal study area the total rainfall for the same year was 1292 mm with mean relative humidity per month from 79 to 88.

In the Transvaal *G. c. umbrosus* were most easily located in the riparian bush found along the watercourses of streams and rivers from the escarpment draining across the lowveld to the east. *Galago c. garnettii* were studied in dense coastal dune forest consisting of subtropical vegetation characterized by a continuous canopy 10–12 m high and a dense understory, as well as in more open temperate forest with a canopy 20–30 m high with a dense understory difficult to penetrate in the summer months.

In the Transvaal *G. crassicaudatus* may be found in the same areas as *G. senegalensis*, *Cercopithecus aethiops* (the vervet monkey), *Cercopithecus mitis* (the samango monkey), and *Papio ursinus* (the chacma baboon). In Natal, both *C. aethiops* and *C. mitis* may be found in the same area as *G. crassicaudatus*.

B.　*Galago senegalensis*

Galago senegalensis are found on the escarpment in open woodland, orchard bush with scrub, and isolated thickets with grassland in the thorn- and bushveld areas of the Transvaal characterized predominantly by flat karroo sandstone in association with *Acacia* thorn trees. They also frequent narrow tracts of vegetation along stream beds and, bordering the escarpment, they even penetrate dense riparian and montane forest where they may be sympatric with *G. crassicaudatus*.

Galago senegalensis tolerate equally marked seasonal and diurnal variations in temperature with a difference between night and day of as much as 17°C, from −6° to 38°C. Mean maximum temperature per month varies from 20.9° to 30.3°C and mean minimum temperature from 3.0° to 17.0°C in the main study area. The rainfall showed a marked annual fluctuation from 599 to 1463 mm in three areas in the same region. It is confined mainly to the summer months from October to April but may be recorded in any month of the year. As for *G. crassi-*

caudatus, winter is characterized by a considerable drying back of the vegetation before leaf growth begins again with the first rains. Relative humidity in the thornveld areas fluctuates monthly between 53 and 70 over the year, considerably less than that which characterizes the areas in which *G. crassicaudatus* are found.

In the area where *G. senegalensis* were studied, *C. aethiops* and *P. ursinus* also occur.

V. Population Densities

It was possible to establish the number of animals within a certain area and thereby calculate the population densities of each species by making direct counts over several consecutive nights in a number of areas in which the animals were studied. Due to the extremely reflective nature of their eyes it was possible to count practically all animals within a particular area in open bush country. Additional clues were gained from knowledge of the movements of individuals, the distribution of loud calls at night, and the number and spacing of sleeping sites located during the day. In forest habitats counting was done along transect lines coinciding with forest paths. Table I shows population densities of the two species.

The figures given in Table I indicate the population densities of bushbabies in areas where vegetation was of a uniform type and where bushbaby groups

TABLE I

Population Densities of *G. crassicaudatus* and *G. senegalensis* in Different Localities Including Areas Where They Are Sympatric[a]

Area	Habitat	Density/km²
G. c. umbrosus		
N. E. Transvaal	Lowveld: riparian bush and savanna	72
N. E. Transvaal	Escarpment: riparian bush and scrub	88
G. c. garnettii		
Kwazulu	Dune forest	125
Kwazulu	Temperate forest	112
G. s. moholi		
N. Transvaal	*Acacia* thornveld	200
N. Transvaal	*Acacia karroo* thickets	500
N. Transvaal	Wooded valley	275
N. E. Transvaal	Lowveld: riparian bush and savanna	103
N. E. Transvaal	Escarpment: riparian bush and scrub	95

[a] From Bearder and Doyle (1974).

were found at regular intervals. Often, however, suitable habitats were interrupted and the distribution of bushbabies was correspondingly patchy.

The population densities calculated indicate that *G. crassicaudatus* are more successful in dense subtropical forest, while *G. senegalensis* thrive in the more open *Acacia* thornveld. In areas where they are sympatric (lowveld and escarpment) the lower population densities suggest that either the habitat may not be optimal for either species or that interspecific competition keeps the population down.

Charles-Dominique (1977) has calculated population densities of five species of prosimians where they are sympatric in Gabon. He reports that the population density of *Perodicticus potto* (the potto), comparable in size to *G. crassicaudatus*, is 8–10 per km², and that the population densities of *Arctocebus calabarensis* (the angwantibo), *G. alleni* (Allen's bushbaby), and *G. elegantulus* (the needle-clawed bushbaby), all of which are comparable in size to *G. senegalensis*, are 7, 15–20, and 15–20 per km², respectively. For *G. demidovii* (the dwarf bushbaby), much smaller than *G. senegalensis*, the population density is 50–80 km². He imputes these relatively low population densities, compared to the South African galagines, to sympatricity and resulting interspecific competition for available space and resources. In addition, he notes that the prosimians in the rain forests of Gabon compete with approximately 120 other sympatric mammalian species.

In Madagascar, which is characterized by a paucity of mammalian fauna, some 15 to 20 mammalian species depending on forest type, population densities of nocturnal prosimians are generally high. Charles-Dominique and Hladik (1971) calculated the population densities of *Lepilemur mustelinus* (the sportive lemur), comparable in size to *G. crassicaudatus*, as 200–450 per km², and *Microcebus murinus*, comparable in size to *G. demidovii*, as 250–360 per km². Charles-Dominique (1977) calculates the population of *Cheirogaleus medius* (the fat-tailed dwarf lemur), comparable in size to *G. senegalensis*, as approximately 250 per km².

VI. Activity Patterns

Both species of bushbaby, like all other galagines, are strictly nocturnal. They begin and end their nocturnal activities with remarkably predictable regularity unless some unusual circumstance determines otherwise. The most important factor determining onset and cessation of activity is the prevailing light intensity. Bushbabies invariably leave cover at sunset and return at sunrise but, in overcast weather, activity begins early and ends late; those sleeping in shady places become active earlier and return later than those sleeping in exposed places. The length of nocturnal activity varies between $9\frac{1}{2}$ hours in summer and $12\frac{1}{3}$ hours in winter; similar responses to light intensity have been demonstrated for other nocturnal prosimians, the lorisids in Gabon (Charles-Domini-que, 1971) and for representatives of the lemurines and cheirogaleines in Mada-gascar (Charles-Dominique and Hladik, 1971; Pariente, 1974). In the laboratory both *G. crassicaudatus* and *G. senegalensis* adapt very readily to arti-ficial changes in illumination.

Total nocturnal activity may be divided into nine periods according to the following chronological scheme:

1. Toilet activities after waking
2. Direct movement to a food tree
3. Active feeding
4. Movement while foraging
5. Rest
6. Movement while foraging
7. Active feeding or rest
8. Direct movement to a feeding tree
9. Toilet activities prior to sleep

A. *Galago crassicaudatus*

Thick-tailed bushbabies always begin their nightly activity after sunset and and return to their nests before sunrise. After waking each morning there is a period of 5 to 30 minutes in which animals groom themselves and one another and urine-wash. Urine-washing is a ritual common to many prosimians in which a few drops of urine are deposited in the cupped hands which then methodically wipe the soles of the feet. In this way the animal is able to deposit its odor at points in its environment (for a fuller description and account of the possible functions of this habit, see Doyle, 1974). Juveniles may play together in the sleeping tree or in trees adjacent to it. The time spent on toilet activities increases to a maximum in the summer months and decreases again toward winter. This is probably related to seasonal differences in ambient temperature and the availability of food.

Initial activity is usually followed by a rapid and direct progression to a food tree. The use of different trees varies during the year as they come into flower or fruit, but generally a tree near to a sleeping place is favored. Particular trees which were productive throughout the year, for example, trees exuding gum, are generally used first on almost every night when the animals sleep close by. Ten minutes to an hour may be spent in this tree during which they may also allogroom, autogroom, and play, depending on the number and age of the animals together. They will then move away from this tree, stopping here and there to feed or groom, visiting favorite spots in a regular circuit night after night. The most characteristic pattern of activity during this time is the stop—go nature of the movement. After spending a period of up to 2 hours in a particular tree or moving slowly between a few trees, the animals will make a sudden spurt of movement before slowing down again or stopping. In this way, during the first third of the night, distance away from the sleeping place is gradually increased.

The middle of the night is a period of comparative inactivity with rest, grooming, and occasional sleep, usually spent at points furthest away from the sleeping place. Particular trees are favored for resting purposes and are used regularly. These are often the largest trees in the home range and presumably give the animals protection against predators while providing a vantage point from which to view the surroundings.

The last third of the night is more or less a repetition of the first third but in reverse. It usually begins with a period of feeding and movement in the general direction of the sleeping place. Many of the food sources used earlier are revisited and, as before, the animals spend long periods in particular places and move quickly between them.

From about 30 to 60 minutes before dawn, the bushbabies move to their preferred feeding or resting trees close to the sleeping place where they engage in social activities or feeding. When it begins to get light they move directly back to the sleeping tree. Group members separated from one another earlier on in the night usually come together at this time. Finally, before moving into the concealment of the sleeping site they will allogroom, autogroom, and juveniles will play.

B.   *Galago senegalensis*

The time at which the lesser bushbabies begin their nightly activities varies within 15 minutes on either side of sunset. Like the thick-tailed bushbaby, the first period of the active night is devoted to grooming, stretching, and urine-washing, after which they leave the sleeping tree generally along a particular path or set route night after night sometimes for many months before choosing a different route. At this point, members of a sleeping group separate, each one going off alone in order to forage. At night the sleeping group, which may vary in composition from time to time, does not function as a unit and approximately 70% of their waking time is spent alone. Initial movements are toward favored feeding trees where the first and longest feeding period of the night begins. Most feeding behavior occurs in the first 4 hours after dark.

Periods of foraging alternate with periods where the animals show no interest in food but groom, rest, or move rapidly from tree to tree. At about dawn, members of a sleeping group gather in one place and may move about together for a while before proceeding purposely toward the sleeping place as light increases. Alternatively, animals may arrive at the sleeping place from different directions at approximately the same time or they may join with others on the way there. Once in the sleeping tree they may groom and settle down well before sunrise.

Although the most active periods of the night in the field are the first 30–60 minutes after nightfall and the last 30–60 minutes before dawn, the general pattern of activity is characterized more by short bursts of activity followed by periods of rest throughout the night. *Galago crassicaudatus*, on the other hand, have a much more distinct bimodal pattern of activity characterized by long spells of relative inactivity followed by bursts of activity. The differences in activity patterns between the two species may be related to differences in diets and feeding habits.

Even in captivity prosimians display a similar circadian activity cycle in which periods of movement and feeding are interspersed with periods of rest, grooming, and perhaps play. This will occur even in the absence of those ecological factors to which it is presumably an adaptation (Pinto *et al.*, 1974).

VII.   Sleeping and Nests

A.   *Galago senegalensis*

A sleeping site may be defined as a particular tree used repeatedly for sleeping purposes by one or more animals. Bushbabies were found sleeping in dense thorny trees, on nests, forks, branches in the midst of clumps of mistletoe, on old birds' nests, and occasionally in cavities or hollow trees. In 40% of the cases, branches or forks were favored as sleeping places and in 52% nests were favored. Nests were more likely to be occupied by two or more animals than by single animals. Sleeping places were found on an average of from 5 to 6 m from the ground with a range of 3 to 11 m, the upper limit being largely determined by the height of the trees.

Sleeping bushbabies without nests are extremely difficult to find due to their preference for dark and confined parts of trees together with their dull gray color and light undersides which provide excellent camouflage. Each individual or group may have two or three favorite sleeping places which are used at certain times of the year. Sometimes these sleeping places will be used on alternative nights with no one place being favored; or one sleeping place may be favored and others used infrequently.

During the year, with changes in temperature and leaf cover, particular types of trees become more or less suitable for sleeping purposes, so that there is a gradual change in the use of sleeping places throughout the year. In winter, when the temperature at night may drop as low as 20°F, it is not uncommon to find bushbabies sleeping at the ends of branches exposed to the direct rays of the sun with little protection or camouflage. At this time of the year bushbabies have been observed to change their sleeping places during the daytime as the original

sleeping place becomes shaded with the movements of the sun. During summer
bushbabies are rarely exposed to the sun except in the early mornings and sleeping trees are chosen with an eye to dense foliage and shade.

Nests are made by females just before and for some time after the arrival of
infants and are probably used as platforms on which to give birth while providing support and camouflage for the offspring. More than one nest may be built
within the home range and the infants may be transferred from one to another
over the first few weeks of life. Old nests may be renewed as they deteriorate or
new nests built in the same tree or in adjacent trees. After nesting material is no
longer available an old nest may still be used for as long as it remains intact. New
nests are only seen between November and May which covers the birth season.
The nest is invariably constructed from flat, soft leaves taken when green and
placed in thorn trees to form an open-topped platform. Leaves and small branches are usually taken from broad-leafed trees or climbers growing in close
association with the thorn trees. Occasionally nests are built on top of a
platform of twigs formed by an old bird's nest.

The nest is of a size to suit the number of animals in the family group. They
may also be built after they are no longer essential for support of the young and
where they may merely provide a comfortable platform in a suitable place
where the whole family group can sleep together without being exposed from
below.

B. *Galago crassicaudatus*

The sleeping places of *G. crassicaudatus* are even more difficult to find than
those of *G. senegalensis*. Nests manage to remain completely hidden from view
and can only be located if the animals using them are followed before dawn.
In three study areas, 35 sleeping places were found.

Galago crassicaudatus usually sleep in a dense tangle of branches and creepers
between 5–12 m from the ground. Certain trees are suitable by virtue of their
supporting dense growths of climbers although it could not be ascertained that
nests were actually constructed within these tangles. On rare occasions when
animals were exposed from below they could be seen sleeping on the tops of
branches or tree forks either alone with the tail curled around the body or in a
huddle with others. Compared to *G. senegalensis*, *G. crassicaudatus* were found
sleeping on branches of forks 91% of the time and on nests only 8% of the time.
When nests were found they consisted of inaccessible platforms built at the ends
of branches using green and leafy twigs and small branches. They were occupied
by a female with small infants. The animals slept on the top, depressing them in
the center. From above they were sheltered by foliage. Nests have been reported
for the same species by Jolly (1966, unpublished paper) in Zambia and by
Haddow and Ellice (1964) in East Africa, where they consist of untidy aggrega-

tions of leaves and twigs often situated in dense foliage or creepers or in the forks of large trees well above the ground.

Like *G. senegalensis*, sleeping places are changed as different trees become more or less suitable with seasonal variations. More than one sleeping place may be used on consecutive days or by different members of the same group. Other species, like genets (*Genetta tigrina*) and vervet monkeys (*Cercopithecus aethiops*) were found sleeping in the same trees, the former in the day and the latter at night.

The adaptive advantage of the use of particular sleeping places may depend on a variety of factors in each species but Bearder (1975) notes that some of these advantages are evidently sacrificed at the expense of others. The sleeping places of *G. crassicaudatus* usually ensured complete concealment as a protection against predators, high temperatures, heavy rain, and hail. *Galago senegalensis* rely for protection on the thorny trees in which their nests are located and will change their sleeping places during the day in response to variations in temperature and other weather conditions. *Galago crassicaudatus*, on the other hand, rely more on concealment and direct protection and were not observed to move during the daytime.

Like *G. senegalensis*, nest building coincides with the birth season and it can be related to the need to provide a platform for the support of the young.

Unlike the lorisines, infant bushbabies of both species require a protective support at least for the first $1\frac{1}{2}$ weeks of life. During this period, the mothers of both species use different sleeping places even in the course of a single night carrying their infants from place to place.

VIII. Locomotion

A. *Galago senegalensis*

The lesser bushbaby has three fundamental methods of locomotion—jumping, hopping, and walking/climbing. Its jump has been analyzed by Hall-Craggs (1965, 1974). Normal standing jumps are performed from a crouched position; the legs are flexed and then brought forward to absorb the force of the landing. During the jump the hands are usually held against the chest and they grip the substrate on landing. On extremely long jumps of $5\frac{1}{2}$ m or more, the arms are brought sharply forward and held straight above the head, an action which presumably provides extra momentum.

Except for fairly rare occasions, when they walk quadrupedally, the lesser bushbaby's method of locomotion is entirely by means of hops and leaps and it jumps to and from vertical supports as often and as easily as it uses horizontal supports, since the natural habitat, unlike that of the typical vertical clinger and leaper, does not require exclusive adaptation to a vertical branch zone. The natural habitat also requires it to be able to move quadrupedally in very thorny trees too dense for jumping or hopping. The long legs enable the body to be arched over thorns and projections or the thorns themselves may be used as a ladder.

Hopping is reserved almost exclusively for movement across open spaces. When they have to descend, *G. senegalensis* check the environment carefully in all directions for up to 20 minutes before dropping to the ground. When on the ground they sit in an upright position facing the tree they have just left while checking once again. Providing there is nothing to alarm them, they turn around and take long rapid jumps, kangaroo fashion, to the next tree, sometimes pausing to look around before jumping into it. Movement along the ground may be performed in a broken series of short hops punctuated by pauses to check the surroundings until reaching the point of no return from which they move rapidly to the next tree.

If the open space to be crossed is very large, any small trees, stumps, fallen branches or fence posts may be used. Individuals were often seen to use particular stretches of barbed wire fencing in order to progress from one tree to another without having to descend to the ground. In doing so they either jump from one post to the next or they walk along the top of the wire, the accompanying tail movements appearing to subserve balance.

Before making long horizontal jumps in either the laboratory or the field, *G. senegalensis* carefully assess the situation moving the head from side to side, checking all directions from a bipedal standing posture apparently gauging the distance from a number of vantage points. Ears move independently to and fro and the head may rotate owl-like through 180°. However, in rapid flight involving a succession of long jumps, in an agonistic situation for instance, a series of long jumps may be made without any hesitation and with complete accuracy from one thorn-covered branch to another.

Galago senegalensis may also move quadrupedally in any position that the substrate requires, upside down on a horizontal branch, upward and downward on vertical surfaces.

B.   *Galago crassicaudatus*

Despite similarities in structure between *G. crassicaudatus* and *G. senegalensis* and their close taxonomic affinity, there are fundamental and important differences in their characteristic modes of locomotion.

Essentially the thick-tailed bushbaby has the same three fundamental methods of locomotion as the lesser bushbaby, namely, jumping, hopping, and walking/climbing. Unlike the lesser bushbaby, however, the quadrupedal walking is the most common form of locomotion. Bearder (1975) describes it as a cat-like prowling along the tops of branches which may increase in speed to a trot or a run, the body swaying from side to side as the hind foot is brought forward to meet the hand on the same side of the body while the other limbs are fully extended resulting in a fluid movement which becomes incomplete as speed increases and the swaying is less noticeable. While moving in this way on horizontal branches, the hands and feet maintain a firm grip, but on smooth or flat surfaces the hands and feet are splayed. The lesser bushbaby has been observed to run, for instance, along the top strand of a barbed wire fence, but has never been observed to run, or, for that matter, walk, for any length of time where it can jump or hop.

Hopping in *G. crassicaudatus* involves both fore and hindlimbs in which both hands move forward on to the substrate and the feet are brought up to meet them. This action is often interspersed with running. Horizontal leaps of up to 2 m are made between trees, providing good support is available on which to land or when jumping into a tree from the ground. The hindlimbs provide the propulsive force while the arms absorb the initial impact on landing. *Galago crassicaudatus* do not generally jump between trees if they are able to climb from one to the other but they will jump rather than descend to the ground.

Galago crassicaudatus occasionally make long downward leaps or drops. The hindlimbs provide the propulsion and the arms are raised sharply above the head at takeoff. The hands and feet are then held forward in such a way that the force of landing is distributed evenly between them. Like *G. senegalensis*, kangaroo-like hops are also used for progression on the ground, but relatively rarely, and the movements are much more clumsy.

The excellent grip of *G. crassicaudatus*, achieved as a result of extreme opposability of the digits and the arrangement of friction pads on the palms and soles, enable them to adopt a wide variety of postures in climbing from place to place. They may hang from one or both legs in order to catch hold of a lower branch before dropping down and often climb beneath a branch or use spiraling movements to climb from a vertical trunk onto a horizontal limb. The most common method of moving between trees is by deployment of weight across the end of adjoining branches. The nearest branch of the next tree is grabbed with one hand, or occasionally both hands, and pulled nearer before stepping across. Perfect balance is maintained with the help of the thick tail and weight is transferred slowly and evenly to the next branch before the other is released. Climbers are also used in order to ascend or descend, always moving head-first and frequently reaching across from one climber to the next. A looping action is sometimes employed in descending tree trunks. The body is bent caterpillar fashion with legs clasped on either side of the support and the hands gripping the bark.

Many of these actions are more characteristic of lorisines than any other galagine, for example, *Perodicticus potto* (Walker, 1969; Charles-Dominique, 1971) which has a specialized quadrupedal form of locomotion.

Although movement by saltation on the ground does occur when alarmed, G. crassicaudatus are far more likely to walk on the ground which they do very cautiously, stopping every now and again to stand upright and survey the surroundings before proceeding. They may also run or hop with equal regularity.

The adaptive advantages and ecological significance of slow and stealthy locomotion in prosimians have been discussed by many authors. Walker (1969) notes that the lorises have in common a slow and stealthy locomotion which distinguishes them from the active leaping galagos. He suggests that this type of movement, which is ideally suited to ensure minimum disturbance of foliage and arboreal supports, is related to their mode of obtaining food. Ehrlich (1968) suggests that this type of locomotion, in the slow loris, for instance, is an adaptation for avoiding predators by moving so slowly and fluidly as to minimize visual disturbance and avoid detection by fast-moving carnivores, such as felids, for example, which have a visual cortex which responds to moving rather than to stationary stimuli. Charles-Dominique (1971) suggests that nocturnal, arboreal predators are guided by auditory stimuli in locating prey, which would emphasize the advantages of the slow and silent locomotion of the potto and angwantibo. He notes too that, in the last resort, lorises will use active defense mechanisms against predators while the fast-moving galago is able to flee and he concludes that these adaptations in the lorises have extensively modified their feeding behavior and diet.

The type of locomotion exhibited by G. crassicaudatus suggests that the distinction between the locomotion of the lorises and the galagos is not as clear out as has been supposed, in that G. crassicaudatus appear to be intermediate between the two. In addition, G. crassicaudatus do not obtain their food by active leaping but by quiet and careful movements similar to those of the lorises yet they remain capable of moving quickly, albeit noisily, and they employ rapid escape as their only means of defense.

Bearder (1975) notes that if a comparison is made between the structure, methods of locomotion, and feeding habits of the slow-moving Perodicticus potto, the extremely agile G. senegalensis, and G. crassicaudatus, the structure of G. crassicaudatus is similar to that of G. senegalensis, their type of locomotion is intermediate between that of the other two species, but their feeding habits are like those of P. potto.

IX.   Feeding

A.   Galago senegalensis

The lesser bushbaby feeds exclusively on insects and acacia gum both of which are available throughout the year to a varying extent. This diet appears to provide for all water requirements since bushbabies have never been seen to drink in the wild or even to lick condensation from the surface of leaves. They do, however, readily drink milk in the laboratory (Doyle and Bekker, 1967) as well as water.

During the winter months the animals subsist largely on gum supplemented by insects. Acacia gum is frequently found in forks, on the undersides of branches, in the axils of fine twigs, or near the base of tree trunks. Typically bushbabies adopt a head-down posture while licking gum but the feeding position may be changed repeatedly while licking at one spot. The bark may be chewed away to expose more gum.

In midwinter, bushbabies spend an average of 12 minutes feeding in every hour. Periods of up to 15 minutes at a time may be spent on the ground looking for insects, worms, and grubs. In spring and summer when insects are plentiful there is a marked decrease in the amount of time spent feeding with very little searching on the ground. Gum is frequently licked but for much shorter periods. There is a noticeable deterioration in the physical condition of bushbabies in winter.

Individual bushbabies may go for periods of up to 3 hours without feeding. In general, the maximum amount of feeding occurs in the first few hours after dark. Periods of foraging alternate with periods when animals show no interest in food but groom, rest, or move rapidly from place to place. When looking for food within a tree the bushbaby flits from one branch to another feeding briefly here and there before moving on to the next tree. Alternatively, it may feed sporadically while moving to a particular tree which may be used repeatedly as a source of food.

In catching insect prey the bushbaby uses one hand which shoots directly from the shoulder at considerable speed hitting the prey and usually pinning it to the substrate. The eyes are shut at the moment of impact and the ears put back in a manner characteristic of the defensive—threat situation. This posture, which probably serves as a protection against insects with flapping wings or spurred legs, is maintained until the prey is brought to the mouth and the head bitten. Large food, too big for one hand to pick up and hold, is held in or down with both hands. Confronted with strange insects or large insects that may inflict injury, like large grasshoppers, *G. senegalensis* display greater precaution in their approach than normally displayed with small familiar insects. The approach to the prey is accompanied by continued movement of the ears and rotation of the head and the occasional assumption of an erect posture while maintaining visual fixation of the prey. Several abortive strikes may be made before finally making a successful strike.

B. *Galago crassicaudatus*

The diet of *G. crassicaudatus* in the natural habitat consists of gum, fruits, flower secretions, seeds, and insects. Feeding habits differ seasonally depending on the availability of each type of food. Gum represents 62% (based on the number of incidences of feeding observed) of the food utilized and is eaten throughout the year, mostly in winter when other foods are in short supply. Fruit represents 21%, mostly in summer, and insects 5%. Exotic fruit, like tropical fruit, is also eaten where plantations of these fruits adjoin natural habitats. While feeding, *G. crassicaudatus* adopt a variety of postures and use hands and mouth in different ways to suit the type of food. They will move to any accessible part of the tree and on to the ground to reach a desired food item.

Galago crassicaudatus adopt a wide range of postures to feed on gum, of the same kind favored by *G. senegalensis,* but their wide reach and the extreme opposability of their digits enable them to feed on gum inaccessible to *G. senegalensis,* for instance, on the undersides of broad branches and trunks.

The gum is usually licked away slowly for up to 20—30 minutes at a time and bark may be chewed away to expose the gum. During long bouts of licking the animals sometimes move away to rest before continuing to feed. It is not uncommon to see two or three bushbabies licking simultaneously at one spot and this rarely leads to fights.

While they are in season, bushbabies return repeatedly to trees bearing edible fruits. Wild figs (*Ficus* sp.) and jackalsbessie trees (*Diospyros mespiliformis*), which carry a large amount of fruit, ripening over a long period, are of particular importance. Various fig trees come into fruit at different times throughout the summer and they are used in succession. At this time, *G. crassicaudatus* consume greater quantities of food than in winter when, like *G. senegalensis*, their condition noticeably deteriorates.

G. crassicaudatus feed noisily and often wastefully on hard-shelled fruits such as wild figs and guavas. Some animals will pick only ripe fruit, others will pick unripe fruit and immediately drop it. Even ripe fruit is dropped accidentally or knocked down in the course of the animal's movements but no interest is shown in it once it has fallen. Fruit is either picked directly with the mouth or first guided to the mouth with one hand and then carried away to be eaten in comfort. Some fruits are held in one hand while part of the peel is chewed away and discarded by spitting or flicking the head. The soft inside is then licked or scraped with the teeth and the remaining shell dropped to the ground. Large fruits are sometimes eaten in a similar way without being picked.

Bushbabies will remain in the vicinity of fruit trees for several hours at a time and return to them frequently during the night, but feeding seldom lasts more than 20 minutes at a time. When several bushbabies are attracted to the same tree at the same time they sometimes quarrel mildly, never seriously.

Bearder (1975) notes that the eating of fruit is probably the most important dietary adaptation of this species. *Galago crassicaudatus* are absent from regions where wild fruit is not abundant for at least half the year and this appears to be a significant factor limiting their distribution.

Bushbabies frequently come across seed pods and woody fruits on the ground or lodged in the forks of trees during their normal movements or while moving slowly in the dense tangle of vegetation close to the ground. Such food is usually consumed on the ground where it is discovered but it may first be carried into a tree in the mouth. Bushbabies search among the leaves on the ground with their hands or nose, pick up seeds with one hand, and chew the husk away noisily. Seeds and dried fruits are eaten more in the winter when fresh fruit is in short supply. Trees are also visited as they come into flower and as much as an hour may be spent searching for nectar. Flowering branches are held in one hand and the flowers either smelled or licked.

Both large and small insects caught in trees or on the ground are eaten when the opportunity arises but not in preference to other foods. No interest is shown in the numerous moths which are attracted to fruit trees in which the bushbabies feed. Insects are picked up in one hand in a rather slow, grabbing motion and then transferred to the mouth.

Although bushbabies are often observed descending to the ground very close to water they have never been seen to drink. They may occasionally lick the surface of leaves, sometimes holding them with one hand or biting them off and licking the petioles. Evidently, *G. crassicaudatus*, like *G. senegalensis*, obtain most of their moisture requirements from food.

Bushbabies have never been observed to eat vertebrates of any kind despite numerous opportunities for them to do so, for example, roosting birds. A pair of doves experimentally introduced into a bushbaby cage 4 m² in size took over one of the next boxes and reared two successive clutches without experiencing any harm. Reports of the predatory habits of *G. crassicaudatus* by other authors may possibly be explained on the basis of local habit.

X.   Group Structure

A.   *Galago senegalensis*

The largest group that the lesser bushbaby forms is what Martin (1968) calls the "family group" to distinguish it from the "social group" which contains more than two sexually mature adults. Bearder and Doyle (1974) report that of 119 sleeping groups, consisting of 238 animals, 60 (85% of the animals) were in groups of two to six, most of which were groups of two to four. Common assortments included an adult pair with or without infants or juveniles, two adult females with infants, and a single adult female with infants or juveniles. This compares favorably with the findings of Haddow and Ellice (1964) for the same species in East Africa. They reported 70% of animals found in sleeping groups of two to six.

These groups do not, however, function as a unit during the night. Individuals spend 70% of their waking time alone and movement through the home range of two or more animals together for any length of time is rare. At sunset, groups separate for individual foraging and they may or may not meet again before dawn. If they do meet they generally move about together before arriving at the sleeping site or they may meet at the sleeping site.

The composition of sleeping groups varies during the year probably due to increasing aggressiveness between animals, particularly males, as young adults mature. Bearder (1969) suggests that this is crucial for pair formation and the establishment of new sleeping sites and new home ranges.

Although it would be misleading to refer to groups in this subspecies as family groups, association between individuals is most likely to be based on familial ties. No natural group configuration occurs in the wild; even sleeping groups change their compostion, but social contact between particular individuals does occur on a regular basis. In the laboratory, male and female with young and juveniles will live together peacefully. Provided there is sufficient space, additional females and subadult males do not increase friction and females may live together peacefully and produce infants in the same nest.

B. *Galago crassicaudatus*

In the three localities in which *G. crassicaudatus* were studied, Bearder (1975) reports that this species, like *G. senegalensis*, sleep either alone or in groups of two to six. In 148 groups, 54% were found in groups of two to six. Large groups would consist of an adult male, an adult female, and the offspring of one or more generations. Animals most likely to be found sleeping alone were males and maturing offspring.

Like *G. senegalensis*, sleeping groups tend to split up at night for individual foraging and come together at dawn, although they do spend more time together at night. The mother and her young tend to remain a cohesive group whereas in *G. senegalensis* they do not, although infants may follow their mothers for short periods. In *G. crassicaudatus* too, the male was more likely to join the group and move about with it than in *G. senegalensis*. Only as the young males begin to mature sexually do they separate themselves from the group and begin to exhibit spacing behavior, such as olfactory marking and vocal advertisement, which serves to maintain distance between themselves and rival conspecifics.

XI.   Home Range

Bushbabies, like all primates, restrict their activities to a measureable, circumscribed, geographical area called a home range, defined in the broad sense of being a composite measure of multiple daily ranges, taking seasonal changes into account, covered by a group in the course of normal feeding and maintenance activities. Sizes of home ranges in primate species vary widely and appear to depend on a variety of factors, such as complexity of social structure, the nature of the terrain or vegetation, dietary preferences of the animals, and whether they are terrestrial or arboreal.

Several other terms are useful in describing the use of space by primates and other animals within their home ranges. "Core areas" are parts of the home range which are used more frequently than other parts (Kaufmann, 1962) and which are usually exclusive to the group using them (Jay, 1965). Core areas contain "fixed points" such as sleeping trees, resting trees, and food sources (Mason, 1968) which are usually connected by "traditional pathways."

A. *Galago senegalensis*

The lesser bushbaby covers an average distance of 2.1. km every night representing movement through a mean number of 500 trees. When moving within the home range, bushbabies habitually use the same generalized pathways. Although the exact route may be highly variable, it is not uncommon to see individuals repeatedly making the same jumps from one particular tree to another or crossing the same open spaces. It is possible for a bushbaby to traverse the home range within 2 hours, although this is rarely done. The animal may make a circular excursion of half a mile or more, starting from a particular point and returning to the same point from the opposite direction some hours later. The flexibility of movement in a directional way along recognizable pathways and the variability of these pathways strongly indicates that the animals have a thorough knowledge of the area in which they live.

The average home range size established for members of a group is 2.8 hectares with a range of 1.25 to 3.95 hectares. The size of this range is much more extensive than that described for many arboreal primates, for example, *Callicebus* (the titi monkey) whose home range is reported by Mason (1968) to be 0.4 hectares. This may be partly due to the nature of the natural habitat—relatively open bush which is likely to be less productive than dense forest. Boundaries of home ranges are not strictly rigid and may shift over the course of a year. Natural boundaries are formed by roads, breaks, or large open spaces, but there is usually a slight overlap between the ranges of adjacent individuals. Figure 2 illustrates the observed movements of individuals forming three major sleeping associations over a period of several consecutive months.

B. *Galago crassicaudatus*

The thick-tailed bushbaby covers an approximate distance of 1 km per night though Bearder (1975) cautions that this is probably an upper estimate. Circular

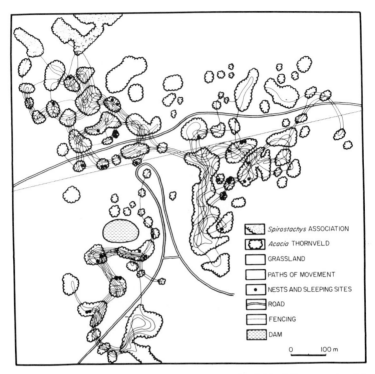

Fig. 2. Map illustrating the observed movements of individual *G. senegalensis* forming three major sleeping associations in the northeast Transvaal.

excursions are typical and animals seldom retrace their steps although they sometimes return by the same route as they went. Adult males tend to move farther than females particularly those with young. When food is abundant in particular localities distance covered may be less particularly if the source of food is near a sleeping tree.

Regular observations of groups of marked animals for more than a year reveal a well-defined range of movements which remain stable over long periods. The home range of one maternal group was definitely established as 7 hectares while that of an adult male was known to be larger and overlapped with the home range of the maternal group about 80%. The size of a home range is compatible with the ability of an animal to move from the center of the range to any point on the periphery and back in a single night. The shape of the home range of *G. c. umbrosus* conforms very closely to that of the dense riverine bush in which this subspecies lives. Figure 3 illustrates the home range of a *G. c. umbrosus* group based on observations of nightly movements over more than a year.

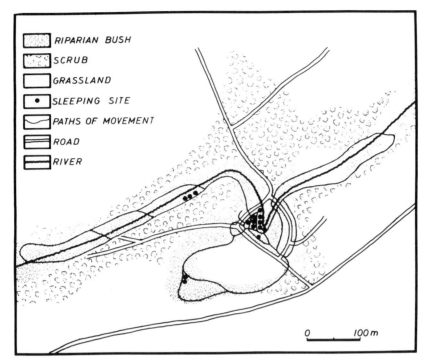

Fig. 3. Map showing the major pathways constituting an annual home range of a *Galago crassicaudatus umbrosus* group in the northeast Transvaal.

Records of the movements of bushbabies over a long period of time show that even localized movements invariably encompass the center of the range which conforms to the previously stated definition of a core area. On the other hand, while similar movements are made on consecutive nights over a period of days, weeks or even months, on an average only 45% of the home range is covered in any one month. After utilizing a particular sector of the home range night after night animals very suddenly alter their movements in order to utilize a new sector. These changes in movement appear to be related to seasonal changes in the productivity of different food trees in different parts of the home range such that the movements of the animals in any one month are almost identical to their movements in the same month of the previous year.

Of five different sectors, into which the home range of one maternal group was arbitrarily divided to show the progressive decrease in their amount of use, the core area was by far the smallest at only 4% of the total with the other sectors ranging from 11 to 39%. However, by far the greatest number of sightings were made in this area, 43% as comapred with a range of 3 to 37% for the others.

It also contained 67% of the sleeping places, compared with a range of 0 to 20% for the other four. Not only did the core area contain a large proportion of suitable sleeping trees, but also a larger proportion of suitable feeding trees.

The home range movements of *G. crassicaudatus* also vary in relation to habitat differences. Like *G. senegalensis*, in relatively open *Acacia* thornveld, *G. c. lönnbergi* in open woodland in Rhodesia regularly descend to the ground to cross open spaces as much as 100 m across. In dense vegetation, the riverine bush where *G. c. umbrosus* were studied, or the dune forest of *G. c. garnettii*, animals seldom descend to the ground.

XII.   Birth Periodicity

A.   *Galago senegalensis*

Doyle *et al.* (1971) have reliably established the gestation period in *G. s. moholi* as ranging from 121 ± 1 day to 124 ± 1 day in 29 births in the laboratory. Females usually produce twins on second and subsequent births and generally give birth twice in a 12-month period, the second conception occurring during an immediate post-partum estrous period. In the laboratory, births tend to occur throughout the year probably due to the absence of seasonal changes.

In the field, however, births occur exclusively between spring and late summer coinciding with that time of the year when rainfall and temperature are highest and when food and nesting material are abundant. Field studies confirm that there are two successive birth seasons in the year. The first infants were found in October and early November but a much higher proportion of infants were found in January and February. Matings were actually witnessed in late September and early October coinciding with the second birth peak and confirming roughly the gestation period established in the laboratory. Evidence from the field also supported the occurrence of an immediate postpartum estrus. Laboratory data showing that females first come into estrus and conceive at 200 days could not be confirmed under field conditions. Figure 4 illustrates the chronology of reproductive activities in the two species of bushbaby.

B.   *Galago crassicaudatus*

The thick-tailed bushbaby has a distinct and restricted breeding season in southern Africa, which is in accordance with the reports of Gerard (1932) and Cansdale (1960). Jolly (1966, unpublished paper) records that four of her semi-captive female *G. crassicaudatus* were in estrus over the same period, suggesting a strict breeding synchrony such as that seen in *Lemur catta* (the ring-tailed lemur, Jolly, 1966). In the northeast Transvaal all infants were born within

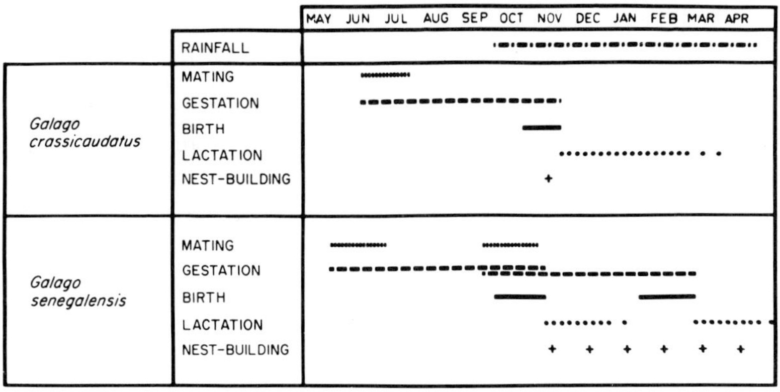

		MAY JUN JUL AUG SEP OCT NOV DEC JAN FEB MAR APR
	RAINFALL	·■·■·■·■·■·■·■·■·■·■·■·■·■·■··
Galago crassicaudatus	MATING	·············
	GESTATION	■·■·■·■·■·■·■·■·■·■·■·■·■·
	BIRTH	▬▬
	LACTATION	················· · ·
	NEST–BUILDING	+
Galago senegalensis	MATING	··········· ···········
	GESTATION	·■·■·■·■·■·■·■·■·■·■·▬▬▬▬·■·■·■·■·■·■·■·■·
	BIRTH	▬▬▬▬ ▬▬▬▬
	LACTATION	········· · ·········
	NEST–BUILDING	+ + + + + +

Fig. 4. Diagram showing breeding and related activities in *Galago crassicaudatus* and *Galago senegalensis* in relation to rainfall and time of year.

a period of 3 weeks in early summer at the beginning of November. Circumstantial evidence suggests the same for the other areas as well. Juveniles were seen in Zululand at the beginning of January while all animals observed in Zululand and Rhodesia between June and August were full grown.

Manley (1966), citing several sources, reports a gestation period for *G. crassicaudatus* of 130–135 days. This would mean that in southern Africa mating takes place in midwinter (June/July). Although mating was not actually witnessed at this time it was a period of intense social activity characterized by many skirmishes and loud calls.

Figure 4 shows that in both *G. senegalensis* and *G. crassicaudatus* the timing of births and the start of nest-building activity coincided with the onset of the rainy season when food supplies increase in abundance and cover becomes denser. This correlation between rainfall and birth periodicity, which has also been reported by Martin (1972) for several species of lemur, probably represents an adaptation to seasonal food availability. Infants born at the beginning of summer are big enough by winter and have accumulated sufficient reserves to tide them over the period of food scarcity which follows. In addition, the availability of suitable nesting material and dense cover further contributes to the survival of small infants at this time.

In *G. crassicaudatus*, although the period of gestation is comparable to that of *G. senegalensis*, infants take longer to develop and are not weaned until much later than in *G. senegalensis*, and there is no postpartum estrus and no second birth season.

Bearder (1975) found sixteen infants which included six sets of twins, one set of triplets, and one singleton, suggesting that twins are at least as common in *G. c. umbrosus* as in *G. s. moholi*.

XIII.   Mother—Infant Relationship and Development

A.   *Galago senegalensis*

Judged by physical characteristics, no infants below the age of about 10 days were seen in the field. This is not surprising since laboratory data show that infants are confined to the nest for the first 10—14 days (Doyle *et al.*, 1969). The presence of infants in a particular home range coincided with the appearance of one or more substantial nests. After the age of 2 weeks, infants were often seen sleeping with the mother on shaded forks or branches.

Nursing was rarely seen and obvious attempts on the part of an infant to suckle at night were usually rejected by the mother. It is probable that most suckling occurs during the daytime in the nest or on a suitable branch or fork with the infant completely hidden beneath her.

Before infants are moving around fully of their own accord, the mother frequently carries them to various places within the home range where they may be left for long periods. They are carried one at a time in the mouth by a dorsolateral fold of the skin. The infant tucks its hands and legs completely under the body and its tail curls passively round the mother's neck or over her back. The mother carries each infant through several trees and then releases it on another nest site or on a branch where it will cling quite safely. It may then be carried further or the mother may return for the second infant. The infants may be placed together or left separately and certain trees may be used repeatedly by the same female for the deposit of her infants. At the beginning of the night's activity the female leaves the sleeping site alone in order to forage. Some time later she returns to the infants and makes the first change to a new nest site or resting place. While still very small she may change them a few more times during the night. If they have been left in different places, which may be as much as 40 m apart, the mother may visit each one sporadically and spend short periods with each during which she will invariably groom them. Short periods of grooming, of not more than 15 minutes, will occur whenever mother and infants are together whether in the process of carrying them from one place to another, when she returns to them after a period of time, or when she settles down to suckle them. She will always collect them both together again before dawn.

Between birth and about 10 days of age little is known of the behavior of infants in the wild. The indications, confirmed by laboratory studies (Doyle *et al.*, 1969), are that they spend the whole time in a nest. From the age of 10 to 21 days they are still entirely dependent on the mother for sustenance and transport. At this stage they are still moved frequently during the night from one nest site to another where they may be left for long periods, of up to 3 hours. At this stage the mother will frequently chase away adult conspecifics which approach the infants.

From 21 to 28 days the infants are still dependent on the mother for transport from tree to tree, but are able to jump around, explore, and lick gum. From about 4 to 6 weeks the infants are able to follow the mother. They may jump up to about 1 m and descend to the ground for the first time taking short hops of about 30 cm and are thus able to go from tree to tree. They follow the mother away from the sleeping tree in the evening to certain trees in the vicinity where they may spend the rest of the night alone, feeding, sleeping, playing, and moving from one tree to another over short distances. The mother returns short-ly before daybreak and utters a soft "coo." The infants approach her giving high-pitched "clicks." She moves slowly toward the sleeping site with the in-fants following her.

After about 6 weeks the infants leave the sleeping place of their own ac-cord and begin independent movement over an ever-increasing range. The family still comes together in the morning before moving to the sleeping site. Infants find and lick gum independently long before they are able to catch insects which they probably learn by watching the mother. The frequent changing of infants from one nest site to another while still young is also characteristic of their behavior in the laboratory (Doyle *et al.*, 1969). While still young and vulnerable, this behavior may serve to reduce the probability of their being found by predators.

It is not known when weaning takes place since infants will frequently spend the day sleeping in a group with their mothers until late adolescence and often beyond. In the laboratory it has been ascertained that weaning is complete by about 11 weeks of age (Doyle *et al.*, 1969).

B.   *Galago crassicaudatus*

Relatively little is known of the mother—infant relationship and development in *G. crassicaudatus* during the first 10 days of life since, like *G. senegalensis*, they do not venture from the nest until more than a week of age. Lactating mothers are recognizable by the distinctive bald patches on the abdomen at the location of the nipples. A female with infants leaves the nest at dusk up to 40 minutes later than other animals after an extended period of grooming the infants. For much of the night infants are left alone in the nest or, like *G. senegalensis*, the mother may carry the infants to different nest sites where they may be left clinging safely for long periods alone or together. Before dawn both infants are retrieved to the sleeping site which may change in successive nights.

By little more than 3 weeks of age the mother may carry both infants with her, one in the mouth and another clinging to the fur on her back (Bearder, 1975), or both clinging to fur on back and abdomen (Davis, 1960; Buettner-Janusch, 1964). Occasionally the infants are left alone on a branch or in a tree fork while the mother moves away to forage. Even with infants clinging to her the mother may make small jumps or climb upside down beneath a branch. By 25 days they are able to follow the mother but they will continue to be carried, at least occasionally, until about 70 days of age, clinging to the mother, and

suckled until about 120 days of age. At 32 days, infants were first observed licking gum, indulging in allogrooming, playing with the mother or each other,
and associating with other adults although they probably indulge in these activities earlier.

As with *G. senegalensis* (Andersson, 1969) they respond to specific maternal
contact calls and from as early as 3 days of age they emit high-pitched squeaks
to attract the mother when in distress. At 9 days of age squeaks are replaced
by clicks and crackles which persist until late adolescence. By 4 weeks their
locomotion is more efficient and they are able to make small jumps. By 40 days
they may move between trees in the absence of the mother following other
animals short distances. They are still reliant on the mother for making difficult crossings between trees and are still frequently carried, though one may
be carried while the other follows.

Between 10 and 16 weeks of age the infants follow the mother at all times,
are able to negotiate difficult passages on their own or find easier ones of their
own accord. For instance, they may cross the ground between trees while the
mother jumps. The mother always initiates movement and waits for the infants
if they are slow to follow. When left behind they emit clicks and buzzes and show
other signs of agitation until they are able to rejoin the mother. By this time they
are extremely active and playful and spend long periods grooming together.
At the end of this period they may still show signs of agitation on being separated from the mother although they are able to move from one food tree to another or to a sleeping place of their own accord.

By 17 weeks their movements continue to be more limited than those of
adults. Although a juvenile may take the lead in group movements the mother
continues to control the timing and general direction. Movements of the group
are more cohesive in unfamiliar parts of the range while in familiar parts individuals may use different routes and feeding places while continuing to move
in the same general direction.

Not until adolescents are 10 months of age do they begin to make completely
independent movements. Subadult males may move away from the parental
range at this time while females maintain a loose association with the mother
until at least 14 months.

XIV.   Relationships with Other Species

A.   *Galago senegalensis*

A small, agile, nocturnal primate like the lesser bushbaby probably has little
to fear while it remains in the thorn trees. Infants less than a month old, and
particularly before the age of 2 weeks up to which time they are still confined to

the nest, are vulnerable to any carnivore that may come across them. The care of the mother, however, ensures that the chances of this occurring are minimal. Although predation has never been witnessed in G. *senegalensis*, possible predators include civets, snakes, owls, genets, jackals, raptorial birds, cats, and man. Of these, the last mentioned causes by far the most serious depredation of populations and the species has disappeared from many regions surrounding towns and villages. Despite protective laws they continue to be caught and kept as pets.

Of the nocturnal predators, only the cats, snakes, and genets are able to climb trees but none are sufficiently agile to catch an alert bushbaby. Owls, raptorial birds, and ground-dwelling predators are no threat while the bushbaby remains in the cover of trees, but may well be in the open. From their behavior it is obvious that bushbabies are extremely wary before descending to the ground and nervous when in small trees. If they are disturbed they either move upward or retreat rapidly to a large tree and if the source of the disturbance is identified movement is accompanied by alarm "yaps" which alert other bushbabies in the vicinity which also move upward. The bushbabies remain alert and may continue yapping until the disturbance is over.

During the day, bushbabies are either completely hidden from view from above by foliage or they are protected from aerial predators by dense thorny branches covering nests and sleeping places. If they are, however, disturbed during the day they will move rapidly away in the trees. On the whole, most predation on bushbabies occurs during infancy while the only serious threat to adults in unprotected areas is man.

The lesser bushbaby shows no active interest in other species apart from potential predators. The only nocturnal, arboreal animals, apart from the genet, seen in the bushbaby habitat are the climbing mouse and the acacia rat. If the bushbaby meets one of these animals they may stop and briefly regard one another but no further interaction occurs. The movement of ground-living animals such as hares, buck, and domestic cattle may cause the bushbaby to look in their direction but they do not induce any alarm. Roosting birds are sometimes inadvertently disturbed and fly away. During the day only birds and bush squirrels are known to share the same trees as bushbabies, but they ignore one another.

B. *Galago crassicaudatus*

Predation has also not been witnessed in G. *crassicaudatus* although there are reports of bushbabies being included in the diets of leopards (*Panthera pardus*) and the black eagle (*Aquila verreauxi*). Most evidence of injury to bushbabies comes from Rhodesia where G. *c. lönnbergi* live in a more open habitat than the other two subspecies and frequently have to descend to the ground in order to cross open spaces between clumps of trees. Bushbabies were known to have been knocked down by cars on roads or killed by domestic dogs and a number were seen with damaged limbs. Dogs are also reported to catch and kill *G. c. umbrosus* in the northeast Transvaal on the ground.

Raptorial birds and snakes are a potential source of danger to both species of bushbaby during the day, particularly if they sleep in exposed positions. Apart from the black eagle, the martial eagle (*Polemaetus bellicosus*) and the crowned eagle (*Stephanoaetus coronatus*) were seen in the same area where *G. c. umbrosus* were studied and where they are known to prey on young vervet monkeys (*Cercopithecus aethiops*), rock hyraxes (*Procavia capensis*), and small species of antelope. Bushbabies were seen to react violently to the presence of snakes and four of the species found in the study area are known to be particularly dangerous—the Egyptian cobra (*Naja naja*), the black-necked spitting cobra (*Naja nigricollis*), the black mamba (*Dendroaspis polylepis*), and the python (*Python sebae*). The latter are now somewhat rare due to exploitation for their valuable skin, but others are encountered frequently.

Galago crassicaudatus are occasionally caught by man for food, but far more rarely than *G. senegalensis* are they caught and sold as pets. There is a superstitious belief among local inhabitants in some areas that the loud cries and whistles made by bushbabies come from a large multicoloured snake which is said to attack and kill people at night by pecking a neat hole in their heads (Astley Maberly, 1967). For this reason this species is generally avoided at night and as a result is afforded some degree of protection.

The reactions of bushbabies to the observer illustrates their response to potentially dangerous species. They usually retreat to a vantage point from which they stare, or they approach hesitantly making a number of characteristic calls. Continued disturbance causes them to move higher in the trees or to move away. If they are then followed they repeatedly change direction until they are invariably lost. Simian primates generally take many weeks before they will allow an observer to approach reasonably close. *Galago crassicaudatus* become habituated to an observer within a few hours and *G. senegalensis* take even less time, which suggests that they have little to fear from large ground-living mammals. The response of one species to another is highly appropriate to its potential as a predator. In Rhodesia, for example, an individual *G. c. lönnbergi* foraged on the ground near two bushbuck but it moved quickly into a tree at the approach of a domestic cat.

Threat of predation is probably responsible for a number of behavioral characteristics of *G. crassicaudatus*. They spend the day well hidden in sleeping places which they do not leave until after sunset and to which they return well before sunrise. At night they remain in the safety of the trees and display great caution when they occasionally descend to the ground. They are able to jump and move rapidly to avoid danger and their characteristic fluid mode of progression under normal circumstances, described above, is such that, like the slow-moving potto and angwantibo, they are often difficult to detect in motion. The lesser bushbaby, on the other hand, relies exclusively on its speed and agility.

Both species of bushbaby have a number of vocalizations which warn conspecifics of the presence, location, and the magnitude of potential danger. *Galago crassicaudatus* are sufficiently large when adult to be unperturbed by owls and small carnivores like genets, and infants which are otherwise vulnerable are adequately protected by the watchfulness of the mother. In addition, the structure of the group apparently ensures that the offspring learn from the mother which species to avoid.

In the northeast Transvaal and in the eastern highlands of Rhodesia, three species of nonhuman primate, in addition to the two species of bushbaby, were found—the chacma baboon (*Papio ursinus*), the vervet monkey (*Cercopithecus aethiops*), and the samango monkey (*Cercopithecus mitis*).

Between the five species there is a varying degree of niche separation. Baboons generally live in higher regions on mountain ledges and open slopes. They sleep along cliff faces and rarely descend to the valley floor. Samango monkeys occupy the dense forested regions at the head of mountain streams, extending downstream to where the dense vegetation narrows to a strip bordering the watercourses. Each main branch of the valley appears to be utilized by a single group. Further downstream where riparian bush adjoins more sparse vegetation and farmland, vervet monkeys are found. Their range overlaps that of the samango monkey but they were not seen together. The samango monkey is found in the dense regions while the vervet monkey is found in greater number in more open areas. In East Africa, Strühsaker (1967) observed intermingling of the two species without any aggression.

A similar separation exists between the nocturnal inhabitants of the same forests, *G. crassicaudatus* and *G. senegalensis*. The former are mainly confined to dense vegetation throughout the length of the valley while the latter are absent from the forest regions at the top and are typically found in open vegetation lower down. Encounters between the two species were not uncommon. They were seen in the same trees feeding on the same type of food but, apart from briefly staring at one another, they ignored one another completely.

Vervet monkeys occasionally used the same trees as *G. crassicaudatus* for sleeping purposes and they were seen together in the mornings and evenings. On one occasion two young monkeys approached an adult bushbaby which ignored them until they came to within 1 m. It then spat at them and they retreated rapidly.

The most numerous of all interspecific interactions which were witnessed were those between bushbabies and genets. The adult *Genetta tigrina* is approximately the same size of the adult *G. crassicaudatus* and they share several ecological characteristics. It may be said that they show a mutual respect for one another and, although the bushbabies displace genets, in most cases their relationship is a complex one. Each species was seen to spit, growl, and lunge at the other on occasions, with females of either species particularly intolerant of the other when they had young. At other times the two species sleep peacefully in the same trees or rest and groom close by at night and they were seen feeding together in the same fig tree. The approach of a genet occasionally induced mild intensity alarm vocalizations from bushbabies, similar to those evoked by a cat.

The nocturnal marsh mongoose (*Atilax paludinosus*) was also encountered by bushbabies but its movements provoked nothing more than a startle reaction typical of their response to any sudden noise.

Interactions between *G. crassicaudatus* and owls were observed in which the former showed little sign of fear. On four occasions a wood owl (*Ciccaba woodfordi*) swooped down on a bushbaby which merely jumped away at the last minute making a brief, low-intensity call. Two juvenile bushbabies were seen to approach a barn owl (*Tyto alba*) which craned its neck and extended its wings forward before flying away. Eagle owls (*Bubo africanus*) were common but they did not appear to cause *G. crassicaudatus* any alarm, unlike the smaller *G. senegalensis* which reacted strongly to them.

Fruit bats (*Epomophorus wahlbergi*) and various species of mice were common food competitors of bushbabies at night but they were always ignored.

XV. Discussion

There are some striking similarities and differences in the ethology and ecological characteristics of the two species of bushbaby. Despite a large difference in size they are morphologically very similar. Both are nocturnal and have a broadly bimodal pattern of activity at night. In captivity they are both more or less omnivorous though in the wild *G. senegalensis* is strictly insectivorous, except for gum, while *G. crassicaudatus* subsists largely on gum and fruit supplemented by insects. Both build nests and both give birth during the wet summer months when food and cover are readily available. They give birth to a small number of offspring which are weaned before the start of the winter. In the initial stages of infancy, mothers of both species move their infants from one location to another, carrying them one at a time in the mouth. Later they may be left clinging alone to a branch while the mother moves away in order to forage, but they are always retrieved before daybreak to sleep with the mother.

Group structure, home range, and the use of space is basically similar in both species. Both have small family groups in which adults show possible territorial behavior in their attitudes toward rival conspecifics. The movements of males are separate from those of females although familiar pairs come into regular contact and they may share a sleeping site at night. Bonds between certain individuals appear to be enhanced through allogrooming. Both species utilize all levels within their habitats and are wary before descending to the ground. A comparison of various social behaviors, not discussed in this chapter, grooming, play, and particularly vocalizations, further emphasizes the similarity between the two species. The adaptability of the two species is indicated by their occurrence in a variety of habitats.

Despite these similarities G. *crassicaudatus* and G. *senegalensis* are best adapted for life in different habitat types for which they show distinct preference. In general, the smaller G. *senegalensis* are found in a greater variety of habitats at higher population densities than G. *crassicaudatus*. They breed more rapidly, require less food of a type which is widespread in its occurrence, and have a smaller home range. In addition, they have greater agility, require less cover, and are able to exploit the smaller branch zone.

While the two species live sympatrically the population densities of one or both are lower than in those regions where they are found alone. This may be due to interspecific competition or because the vegetation does not represent an optimum habitat. In fact, competition appears to be minimal; each species typically occupies different parts of a shared habitat and no antagonism or fear was seen during interspecific encounters. G. *senegalensis* spend more time in the open bush or at the periphery of riparian bush or forest where G. *crassicaudatus* spend most of their time. Only in Rhodesia where G. *crassicaudatus* have adapted to a more open habitat than elsewhere, were there indications of some competition for food and where the population density of G. *senegalensis* was far lower than elsewhere.

The highest population densities of G. *crassicaudatus* were found in dense subtropical evergreen forest while G. *senegalensis* are most abundant in open *Acacia* thornveld savanna. It is possible that G. *senegalensis* are least liable to predation in open areas where they are able to spot danger from a distance and react appropriately. *Galago crassicaudatus*, being larger and slower, have greater requirements for cover while sleeping during the day. Their relatively quiet movements in dense vegetation may not only enable them to avoid detection, but also ensure that they are not taken by surprise, since arboreal predators are more abundant and better concealed in forest areas than in open woodland. The fact that G. *crassicaudatus* may also be found in open regions, however, indicates that their type of locomotion is not a major factor limiting their distribution. Given sufficient cover for a daytime retreat the most important ecological factor which influences distribution appears to be the availability of trees bearing edible fruits. Where bushbabies are absent from suitable habitats, which are not separated by major physical barriers, their present distribution may partially be explained by considering the influence of disease, epidemics, veld fires, or the destruction of natural vegetation for development schemes, which may be disastrous in their effects.

As for the other primates indigenous to southern Africa, man is the greatest threat to both species of bushbaby. Although neither species of bushbaby is yet being used to any great extent in biomedical research, unlike the chacma baboon, which is being used in research of a terminal nature and which is also being exterminated as vermin in many parts of the country, both species of bushbaby, seldom seen and relatively unknown, are at a disadvantage in that no thought is given to the possible presence of bushbabies in areas earmarked for housing or agricultural developments.

Acknowledgements

This chapter is based largely on the work of the second author while a candidate successively for the M.Sc. and Ph.D. degrees under the supervision of the first author. The research was made possible by ad hoc grants from the National Geographic Society, Washington, D.C., a Larger Grant from the Human Sciences Research Council of the Republic of South Africa, and by continuing support from the University Council, University of the Witwatersrand, Johannesburg, to each of which the authors wish to express their indebtedness. Most of the field research on which this chapter is based was carried out on the farms "Mosdene" and "Dindinnie," owned respectively by Mr. E. A. Galpin and Mr. Gordon NcNeil, two people among a host of far-sighted N.Transvaal farmers who have a vivid appreciation of wildlife in all its forms and who have devoted a large part of their lives, energies, and resources to both preserving and perpetuating it.

References

Andersson, A. B. (1969). "Communication in the lesser bushbaby (*Galago senegalensis moholi*)." Unpublished M.Sc. dissertation. University of the Witwatersrand, Johannesburg.

Astley-Maberly, C. T. (1967). "The Game Animals of Southern Africa." Nelson, Johannesburg.

Bearder, S. K. (1969). "Territorial and intergroup behaviour of the lesser bushbaby, *Galago senegalensis moholi* (A. Smith), in semi-natural conditions and in the field." Unpublished M.Sc. dissertation, University of the Witwatersrand, Johannesburg.

Bearder, S. K. (1975). "Aspects of the ecology and behaviour of the thick-tailed bushbaby, *Galago crassicaudatus*." Unpublished Ph.D. thesis. University of the Witwatersrand, Johannesburg.

Bearder, S. K. and Doyle, G. A. (1974). *In* "Prosimian Biology" (R. D. Martin, G. A. Doyle, and A. C. Walker, eds.), pp. 109–130. Duckworth, London.

Buettner-Janusch, J. (1963). *In* "Evolutionary and Genetic Biology of the Primates" (J. Buettner-Janusch, ed.), Vol. 1, pp. 1–64. Academic Press, New York.

Buettner-Janusch, J. (1964). *Folia Primatol.* **2**, 93–110.

Cansdale, G. S. (1960). "Bushbaby Book." Phoenix House, London.

Charles-Dominique, P. (1971). *Biol. Gabon.* **7**, 121–228.

Charles-Dominique, P. (1977). "Ecology and Behaviour of Nocturnal Primates." Duckworth, London.

Charles-Dominique, P. and Hladik, C. M. (1971). *Terre Vie* **25**, 3–66.

Davis, M. (1960). *J. Mammal.* **41**, 401–402.

De Boer, L. E. M. (1973). *Genetica* **44**, 330–367.

Doyle, G. A. (1974). *In* "Behavior of Nonhuman Primates" (A. M. Schrier and F. Stollnitz, eds.), Vol. V, pp. 155–353. Academic Press, New York.

Doyle, G. A. (1977). *In* "The Study of Prosimian Behavior". (G. A. Doyle and R. D. Martin, eds.) in press. Academic Press, New York.

Doyle, G. A. and Bekker, T. (1967). *Folia Primatol.* **7**, 161–168.

Doyle, G. A., Andersson, A., and Bearder, S. K. (1969). *Folia Primatol.* **11**, 215–238.

Doyle, G. A., Andersson, A., and Bearder, S. K. (1971). *Folia Primatol.* **14**, 15–22.

Ehrlich, A. (1968). *Folia Primatol.* **8**, 72–76.

Gerard, P. (1932). *Arch. Biol.* **43**, 93–151.

Haddow, A. J. and Ellice, J. M. (1964). *Trans. Roy. Soc. Trop. Med. Hyg.* **58**, 521–538.

Hall-Craggs, E. C. B. (1965). *J. Zool.* **147**, 20–29.

Hall-Craggs, E. C. G. (1974). *In* "Prosimian Biology" (R. D. Martin, G. A. Doyle, and A. C. Walker, eds.), pp. 829–845. Duckworth, London.

Hill, W. C. O. (1953). "Primates: Comparative Anatomy and Taxonomy. I. Strepshirini." Edinburgh U.P., Edinburgh.

Hill, W. C. O. and Meester, J. (1969). *In* "Smithsonian Institution Preliminary Identification Manual for African Mammals" (J. Meester, ed.), pp. 1–8. Smithsonian Institution, Washington, D.C.

Jay, P. (1965). *In* "Behavior of Nonhuman Primates" (A. M. Schrier, H. F. Harlow, and F. Stollnitz, eds.), Vol. II, pp. 525–591. Academic Press, New York.

Jolly, A. (1966). "Lemur Behavior: A Madagascar Field Study." University of Chicago Press, Chicago, Illinois.

Kaufman, J. H. (1962). *Univ. Calif. Pub. Zool.* **60**, 95–222.

Manley, G. H. (1966). *Symp. Zool. Soc. Lond.* **15**, 493–509.

Martin, R. D. (1968). *Z. Tierpsychol.* **25**, 409–495, 505–532.

Martin, R. D. (1972). *Phil. Trans. Roy. Soc. Lond.* **B.264**, 295–352.

Mason, W. A. (1968). *In* "Primates: Studies in Adaptation and Variability" (P. C. Jay, ed.), pp. 200–216. Holt, Rinehart and Winston, New York.

Pariente, G. (1974). *In* "Prosimian Biology" (R. D. Martin, G. A. Doyle, and A. C. Walker, eds.). pp. 183–198. Duchworth, London.

Pinto, D., Doyle, G. A., and Bearder, S. K. (1974). *Folia Primatol.* **21**, 135–147.

Roberts, A. (1951). "The Mammals of South Africa." Central News Agency, Johannesburg.

Pringle, J. A. (1974). *Ann. Natal. Mus.* **22**(1), 173–186.

Shortridge, G. C. (1934). "The Mammals of South West Africa," Vol. 1. Heinemann, London.

Smithers, R. H. N. (1968). "A Check List and Atlas of the Mammals of Botswana." The trustees of the National Museums of Rhodesia, Salisbury.

Strühsaker, T. T. (1967). *Ecology* **48**(6), 891–904.

Walker, A. C. (1969). *E. Afr. Wildl. J.* **33**, 249–261.

2

The Aye-Aye

JEAN-JACQUES PETTER

The aye-aye, *Daubentonia madagascariensis* (Fig.1) originally described by the naturalist Gmelin in 1788, is one of the most curious of primates. The aye-aye, due to its peculiar diet and ecology, is an exceptional example of adaptation among mammals. It is, in fact, because of this specialization that this species is at present one of the most endangered.

Of particular interest is the fact that the aye-aye is a result of an evolutionary radiation which produced a great variety of forms in Madagascar. This speciation was made possible by the early separation of Madagascar from the African continent, the availability of a multitude of ecological niches due to the varied geographical and climatological factors, and a lack of predators.

A great deal of the information obtained on the aye-aye resulted from a recent

Fig. 1. Aye-aye *Darebentonia madagascariensis.*

study designed for the conservation of the species. A number of captures were undertaken for release on an island near the coast of Madagascar, which has been established as a special reserve.

I. Systematics and Description

Daubentonia madagascariensis is the sole living representative of the genus *Daubentonia,* which also comprises a special family, Daubentoniidae, within the group of Malagasy lemurs. Its peculiar characteristics, even from the time of its discovery, astonished naturalists.

The aye-aye is comparable in size to a lemur; its body measures 40 cm in length and has an equally long tail. The pelage, for the most part black, is made up of short hairs, as well as long stiff hairs. The head is short, and the ears are extremely large. The incisors are well developed and have continuous growth. The third digit of the hand is long, thin and very mobile. The only mammary glands are inguinal in *Daubentonia*.

The aye-aye was first discovered by the naturalist Sonnerat in 1780. The genus *Daubentonia* was originally created by Geoffroy St-Hilaire in 1795. The name should be preserved because it precedes that of *Chiromys* proposed by Cuvier in 1800. Initially its systematic position was uncertain, and remained so for 100 years. It was originally believed to be a type of squirrel, and, in fact, Gmelin classed it with rodents. It was not until 1859, when Sandwith sent a specimen to Owen, that its taxonomic position was clarified. Owen clearly demonstrated, in a careful anatomic study published in 1866, the fundamental differences between *Daubentonia* and rodents. The taxonomic difficulties explain the numerous names given by different authors. At present, only one species is recognized within the genus: *Daubentonia madagascariensis* (Gmelin 1788). The skeletal remains of a larger form from the southwest of Madagascar, which apparently disappeared less than 1000 years ago, have been described as a separate species, *Daubentonia robusta* (Lamberton, 1934).

II.   Description—Variability

The absence of specimens prevents the determination of the existence of two subspecies of *Daubentonia madagascariensis* as some authors hypothesize. The slight variations of the pelage would be due simply to the differences in age of the captured individuals.

The body is in general almost entirely black or dark brown. Only the face and the underparts are slightly whitish in color. The long body hairs are lightened at the extremity.

A subspecies, *D. laniger*, whose hairs are more woolly, has been described by Grandidier (1928). The pelage is light coffee colored, with the white tips less marked, and the ears are longer. Schwarz (1931) believed that the specimen was in molt, but Lavauden (1933) stated that the tail was relatively short, and the woolly pelage did not resemble a molt. He concluded that two forms probably exist: one living in the coastal forest and a second woolly form living at higher altitude (500—800 m). According to Lavauden, certain wood-cutters recognize the existence of two forms: one in the primary forest and the other inhabiting the secondary forest. We have not been able to obtain any additional information that would clarify the situation; unfortunately, it is improbable that a second form still exists. It is, however, very possible that a subspecies does exist due to localized areas of isolation within the habitat, as is the case for other lemurs.

A series of captures undertaken in 1966 and 1967 for the purpose of introducing endangered species in a special reserve did not reveal any great variation in the pelage color between individuals. However, all the captures were made along the coast. At present, no information has been obtained on the existence of an aye-aye living in forests at higher altitudes.

All of the captured adults showed little variation in pelage color. During nighttime observations of the animals, the head may appear lighter than it actually is. However, this is simply due to the fact that the white hairs reflect the light of a flashlight with greater luminosity.

One old female with prominent teeth appeared slightly darker. The pelage contained fewer long white hairs than those of the other. Young males have a lighter colored head than the females, especially the adults.

A male in captivity lost a large part of its pelage within a few days during June, 1967, although there was no change in the diet. Within 15 days, it was replaced.

At the time of their capture, the animals had a pelage in relatively good condition. An adult female lacked almost all the long hairs, and it is possible that a sort of molt affecting these hairs occurs periodically every 1 or 2 years.

Unlike other lemurs, the tail is always fluffed, and appears larger than it actually is. The hairs of the tail are approximately 10–12 cm in length.

III. Anatomy

The anatomic study of the aye-aye, whose strangeness excited the curiosity of early anatomists, has progressed considerably since the work of Owen. It has been shown that most of its anatomy clearly resembles that of other lemurs, except for certain characteristics that are clearly different. Most important of these are the peculiar dentition, which originally resulted in its classification with rodents; the third finger of the hand which is long, thin, and very mobile; the presence of claws on all the fingers and toes except the hallux; the globular skull with a primitive brain; and the absence of a seminal vesicle and the inguinal position of the mammary glands.

The head is short and large, the chin reduced, the neck short but powerful, and the large tail is slightly longer than the body. The eyes are prominent and oriented toward the front. The well-developed ears form two large lobes which are very mobile. The posterior appendages are longer and stronger than the anterior ones. The skin is uniformly covered with hair except for the tip of the snout, the extremity of the appendages, the scrotum, and the interior of the ears. On the palm, only the interdigital spaces possess small-scaled flattened pads. The surface of the soles of the feet is covered with warts without crests except for the interdigital spaces.

The slight development of the pads of the palm and the sole, which clearly distinguishes it from other lemurs, is probably related to the well-developed claws, whose presence results in a lessened aptitude for leaping and a modification in the climbing technique.

The skull of the aye-aye is remarkable for its width and the reduction of the facial part in front of the orbit. The premaxillary, however, is well developed. The frontal and sphenoidal sinuses are very large and of greater dimension than those of other lemurs. The condyle of the mandible is lowered, an adaptation to longitudinal movements.

The dentition resembles that of the rodents, and the dental formula is difficult to interpret. It may be designated with some uncertainty as $\dfrac{1 \quad 0}{1 \text{ or } 0 \quad 0 \text{ or } 1}$ $\dfrac{1 \; 3}{0 \; 3}$. A large space separates the incisors from the other teeth, which are very small. The crown of the molars is flat without a distinct tubercule, which is explained by the shape of the condyles. The anterior teeth resemble those of a rodent: long and solid; growth is continuous with permanent pulp and open root. Enamel exists only on the anterior surface. The first superior is tooth I_1 or I_2. The lower is perhaps a canine. According to Peters (1866), who studied tooth replacement in a young female, the temporary dental formula would be $\dfrac{2 \; 1 \; 2}{2 \; 0 \; 2}$, and the specialized tooth is an incisor which appears at birth between two temporary incisors. Only three of the temporary teeth are replaced definitely. The two segments of the inferior mandible are not fused along the median, but are attached by ligaments which permit a certain independence of these two parts during use of the incisors.

One of the most curious traits of the aye-aye is the existence of a specialized third finger. It is shorter in length than the fourth digit, and also smaller in diameter. It is extremely dexterous, due to the development of the metacarpal which is extended beyond the palm and forms a sort of supernumerary phalange, and is particularly long and dextrous. However, it is weak and is incapable of gripping an object. Thus, it is not used in the normal functions of the hand, but is used only as a probe or as an instrument to feel or scratch. The finger is clawed as are all the other fingers of the hand.

The foot is similar to that of *Lemur*; the well-developed thumb is opposable to the rest of the hand which results in a firm grip. The thumb has a flattened nail, while the other fingers are clawed.

IV.   Ecoethology

Although the anatomy of the aye-aye is well known, its eco-ethology is still little understood. The few observations available are from the studies of the early naturalists, principally Winson (1855), Sandwith (1859), Lavauden (1933), and Rand (1935), and from a study I first undertook in 1957, and continued in 1967 with A. Peyrieras to conserve the species.

A.   Posture

The immobile posture is similar to that of *Lemur*. At rest, *Daubentonia* is generally rolled up into a ball, as is the *lemur*, with its large tail covering its body. In the nest, it frequently sleeps on its side. During feeding, it can assume various postures, clinging to a vertical trunk or remaining suspended underneath a large branch.

B.   Locomotion

According to our observations, locomotion of *Daubentonia* in branches is similar to that of *Lemur*. However, it is less adept in horizontal locomotion, although its claws permit greater possibilities for clinging to a large vertical trunk, for example, in search of xylophagic larvae. A film "The Aye-Aye of Madagascar" (Petter 1967) taken with the help of Peyrieras and Gobert during several missions to Madagascar, and notably the release of several captured aye-aye on the island of Nosy Mangabe, illustrates well locomotion under different conditions.

The climbing of a vertical trunk is performed by rapid, successive leaps, the four appendages gripping the trunk at almost the same moment. The claws permit clinging to very thin branches. *Daubentonia* are even capable of hanging by their rear legs in a rather unique manner by hooks formed by their claws.

Their movement in branches is clearly less rapid than that of *Lemur* and the leaps are shorter. Often, especially in moving from the ends of one branch to another, the animal seems to take great care not to fall, and only releases one branch when it has hold of another.

As our captures have shown, their lack of fear of man, as well as their slow locomotion allows an agile climber to easily trap them in the branches.

When an aye-aye leaps to the ground, the four legs touch at the same time. A young animal observed in captivity sometimes touched the ground with its teeth due to the weight of its large head. Locomotion on the ground appears very mechanical because the arms make long forward movements. This peculiarity is caused by the presence of the third finger which must remain folded underneath the hand. The fingers of the hand are raised and do not touch the ground.

Thus, pressure is exerted, especially on the palm, or the extremity of the palm, which prevents rapid displacement.

This clumsiness, however, does not prevent the aye-aye from frequently descending to the ground, as we have observed in the field. They are even capable of making long trajets on the ground in degraded areas. They prefer to follow a path or a clearing, but upon reaching a tree or vertical object, their first reflex is to climb. Their nests are often found in trees near a path, and an animal was even observed marching on a path near a village early one morning in August.

C. Manipulation

The anatomic specializations of the aye-aye in relation to the diet (form of the third finger and the development of the incisors) are remarkable. The manner in which the third finger and the incisors are used to search for and eat insect larvae, as well as to empty coconuts, or to drink are described later in the chapter. Generally, the principal function of the third finger is as a probe, for example, to scratch and clear an interesting area such as the entrance to the hole of a larva. The rapid and precise movements of this extremity resemble that of an engraver during a delicate operation. We have never observed, however, the third finger used as percussion instrument, as has been described by early observers.

The third finger is a very specialized instrument which, as we have stated above, handicaps the prehensile and locomotive functions of the hand. The aye-aye can grip and hold small objects only with great difficulty. In captivity, for example, he does succeed in blocking an egg on the ground with the palm on one of the hands in order to empty it with the third finger of the other hand.

V. Rhythm of Activity

The aye-aye leaves its nest at nightfall (6:00 PM during November, 6:30 PM at the end of December) if it does not rain too hard. During the first few minutes it is easy to follow. If there is no moonlight, however, observation becomes nearly impossible, because the animal is capable of rapid and silent movement in treetops. An electric flashlight is useful, but continued use causes the animal to flee. In bright moonlight, observation is relatively easy because the animal can be followed without too much difficulty, the flashlight being used only periodically to observe particular activities.

Return to the nest occurs with the same regularity as departure, that is, just before sunrise. At the beginning of November, this occurs at 4:00 AM, during

December at 3:30 AM, even before the least amount of light is detectable by the human observer.

Except during heavy rainfall, we have never observed the aye-aye to remain immobile for any period of time, as is the case for other nocturnal lemurs. Upon leaving the nest, it often remains for several minutes in nearby branches, and, suspended by its legs, scratches itself with its foot. It may also collect, using its teeth, some branches from a nearby tree for nest material, to add to the existing nest. The animal then proceeds to the fruit trees (coconut and mango) where it begins to feed. After 1 or 2 hours, it begins to search for insect larvae in the dead branches.

The general impression of the nocturnal activity of the aye-aye is one of continual inspection. One hour may be required to progress 100 m. However, it is capable of much more rapid movement, and it is nearly impossible to follow an aye-aye which flees into obscurity even if there are no obstacles in the underbrush.

VI. The Nest

In the border zones of villages where the aye-aye still exist, the trees chosen as nest sites are most often coconut trees. An aye-aye often constructs and utilizes several nests (two to five), situated at a height of between 10 and 15 m. The tree in which the nest is located often contains a large number of lianas. Old nests are often reoccupied, in which case they are in part repaired, and fresh branches and leaves can be distinguished.

The nests are also found in "Badamiers" (*Terminalia catappa* Linn.), "Hintsyna" (*Afzelia bijuga* A. Gray), "Copaliers" (*Trachylobium verrucosum* Gaertn.), as well as in mango and litchis. Three nests have been observed in epiphytic ferns in the form of a cup attached to vertical trunks. Two nests have been found in coconut trees, one of which was constructed entirely with the fronds of the coconut palm.

The nest is most often in the form of an oval. If the tree is very dense, the upper part of the nest may be only loosely constructed. A clear opening of about 15 cm in diameter is always present on one side of the nest. The entrance is blocked by the animal when it is in the nest, and there is no other opening for an exit. The orientation of the entrance varies and depends on the orientation of the branch which is most favorable for entry.

The nest is often constructed in the fork of a tree or in a dense tangle of lianas. Construction material is generally collected from the same tree in which the nest is situated, or from nearby trees, which results in a mixture of material. A recent nest constructed on an old nest was composed entirely of branches 50 cm in length of "Longozo" (*Ammomum danielli* Hook) which the animal must have collected on the ground. It was completely surrounded by urticant lianas, genus *Figus* (Urticaceae) and *Smilax* (Liliaceae) (Avetro in Betsimisaraka), which were not utilized for the nest.

The walls are made of branches with leaves, more or less interlaced. The roof may be formed from separate leaves, simply laid flat and held between the branches that form the walls. The ordinary rat (*Rattus rattus*) often found near the villages, may also construct round nests in dense lianas, which may be mistaken for those of the aye-aye. However, these nests are made of leaves only. We found, however, a nest constructed entirely of the large leaves of the tree by an aye-aye in a Badamier. The interior of the nest did not contain a branch, which, according to certain Malagasy legends, serves as a "pillow."

During construction of the nest, the animal arranges the branches without trying to interlace them. The lateral walls are constructed around the platform on which the animals interlace the branches around themselves. New branches are carried to and added to the nest on the same side. Some are simply placed on the nest, while others are forced in between existing branches. The branches, which may be as much as 2 cm in diameter, are rapidly sectioned by the incisors, and transported between the teeth to the nest. The branches cut for the nest are primarily from 2–4 m, or less from the nest site, rarely further away. About sixty branches may be used to construct the nest. The construction may take 1 hour, during which time the aye-aye is constantly active, collecting and transporting material. The construction of the nest generally takes place in the morning, before sunrise. This is also the time during which fresh branches are added to existing nests.

Contrary to certain reports, the nest is occupied for several days in succession. It may be abandoned momentarily, then reutilized. An accident may result in the nest being abandoned. For example we observed an aye-aye in its nest in a coconut palm for 2 days. The second day a large branch fell from the tree just at the moment when the aye-aye left the nest. The next morning it did not enter the nest. In another incident, however, a nest that we disturbed to force the animal to leave during the day was repaired and reoccupied the next morning.

An abandoned aye-aye nest may serve as a refuge for a variety of animals. Often ants, which are particularly abundant, may invade the nest. We also found, on separate occasions, rats, a *Cheirogaleus*, and a Tenrecidae.

VII. Diet

The diet of the aye-aye consists of various fruits and larvae. All of its nocturnal activity is concerned with the collection of food. A surprising amount of energy, unusual for their size, is expended in the search for larvae.

Fruits constitute an important part of the diet; coconuts especially, which are always present near the villages, replace other fruits no longer in season. The coconut is always attacked while it is still on the tree. The aye-aye chooses those which are not completely ripe (more or less green). At this time, it is full of "milk," and the soft pulp is about 0.5 cm thick. The coconut is opened with the teeth, accompanied by violent movements of the head, during which the fibres on one side are torn open. After this, a hole of 3—4 cm is made with the incisors. This noisy operation requires about 2 minutes. The juice and the pulp are extracted almost completely in spite of the smallness of the opening. In order to accomplish this task, the third finger is extended into the opening and rapidly withdrawn to the mouth. The movements are extremely rapid, and in this manner all is extracted except for a small amount of pulp, just opposite the hole. The emptied coconut is abandoned on the tree where it may remain for several days before it falls. At the base of these trees, a large number of coconuts may be found in which the outer fiber has been torn open in characteristic fashion, and with a small round hole in the nut.

The aye-aye also eats litches and mangoes when these are in season; of the latter only the ripened part of the fruit is consumed. The mango is often carried between the teeth for several meters, and then, while suspended by the feet, the aye-aye will leisurely consume the fruit which is held and turned between the hands.

The search for insect larvae in dead branches occupies an important part of the nocturnal activity. The animal advances slowly along the branches, sometimes suspended from a horizontal branch above, and carefully examines every centimeter. It is difficult to determine whether detection is by sound or odor. The nose is positioned close to the branch during the search, but during close inspection either the snout or the ears may be oriented toward the bark. The search along the dead branches may lead the animal almost to ground level. One night we observed an aye-aye on a fence 1 m above the ground in the search of larvae.

When a larva is detected, the bark is forcefully attacked with incisors (Fig. 2). Once the hole is made, the third finger is introduced as a probe and agitated several times in a turning, back and forth motion before being withdrawn in a rapid movement to the mouth (Fig. 3). This may be repeated several times during which the larva seems to be reduced to a pulp, rather than extracted as a whole.

Another important food source is found in the larvaes which parasitize the nut of the Badamier fruits. In January, when it does not rain too hard, at least one aye-aye was found every night in a Badamier in which the fruits (the size of a small apricot) were still green. At first, we thought we were observing an animal consuming an enormous quantity of these fruits, but, after examining the remains on the ground, we found that the fruit and nut had been opened simply to extract from the interior, a small larva measuring 7 mm.

Fig. 2. Aye-aye biting the bark of a branch in search of the hole of a larva.

The animal collects the fruit with its teeth, then sitting or suspended by the feet, begins to open the fruit. Holding it in its hands, the animal removes the green pulp from one end, and then opens the hard nut on one side. This may require 20–30 seconds, and the noise may be heard from quite a distance. The fruit is then held in one hand and the third finger of the other hand is introduced to rapidly extract the larva (Fig. 4) which is passed laterally to the mouth, in the free space just behind the incisors. This is repeated two or three times after which the fruit is dropped to the ground and the animal proceeds to search for another. We observed an aye-aye for 1 hour in a Badamier during which about forty larvae were consumed, with an expenditure of energy that seemed out of proportion to the amount of food collected.

We also observed the aye-aye on several occasions in the "bread-fruit tree" (*Artocarpus*). It seemed to lick the terminal buds of the branches, however, we were not able to determine if the animal was interested in the plant itself or in a parasitic larva as in the Badamier fruits. None of the buds that we collected contained insect larvae, but the time and care spent in inspecting the tree gave the impression that the animal searched for prey.

The incisors are utilized primarily to tear open bark, or the fibers of the coconut. The muscles of the neck are well developed, and provide for great force in tearing. A young aye-aye maintained in captivity began to tear open coconuts at

Fig. 3. Aye-aye using his slender finger to collect larvae under the bark.

Fig. 4. Aye-aye searching for larvae in a fruit with his slender finger.

the age of approximately 5 months. The teeth are also used to attack wood or to make a hole in a nut. Gouging is accomplished by the inferior incisors; the superior incisors serve as a point of leverage. As we stated above, the two inferior demi-jaws are not fused by a median symphisis as in other lemurs. This characteristic allows the inferior incisors to operate separately on two different levels, which enables the animal to adapt more efficiently to the form of the object.

The incisors are also used to consume fibrous fruits, such as the mango, which are scraped in a fashion similar to that of other lemurs, whose lower front teeth form a "tooth comb." The sap of Hintsyna is collected in the same manner; after the bark is removed the sap is delicately scraped, with the lower incisors, from the galls which are formed on certain trees. It is with the incisors of both jaws, as well, that a large larva is eaten while being held in both hands.

In order to eat the majority of fruits, however, the third finger is used as often as the incisors. A banana is scraped using the third and fourth fingers in successive, rapid motions to transport food to the mouth.

In captivity the aye-aye can be maintained in good health with a varied diet. They will accept coconuts, eggs, rice with bouillon and chicken blood, various fruits, sugarcane, honey, and insect larvae [various species of coleoptera]. A colony of coleoptera larvae breeding in sawdust provided us with a dozen larvae every night. This seems to be a preferred food item of the aye-aye; and they will easily consume two dozen or more, and will even accept dead ones. In fact, all types of larvae presented were accepted.

When an aye-aye finds an egg, it is able to transport it by holding it in its mouth, behind the incisors. To consume an egg, the same technique is used as for a coconut. The egg is held in one hand, and after a hole is made with the teeth, it is emptied using the third finger. The movement of the third finger occurs about twice a second to extract the egg contents toward the tongue which is extended just beyond the incisors. After about fourty movements, the aye-aye changes hands. One adult used the fourth finger of the left hand to extract the contents of an egg.

The aye-aye has never been observed to accept spontaneously an adult insect. A young in captivity ate several "hannetons," but it was necessary to hold the insect for the animal. In fact, the aye-aye would flee from large insects which were attracted to a nearby lamp.

VIII.   Reproduction

In spite of our efforts, no observations of reproduction has even been recorded for the aye-aye in captivity. The aye-aye may reproduce only every 2 or 3 years. According to local woodcutters and our own observations, there is only one young, which is born in October—November.

Among our captured animals, a young male about 1 year old had only small, partially developed testicles, about 1.5 cm long. The mammary glands of a young 2-year-old female were not yet apparent. These two young were still with their mother: The young maintains the habit of sucking for a long period of time, and often makes these movements during its sleep. When a young sleeps with its mother, they often have the habit of mutually sucking the clitoris and penis of each other while sleeping.

A couple of aye-aye, maintained in captivity for nearly 3 years with constant care and a carefully varied diet, never demonstrated any sexual behavior. They slept together in a nest constructed in a large box. The testicles of the adult male were well developed all year long (2.5 cm × 1.5 cm). The skin of the scrotum was black and hairless. The female was younger, probably in its third year at the time of capture, but certainly reproductively mature. In the nest the female always occupied the deeper position, and the two animals slept side by side, the head of one against the genital region of the other. We have noticed this position, which allows the young to find the inguinally positioned mammary glands without difficulty, to be typical.

IX.   Communication

We have been able to distinguish two types of sound characteristic of the aye-aye: a sort of "Ron-tsit" repeated in succession, which indicates alarm or danger, and a sonorous "creee," which is rarely heard in captivity. According to recent observations made during capture and release operations on the Island of Nosy Mangabe, the latter cry seems to be a signal of contact between the mother and the young. The young especially emits this cry when it is isolated from the mother. During an observation at Nosy Mangabe, this was clearly illustrated: a young, released with its mother, emitted this cry every 5 seconds when it was separated, until the mother responded with the same cry. The young descended immediately to rejoin the mother.

Both male and female aye-aye frequently urine-mark the ground or various supports. A young captive female frequently marked after a period of anger or aggression. She advanced while rubbing the clitoris against the ground, leaving a trace of several meters. A captive male marked an inclined trunk in the same manner after being disturbed. This type of marking also occurs when two animals come in contact. All the captured animals, as well as a young raised in captivity, release the same acid odor, more or less intense according to the individual, but very distinct from that of other lemurs.

The young captive female occasionally marked a branch by tearing away bark with its incisors. In the same way she could leave painful tooth marks on the arm of a caretaker. This behavior in association with urine-marking probably permits the adult to leave a lasting mark on the bark of a tree.

The aye-aye seems to be a nonsocial animal. We have only observed isolated individuals. Woodcutters have confirmed this finding. This isolation is relative, because in the forest of Mahambo three individuals were found within a diameter of 50 m, and two (probably a mother and its young) were often found close together. It is therfore possible that the social structure is primitive and similar to that of *Microcebus murinus*.

The rarity of the aye-aye and the destruction of the biotype prevent a precise determination of the size of the territory, which may vary depending on the richess of the milieu. According to our observations, however, an animal may be found up to 5 km from where it was first observed. In certain areas we found a male, a female, and 2 young within an area of about 5 km in diameter.

During the middle of the day, as we observed during the captures, the animals are nearly insensitive to noise or even human presence. It is necessary to touch the nest or even the animal to force it to leave the nest. The return to sleep of a female with her young after having seen a climber next to the nest is quite surprising.

The aye-aye, similar to other nocturnal lemurs, probably undergoes a cyclic lowering of the body temperature during the day. In fact, the animals are very lethargic at this time of day, and an animal removed from the nest generally begins to tremble. Those captured at the end of the day or at night are usually much more aggressive.

The aye-aye, with its claws and large incisors, is powerfully armed, but due to the absence of predators during its evolution in the Malagasy forest, has not developed an instinctive defense behavior. An arboreal carnivore, such as *Cryptoprocta ferox*, is certainly capable of killing an aye-aye at night.

During the day, when a climber arrives near a nest, the animal may leave the nest rapidly, then return tranquilly to it if one remains motionless. Most often, the usual behavior is to suddenly present the head at the entrance of the nest, accompanied by a hissing noise. The suddenness of this movement can easily catch a climber unaware and cause him to lose his balance, an incident that occurred to a technical aide during a routine insepection of an old nest which was thought to be empty.

When the animal is disturbed by a climber, he may emit a cry of danger, the "Ron-tsit" described above, which should suffice to scare off an eventual predator. This cry is often emitted at night when the animal is disturbed by the light of an electric flashlight pointed in its direction. The aye-aye is certainly rarely attacked by a predator while it is in its nest, in which it occupies a strong defensive position. A male attempting to enter an occupied nest was brutally rejected, which illustrates the presence of a certain aggressiveness among adult individuals.

All the young, captured in the nests except one, did not exhibit any aggression, and were taken without difficulty. Six adults were kept in the same cage for several days without any aggressive interactions. The young tamed female and a subadult male placed together for the first time did not display any particular reaction. The aggressiveness of the aye-aye is manifested by an aggressive posture and a hissing noise (similar to that of a cat) which serves to intimidate potential enemy.

Agressive and play behavior are very often similar as observed in the young captive female. At the age of 3 months, the animal passed long periods of time in play, which suddenly would change into a demonstration of aggression often occuring after a brusque movement during a play session with another animal. This involved an aggressive posture and hissing sound (or "Ron-tsit") as well as urine-marking on the ground. Urine-marking also occurred whenever someone suddenly entered the room. We would like to stress the sudden, abrupt and mechanical nature of the change between play and aggressive behavior. This transition is not at all comparable to that of any other lemur, and the impression is that the animal simply does not possess any intermediate behavior patterns.

X.   Geographic Distribution, Habitat, Protection

Despite the observations that have been made to date on the aye-aye, it is evident that the eco-ethology is still far from complete. The difficulties involved in observing these animals and the failure to obtain reproduction in captivity do not bode well for the survival of the species. For these reasons we have tried to provide for the conservation of the aye-aye in its natural habitat.

In order to accomplish this task, a bibliographic and field study was undertaken before trying to initiate the practical measures needed for conservation. According to the literature, the aye-aye existed on the east coast in the forested zones between Antalaha (north of Maroantsetra) and Mananjary and on the northwest coast between Ambilobe and Analalava. Kaudern, in 1914, suggested their possible existence in the forest of Ankarafantsika, north of the Betsiboka; but Lavauden, in 1933, was certain that they no longer existed there. Decary communicated to us his finding of an aye-aye south of Farafangana on the east coast, which would extend the range along nearly the entire east coast of Madagascar. Very little information is available from collections, since almost all the specimens only indicate "Madagascar." The few precise localities fall within the limits stated above. The specimen described by Grandidier in 1928 was captured (according to Lavauden) above Beforona, between Moramanga and Andevoranto (among the collections of the "Académie malgache").

Since the above information on the distribution of the aye-aye is relatively out-of-date, I attempted, with A. Peyrieras, to define its present status. Our search was carried out by periods of observation (day or night) and by interviewing woodcutters. The results may be summarized as follows. The number of aye-aye has been reduced dramatically in the last 30 years. Only rarely are the animals acutally killed by the woodcutters, but as any specialized forest-dwelling species the aye-aye disappears following destruction of the forest. In 1975, due to demographic pressure, the destruction of forests increased tremendously.

On the east coast, the costal forest north of Tamatave has disappeared almost entirely, either through exploitation or burning, sometimes replaced by imported tree species. The only remains are several isolated and partially destroyed areas such as Mahambo. In 1975, the aye-aye still exists, but the reserve itself is being increasingly destroyed and their future in this area is threatened.

The reactions of the local Malagasy people to the aye-aye are varied. In the region between Mananara and Rantabe, the aye-aye is strictly "fady" (taboo). Near Maroansetra, where the inhabitants are in general more civilized it is simply considered an animal of "bad luck."* If an aye-aye is found in a village, it may be necessary to kill it to remove the evil. The animal is killed, exposed for several hours in public, then buried with various rituals according to the area. It is never eaten. This actually occurred in January 1957 at Mariarano and March 1957 at Antanidambo. At Antanidambo, by a coincidence which was easily explained by the villagers, an old woman died the day after the aye-aye was killed. These villages are situated in the littoral forest which has been cleared by cutting and plantation to resemble more or less sparsely treed savanna.

In the Maroantsetra region, we have investigated up to 25 km inland, the forests of Rantabe to Beanana. We did not find any aye-aye, but according to the woodcutters, they are most often found near 3—to 6—years-old areas of "Tavy" (portions of the forest burned for plantation). They are attracted by trunks which contain a large number of larvae.

With A. Peyrieras (1970), we have also searched, to the south of Tamatave without success, to the limit of the interior forest. The woodcutters that we interviewed in the region were also very reluctant to speak of the "fady." According to them, the aye-aye exists in the forest, but it was impossible to convince them to show us any, and we did not remain in the area long enough to make any detailed search. It was in this area that Ursch (in Millot, 1952) obtained several specimens.

Although we have not been able to obtain substantial evidence, it is possible that the aye-aye existed at elevations of 1000—1200 m. However, this would be unusual, and the word of observers must be viewed with caution. For several stays in the region of Perinet, Fanovana, Ambodiriana, near Moramanga, we inter-

---

*The difference in the "fady" is mostly a matter of location rather than race. The Tsimehety and Betsimisaraka, who inhabit these regions, agree.

viewed as many woodcutters as possible. As is the case elsewhere, most are afraid of these animals, but several old villagers said they had seen the aye-aye and observed their nests. According to them, the aye-aye is found in the deep forest and is rare. In this region as well, the appearance of an aye-aye in a village is a prophecy of death.

We have not been able to obtain any information on their range in the regions of the south. Near Ranomafana and Ifanadiana, the people are not familiar with the aye-aye, even though they seemed to know the local fauna well. In the north, the aye-aye has been said to exist to the east of the Mountain d'Ambre. We found one that had been killed, and exposed at the entrance of a village near Sahafany. In the west, we have on several occasions visited the forest of Ankarafantsika (100 km east of Majunga) where Kaudern indicated the possible existence of the aye-aye. We have never found any trace of their presence.

It is certain, however, that the aye-aye still exists in the north in the "Baie of Ampasindava," where, in 1931 a Franco-Anglo-American expedition collected a single specimen. An animal was killed in February 1957 near the village of Ambaliha, and seems to be known by the Sakalava woodcutters of the region, who also consider the animal as "fady." As elsewhere, it is certainly very rare, and the rapid disappearance of the forests in the region of Sambirano, and even the mangrove forests, leaves little hope for the future.

At present the regions of the west where the aye-aye may still exist are difficult to penetrate, since they are far from the villages and roads. The same is true for the east, where, if it still exists, it is in the vast areas between 200−600 m where the forest is still intact. Difficult accessibility, humid climate, and tortured topography of this forest prevents penetration and research in this area.

Following our observations in the region of Mahambo, the Forest Service as early as 1957, agreed to conserve the corresponding zone of the forest as a reserve, and received the aide and encouragement of the WWF. However, fearing for the future of a conservation effort in this too frequented area (as the present situation confirms), additional steps for conservation were taken.

During the course of careful searches along the oriental coastal forest, 20 animals were located for eventual capture in the vicinity of certain villages. These animals were most likely condemned to extinction in the near future. It was decided that a special reserve would be created on the Island of Nosy Mangabe, situated in the Antongil Bay, in the northeast of Madagascar.

The island, which has a surface of 520 hectares was judged to be ideal. Approximately 6 m from Maroantsetra, in the northeast of Madagascar, it is entirely covered by dense forest. Thanks to the efforts of M. Ramanantsoavina, director of the Forest Service, the reserve was officially established in December, 1966. A path was established as well as a plantation of coconut and mango in a zone near the beach to provide a food supplement for the animals, After all arrangements were carried through, the captures were undertaken.

The first capture was made in October, 1966; thirteen other individuals were captured near various coastal villages thanks to observation and efforts of more than 1 year. Nine animals were released on Nosy Mangabe, in the presence of the local administration, and a film recording was made.

XI. Conclusion

The aye-aye, because of its numerous anatomic and behavioral traits, appears to be quite a grotesque representative of the Primates. Its specialized diet gives this animal a distinct advantage in interspecies competition. The ecological niche of the aye-aye is practically unoccupied by other species in Madagascar, where woodpeckers (Piciformes) the most notable predator of xylophagic larvae, are absent.

To our knowledge the only predators of larvae which may enter into slight competition are *Hypositta corallirostris* [a sort of small nuthatch in convergent form whose taxonomic placement is still uncertain], which is rare; and *Falcules pallista* (a Vangidae with a large recurved beak), which is abundant in the west and northeast, very efficient, but probably represents only a small incursion into the insect larvae resources of the Malagasy forest.

The abundance of the forest, which covered the entire island, permitted, however, the development of two forms of aye-aye, the larger of which has disappeared.

The extreme specialization which has developed in conjunction with the success of the species also imposes certain inconveniences, and the aye-aye is apparently handicapped in other areas which are necessary for survival. Thus, the specialized third finger results in decreased possibilities for prehension, and their encumbered, slow locomotion greatly hinders escape from a predator. The specialization of the incisors also limits the range of their use.

In compensation, the animal has conserved its nocturnal habit, and is well protected during the day in a solid nest that can be effectively defended. The teeth can inflict serious wounds. An aye-aye trapped by hand, could easily section a finger in one movement. However, the cryptoprocte (*Cryptoprocta ferox*), the largest Malagasy Viverridae, is capable of killing an aye-aye at night in the trees. The species has managed to maintain itself despite this slight predation. In fact, the increase in body size during the course of its evolution (Grandidier, 1928) probably contributes to its success by removing predation by the smaller predators.

The existence of claws on all the fingers (except the hallux) probably evolved in relation with the development and utilization of the incisors. They permit not only a solid hold during use of the teeth, but also a greater efficiency (compared to other lemurs) for climbing large trunks in a variety of positions.

Despite the process of specialization during its evolution, the species has not lost all its plasticity. Thus, the aye-aye which survives easily in zones lacking coconuts, is also perfectly adapted to this imported tree. In addition, it accepts a large variety of foods in captivity.

The nest may be constructed at various heights, in different species of trees, and with a variety of materials. It is capable of movement on a variety of supports, and survives in almost totally degraded forests provided that several large trees remain for nest construction.

The solid nest as well as the large thick tail are protections against external variations in temperature. Nevertheless, the aye-aye is limited to the coastal zone and is rarely encountered at elevations above 200 m.

The shelter of the nest allows the aye-aye to pass the day in a sort of lethargic state, which permits a certain reduction in thermal expenditure.

Present knowledge, unfortunately still fragmentary, of this extraordinary species illustrates its close relationship to other living lemurs. The chromosomic formula ($2N = 30$) permits the supposition of a cytogenetic evolution from a more primitive form similar to the genius *Microcebus*. The intermediate forms which would permit verification of this hypothese are, unfortunately, unavailable. This would have occurred during the Quaternary Period with the evolution of forms comparable to those of Eocène in Europe.

Future studies will permit a better understanding of the history and biology of the aye-aye, if its certain extinction can be avoided by a rational and durable protection of at least part of its environment.

References

Baron, L. (1882). Notes on the habits of the Aye Aye of Madagascar in its nature state. *P.Z.S. London,* 639−640.

Grandidier, G. (1928). Une variété de *Cheiromys madagascariensis* actuel et un nouveau Cheiromys fossile. *Bull. Acad. Malgache* **11**, 101−107.

Kaudern, W. (1915). Saügetiere aus Madagaskar. *Ark. Zool.* **9** (18), 1.

Lamberton, C. (1911). Contribution à l'étude des moeurs du Aye Aye. *Bull. Acad. Malgache* **8**, 129−140.

Petter, J. J. (1959). L'observation des Lémuriens nocturnes dans les forêts de Madagascar. *Natur. Malgache* **11** (1−2), 165−173.

Petter, J. J. and Petter-Rousseaux, A. (1959). Contribution a l'étude du Aye Aye. *Natur. Malgache* **11** (1−2), 153−164.

Petter, J. J. (1962). Recherches sur l'Ecologie et l'Ethologie des Lémuriens malganches. *Mem. Mus. Nat. Hist. Nat.* (A) **27**, 1−146.

Petter, J. J. (1962). Ecological and behavioral studies of Madagascar Lemurs in the field. *Ann. N.Y. Acad. Sci.* **102** (Art 2), 267−281.

Petter-Rousseaux, A. (1962). Recherches sur la biologie de la reproduction des Primates inférieurs. *Mammalia* **26** (Suppl. 1), 1−88).

Petter, J. J. and Peyrieras, A. (1970). Nouvelle contribution à l'étude d'un Lémurien malgache, le Aye Aye (*Daubentonia madagascariensis* E. Geoffroy). *Mammalia* **34** (2), 167–193.

Rand, A. L. (1953). On the habits of the Madagascar mammals. *J. of Mammal.* **16** (2), 89–104.

Sandwith, H. (1863). in Own R. On the Aye Aye. *Trans. Zool. Soc. London,* 33–101.

3

Conservation of the Brazilian Lion Tamarins (*Leontopithecus rosalia*)

ADELMAR F. COIMBRA-FILHO
and RUSSELL A. MITTERMEIER

I. Introduction

The lion tamarins* of the genus *Leontopithecus* are unquestionably the most endangered primates in the New World, and also rank high on the list of the world's most endangered mammals. They are restricted to patches of forest in southeastern Brazil and their habitat is rapidly being destroyed by the advance of civilization. In the past few years, these animals have received much international attention, as concerted attempts have been made to ensure their survival in the wild and in captivity (e.g., Bridgwater, 1972a; Hampton, 1972).

Although very little is yet known about the behavior and ecology of *Leontopithecus* in the wild, we have in several recent papers attempted to piece together the available data in the hopes that it will be of some use in future conservation efforts (Coimbra-Filho, 1969, 1970a, 1972, 1977; Coimbra-Filho and Mittermeier, 1973). Other papers have outlined the requirements for captive breeding of *Leontopithecus* (Coimbra-Filho, 1965; Hill, 1970; DuMond, 1971, 1972; Coimbra-Filho and Magnanini, 1972). The literature also includes a number of studies of the behavior of captive *Leontopithecus* (Ditmars, 1933; Frantz, 1963; Altmann-Schönberner, 1965; Epple, 1967; Snyder, 1972, 1974). Finally, several papers have attempted to advertise the situation of *Leontopithecus* and attract international attention to the disappearance of these unique animals (Perry, 1971, 1972; Magnanini, 1973; Mittermeier and Douglass, 1973; Magnanini *et al.*, 1975; Coimbra-Filho *et al.*, 1975; Kleiman, 1976).

In this paper, we briefly discuss taxonomy and the available data on distribution and ecology, review past conservation attempts on behalf of these animals, discuss the current status of wild populations in Brazil and what is being done to ensure their survival, give a progress report on captive breeding projects in the United States and in Brazil, and make several recommendations for future efforts on behalf of these monkeys.

---

*The vernacular names "marmoset" and "tamarin" may cause some confusion. The family Callitrichidae can be divided into two major subgroups on the basis of lower canine to lower incisor relationships, and some authors (e.g., Napier and Napier, 1967) use this difference to assign vernacular names. *Cebuella* and *Callithrix*, with lower canines roughly equal in length to lower incisors ("short-tusked"), are referred to as marmosets, whereas *Leontopithecus* and *Saguinus*, with lower canines much longer than lower incisors ("long-tusked"), are called tamarins. However, many recent authors refer to *Leontopithecus rosalia rosalia* as the golden lion marmoset (e.g., articles in Bridgwater, 1972a), and there is a tendency, especially among biomedical researchers, to refer to all callitrichids as marmosets. In this paper, we continue to call the members of the genus *Leontopithecus* lion tamarins, since we find it useful to distinguish between the two subgroups of the Callitrichidae by the vernacular names.

II.   Taxonomy, Distribution, and Ecology

A.   Taxonomy

The genus *Leontopithecus* belongs to the New World superfamily Ceboidea, family Callitrichidae. It contains one species, *Leontopithecus rosalia*, which is divided into three distinct subspecies: *Leontopithecus rosalia rosalia*, the golden lion tamarin (Fig. 1A); *Leontopithecus rosalia chrysomelas*, the golden-headed lion tamarin (Fig. 2B, C, D); and *Leontopithecus rosalia chrysopygus*, the golden-rumped lion tamarin (Figs. 1B and 2A). The characters distinguishing these three subspecies are discussed in Coimbra-Filho and Mittermeier (1972). The same paper also provides a detailed review of the taxonomic history of the genus, a list of vernacular names, measurements of museum specimens of *L. r. chrysomelas* and *L. r. chrysopygus*, and synonymies for all three subspecies.

The golden lion tamarin (*L. r. rosalia*) is unquestionably the best known member of the genus and has frequently been exhibited in zoos in many parts of the world. It was described by Linnaeus in 1766, but had been known to science for more than 250 years prior to his description. The first reference to *L. r. rosalia* is by the Jesuit Pigafetta, chronicler of the voyage of Magellan. Pigafetta observed the animals in the wild in December, 1519 and called them "beautiful simian-like cats similar to small lions" (Feio, 1953). In later centuries, *L. r. rosalia* was a popular pet among European royalty and it is even reported that one was owned by Madame Pompadour (Hill, 1970).

The golden-headed lion tamarin (*L. r. chrysomelas*) was discovered by Prince Maximilian zu Wied (1826—Fig. 3A) and described by Kuhl in 1820. This subspecies is very poorly known and was only once exhibited outside Brazil (in London in 1869) and once in Brazil (in Rio de Janeiro in 1961) prior to December, 1969 (Coimbra-Filho and Mittermeier, 1972).

The golden-rumped lion tamarin (*L. r. chrysopygus*) was discovered by Natterer in 1819 and described by Mikan in 1823 (Fig. 3B). This animal is the least known member of the genus *Leontopithecus*. It has never been exhibited outside Brazil and, prior to 1973, had never been kept in captivity.

B.   Distribution

The ranges of *L. r. rosalia* and *L. r. chrysopygus* have diminished tremendously in the past 150 years (Fig. 4A). The total range of *L. r. chrysomelas* was always quite small and has not changed considerably since the animal was first discovered. However, there has been substantial forest destruction within its small range, so that the actual forest habitat available to this subspecies has also decreased.

In the last century, *L. r. rosalia* was found from the western part of the former

Fig. 1. (A) Family group of *L. r. rosalia*. The male is on the left, the female on the right. The male was born in captivity to a wild-born male and a captive-born female; the female was wild born and captured in 1973. The infants were 40 days old when this picture was taken. (B) *L. r. chrysopygus* male with two 60-day-old infants born in the Tijuca Bank in 1974.

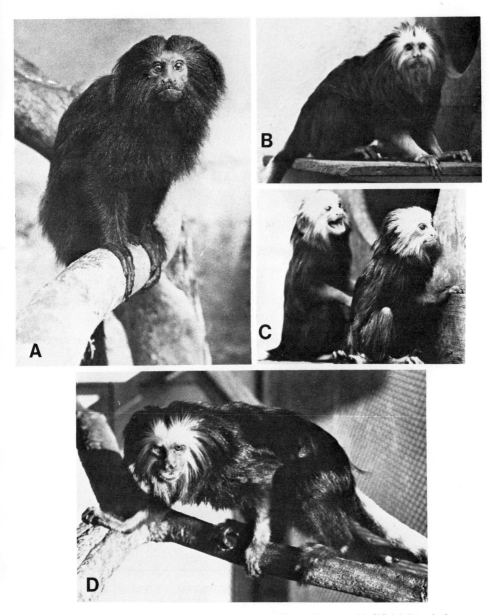

Fig. 2. (A) Young adult male *L. r. chrysopygus* about 2 years old. (B) Adult male *L. r. chrysomelas*. (C) Two 70-day-old *L. r. chrysomelas* juveniles born in the Tijuca Bank on September 19, 1974. (D) Adult female *L. r. chrysomelas* obtained in December, 1969 from the vicinity of Itabuna, Bahia. This female gave birth to three sets of hybrid *L. r. rosalia* × *L. r. chrysomelas* infants prior to the pure *L. r. chrysomelas* twins in Fig. 2C.

Fig. 3. (A) Plate of *L. r. chrysomelas* from Wied-Neuwied (1826). (B) Plate of *L. r. chrysopygus* from the original description (Mikan, 1823). Note the extent of light color pattern on the rump and legs and compare with Figs. 1B and 2A. The coloration of this part of the body is variable in *L. r. chrysopygus*.

state of Guanabara* to the southeastern part of the state of Espírito Santo, in seasonal tropical forest (Veloso, 1966) of the Atlantic coast (Fig. 4A, B). The geographic coordinates of the former range are roughly 23°S, 44°W to 20.5°S, 40.5°W. *Leontopithecus r. rosalia* was always most abundant in the Rio São João basin of the state of Rio de Janeiro. Habitat destruction has been so widespread that the animal is now entirely restricted to the remaining forests in the São João basin. Populations still exist in the municipalities of Silva Jardim, Casimiro de Abreu, Cabo Frio, Araruama, Sao Pedro da Aldeia, and possibly in forest

---

Fig. 4. (A) Map of southeastern Brazil showing the past and present ranges of the three subspecies of *L. rosalia*. (B) Map showing the present distribution of *L. r. rosalia* and the location of the Poço das Antas Biological Reserve in the state of Rio de Janeiro.

---

*The state of Guanabara, in which the *city* of Rio de Janeiro was located, united with the larger *state* of Rio de Janeiro in 1975. The resulting state is known as Rio de Janeiro.

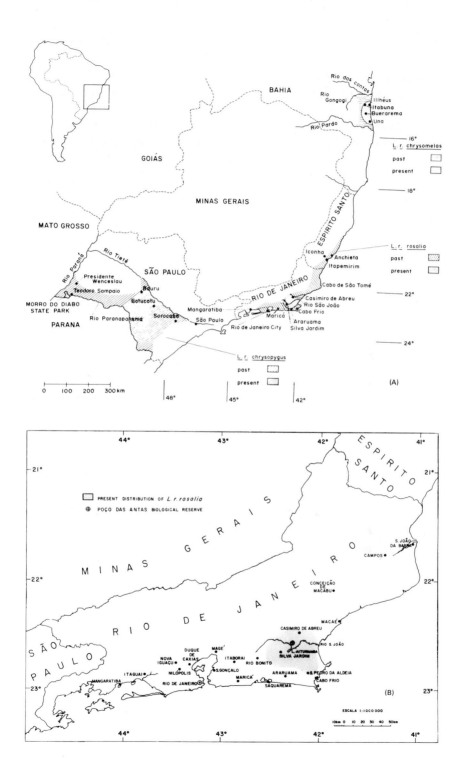

patches in Rio Bonito and Saquarema. However, the forests in all these municipalities are in most cases widely scattered and have already been subjected
to varying degrees of exploitation. The open areas of pasture and agricultural
land separating the remaining forests make it impossible for gene flow among
the few remaining *Leontopithecus* to occur. In 1973, we reported that the remaining forest habitat of *L. r. rosalia* was no more than 900 km² (Coimbra-Filho and
Mittermeier, 1973). Since that time, uncontrolled forest destruction has continued and the present habitat is probably considerably less (Fig. 5).

Leontopithecus r. chrysomelas inhabits Atlantic rain forest (Rizzini, 1963, 1967)
in the southern part of the state of Bahia (Fig. 6). The northern limit of the range
is the Rio das Contas (14°S), the southern limit the Rio Pardo (15.5°S), and the
western limit apparently the Rio Gongogi and its headwaters (40°W) (Fig. 4A).
The forest belt making up the range of this subspecies is only about 50 km wide
and originally stretched unbroken from the central Bahian plateau to within
a few kilometers of the coast. However, little of the original forest remains, since
the region is being heavily logged and has long been used for cocoa and rubber
plantations (Figs, 6B and 7). Cocoa, however, is grown by the "shading method"
(*cacaual cabrocado*) in which many large forest trees are left standing to protect
the cocoa plants from direct sunlight (Fig. 7). Consequently, there is some continuation of the forest canopy even within plantations, and *L. r. chrysomelas*
can sometimes be observed in or near this shade tree complex. *Leontopithecus
r. chrysomelas* can still be found in forests in the municipalities of Una, Buera-
rema, and Itabuna, and perhaps in Ilheus.

We previously reported that a population of this subspecies had been discovered in 1971 in the northern part of the state of Espírito Santo (A. Ruschi,
personal communication), near the border between Espírito Santo and Bahia
(roughly 18°S) (Coimbra-Filho and Mittermeier, 1973). However, this record
has been carefully checked and has unfortunately proved erroneous.

Leontopithecus r. chrysopygus once occurred in a large part of the state of São
Paulo, primarily in riparian forest that represented an extension of Atlantic
coastal forest inland along the major rivers (Rizzini, 1963, 1967). Its range was
bordered on the north by the Rio Tietê, on the south by the Rio Paranapanema,
and on the west by the Rio Paraná (Fig. 4A). However, forest destruction for
lumber, agriculture, and pasture land has proceeded even more rapidly in São
Paulo than in other parts of Brazil, especially since the turn of the century
(Figs. 8 and 9). Today, very little forest remains in São Paulo (Victor and
Montagna, 1970). As a result, the range of *L. r. chrysopygus* has been reduced to
a single 37, 147-hectare tract of forest, the Morro do Diabo State Park in the
westernmost part of the state (Figs. 4A, 8A, and 9A) (Coimbra-Filho and
Mittermeier, 1973).

Leontopithecus r. chrysopygus has always been a rare animal. After its discovery by Natterer in 1819, it was not heard of again for more than 80 years.

Fig. 5. (A) Forest habitat of *L. r. rosalia* in the municipality of Silva Jardim. This photograph was taken in 1968. The forest has since been cut down. (B) Destruction of *L. r. rosalia* forest habitat in Silva Jardim. This photograph was taken in 1974.

Fig. 6. (A) Forest habitat of *L. r. chrysomelas* in the municipality of Una, state of Bahia. This is one of the last tracts of primitive forest in the area. A similar tract of forest was offered to the IBDF in 1973. The forests in this region contain much valuable timber, including species like Brazilian rosewood (*Dalbergia nigra*), and are being cut down for rubber and cocoa plantations. (B) Destruction of primitive forest in Una. Note the areas where forests have been cleared to make way for rubber plantations. This photograph was taken in 1973.

Fig. 7. (A) Cocoa plantation in the municipality of Itabuna, state of Bahia. Note that many of the larger forest trees are left standing to provide shade for cocoa plants. *Leontopithecus r. chrysomelas* is sometimes found in the vicinity of such plantations. (B) Schematic representation of the "shading method" (*cacaual cabrocado*) used in cocoa plantations in the state of Bahia. The dotted lines indicate where trees from the original forest were removed.

69

Fig. 8. (A) Forest habitat of *L. r. chrysopygus* in Morro do Diabo State Park, municipality of Teodoro Sampaio, state of São Paulo. (B) Forest destruction caused by herbicide (2,4,5-T and 2,4-D). This photograph was taken in 1973 very close to the Morro do Diabo State Park. (Photo courtesy of O. Globo.)

Fig. 9. (A) Forest in Morro do Diabo State Park. Shown here is the road which was opened in 1970 and cuts the park in half. This road is now paved. (B) The fate of São Paulo's forests. The forest in this area was burned down to make way for pasture land.

Between 1900 and 1905, the São Paulo museum obtained several specimens. From 1905 to 1970, there was no further information on the animal and some authorities considered it extinct or nearly extinct (P. Hershkovitz, in litt., 1965 Coimbra-Filho, 1970a). However, in 1970, Coimbra-Filho rediscovered the species in the Morro do Diabo State Park and, in 1973, obtained seven live adults, six of which are currently in the Tijuca Biological Bank (the seventh died).

For a more detailed discussion of the distribution of the genus *Leontopithecus*, see Coimbra-Filho and Mittermeier (1973).

C.   Ecology

1.   Habitat

Leontopithecus seem to prefer primary forest habitat in areas where such habitat remains. *Leontopithecus r. chrysopygus* has thus far only been observed in the least disturbed forest in the Morro do Diabo State Park and, in the only study of primates within the range of *L. r. chrysomelas*, Laemmert *et al.* (1946) found this subspecies to be partial to old forest as well (Coimbra-Filho and Mittermeier, 1973). However, both *L. r. rosalia* and *L. r. chrysomelas* have also been seen in various kinds of secondary forest—perhaps because little primary forest remains within their ranges.

Within the forest, *Leontopithecus* are usually found between 3 and 10 m above the ground (Fig. 10A). This part of the forest is frequently characterized by heavy vine and epiphyte growth and interlacing branches, which provide excellent cover for the tamarins and their prey.

Leontopithecus frequently sleep in tree holes (Fig. 10B; Coimbra-Filho, 1977). If tree holes are not available (e.g., in some secondary forests), dense vine tangles or epiphytic communities can probably be used in their place.

2.   Diet

Leontopithecus are primarily insectivorous and frugivorous. Their diet includes a variety of insects (including members of the orders Blattariae, Orthoptera, Homoptera, Lepidoptera, some kinds of Hemiptera, and Coleoptera,

---

Fig. 10. (A) Alternate tree routes used by *L. r. rosalia* during foraging. (B) Tree hole occupied by two *L. r. rosalia* in a forest near Guapi, south of Lagoa de Juturnaíba, state of Rio de Janeiro. The hole was about 15 m up in the tree and about 8 cm in diameter at its widest point (indicated by an arrow in the inset). The forest in which this observation was made has since been largely destroyed.

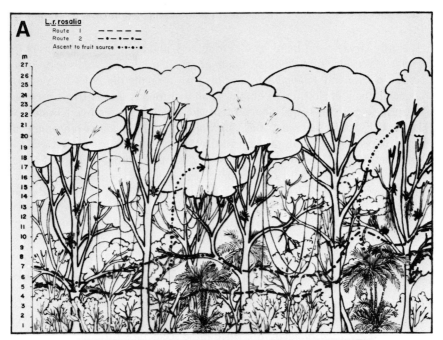

especially larvae), and many species of spiders, and even araneomorphs of the families Lycosidae and Ctenidae. Captive specimens also eat snails (e.g., *Brady-baena similaris* and *Bulimulus tenuissimus*), occasionally dig holes in the earth floors of their enclosures to obtain small arthropods and earthworms, and sometimes chew on strips of bark. *Leontopithecus* also eat small vertebrates like lizards (e.g., *Anolis punctatus, Anolis fuscoauratus,* and *Tropidurus torquatus*) and captive specimens readily accept mice. We have not observed predation of birds, but Ruschi (1964) reported a case of *L. r. rosalia* eating the eggs of *Turdus r. rufiventris* and saw wild *Leontopithecus* searching for hatchlings of other passerine species. Eggs of domestic quail (*Corturnix* sp.) are part of the regular diet of captive *Leontopithecus* kept in the Tijuca Bank.

Many kinds of fruit are also important in the diet of *Leontopithecus*. In the periodically flooded river margins of the Rio São João basin, *Ficus* spp. (Moraceae) and *Inga* spp. (Leguminosae Mimosaceae) are dominant species and provide food sources for *L. r. rosalia*. Fruits of *Tapirira guianensis* (Anacardiaceae), *Marlierea* sp. (Myrtaceae), various species of Passifloraceae, and *Posqueria latifolia* (Rubiaceae) are also eaten. *Leontopithecus r. rosalia* probably serves as a dispersal agent for the small, closed seeds of the Passifloraceae and the relatively large seeds of *Tapirira guianensis,* since these seeds pass through the animal's digestive tract without losing the ability to germinate. *Tabebuia cassinoides* and *Tabebuia obtusifolia* (Bignoniaceae) and, in somewhat drier, more elevated areas, *Platymenia foliolosa* (Leguminosae Mimosaceae) are characteristic forest species within the range of *L. r. rosalia,* but are not a source of fruit for the animals.

Wild *L. r. chrysopygus* have been observed feeding on fruits of *Ficus* (*Urostigma*) sp., *Eugenia sulcata* and two other species of *Eugenia* (Myrtaceae), *Campomanesia guabiroba* (Myrtaceae), *Esenbeckia* sp. (Rutaceae), one species of Annonaceae, and several species of Sapotaceae. *Ficus* sp. and *Arecastrum romanzoffianum* (Palmae) are common trees in the habitat of this subspecies; the *Ficus* probably also serves as a fruit source. *Aspidosperma* sp. (Apocynaceae), *Cedrela* sp. (Meliaceae), and *Gallezia integrifolia* (Phytolaccaceae) are characteristic species of the region, but do not provide fruit for the monkeys.

Few data are available on the feeding habits of *L. r. chrysomelas.* It has been observed foraging in bromeliads.

3. Group Size

Group size for *L. r. rosalia* varies from two to eight, with three to four being the numbers most frequently encountered. One "group" of fifteen to sixteen was seen in a fruiting tree (*Inga* sp.), but this was probably a temporary, unstable association of several groups at a favorable food source. Wied-Neuwied (1826) saw *L. r. chrysomelas* in groups of four to twelve. Three *L. r. chrysopygus* groups in the Morro do Diabo State Park consisted of two animals (an adult pair), three animals (an adult pair and one subadult), and five animals (one adult pair, one young adult male, and two young adult females).

4.   Reproduction

Leontopithecus r. rosalia births take place from September to March in southern Brazil. These are the warmest and wettest months and correspond to the period from early spring to late summer in the Northern Hemisphere. As with other Callitrichidae, twin births are the rule in *Leontopithecus*, although single births or triplets sometimes occur. Previous studies indicated that the gestation period of *L. r. rosalia* ranged from 132 to 134 days (Rabb and Rowell, 1960; Ulmer, 1961; Walker, 1964). However, recent investigations have shown a range of 126 to 132 days and an average of 128 (Kleiman, 1977b; Wilson, 1977).

III.   Past Conservation Measures on Behalf of *Leontopithecus*

The major problem in the conservation of the lion tamarins is that they occur in the most densely inhabited part of Brazil. The first Portuguese settlers arrived in southeastern Brazil more than 450 years ago and the area is today the cultural, industrial, and agricultural center for the vast Brazilian nation. Modern Brazil is rapidly taking its place among the developed nations of the world and is greatly concerned with economic progress and expansion. Naturally, such progress manifests itself most strongly in the advanced southeast. This factor, together with the burgeoning human population of this part of Brazil, does not bode well for the remaining tracts of forest that make up the habitat of *Leontopithecus* and other interesting primate species (*Brachyteles arachnoides, Alouatta fusca, Callicebus personatus*, several species of *Callithrix*, and *Cebus apella*). Forests are usually regarded as unwanted barriers to progress which should be stripped of their valuable lumber and then clear cut or burned down to make way for cattle or agricultural projects. The fate of the unique wildlife heritage inhabiting these forests has, in general, been of little or no concern. The few devoted Brazilian conservationists have faced an extremely difficult, uphill battle in their attempts to save even the smallest remnants of southeastern Brazil's flora and fauna from the bulldozer and chain saw of progress.

The lion tamarins have always been relatively uncommon and more restricted in range than most other inhabitants of the southeastern coastal forests. Consequently, although their situation is only an early warning of what is happening to the entire region, they have shown the effects of civilization's advance sooner than many other animals of the region.

The strikingly beautiful golden lion tamarin has also long been a popular zoo exhibit in the United States and Europe, was occasionally used as a research animal in biomedical laboratories, and was sold in the pet trade. It is estimated that between 200 and 300 of these animals were shipped out each year in the period 1960–1965 (Hill, 1970). Consequently, in addition to suffering the effects of widespread habitat destruction, this subspecies was subjected to considerable trapping pressure for export to foreign markets.

A. International Conservation Measures up to 1972

Clyde Hill of the San Diego Zoological Society was the first to call international attention to the endangered status of *L. r. rosalia*. In 1965, he proposed to the President of the American Association of Zoological Parks and Aquariums (AAZPA) that the association "blacklist" the importation of *Leontopithecus*. In 1966, the AAZPA formally recognized *L. r. rosalia* as an endangered species and recommended to the International Union for the Conservation of Nature (IUCN) that *L. r. rosalia* be included in the "Red Data Book," the major international reference on the world's rare and endangered species. In 1967, the International Union of Directors of Zoological Gardens (IUDZG) pledged that its members would not import *L. r. rosalia* and publicized the animal's endangered status (J. Perry, personal communication). These actions, together with measures included in the 1969 U.S. Rare and Endangered Species Act (U.S. Public Law 90–135), ended the legal importation of this subspecies into the United States (Hill, 1970; Bridgwater, 1972b). Through the efforts of the IUCN, zoos in other importing countries were also encouraged to end purchases of *Leontopithecus* (Perry, 1972).

A Wild Animal Propagation Trust* Golden Marmoset Committee, originally organized in 1966, was reorganized in 1969 and placed under the direction of Donald D. Bridgwater (then General Curator of the National Zoological Park, now Director of the Minnesota Zoological Garden). Under his guidance, the committee began to "monitor captive population status, develop maintenance policies, effect agreement documents, act as a central clearinghouse for activities on behalf of the marmoset, stimulate various conservation actions, translate papers, and effect animal exchanges" (Bridgwater, 1972b). Since the last wild-caught *Leontopithecus* entered the United States in 1969 (other countries may have imported specimens after 1969), the major goal of this committee was to ensure the continued survival of the United States colonies. One of the first things the committee did was direct efforts to consolidate all *Leontopithecus* in the United States into several major breeding centers, since the 70 + animals

*The Wild Animal Propagation Trust was an independent organization that sought cooperative breeding arrangements through special species committees. It has now been disbanded (J. Perry, personal communication).

were widely dispersed in nineteen zoos. Most of the zoos had less than four animals each, and four had only single animals (Perry, 1972 and personal communication). Consolidation has now been for the most part achieved, but largely because some zoos have specialized and experienced breeding success, whereas others have experienced losses (Table I, Kleiman, and also J. Perry, personal communication).

The committee also began to organize the available data on husbandry of *Leontopithecus* and other Callitrichidae. With this goal in mind, Bridgwater, with the help of John Perry (Assistant Director, National Zoological Park), William G. Conway (Director, New York Zoological Society), and others, organized a special Golden Lion Marmoset Conference. This conference, which was held at the National Zoo in Washington, D.C. from February 15–17, 1972,

TABLE I

Total number of Captive *Leontopithecus r. rosalia* as of August, 1975 [a]

Institution	Total number of specimens		
	♂	♀	Undetermined
National Zoological Park, Washington, D.C.	12	10	
Oklahoma City Zoo, Oklahoma	5	7	1
Los Angeles Zoo, California	5	3	2
Monkey Jungle, Miami, Florida	7	1	
Brookfield Zoo, Chicago, Illinois	2	3	
Omaha Zoo, Nebraska	2	1	
San Antonio Zoo, Texas	1	2	
Fort Worth Zoo, Fort Worth, Texas	1	1	
Houston Zoo, Texas	1	1	
San Diego Zoo, California	2	0	
Glen Roskilly (private)	1	0	
Japan Monkey Centre	2	1	
Tokyo Ueno Zoo, Japan	1	0	
Fukuoka Zoo, Japan	1	1	?
Pretoria Zoo, Pretoria, South Africa	<	10	>
Private collection, Pretoria, South Africa	<	7	>
London Zoo	1	0	
Location	♂	♀	Undetermined	Totals
Totals				
United States	39	29	3	71
Other				24
				95

[a]The animals in the Tijuca Biological Bank are not included in this table. Information courtesy of Devra G. Kleiman, National Zoological Park, Washington, D.C.

brought together two dozen North American, European, and Brazilian marmoset specialists. The major contribution of this conference was a list of recommendations for a basic *Leontopithecus* husbandry program (DuMond, 1972). This list outlined the requirements for successful captive breeding of the genus and has since served as a valuable guideline for breeding programs. A number of other important papers on various aspects of callitrichid biology were also presented and are included in a volume entitled "Saving the Lion Marmoset" (Bridgwater, 1972a), published by the Wild Animal Propagation Trust.

B.   Past Conservation Measures in Brazil

Coimbra-Filho and Alceo Magnanini (Tijuca National Park, Rio de Janeiro) were among the first Brazilians to realize the precarious situation of the genus *Leontopithecus*. In 1962, they tried to save part of the natural habitat of *L. r. rosalia* by organizing the Biological Reserve of Jacarepagúa in the state of Guanabara. Although the monkey had already disappeared from Jacarepaguá (because of forest clearance for agriculture and heavy trapping), Coimbra-Filho and Magnanini planned to reintroduce it once the reserve had been established. Jacarepaguá would also have been a superb sanctuary for a great variety of aquatic birds and a number of other mammals (Coimbra-Filho and Magnanini, 1962). Unfortunately, attempts to establish the reserve failed because of lack of government interest.

Several years later, at just about the time that conservation measures on behalf of *Leontopithecus* were initiated in the United States, Brazilians began to show concern for the fate of some of their unique wildlife. In 1965, the Brazilian government ratified the Convention for the Protection of the Flora, Fauna and Natural Scenic Beauties of the Countries of the Americas, which recognized the importance of saving species endangered with extinction. In early 1967, the Brazilian Nature Conservation Foundation (FBCN) asked the Brazilian Forestry Development Institute (IBDF) to provide special protection for *L. r. rosalia* and *Brachyteles arachnoides*. On May 22, 1967, the IBDF issued a law (Portaria No. 18), which prohibited the hunting, capture, purchase, and exportation of these two species. In 1968, the FBCN asked the IBDF to enact official protection for all Brazilian endangered species and supplied a list of them. This list (Carvalho, 1968), officially released as Portaria No. 303 on May 29, 1968, included the three members of the genus *Leontopithecus* and a number of other rare primates, including *Brachyteles arachnoides*, *Callimico goeldii*, and *Cacajao* spp. In combination with several previously enacted laws (Brazilian Fauna Protection Law, Lei no. 5.197, January 3, 1967; Decreto-Lei no. 289, January 3, 1967; Decreto-Lei no. 289, February 28, 1967), it forbid the capture, hunting, purchase, sale, and exportation of all endangered species and products made from them. Portaria 18, Portaria 303, and the later Portaria 3.481 (May 31, 1968) ended legal exportation of *Leontopithecus* from Brazil, but unfortunately did nothing to prevent habitat destruction which is, in reality, the major problem.

At a Conservation Symposium held in Rio de Janeiro in 1968, Coimbra-Filho (1969) presented the results of some preliminary field studies of the behavior, ecology, and distribution of *L. r. rosalia* that he had been conducting on and off since 1966. He pointed out that the range of this animal was but a fraction of what it once had been and that even the small remaining area was being destroyed. A reserve was clearly needed if there was to be any hope that the golden lion tamarin would survive in the wild.

Based on Coimbra's initial studies and two subsequent surveys (one by helicopter, one on foot and in a small plane) conducted by Coimbra-Filho, Magnanini, and José Candido de M. Carvalho (Museo Naçional, Rio de Janeiro), two excellent reserve sites were selected in the state of Rio de Janeiro. When John Perry visited Brazil in early 1970, the major issue was the acquisition of land for a reserve (Perry, 1971). The land was valued at roughly $500 (United States) per hectare and Perry pointed out it was unlikely that the sum needed to purchase it could be obtained from international funding agencies. Expropriation of the land was therefore deemed the best possible solution. Under Brazilian law, an expropriation decree establishes a 5-year governmental option. During this time the government may proscribe actions which would adversely affect the property, but must also establish a price which must be paid prior to the end of the 5-year period. If a price is not established, the option lapses after the 5 years have ended.

Unfortunately, the two prime reserve sites were deforested in the year following the surveys. A third site, known as Poço das Antas* ("pool of the tapirs"), in the municipality of Silva Jardim, state of Rio de Janeiro (Fig. 4b), was selected in March, 1971, as the best remaining locality for a *L. r. rosalia* reserve. The Poço das Antas area, though inferior to the other two sites, comprised some 3000 hectares and still contained a reasonable number of tamarins.

In May, 1971, Magnanini and Coimbra-Filho submitted two project proposals to the IUCN (IUCN No. 24–1; IUCN No. 16–3, WWF Project 793). The first of these was for the Poço das Antas Experimental Center for Wildlife, but did not include a request for funds. The second was for the Tijuca Biological Bank (or the "Tijuca Bank of Lion Marmosets," as it is sometimes called). The Experimental Center for Wildlife was to be part of the Poço das Antas Reserve, was

---

*The locality which gave rise to the name of this site appears as Poço d'Anta on maps. However, in the original IUCN proposal, the name Poço das Antas was used, and this is the spelling now associated with the *Leontopithecus* reserve.

modeled after the Patuxent Wildlife Center in the United States, and was to be dedicated to captive breeding of endangered Brazilian species. The Tijuca Bank was clearly intended as a safety measure. At first, a *"Leontopithecus* bank" was to be part of the Experimental Center at Poço das Antas. However, Magnanini, Coimbra-Filho, and A. D. Aldrighi (Director, Tijuca National Park) foresaw some of the difficulties involved in establishing the Poço das Antas Reserve and decided to set up an emergency *Leontopithecus* breeding station in Tijuca National Park instead.

The Tijuca Bank was to consist of 22 cages. It was intended to serve as a breeding colony for the three subspecies of *Leontopithecus*, and also a holding site for animals translocated from areas where habitat destruction was inevitable and ones confiscated from illegal dealers. It was hoped that many of the translocated animals could eventually be released into Poço das Antas Reserve once it was officially established.

There were a number of advantages to establishing a captive colony in Tijuca National Park. Permission to establish the "bank" in Tijuca had already been granted by IBDF, thus avoiding the delay involved in expropriation. Also, Tijuca is a 3300-hectare national park, entirely surrounded by the city of Rio de Janeiro, and is relatively well protected. The Tijuca area was once part of the natural habitat of *L. r. rosalia* as well, so it was hoped that this subspecies could eventually be reintroduced into the park. Furthermore, both Magnanini and Coimbra-Filho worked right next to the park and could supervise the activities of the "bank" as part of their daily schedule.

The proposal for the Tijuca Bank included a request for $47,500. In July, 1971, the IUCN replied that it would be more realistic to aim for $10,000. Magnanini cut as many corners as possible and came up with $22,400 as an absolute minimum for first year operations, to be followed by $10,000 for the second year, $7,000 for the third year, and $5,500 per annum thereafter (Magnanini and Coimbra-Filho, 1972). The large first year investment took into account construction of cages (at $400 each) and costs involved in translocation of animals from forests slated for destruction. The new, reduced proposal was submitted to IUCN in August, 1971. In the meantime, work on the Tijuca Bank had already begun, aided by a grant of $3,000 from the IBDF.

At the Golden Marmoset Conference in February, 1972, the U.S. National Appeal of the World Wildlife Fund presented Coimbra-Filho and Magnanini with $5,000 to support further work on the Tijuca Bank. The participants in the conference also drafted several letters to encourage further action on behalf of the genus *Leontopithecus*. One letter was sent to the President of Brazil, asking him to act rapidly and favorably regarding the creation of the Poço das Antas Biological Reserve. Other letters were sent to the Director General of the IUCN and the President of the World Wildlife Fund, requesting support for the Tijuca Bank and the Poço das Antas Reserve.

IV. Conservation Measures on Behalf of *Leontopithecus* since the 1972 Washington Conference

A. Captive Breeding in Brazil—The Tijuca Biological Bank

After the initial $5,000 grant from the U.S. National Appeal of the World Wildlife Fund and the $3,000 from the IBDF, there were a number of other contributions to the Tijuca Bank. By December, 1972, the World Wildlife Fund had given an additional $10,000, bringing their total contribution to $15,000. By December, 1973, a further $7,800 had been received from the IBDF's allocation for the Tijuca National Park. With these funds, it was possible to complete the 22 cages of the "bank" and to construct a kitchen, an office, and facilities for visiting researchers.

Since the completion of the "bank" in January, 1974, the IBDF has contributed approximately $3,900 and the Brazilian Academy of Sciences $5,200 to cover the costs of hiring two experienced animal keepers and food for the captive *Leontopithecus*.

The 22 cages of the Tijuca Bank range in size from 4 m × 3 m × 2.5 m to 9 m × 3 m × 2.5 m. They all have wooden nest boxes, horizontal wooden perches, and natural dirt floors. Neither the wooden fixtures nor the dirt floors are ever cleaned, but the cages are open to the sun and periodically hosed down. They have two adjacent concrete walls, two wire walls, and slanting roofs of translucent, heavy duty plastic, and are arranged so that the two walls of one cage face the open wire sides of neighboring cages (Fig. 11). This ensures that monkeys in different cages cannot see one another, thus reducing tension resulting from constant visual presence of extra-group conspecifics—an important consideration with aggressive primates like *Leontopithecus* (Magnanini *et al.*, 1975).

The daily diet of the animals in the Tijuca Bank consists of whole wheat bread and vitaminized milk (given in the morning), bananas, pears, apples, and grapes (given in the afternoon), water (given twice daily), domestic quail eggs (one per monkey every fourth day), and grasshoppers (one per monkey every third day). During the summer months, cages are sprayed daily between 11.00 and 15.00 hours to keep the monkeys from becoming overheated (Magnanini *et al.*, 1975).

As of September, 1975, the Tijuca Bank housed a total of 48 *Leontopithecus*, including thirty-one *L. r. rosalia*, eight *L. r. chrysomelas*, and nine *L. r. chrysopygus*. Of these, two *L. r. rosalia*, two *L. r. chrysomelas*, and three *L. r. chrysopygus* were born in the "bank" during 1974. A list of all specimens and their origins is given in Table II.

The six adult *L. r. chrysopygus* currently in the Tijuca Bank are the first members of this subspecies ever kept in captivity anywhere, and they have now successfully raised three offspring. Two infants were born in September, 1974, and both are doing well. Two others were born to another pair in October, 1974.

Fig. 11. (A) Cages in the Tijuca Bank in the city of Rio de Janeiro. Note how the concrete walls block animals in neighboring cages from view (B).

TABLE II

Leontopithecus in the Tijuca Biological Bank as of September, 1975[a]

L. r. rosalia	31 animals (8 adult males, 14 adult females, 4 subadult males, 3 subadult females, 1 juvenile male, 1 juvenile female)

0/1 adult; captive born
0/1 adult; captive born; donated by Rio Grande do Sul Zoo
1/0 subadult; wild born; donated by Rio Grande do Sul Zoo
1/1 subadults; wild born; donated by Rio Grande do Sul Zoo
1/1 adults; wild born; donated by Rio Grande do Sul Zoo
1/1 adults; male captive born, female wild born
1/1 adults; wild born; captured in 1974 in forest in Rio São João basin
1/1 adults; wild born; captured in 1974 in forest in Rio São João basin
0/2 adults; wild born; captured in 1974 in forest in Rio São João basin
1/2 adults; wild born; captured in 1974 in forest in Rio São João basin
1/1 adults; 1/0 juvenile adult male captive born; juvenile and adult female wild born
1/1 adults; 1/1 subadults male captive born to wild-born male and captive-born female; female wild born; subadults born in the Tijuca Bank in October, 1974
1/2 adults; 1/1 subadults; 0/1 juvenile wild born; entire group captured in 1975 in forest in Rio São João basin

L. r. chrysomelas 8 animals (4 adult males, 2 adult females, 2 subadult males)

1/0 adult; wild born, but obtained when infant and reared in captivity
1/0 adult; wild born, but obtained when infant and reared in captivity
1/1 adults; wild born, but obtained when infants and reared in captivity. This pair had two infants in 1974, but both were abandoned right after birth by both parents and did not survive
1/1 adults; 2/0 subadults adult male wild born, obtained in 1973 from Itabuna, Bahia; female wild born, but hand-reared in captivity, and obtained from the vicinity of Itabuna, Bahia in December, 1969. The subadults were born in the Tijuca Bank in October, 1974. This same female also gave birth to three sets of hybrid *L. r. rosalia* × *L. r. chrysomelas* which we have discussed elsewhere (Coimbra-Filho and Mittermeier, 1976)

L. r. chrysopygus 9 animals (3 adult males, 3 adult females, 1 subadult male, 2 subadult females)

1/1 adults; wild born; captured in November, 1973 in the Morro do Diabo State Park.
1/1 adults; 0/1 subadult adults wild born; captured in November, 1973 in the Morro do Diabo State Park. The subadult was born in the Tijuca Bank in October, 1974; its twin died shortly after birth
1/1 adults; 1/1 subadults adults wild born; captured in November, 1973 in the Morro do Diabo State Park. The subadults were born in the Tijuca Bank in September, 1974

[a] Each line indicates the animals being kept in a single cage.

One of these died shortly after birth, but the other is in fine condition. We originally intended to capture only one pair of *L. r. chrysopygus* to make behavioral comparisons with the other two subspecies. However, several incidents pointed out the vulnerability of the Morro do Diabo population (see below), and we decided that it would be wise to set up a small captive breeding nucleus in case something disastrous happened to the Morro forest (Magnanini *et al.*, 1975).

B. Captive Breeding in the United States and Elsewhere

As of August, 1975, a total of 95 *L. r. rosalia* existed in fourteen institutions and two private collections outside Brazil. Eleven of the colonies (71 animals) are in the United States, three in Japan (six animals), two in Pretoria, South Africa (seventeen animals), and one in London (one animal). Major breeding colonies (10+ animals) are maintained by the National Zoological Park, the Oklahoma City Zoo, the Los Angeles Zoo, and the Pretoria Zoo (Table I; Kleiman, personal communication).

Since 1969, no further wild specimens of *L. r. rosalia* have been exported from Brazil. Table III shows the changes in the total reported non-Brazilian *L. r. rosalia* captive population in the period 1966—1974. After hitting a peak in 1968 (probably the result of last minute buying before the imposition of the import ban), the population has been slowly but steadily declining.

Initially, the most significant trend was the decline in the number of wild-caught specimens, many of which simply died of old age. One of the major problems in breeding *Leontopithecus* in captivity was the difficulty of breeding from captive-born parents. Births from wild-caught parents or from a wild-caught parent and a captive-born parent were not uncommon, but births from two captive-born parents were quite rare.

However, in the past few years researchers have begun to understand why successful breeding from captive-born parents was generally unsuccessful. It seems that learning plays an important role in the acquisition of parental behavior patterns in *Leontopithecus*. In the wild, offspring usually remain with their parental group through the birth and rearing of at least one set of younger siblings. Older offspring frequently perform what has been called "aunting behavior" or "allomaternal behavior" (Hrdy, 1976), carrying, playing with, and in other ways helping to raise younger siblings. By performing such "helping behavior," they apparently gain considerable experience in raising infants. This experience is quite important when the time comes for them to raise their own infants. In addition, by remaining with their family group until adulthood, young tamarins also experience the full process of being raised by their own parents.

In many captive situations in the past, offspring were removed from their parental group for one reason or another and failed to learn all the behavior

TABLE III

Trends in the Captive Colonies of *Leontopithecus r. rosalia*[a]

	1964	1965	1966	1967	1968	1969	1970	1971	1972	1973
Total population	ND[b]	ND	72	99	102	96	84	76	75	71
Number of zoos reporting	ND	ND	23	27	28	24	23	20	17	14
Change	—	—	—	+27	+3	−6	−12	−8	−1	−4
Number per collection	—	—	3.1	3.7	3.6	4.0	3.7	3.8	4.4	5.1
Captive bred	ND	ND	6	8	19	22	34	39	43	44
Percentage of total	—	—	8.	8.	19.	23.	40.	51.	57.	62.
Births surviving[c]	5	7	5.	10	10	18	11	21	11	—
Births 1966–1972	—	—	—	—	—	—	—	—	86	54
Increase in captive-bred	—	—	—	—	—	—	—	—	—	32
Disappearance	—	—	—	—	—	—	—	—	—	—
Births not surviving[d]	0	2	3	0	12	11	6	16	14	—
Wild caught population	ND	ND	64	91	83	74	50	37	32	27
Change	—	—	—	+27	−8	−9	−24	−13	−5	−5
Percentage of change	—	—	—	+42	−9	−11	−32	−26	−14	−16

[a]Data compiled from *Zoo Yearb.* 5–14 and reproduced with permission of John Perry (National Zoological Park). Animals from the Tijuca Biological Bank and South African colonies not included in this table.

[b]ND, no data.

[c]Born in calendar year; surviving on subsequent reporting date.

[d]Born in calendar year; not surviving on subsequent reporting date.

patterns associated with successful rearing of infants. Consequently, when they reached adulthood, these improperly socialized individuals either failed to breed or, if infants were born to them, the infants were either ignored, mistreated, or killed. This problem and the possibilities for corrective learning or "recuperation" are discussed in more detail in Coimbra-Filho and Mittermeier (1976).

Recently, however, a number of captive-born pairs have bred successfully. With improved knowledge of the behavioral requirements of *Leontopithecus*, it seems that the problem of breeding from captive-born pairs has now been largely overcome.

For several years, a major concern was the decrease in number of captive females. Apparently by pure chance, there was a higher number of male births in the colonies (e.g., the Monkey Jungle in Miami, Florida had so many male births that it now has seven males and only one female). Since *Leontopithecus* is a monogamous species, many males were without mates. Fortunately, the sex ratio now appears to be evening out (D. Kleiman, personal communication).

Several other problems face captive colonies. Outbreak of disease is always a potential threat (and perhaps the most serious one at present). Several small zoo colonies lost most of their *Leontopithecus* to unknown diseases of viral origin (Bridgwater, 1972b). If an epidemic were to strike one of the larger colonies, a substantial portion of the captive breeding stock could be lost. In the past diet was a problem. Captive *Leontopithecus* were given primarily fruit, which was not adequate for long-term survival. This deficiency has now, for the most part, been corrected and no longer presents a problem in the major colonies (Kleiman, 1976).

C.   Survival of *Leontopithecus* in the Wild

In spite of the international concern for the fate of the genus *Leontopithecus*, the governmental agencies responsible for the preservation of Brazil's fauna have still done very little to ensure the survival of these unique animals in their natural habitat. The situation of wild populations continues to deteriorate.

1.   *Leontopithecus r. rosalia*

Leontopithecus r. rosalia, the best known and formerly the most abundant subspecies, is presently in the greatest danger of extinction. For more than 2 years, establishment of the Poço das Antas Reserve was delayed because of lack of interest on the part of a minor Ministerio da Agricultura official in whose hands the document had fallen (Magnanini *et al.*, 1975). On March 11, 1974, the President of Brazil finally signed the laws establishing the reserve (Decreto 73.791) and allowing for expropriation of the roughly 3000 hectares which are to make up the reserve (Decreto 73.392). The task of land expropriation was placed in the hand of the IBDF and the National Institute for Colonization and Agrarian Reform (INCRA).

Unfortunately, at the present time—more than a year after the official establishment of the reserve—virtually nothing has been done to implement the existing protective legislation. Habitat destruction both near and in the area of Poço das Antas Reserve continues (Fig. 12). Irreplaceable forests are being burned for charcoal production and to make way for farmland.

Very little time remains. The entire basin of the Rio São João (some 210,000 hectares) is considered excellent farmland and is currently slated for a large-scale agricultural project entitled "Projeto do Vale do Rio São João." This project will include drainage of the rivers making up the São João basin, development of elaborate irrigation systems, and construction of a dam to expand the capacity of Lagoa de Juturnaíba, a large lake in the area. The goal of this project is to make the São João area one of the major sources of farm produce for the city of Rio de Janeiro, while the Lagoa de Juturnaíba will serve as an important reservoir (Anonymous, 1973). The municipalities most affected will be Casimiro de Abreu, Cabo Frio, São Pedro da Aldeia, Araruama and Silva Jardim—precisely the areas in which the last remaining populations of *L. r. rosalia* are to be found. Needless to say, the Rio São João project, which will

Fig. 12. Forest in the Poço das Antas National Biological Reserve. Since no action has been taken to protect the land in this reserve, the owners continue to plant crops, hunt, and cut down the forest. Note the banana plantation in the foreground.

take roughly 5—6 years to complete, will ensure the destruction of what little remains of *L. r. rosalia*'s forest habitat. If adequate protection is not quickly provided for the Poço das Antas Reserve, it may well be destroyed along with the rest of the forests in the São João basin.

At present, only about 40 to 50% of Poço das Antas still has stands of forest that could serve as *Leontopithecus* habitat. A helicopter survey conducted by Magnanini and Coimbra-Filho in May, 1975, indicated that so much of the existing reserve area had been destroyed that an addition of some 2500 hectares would be necessary if there was to be any hope that the reserve would be of use in saving *Leontopithecus* from extinction. In addition, it was decided that a certain portion of the currently delimited reserve area was in such bad shape that it should be dropped from the reserve. A new proposal, which would increase the size of the Poço das Antas Reserve to some 5000 hectares, has now been prepared and submitted.

In any case, considerable reforestation will be necessary before most of Poço das Antas can again be considered suitable habitat for the lion tamarins. Even if Poço das Antas can be saved in time, it is questionable that this minute patch of forest can long exist as an ecological island in the midst of hundreds of thousands of hectares of cultivated land.

2. *Leontopithecus r. chrysomelas*

The status of the remaining habitat of *L. r. chysomelas* is not much better than that of *L. r. rosalia*. Thus far, nothing has been done to ensure the survival of this subspecies in the wild. However, a tract of 4535 hectares of almost virgin forest exists in the municipality of Una and would make an ideal reserve for *L. r. chrysomelas*. This land is in private hands, but was offered to the IBDF for approximately $160 per hectare in October, 1973. The forests on this land are rich in *jacarandá* or Brazilian rosewood (*Dalbergia nigra*, Leguminosae Faboideae) which currently has a very high market value, and plans have already been proposed to turn 1500 hectares into rubber plantation and 400 hectares into cocoa plantation. Consequently, if the land is not purchased or expropriated very quickly, the last major tract of *L r. chrysomelas* habitat will also be destroyed. As in the case of *L. r. rosalia*, IBDF has not yet acted.

3. *Leontopithecus r. chrysopygus*

Always the rarest and least known member of the genus, *L. r. chrysopygus* is ironically now the least endangered. The remaining population of this subspecies occurs in the 37,147-hectare Morro do Diabo State Park, and therefore has a better chance of survival in the wild than either of its relatives.

Nonetheless, *L. r. chrysopygus* is definitely an endangered animal. The Morro do Diabo forest is nothing more than a minute remnant of the vast forests which once covered much of the state of São Paulo and is today almost entirely surrounded by pasture and farmland. Its vulnerability was clearly pointed out in 1973. Some local farmers, experimenting with herbicides (2,4,5-T and 2,4-D) near the park, destroyed many hectares of what little forest remains in close proximity to the park. Furthermore, Morro do Diabo is cut by a railroad and, since 1969, by a highway which is now paved and divides the park into two parts (Fig. 9A).

It should be pointed out that the reserves that have been established or proposed for the three subspecies of *Leontopithecus* would not only save populations of these animals. Many other unusual southeastern Brazilian species exist in these forests and the forests themselves are unique and worthy of preservation. Table IV lists a number of endangered mammal and bird species found within the present ranges of the three *Leontopithecus*.

D. The 1975 Conference on the Biology and Conservation of the Callitrichidae

Interest in *Leontopithecus* and many other callitrichids has grown rapidly since the 1972 conference, especially because several callitrichid species (e.g., *Callithrix jacchus*, *Saguinus fuscicollis*, and *Saguinus mystax*) are very important in biomedical research and are becoming more and more difficult to obtain. As a result, a conference on the biology and conservation of all Callitrichidae was organized by Devra G. Kleiman (National Zoological Park, Office of Zoological Research) and held from August 18–20, 1975, at the National Zoological Park's Conservation and Research Center at Front Royal, Virginia. This conference brought together some 50 scientists from the United States, South America, and Europe, served as a follow-up to the February, 1972 conference, and promoted the exchange of new data and ideas within the diverse group of researchers working with callitrichids.

The 1975 conference showed that our knowledge of callitrichids has come a long way in 3 short years. The only disappointing note was that the situation of the lion tamarins, which was the main focus of the 1972 conference and a major concern at Front Royal, was worse than ever before. In spite of continuing international interest in the fate of these animals and the fact that the Poço das Antas Reserve had finally been established (at least on paper), there were less lion tamarins and less lion tamarin habitat in 1975 than in 1972.

V. Recommendations for the Future

In 1973, we estimated that the remaining wild population of *L. r. rosalia* did not exceed 400 animals, while those of *L. r. chrysomelas* and *L. r. chrysopygus* were between 200–300 and 100–500, respectively (Coimbra-Filho and Mittermeier,

TABLE IV

Rare and Endangered Species of Mammals and Birds
Sympatric with the Three Subspecies of *Leontopithecus*

L. r. rosalia   state of Rio de Janeiro
  Mammals
    Alouatta fusca
    Lutra sp.
    Bradypus torquatus

L. r. chrysomelas   state of Bahia
  Mammals
    Alouatta fusca
    Lutra sp.
    Bradypus torquatus
    Chaetomys subspinosus

  Birds
    Tinamus s. solitarius
    Crypturellus noctivagus zabele
    Spizaetus t. tyrannus
    Crax blumenbachii
    Neomorphus geoffroyi dulcis
    Pyrrhura cruentata
    Amazona brasiliensis rhodocorytha
    Xipholena atropurpurea

L. r. chrysopygus   state of São Paulo
  Mammals
    Panthera onca
    Chrysocyon brachyurus
    Lutra platensis
    Myrmecophaga tridactyla

  Birds
    Tinamus s. solitarius
    Spizaetus ornatus
    Spizastur melanoleucus
    Pipile jacutinga

1973). More recent population estimates indicate that *L. r. rosalia* is down to 100–200 animals, whereas *L. r. chrysomelas* and *L. r. chrysopygus* are in the vicinity of 200 animals each. If adequately protected reserves are not established quickly, *L. r. rosalia* and *L. r. chrysomelas* could be extinct in the wild within a few years.

The Poço das Antas reserve is the last hope for the survival of *L. r. rosalia* in its natural habitat. The forests in the rest of its range are going to be destroyed by the huge agricultural project planned for the Rio São João basin. Poço das

Antas has already been officially established as a reserve and the funds are available for the payment of indemnities to the owners of the land. Hopefully, the size increase for the reserve will be rapidly approved by the President of Brazil.* Once this has been done, it will be up to IBDF and INCRA to pay indemnities and enforce the existing protective legislation.

The 4535-hectare tract of private land in the state of Bahia is the last remaining area within the range of *L. r. chrysomelas* that is large enough to be a significant reserve. The cost of the land is really very low, especially when one considers how much is spent each year in forest destruction and "reforestation" with exotic species like *Eucalyptus*. The possibility exists for IBDF to effect an agreement with CEPLAC (Center for Cocoa Experimentation in Bahia) to obtain the land in Una, but again almost nothing has been done.

The future of *L. r. chrysopygus* in the wild depends entirely on protecting the Morro do Diabo State Park. Given the size of Morro do Diabo, the remaining wild population of this subspecies is much more secure than that of either of its two relatives. The combined area of Poço das Antas Reserve and the proposed reserve in Una does not amount to more than 20% of the total area of Morro do Diabo State Park.

The future of captive populations depends on successful breeding from captive-born parents and control of disease outbreak. Hopefully, the new data on *Leontopithecus* biology and captive breeding requirements that have been gathered over the past few years will make this possible.

The energies of all those interested in the survival of the genus *Leontopithecus* should now be devoted to establishing well-protected reserves for *L. r. rosalia* and *L. r. chrysomelas*, since the conservation of wild populations of these animals is of primary concern. At the same time, everything possible should be done to ensure that captive stocks of all three subspecies survive and breed successfully—just in case attempts to save wild populations fail.

Acknowledgments

We would like to thank the following people for their efforts on behalf of the lion tamarins: A. Magnanini, A. D. Aldrighi, M. P. Autuori, J. L. Belart, D. D. Bridgwater, W. G. Conway, R. W. Cooper, F. V. DuMond, M. da Frota Moreira, C. Hill, M. L. Jones, D. G. Kleiman, J. C. de Mello Carvalho, M. A. Moraes Victor, P. Nogueira-Neto, A. de A. Pacheco Leão, and J. Perry. Advance drafts of this paper were read by D. D. Bridgwater, N. Duplaix-Hall, M. L. Jones, C. Hill, D. G. Kleiman, and J. Perry, and we thank them for their comments. We are especially grateful to D. G. Kleiman and J. Perry for pro-

---

*On November 3, 1975, the president of Brazil approved the legislation to increase the size of Poço das Antas to 5000 hectares (Decreto No. 76.583, Decreto No. 76.584).

92   Adelmar F. Coimbra-Filho and Russell A. Mittermeier

viding us with unpublished information on the current status of captive *Leontopithecus* colonies. A.F.C.-F. is a fellow of the Brazilian National Research Council. R.A.M. studies in South America were supported in part by a National Science Foundation Graduate Fellowship and in part by the New York Zoological Society.

References

Altmann-Schönberner, D. (1965). Beobachtungen über Aufzucht und Entwicklung des Verhaltens beim grossen Löwenäffchen, *Leontocebus rosalia. Zool. Gart.* **31**, [N.F.], 227–239.
Anonymous (1973). Vale do São João: uma nova realidade, *Bol Saneamento (Rio de Janeiro)* **27**(47), 64.
Bridgwater, D. D. (1972a). "Saving the Lion Marmoset." Wild Animal Propagation Trust, Wheeling, West Virginia.
Bridgwater, D. D. (1972b). Introductory remarks with comments on the history and current status of the golden marmoset. *In* "Saving the Lion Marmoset" (D. D. Bridgwater, ed.), pp. 1–6, Wild Animal Propagation Trust, Wheeling, West Virginia.
Carvalho, J. C. de M. (1968). "Lista das Espécies de Animais e Plantas Ameaçadas de Extinção no Brasil." Fundação Brasileira para a Conservação da Natureza, Rio de Janeiro, Brazil.
Coimbra-Filho, A. F. (1965). Breeding lion marmosets, *Leontideus rosalia*, at Rio de Janeiro Zoo. *Int. Zoo Yearb.* **5**, 109–110.
Coimbra-Filho, A. F. (1969). Mico-leão, *Leontideus rosalia* (Linnaeus, 1766), situação atual da espécie no Brasil (Callithricidae, Primates). *An. Acad. Brazil. Cienc Suppl.* **41**, 29–52.
Coimbra-Filho, A. F. (1970a). Considerações gerais e situação atual dos micos-leões escuros, *Leontideus chrysomelas* (Kuhl, 1820) e *Leontideus chrysopygus* (Mikan, 1823). *Rev. Brasil. Biol.* **30**, 249–268.
Coimbra-Filho, A. F. (1970b). Acêrca da redescoberta de *Leontideus chrysopygus* (Mikan, 1823) e apontamentos sobre sua ecologia. *Rev. Brasil. Biol.* **30**, 609–615.
Coimbra-Filho, A. F. (1972). Mamíferos ameaçados de extinção no Brasil. *In* "Espécies da Fauna Brasileira Ameaçadas de Extinção," pp. 13–98. Academia Brasileira de Ciencias, Rio de Janeiro, Brazil.
Coimbra-Filho, A. F. (1977). Natural shelters of *Leontopithecus rosalia* (Linnaeus, 1766) and some ecological implications (Callitrichidae, Primates). *In* "Conference on the Biology and Conservation of the Callithrichidae," in press.
Coimbra-Filho, A F and Magnanini, A. (1962). "Aves da restinga." Centro de Pesquisas Florestais e Conservação da Natureza, Rio de Janeiro.
Coimbra-Filho, A. F. and Magnanini, A. (1972). On the present status of *Leontopithecus* and some data about new behavioral aspects and management of *L. rosalia rosalia*. *In* "Saving the Lion Marmoset" (D. D. Bridgwater, ed.), pp. 59–69, Wild Animal Propagation Trust, Wheeling, West Virginia.
Coimbra-Filho, A. F. and Mittermeier, R. A. (1972). Taxonomy of the genus *Leontopithecus*

Lesson, 1840. *In* "Saving the Lion Marmoset" (D. D. Bridgwater, ed.), pp. 7–22, Wild Animal Propagation Trust, Wheeling, West Virginia.

Coimbra-Filho, A. F. and Mittermeier, R. A. (1973). Distribution and ecology of the genus *Leontopithecus* Lesson, 1840 in Brazil. *Primates* **14**(1), 47–66.

Coimbra-Filho, A. F. and Mittermeier, R. A. (1976). Hybridization in the genus *Leontopithecus* (*L. r. rosalia* (Linnaeus, 1766) × *L. r. chrysomelas* (Kuhl, 1820))—Callitrichidae-Primates. *Rev. Brasil. Biol.* **36**(1), 129–137.

Coimbra-Filho, A. F., Magnanini, A., and Mittermeier, R. A. (1975). Vanishing gold. *Anim. Kingdom,* **78**(6), 20–27.

Ditmars, R. L. (1933). Development of the silky marmoset. *Bull. N. Y. Zool. Soc.* **36**(6), 175–176.

DuMond, F. V. (1971). Comments on minimum requirements in the husbandry of the golden marmoset (*Leontideus rosalia*). *Lab. Prim. Newsl.* **10**(2), 30–37.

DuMond, F. V. (1972). Recommendations for a basic husbandry program for lion marmosets. *In* "Saving the Lion Marmoset" (D. D. Bridgwater, ed.), pp. 120–139, Wild Animal Propagation Trust, Wheeling, West Virginia.

Epple, G. (1967). Vergleichende Untersuchungen über Sexual- und Sozialverhalten der Krallenaffen (Hapalidae). *Folia Primatol.* **7**, 37–65.

Epple, G. (1972). Social communication by olfactory signals in marmosets. *Int. Zoo Yearb.* **12**, 36–42.

Feio, J. L. de A. (1953). Contribuição ao conhecimento da historia da zoogeografia do Brasil. *Publ. Avulsas Mus. Nac. (Rio de Janeiro)* **12**, 1–22.

Frantz, J. (1963). Beobachtungen bei einer Löwenäffchenaufzucht. *Zool. Gart.* **28**, [N.F.], 115–120.

Hampton, S. (1972). Golden lion marmoset conference. *Science* **177**, 86–87.

Hill, C. A. (1970). The last of the golden marmosets. *Zoonooz* **43**(1), 12–17.

Hrdy, S. B. (1976). Care and exploitation of nonhuman primate infants by conspecifics other than the mother. *In* "Advances in the Study of Behavior," Vol. 6, pp. 101–158. Academic Press, New York.

Jones, M. (1973). "Studbook for the Golden Lion Marmoset, *Leontopithecus rosalia*." American Association of Zoological Parks and Aquariums, Wheeling, West Virginia.

Kleiman, D. (1976a). Will the pot of gold have a rainbow? *Anim. Kingdom,* **79**(1), 2–6.

Kleiman, D. (1977). The reproductive cycle and sociosexual interactions in pairs of golden lion marmosets (Primates, Callitrichidae). *In* "Conference on the Biology and Conservation of the Callitrichidae," in press.

Kuhl, H. (1820). "Beitrage zur Zoologie und vergleichendem Anatomie." Frankfurt a. M.

Laemmert, H. W., Jr., Ferreira, L. C., and Taylor, R. M. (1946). An epidemiological study of jungle yellow fever in an endemic area in Brazil. Part II. Investigation of vertebrate hosts and arthropod vectors. *Amer. J. Trop. Med. (Suppl.)* **26** (6), 1–69.

Magnanini, A. (1973). Uma espécie ameaçada de extinção no Brasil. *Bol. Inform. No. 8/73, Fund. Brasil. Conserv. Natur.,* pp. 21–33.

Magnanini, A. and Coimbra-Filho, A. F. (1972). The establishment of a captive breeding program and a wildlife research center for the lion marmoset *Leontopithecus* in Brazil. *In* "Saving the Lion Marmoset" (D. D. Bridgwater, ed.), pp. 110–119. Wild Animal Propagation Trust, Wheeling, West Virginia.

Magnanini, A., Coimbra-Filho, A. F., Mittermeier, R. A., and Aldrighi, A. D. (1975). The Tijuca Bank of Lion Marmosets—a progress report. *Int. Zoo. Yearb.* **15**, 284—287.

Mikan, J. C. (1823). "Delectus Florae et Faunae Brasiliensis." Fol. Vindobonae.

Mittermeier, R. A. and Douglass, J. F. (1973). The plight of the lion marmosets. *Lab. Prim. Newsl.* **12**(3), 12—13.

Napier, J. R. and P. R. Napier (1967). "A Handbook of Living Primates." Academic Press, London and New York.

Perry, J. (1971). The golden lion marmoset. *Oryx* **11**(1), 22—24.

Perry, J. (1972). In danger—golden lion marmosets. *Smithsonian* **3**(9), 49—53.

Rabb, G. B. and Rowell, J. E. (1960). Notes on reproduction in captive marmosets. *J. Mammal.* **41**, 401.

Rizzini, C. T. (1963). Nota prévia sobre a divisão fitogeográfica do Brasil. *Rev. Brasil. Geogr.* **25**(1), 1—64.

Rizzini, C. T. (1967). Delimitação, caracterização e relações da flora silvestre hileiana. *Atas Simp. Sobre Biota Amazon.* **4**, 13—36.

Ruschi, A. (1964). Macacos do Espírito Santo. *Bol. Mus. Biol. Mello Leitão Zool.* **23 A**, 1—23.

Snyder, P. (1972). Behavior of *Leontopithecus rosalia* (the golden lion marmoset) and related species: a review. *In* "Saving the Lion Marmoset" (D. D. Bridgwater, ed.), pp. 23—49. Wild Animal Propagation Trust, Wheeling, West Virginia.

Snyder, P. (1974). Behavior of *Leontopithecus rosalia* (golden-lion marmoset) and related species: a review. *J. Hum. Evol.* **3**, 109—122.

Ulmer, F. A. (1961). Gestation period of the lion marmoset. *J. Mammal.* **42**, 253—254.

Veloso, H. P. (1966). "Atlas Florestal do Brasil." Serv. Inform. Agr., Ministerio da Agricultura, Rio de Janeiro.

Victor, M. A. M. and Montagna, R. G. (1970). Analise panorámica da situação florestal e efeito da Lei dos Incentivos Fiscais em São Paulo. *Silvicultura São Paulo* **7**, 7—18.

Walker, E. P. (1964). "Mammals of the World," Vol. I. Johns Hopkins Univ. Press, Baltimore, Maryland.

Wied-Neuwied, M. (1826). "Beiträge zue Naturgeschichte von Brasilien," Vol. II. Weimar.

Wilson, C. (1977). Notes on the gestation and reproduction of *Leontopithecus* at Oklahoma City Zoo. In "Conference on the Biology and Conservation of the Callitrichidae," in press.

Note Added in Proof

In 1976, an area of 5000 hectares was purchased in the municipality of Una, state of Bahia, for a *Leontopithecus rosalia chrysomelas* reserve. Once fully implemented, this reserve will provide at least some protection for the golden-headed subspecies.

In addition, a second population of *Leontopithecus rosalia chrysopygus* was discovered in 1976 on a privately owned tract of forest in Galia, west of Bauru, state of São Paulo. The population there is thought to be very small, but no estimates have yet been made.

4

Rediscovery and Conservation of the Peruvian Yellow-Tailed Woolly Monkey (*Lagothrix flavicauda*)

RUSSELL A. MITTERMEIER, HERNANDO DE MACEDO-RUIZ,
B. ANTHONY LUSCOMBE, AND JOHN CASSIDY

I. Introduction

The woolly monkeys of the genus *Lagothrix* are large, robustly built, prehensile-tailed monkeys found primarily in the Upper Amazon region of northwestern South America. The genus consists of two species, *Lagothrix lagothricha* (Fig. 1A), the common woolly monkey, and *Lagothrix flavicauda* (Figs. 1B, 2, and 3), the Peruvian yellow-tailed woolly monkey. *Lagothrix lagothricha* is divided into four subspecies and is fairly widespread in the rain forests of Brazil, Colombia, Peru, Ecuador, and Bolivia (Fooden, 1963). It is frequently exhibited in zoos and, until recently, was commonly sold in the pet trade. *Lagothrix flavicauda*, on the other hand, is known only from a very restricted area in the Andes of northern Peru and is among the rarest and least known of New World monkeys. Although first described more than 160 years ago, *L. flavicauda* was, as of 1974, represented by a total of only five museum specimens, the last of which was collected in 1926.

In this paper, we will briefly describe *L. flavicauda*, discuss its history, report the results of an expedition intended to determine if it still existed, review its conservation status, and discuss measures needed to ensure its survival. In addition, we provide plans which briefly outline the requirements for a captive breeding colony of this rare monkey (Section VIII).

II. Description

Lagothrix flavicauda differs from *L. lagothricha* in several features, including the mahogany-colored pelage, the white or buffy circumbuccal patch, the long yellow tuft of scrotal hair, the yellow band on the posteroventral surface of the tail (which originally gave rise to the specific name *flavicauda*), the texture of the hair, and certain features of the skull. The lower back, dorsal and ventral surfaces of the tail, and adjacent parts of the legs are mahogany brown. This color darkens anteriorly to deep mahogany brown on the upper back, nape, crown, sides of the face and arms, and laterally to the same color on the ventral surface of the body. The arms and especially the legs darken distally, becoming very dark brown or almost black on the hands and feet. In both males and females, a long, conspicious tuft of yellow or straw-colored pubic hair is present and varies in length from 100 to 200 mm. This tuft may be somewhat longer and thicker in males (Thomas, 1927a; Hill, 1962; Fooden, 1963), but on the basis of the small sample of specimens currently available, no definite statement can be made concerning dimorphism in this character. The yellow band on the posteroventral surface of the tail extends about one-third of the way up the tail. It flanks the volar skin at the end of the prehensile tail and extends 50–260 mm past the anterior border of the volar skin. Thomas (1927a) noted that a juvenile he examined lacked both the pubic tuft and the yellow band on the tail. Two juvenile specimens examined by us also lacked the yellow pubic tuft and one (the smaller) lacked the yellow tail band as well. The pubic tuft is probably a secondary sexual characteristic that only appears later in development. The tail

---

Fig. 1. (A) Juvenile *Lagothrix lagothricha* from Peruvian Amazonia (B) Juvenile male *Lagothrix flavicauda* obtained in Pedro Ruiz Gallo. Note the white circumbuccal facial patch and the texture of the hair and compare with Fig. 1A.

Fig. 2.   Juvenile male *Lagothrix flavicauda* obtained in Pedro Ruiz Gallo.

band may also appear late in development in some individuals, or may be considerably reduced or absent in a certain percentage of the population. The white to buffy circumbuccal patch extends from the chin to between the eyes and is the most distinctive external character of *L. flavicauda*. (Indeed, a more appropriate common name for this species would be Peruvian "white-nosed" woolly monkey). The fur of *L. flavicauda* is longer and thicker than in *L. lagothricha* (Fig. 1A,B), perhaps as an adaptation to colder temperatures in the montane

Fig. 3.   Juvenile male *Lagothrix flavicauda* and former owners in Pedro Ruiz Gallo.

forests in which *L. flavicauda* lives. Thomas (1927a) describes *L. flavicauda* fur as more "normal" in texture than the hair of other *Lagothrix*. Thomas (1927a, c), Hill (1962), and Fooden (1963) discuss the skull of *L. flavicauda* and point out some ways in which it differs from the skull of *L. lagothricha*. Of particular note are the more narrow braincase of *L. flavicauda* and certain morphological differences in the deciduous dentition. Field measurements are available for five specimens and are given in Table I.

III. History

Lagothrix flavicauda was discovered by Alexander von Humboldt in 1802 and was described by him as *Simia flavicauda—"le choro de la Province de Jaen"—* in 1812 (Humboldt and Bonpland, 1812). Humboldt, however, never saw a live specimen of this monkey. His description was based on nothing more than flat, trimmed skins used as saddle covers by Peruvian muleteers in the vicinity of Jaen (Dept. of Cajamarca, Peru). No type specimen was preserved and Humboldt knew so little about the animal that he considered it a new species of howler monkey, rather than a woolly monkey (Fooden, 1963).

For more than 110 years following Humboldt's description of *L. flavicauda*, almost nothing was heard of the animal. Poeppig (1832) mentioned the occurrence of a *rothe choro* ("red choro"—choro being the most frequently used name

TABLE I

Locality Data and Field Measurements of Known Museum Specimens of *Lagothrix flavicauda*

Catalog no.[a]	Sex[b]	Locality, collector, date collected	Field measurements (mm)
AMNH 73222	M	La Lejia (06°07'S, 77°28'W, 2300 m), Dept. of Amazonas, Peru; collected by Watkins, April 8, 1925, skin and skull	Head-body, 515; tail, 605
AMNH 73223	F	Same as AMNH 73222, skin and skull	Head-body, 535; tail, 610
BMNH 1927.1.1.1	M	Pucatambo (06°09', 77°11'W, 1555 m), Dept. of Amazonas, Peru; collected by Hendee, January 27, 1926, skin and skull	Head-body, 520; hind foot, 145; tail, 560; ear, 32.5
BMNH 1927.1.1.2	Juv. M	Same as BMNH 1927.1.1.1, skin and skull	Head-body, 260; hind foot, 86; tail, 286; ear, 30
BMNH 1927.1.1.3	F	Same as BMNH 1927.1.1.1, skin and skull	Head-body, 520; hind foot, 148; tail, 630; ear, 34
MNHJP 41	F	Forest above Alva (05°56'S, 77°56'W, 1000—2000 m), near Pedro Ruiz Gallo, Dept. of Amazonas, Peru; collected by Mittermeier, Macedo—Ruiz, and Luscombe, skin and skull	No field measurements
MNHJP 42	Juv.	Same as MHNJP 41, skin and skull	No field measurements
MNHJP 43	Adult (?)	Same as MHNJP 41, skin and skull	No field measurements
MHNJP 45	Adult (?)	Same as MHNJP 41, skin only	No field measurements
No number	Adult (?)	Same as MHNJP 41, skin only	No field measurements
MVZ 148038	M	Purchased in market of Huamachuco, Dept. of La Libertad, Peru; obtained by Jones, Nov. 21, 1974, skin only	No field measurements

Reported localities	
Aguas Verdes, near Pedro Ruiz Gallo, Dept. of Amazonas, Peru	Locality reported for live juvenile specimen obtained in Pedro Ruiz Gallo by Mittermeier, Macedo-Ruiz, and Luscombe
Garcia, on the *Marginal de la Selva* road, near km 400 on the Ingénio-Rioja section, Dept. of Amazonas	Information from local people
Laurel, near La Lejía, Dept. of Amazonas	
Region of Upper Río Mayo, along Ingénio-Río Nieva-Rioja road, Depts. of Amazonas and San Martin	Information from Local people Letter in *La Prensa*, June 16, 1974
Forests in vicinity of Moyobamba (very rare), Dept. of San Martin	Freese (1975)
Vicinity of Jaen, Dept. of Cajamarca	Humboldt and Bonpland (1812)

[a] AMNH, American Museum of Natural History, New York; BMNH, British Museum of Natural History, London; MVZ, Museum of Vertebrate Zoology, Berkeley, California; MHNJP, Museo de Historia Natural "Javier Prado," Lima, Peru. Localities reported in the literature or obtained from local people and listed in the second part of the table.

[b] M, male; F, female; Juv., juvenile.

TABLE II

Vernacular Names for *Lagothrix flavicauda* and Some Other Monkeys in the Department of Amazonas, Peru[a]

Lagothrix flavicauda	Maquisapa[b] (Alva, Pedro Ruiz Gallo)
	Maquisapa chusco (vicinity of Mendoza)
	Upa (Aguas Verdes, near Pedro Ruiz Gallo)
	Puca-runtu (= red testicles; Pedro Ruiz Gallo)
	Choro, Chore (vicinity of Chachapoyas)
	Quille Corote (= yellow testicles; vicinity of
	Pucatambo, according to Hendee's 1927 letter)
Ateles belzebuth belzebuth	Choba (Alva)
	Maquisapa fino (near Mendoza)
Aotus trivirgatus	Mono nocturno (Alva, Pedro Ruiz Gallo)
Cebus albifrons	Mono del día (Pedro Ruiz Gallo)

[a]Areas in which the different names are used are given in parentheses.

[b]"Maquisapa" is the name most frequently used in the Dept. of Amazonas to refer to *L. flavicauda*. However, in the rest of Peru, and especially the Peruvian Amazon, "maquisapa" is used for *Ateles*, whereas "choro" is used for *Lagothrix lagothricha*. To avoid confusion, we use the name "mono choro peruano de cola amarilla," a direct translation of Peruvian yellow-tailed woolly monkey, as the vernacular name of *L. flavicauda* in all Spanish publications on the animal.

for woolly monkeys in Peru; see Table II) from Yurimaguas (Province of Alto Amazonas, Dept. of Loreto, Peru), but the identity of this animal is not clear. In 1925, a professional animal collector named Watkins collected two specimens of *L. flavicauda* at La Lejía (Dept. of Amazonas). These were deposited in the American Museum of Natural History in New York, but were not recognized as *L. flavicauda* until 1963 (Fooden, 1963). In 1926, R. W. Hendee, a collector for the Godman-Thomas Expedition to Peru, collected three more specimens at Pucatambo (a tiny rest stop located at the border between the Depts. of Amazonas and San Martin) for the British Museum of Natural History in London. Unlike Watkins' specimens, these animals were immediately recognized as something unusual and were described by the zoologist Oldfield Thomas (1927a) as *Lagothrix (Oreonax) hendeei*, a new species and new subgenus of wooly monkey. Later the same year, Thomas (1927c) raised the new subgenus to full generic status as *Oreonax hendeei*, basing his decision on several features of the deciduous dentition of one juvenile specimen. Thomas (1972b), incidentally, compared his "new" species with Humboldt's description of *Simia flavicauda*, but concluded that Humboldt's animal was probably nothing more than "a local *Lagothrix*, perhaps *L. lagothricha*." Cabrera (1958) considered Thomas' *Oreonax hendeei* to be no more than specifically distinct from other woolly monkeys, and considered Humboldt's *Simia flavicauda* to be indeterminable. Finally, Fooden (1963) carefully checked Humboldt's and Thomas' descriptions, compared them with the two Watkins specimens in the American Museum, and concluded that *Oreonax hendeei* was synonymous with Humboldt's *Simia flavicauda*. He also agreed with Cabrera that the animal was only specifically and not generically distinct from other *Lagothrix*, the correct name thus being *Lagothrix flavicauda* (Humboldt, 1812).

IV. The Rediscovery of *Lagothrix flavicauda*

For nearly 50 years following the collection of the Hendee specimens, nothing was heard of *L. flavicauda*, even though at least one expedition went into the area to locate the animal.* In order to determine if *L. flavicauda* indeed still existed and, if it did, to obtain data on its biology and conservation status, we organized a 12-day expedition (April 26–May 7, 1974) to the region in which the Watkins and Hendee specimens had been found. Included in the survey were forested areas in the vicinity of Chachapoyas, Mendoza, and Pedro Ruiz Gallo (near a military engineering base and about 2 hours north of Chachapoyas by bus), all in the Dept. of Amazonas (Fig. 4). While we were still on the road to Chachapoyas, we met a hunter who had with him a stuffed adult specimen (skin and skull) of *L. flavicauda* that he had shot in montane rain forest only 6 days prior to our arrival. We later obtained from this same hunter three additional skins and two skulls of *L. flavicauda* he had shot for food, and traveled with him to the forests from which he had obtained these animals. (The skins and skulls we obtained represent the first *L. flavicauda* specimens for a Peruvian museum and are now deposited in the Museo de Historia Natural "Javier Prado" of the Universidad Nacional Mayor de San Marcos in Lima.)

By questioning local people in the towns we visited, examining the collections of several amateur taxidermists (a popular hobby in Peru), and visiting the forests in which *L. flavicauda* occurs, we obtained a picture of the mammal and bird faunas of the area and learned something of the habitat and status of *L. flavicauda* and other rare and unusual animals sympatric with it.

Finally, on the day before our return to Lima, local children in Pedro Ruiz Gallo directed us to the home of a soldier who had a juvenile *L. flavicauda* as a pet. We managed to obtain this unique find from him and took it back with us. The animal, a young male, is the first living representative of its species to be seen by the scientific world (Figs. 1B, 2, and 3). (It has now been donated to the Museo de Historia Natural "Javier Prado" of the Universidad Nacional Mayor de San Marcos, where it continues to thrive.)

*In 1940, Dr. Oliver P. Pearson (Museum of Vertebrate Zoology, Berkeley, California) went to northern Peru to collect *L. flavicauda*, but was unsuccessful (O. P. Pearson, personal communication, 1974).

V. Habitat and Range of *Lagothrix flavicauda*

On the basis of available data, *L. flavicauda* appears to be restricted to montane rain forest of the broken, intermountain plateau on the eastern slope of the Andes in northern Peru. Its range includes the southern part of the Dept. of Amazonas and the mountainous, western part of the Dept. of San Martin (see Fig. 4). We have been unable to determine if the animal actually occurs in the

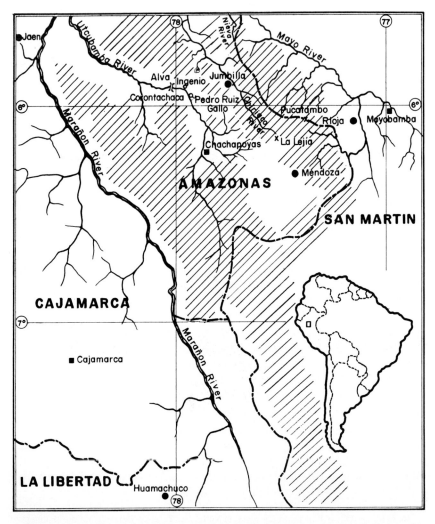

Fig. 4.   Map showing the range of *Lagothrix flavicauda* in the Peruvian Departments of Amazonas and San Martin.

immediate vicinity of Jaen, Dept. of Cajamarca. Humboldt observed his skins in Jaen, but the animals could well have come from the Dept. of Amazonas, on the eastern side of the Río Marañon. In November, 1974, R. Jones (University of California, Berkeley, California) obtained a skin of *L. flavicauda* in the market of Huamachuco (Dept. of La Libertad). However, this specimen lacked precise locality data and almost certainly came from forests north or east of Huamachuco. The immediate vicinity of Huamachuco has no forests that could support *L. flavicauda*.

Localities from which *L. flavicauda* has been collected or reported are listed in Table I and shown in Fig. 4. We have been unable to determine the limits of distribution of the species. It may range somewhat more widely than the list of localities presently available indicates and could conceivably extend into Ecuador (but this is unlikely). Thus far no specimens have been collected outside Peru.

The region in which *L. flavicauda* occurs is characterized by steep, grass or forest-covered mountains, cut by valleys and deep ravines (Fig. 5). To the north, south, and east, there is a transition from montane rain forest to low-altitude Amazon rain forest. To the west, the typical "sierra" of the Andes takes over.

Fig. 5. (A) Montane forest habitat of *L. flavicauda* at Alva, near Pedro Ruiz Gallo, Dept. of Amazonas, Peru. (B) Same as Fig. 5A.

The altitudinal range of *L. flavicauda* is roughly 500—2500 m, and might reach as high as 3000 if adequate forest is available at this altitude. Fooden (1963) estimated that Watkins' specimens came from 3000 m, but Stübel (in Papavero, 1973) gives the altitude of La Lejía as 2300 m.

Few data are available on the behavior and ecology of *L. flavicauda*. Hendee, in a letter written to Oldfield Thomas on February 27, 1927, reported that the group from which he collected the three British Museum specimens was made up of about twenty animals and that they were not particularly timid.

A number of other interesting animals are sympatric with *L. flavicauda*. Among them are at least three other species of monkeys (*Ateles belzebuth belzebuth, Cebus albifrons, Aotus trivirgatus,* and possibly *Alouatta seniculus*), spectacled bear (*Tremarctos ornatus*), possibly mountain tapir (*Tapirus pinchaque*), cock of the rock (*Rupicola peruviana*), marvelous spatule-tail hummingbird (*Loddigesia mirabilis*—another endemic that is apparently even more restricted in range than *L. flavicauda*), and a variety of other birds and mammals (Table III).

VI. Status of *Lagothrix flavicauda*

The information that we gathered during the course of our expedition led us to the conclusion that *L. flavicauda* is an endangered species. It is threatened by habitat destruction and especially by hunting within its restricted range. Prior to 1960, no roads went into the region in which it occurs. However, since that time, the Peruvian army has been constructing the Marginal de la Selva road through the area and has been opening previously isolated localities for colonization and defense purposes. Although a good deal of montane forest still remains, destruction is taking place at an ever-increasing rate. Forests are being cleared to make way for agricultural plots and cattle, although in some places the hillsides are so steep that cattle can barely walk on them.

Hunting, however, is having the most destructive immediate effect on the fauna of the area. Monkeys, bears, and many other mammals and birds are being shot for food and for their skins, some of which are stuffed and sold to townspeople. In the house of the hunter from whom we obtained the four *L. flavicauda*, we also saw skins of a spider monkey (*Ateles belzebuth belzebuth*), a tayra (*Eira barbara*), four coatis (*Nasua nasua*), the skull of a capuchin monkey (*Cebus albifrons*), and a bag of feathers containing the remains of roughly 12 cock of the rock (*Rupicola peruviana*) and a lyre-tailed nightjar (*Uropsalis lyra*). Skins of spectacled bear (*Tremarctos ornatus*) sell for 600 soles (approx. $15.00) each and bear grease is valued at 300 soles per liter. Bear meat is so highly esteemed that it is usually eaten by the family of the hunter and not sold. Monkey meat is also appreciated and *L. flavicauda* is apparently especially sought after because of its size and the quality of its flesh.

TABLE III

Some Birds and Mammals Recorded from the Range of *Lagothrix flavicauda* in the Peruvian Department of Amazonas[a]

Birds

Turkey vulture	*Cathartes aura*
Black-and-chestnut eagle	*Oroaetus isidori*
American kestrel	*Falco sparverius*
Andean guan	*Penelope montagnii*
Sun bittern	*Eurypyga helias*
Scarlet-fronted parakeet	*Aratinga wagleri*
Squirrel cuckoo	*Piaya cayana*
Lyre-tailed nightjar	*Uropsalis lyra*
Sparkling violetear	*Colibri coruscans*
Marvelous spatuletail	*Loddigesia mirabilis*
Cock of the rock	*Rupicola peruviana*
Tropical kingbird	*Tyrannus melancholicus*
Green jay	*Cyanocorax ycas*
Violaceous jay	*Cyanocorax violaceus*
Scarlet-rumped cacique	*Cacicus uropygialis*
Carbonated flower-piercer	*Diglossa carbonaria*
Blue-necked tanager	*Tangara cyanicollis*
Silver-beaked tanager	*Ramphocelus carbo*
Hepatic tanager	*Piranga flava*
Black-beaked grosbeak	*Pheucticus aureoventris*

Mammals

Opossum	*Didelphis marsupialis*
Night monkey	*Aotus trivirgatus*
White-fronted capuchin monkey	*Cebus albifrons*
Spider monkey	*Ateles belzebuth belzebuth*
Spectacled bear	*Tremarctos ornatus*
Coati	*Nasua nasua*
Long-tailed weasel	*Mustela frenata*
Tayra	*Eira barbara*
Jaguarundi	*Felis yagouaroundi*
White-lipped peccary	*Tayassu pecari*
White-tailed deer	*Odocoileus virginianus*
Paca	*Agouti paca*
Prehensile-tailed porcupine	*Coendou bicolor*

[a] Based on the author's field observations, specimens in local taxidermy collections, and information from local inhabitants.

Added to hunting pressure from local people is an even greater threat from the army. The army is reportedly employing professional hunters (called "mitayeros" or "montaraces") to shoot game to feed construction crews working on the road network. This highly wasteful procedure is unnecessary since domestic animals are available (although somewhat more expensive than living off the land).

The army's destructive effect on the fauna has even caused concern among a number of residents of the area. Several weeks after an article announcing our rediscovery of *L. flavicauda* appeared in *La Prensa*, one of Lima's biggest daily newspapers, the following letter was written to the editor of *La Prensa* by a schoolteacher in Pedro Ruiz Gallo:

> Having read your article on *Lagothrix flavicauda* in the May 19 supplement, I I feel that I should inform you that this species of monkey inhabits the region of the Upper Mayo (Fig. 4), where it is being eliminated without compassion by the workers constructing the Ingénio—Río Nieva—Rioja road because it constitutes one of their food resources and is highly appreciated for its size and excellent meat. I ask you to initiate a campaign to make the government take the necessary measures to prevent the extinction of this monkey. It is very easy to hunt since it travels in "herds" and does not try to escape when it sees hunters. When one of the members of a group is injured, the others reunite around it, which facilitates the slaughter of the entire group. This is why one no longer finds *L. flavicauda* in areas where established human communities exist. If protection is not provided very soon, the species *Lagothrix flavicauda* will have been wiped out by the time the *Marginal de la Selva* road is completed. (Translated from letter in Spanish in *La Prensa* Sunday Supplement, Lima, Peru, June 16, 1974, p. 2).

VII. Conservation and Recommendations for the Future

Lagothrix flavicauda is an important part of Peru's faunal heritage since it is the only species of larger mammal that is endemic to Peru. Nonetheless, because of its rarity and the relative isolation of its habitat, it is unknown to the vast majority of Peruvians. In order to publicize the animal, we held a press conference to announce its rediscovery and were aided considerably by an outstanding performance on the part of the juvenile male *L. flavicauda* obtained in Pedro Ruiz Gallo. We obtained coverage from the major Lima dailies and local television, and longer articles also appeared in two Peruvian magazines (Anonymous, 1974; Mittermeier *et al.*, 1974). Articles on *L. flavicauda*, have also been written for several Peruvian and international conservation journals (Mittermeier, *et al.*, 1975a,b; Macedo-Ruiz, *et al.*, 1976; Marcus, 1975) and we have recommended that it be included in a new series of endangered species postage stamps currently being issued by the Peruvian government [other species in this series include the vicuña (*Vicugna vicugna*), the chinchilla (*Chinchilla laniger*), the giant otter (*Pteronura brasiliensis*), the bush dog (*Speothos venaticus*), the spectacled bear (*Tremarctos ornatus*), and the red uakari (*Cacajao calvus rubicundus*)].

In order to provide legal protection for *L. flavicauda*, we have recommended that it be given full protection within Peru and that it be included as a red sheet, endangered species in the International Union for the Conservation of Nature's "Red Data Book." The latter has already been done (Red Data Book, 1974).

In order to ensure the survival of *L. flavicauda* within its natural habitat, we have recommended that a national park, national reserve, or national sanctuary be established in its range. If properly selected, such a protected area would also include populations of other rare and endangered species of the region. In order to locate a site for a park or reserve, we are planning longer and more detailed surveys, to be conducted under auspices of the *Museo de Historia Natural "Javier Prado"* and the *Servicio Forestal y de Caza* (Wildlife Service) of the Peruvian Ministry of Agriculture. Once a suitable site has been located, a long-term study should also be conducted to provide data on the behavior and ecology of the species.

Finally, we are planning a captive breeding program for *L. flavicauda* in order to ensure the survival of some representatives of the species should it disappear in the wild. (A brief outline of the requirements for a captive colony is given in Sections VIII.)

The letter of the concerned resident of Pedro Ruiz Gallo clearly points out the need for rapid action on behalf of *L. flavicauda*. In recent years, much has been accomplished in Peruvian conservation, thanks both to the efforts and initiative of Peruvian conservationists and to international cooperative endeavors dedicated to saving important parts of Peru's wildlife heritage. We therefore have every hope that it will be possible to direct national and international attention to *L. flavicauda* and that everything feasible will be done to save this unique and attractive monkey from extinction.

VIII. Plans for a *Lagothrix flavicauda* Breeding Colony

With a life expectancy in captivity of only about 15 months, *Lagothrix* was previously considered a very difficult genus to maintain (Crandall, 1964). However, the major problems have now essentially been solved. The first successful colony birth took place in 1966 (Williams, 1967a) and there are now at least four captive breeding colonies in Europe: Murrayton Woolly Monkey, Cornwall, England; Apenheul Woolly Monkey Colony, Apeldoorn, Netherlands; Basel Zoo, Basel, Switzerland; and Banham Zoo, Norfolk, England.

The following plans for a *Lagothrix flavicauda* breeding colony are based on the experiences of the successful European colonies (especially Murrayton and Apenheul, where J. C. spent 2 years studying behavior and designing the physical environment), on several publications by Williams (Williams, 1965, 1967a,b, 1972, 1974), and to a lesser extent on a few recent field studies of *L. lagothricha* (Durham, 1975; Kavanagh and Dresdale, 1975; Nishimura and Izawa, 1975; Hernandez-Camacho and Cooper, 1976; Cassidy, in preparation).

A. Size and Composition of Initial Captive Group

The only information on group size for *L. flavicauda* comes from the 1927 Hendee letter, which mentions about twenty animals. The most recent field studies of *L. lagothricha* indicate a wide group size range of 4 to 35 individuals, with an average of about 13 at an altitude of 600 m and 7 at 1500 m (Kavanagh and Dresdale, 1975; Durham, 1975; J. Cassidy, personal observation). Based on this limited field data and the experience of the European colonies, we feel that a group of 9 to 12 individuals would be optimal for starting a captive colony of *L. flavicauda*. Group composition should be 1 adult male; 1−2 subadult males; 3 adult females; 2−3 subadult females; and 2−3 juveniles. Preferably, the animals should come from a single group or from as small a number of groups as possible.* Ideally one complete group should be captured, thus maximizing group cohesion from the outset. The animals should be captured with tranqulizer darts shot from either a CO_2 rifle or a blowgun (Brockelman and Kobayashi, 1971). The blowgun is preferable since it is safer and its silence facilitates the capture of several individuals from a single group.

B. Physical Features of the Colony

1. Location

Since the colony will be located in Peru, it should, if possible, be situated in a forested area similar to the natural habitat in climate and vegetation. This is not absolutely necessary, since the European colonies are located in areas completely different from *Lagothrix* habitat, but would reduce the need for construction of monkey buildings and planting of food trees.

2. Spatial Requirements

During the initial stages of colony formation, only one compound will be necessary. With increase in numbers, either through reproduction or the introduction of additional wild-caught specimens, two more compounds should be added to form a large complex. The final colony area, excluding research buildings, should be between 12 and 20 hectares, divided into three separate but linked compounds of 4 to 7 hectares each (Fig. 6).

Some trees should be removed from the colony area to facilitate observation, but all trees which can serve as food sources should be left standing.

Compounds should be surrounded by a moat or a wall, or some combination

---

*Not everyone agrees with this method for starting a colony. Some specialists feel it is too difficult for most adult *Lagothrix* to adjust to captivity and that it is better to start a colony with behaviorally more flexible juveniles and subadults (but not infants or young juveniles) (M.G.M. van Roosmalen, personal communication).

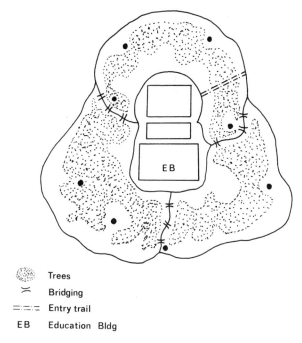

Trees

Bridging

Entry trail

EB   Education Bldg

Fig. 6.   Plan for a *Lagothrix flavicauda* breeding colony. The education building and monkey buildings are located at the center of the colony area; bridging connects the three compounds. The solid black circles indicate the location of the observation points.

of the two—depending on the physical features of the site, availability of water, and other factors. The moat should be 2–5 m wide, with a minimum depth of 50 cm. The wall should be made of smooth, corrugated, unclimbable, weather-resistant material. Buried at least 30 cm below the surface, it should extend 2.5–3.5 m above the ground and incline inward at an angle of 15 to 20°. All trees and branches should be cleared from an area of at least 5 m on each side of the perimeter, and there should be no vegetation other than grass or very low bushes within 3 m of the wall or moat.

When the second and third compounds are added, "bridges" should be constructed between them in order to study group encounters and possible seasonal changes in group composition (Fig. 6). This observation device is already in use at the University of Puerto Rico Primate Center on La Cueva Island (Neurater and Goodwin, 1972).

3.   Building

Three kinds of indoor space are needed in the colony: quarters for monkeys (including hospital area); research facilities (including office space, library, laboratory facilities, video tape studio); and living quarters for animal keepers and researchers (including sleeping space and kitchen). If the colony is located in an area where protection from adverse weather conditions is necessary, indoor monkey rooms are a must. Two 6 m × 6 m rooms per compound, each well furnished with natural materials like ropes, rope nets, wooden bars, and trimmed trees, are adquate. Indoor enclosures should have sealed concrete floors and walls of some washable, scratch-resistant material. Temperature of these rooms should be maintained at about 17°C, preferably using solar heating.

4.   Observation Facilities

The most important function of any breeding colony, aside from its primary purpose of conservation, is to serve as a site for long-term captive studies which provide an invaluable supplement to field studies. Mention has already been made of intercompound bridges. To further facilitate study, several observation towers should be constructed at strategic points within the compounds. The number of towers depends on the site, but a sufficient number should be provided to allow observation of most of the compound from a distance of 50 m or less. Towers 6 m in height should be adequate in most cases, but they can be as high as 8 m if the natural vegetation in the colony makes this necessary. They should include facilities for video tape cameras which serve as a useful aid to behavioral research. (Video tape is recommended since it is economical and easy to use. The tape can be used again and again and instant replay or monitoring is possible.)

5.   Nutrition

Captive studies and field observations indicate that *Lagothrix* sp. are frugivores, with a secondary specialization in folivory (Kay, 1973). The diet of wild *Lagothrix* sp. includes the following (Durham, 1976; Kavanagh and Dresdale, 1975; Nishimura and Izawa, 1975; Cassidy in preparation.):

> *Anacardium excelsum* (Anacardiaceae)—leaves, flowers, buds
> *Attalea regia* (Palmae)—fruit
> *Cecropia* sp. (Moraceae)—flowers, buds, fruit
> *Chrysochlamys* sp. (Moraceae)—fruit, pedicels
> *Ficus* sp. (Moraceae)—flowers, buds, fruit
> *Gustavia* sp. (Lecythidaceae)—fruit
> *Jessenia polycarpa* (Palmae)—fruit
> *Lucuma* sp. (Sapotaceae)—fruit, flowers, buds
> *Mauritia flexuosa* (Palmae)—fruit
> *Protium* sp. (Burseraceae)—fruit
> Annonaceae, several sp.—leaves
> Clusiaceae, several sp.—leaves

Palmae, "palmetto"—fruit, flowers, buds
Sapotaceae, several sp.—fruit

Wild specimens also eat bark, twigs, and some insects (Kavanagh and Dresdale, 1976; Cassidy, in preparation).

Animals in the European colonies are given or browse on the following temperate zone species in their compounds:

Epilobium sp. (Onagraceae)—leaves, common name: willow herb
Rumex sp. (Polygonaceae)—leaves, common name: dock
Taraxacum sp. (Compositae)—leaves, common name: dandelion
Fagus sp. (Fagaceae)—leaves and buds, common name, beech
Ulmus sp. (Ulmaceae)—leaves and buds, common name: elm
Urtica sp. (Urticaceae)—leaves, common name: nettles

In addition to these foods, captive animals are also given a wide range of fruits and green vegetables, onions and garlic, carrots, insects (including mealworms, moths, locusts, stick insects), raw eggs, wheat germ, honey, molasses and brewers' yeast, concentrated fruit juice, vitamin extract, vitamin D drops, and chicken, lightly cooked and eaten with leaves. Occasionally sisal rope or wood is chewed and ingested with chicken. Captive animals also catch insects and occasionally birds in the trees within the colonies.

The following foods, listed in Williams (1967a), are no longer recommended: cooked eggs, cheese and other milk products, condensed or fresh milk, cereals, red meat, coconut, and semolina.

The supplementary foods listed above should be given to colony animals at several feeding platforms. Since provisioning can lead to high intraspecific aggression, foods should be introduced simultaneously at several widely spaced platforms.

Acknowledgments

We should like to express special thanks to Marc Dourojeanni Ricordi (Director General de Forestal y Caza), Carlos Ponce del Prado (Director de Fauna Silvestre) and Antonio Brack Egg (Sub-Director de Conservacion de la Fauna Silvestre) of the Peruvian Ministry of Agriculture for their cooperation in all phases of this project. Thanks also to J. Frooden (Field Museum of Natural History Chicago) for reading and commenting on the manuscript, to G. Corbet (British Museum of Natural History) for providing information on the Hendee specimens, to Ned K. Johnson (Museum of Vertebrate Zoology, Berkeley, California) for information on the Huamachuco skin, to Enrique Zileri (*Caretas* magazine, Lima, Peru) and Lima television channels 4 and 5 for publicizing the situation of *L. flavicauda*, to Hugo Vela (Pedro Ruiz Gallo, Peru) for data on the range and status of *L. flavicauda*, and to our guides Agliberto and Javier Vigo Bardales for helping us rediscover the animal in the first place. We are also grateful to the following people for their help with the captive *L. flavicauda* juvenile: Alfredo

Camacho Pozzo for providing materials and designing the animal's cage, to Dr. Jose N. Pinelo and Luz Sarmiento for polio vaccine and parasitological examination, and to Aurora Ruiz de Macedo, Gloria Pinelo de Macedo, and Lolie Zúniga for providing facilities and caring for the animal. R.A.M.'s studies in South America were partly supported by a National Science Foundation Predoctoral Fellowship and partly by the New York Zoological Society.

References

Anonymous (1974). La superviviente. *Caretas* (*Lima*) **500**, 50—51.

Brockelman, W. Y. and Kobayashi, N. K. (1971). Live capture of free-ranging primates with a blowgun. *J. Wildl. Manage.* **34**(4), 852—855.

Cabrera, A. (1958). Catálogo de los mamíferos de America del Sur. *Rev. Mus. Argent. Cienc. Natur. Bernardino Rivadavia Inst. Nac. Invest. Cienc. Natur. Cienc. Teol.* **4**(1), pp. iv + 307.

Cassidy, J. (1976). The behavior and ecology of the genus *Laghothrix*. In preparation.

Crandall, L. S. (1964). "Management of Wild Animals in Captivity." Univ. Chicago Press, Chicago, Illinois.

Durham, N. M. (1975). Some ecological, distributional, and group behavioral features of Atelinae in southern Peru, with comments on interspecific relations. *In* "Socioecology and Psychology of Primates" (R. A. Tuttle, ed.). Aldine, Chicago.

Fooden, J. (1963). A revision of the woolly monkeys (genus *Lagothrix*). *J. Mammal*, **44**(2), 213—247.

Hill, W. C. O. (1962). "Primates, Comparative Anatomy and Taxonomy", Vol. V. Cebidae, Part B. pp. xxi + 537. Edinburgh Univ. Press, Edinburgh.

Hernandez-Camacho, J. and Cooper, R. W. (1976). The nonhuman primates of Colombia. *In* "Neotropical Primates: Field Studies and Conservation" (R. W. Thorington, Jr. and P. G. Heltne, eds.), pp. 35—69. National Academy of Sciences, Washington, D.C.

Humboldt, A. and Bonpland, A. (1812). "Recueil d'Observations de Zoologie et d'Anatomie Comparée," Vol. I, pp. viii + 368. F. Schoell and G. Dufour, Paris.

Kavanagh, M. and Dresdale, L. (1975). Observations on the woolly monkey (*Lagothrix lagothricha*). *Primates*, **16**(3), 285—294.

Kay, R. F. (1973). Mastication, molar tooth structure and diet in primates. pp. xi + 376. Unpubl. doctoral dissertation, Yale Univ., New Haven, Connecticut.

Macedo-Ruiz, H. de, Mittermeier, R. A., and Luscombe, B. A. (1976). Redescubrimiento del primate peruano *Lagothrix flavicauda* (Humboldt, 1812) y primeras observaciones sobre su biologia. *Rev. Cienc.* **71**.

Marcus M. (1975). The tale of Flavi. *Harvard Mag.* **78**(4), 51—53.

Mittermeier, R. A., Macedo-Ruiz, H. de, and Luscombe, B. A. (1974). The search for the Peruvian yellow-tailed woolly monkey. *Andean Times* (Lima), **34**(1750), 13—14.

Mittermeier, R. A., de Macedo-Ruiz, H, and Luscombe, B. A. (1975a). A woolly monkey rediscovered in Peru. *Oryx* **13**(1), 41—46.

Mittermeier, R. A., de Macedo-Ruiz, H., and Luscombe, B. A. (1975b). Mystery monkey. *Anim. Kingdom* **78**(3), 2—7.

Neurater, L. J. and Goodwin, W. J. (1972). The development and management of macaque breeding programs. *In* "Breeding Primates" (W. B. Beveridge, ed.), pp. 60—71. Karger, Basel.

Nishimura, A. and Izawa, K. (1975). The group characteristics of woolly monkeys in the Upper Amazonian basin. *In* "Contemporary Primatology" (S. Kondo, M. Kawai, and A. Ehara, eds.) pp. 351—357. S. Karger, Basel.

Papavero, N. (1973). "Essays on the History of Neotropical Dipterology," Vol. 2. Museu de Zoologia, Univ. Sao Paulo, Sao Paulo.

Poeppig, E. (1832) Doctor Poeppig's naturhistorische Reiseberichte. *Notiz. Gebiete Natur-Heilk.* **33**(7), 97—106.

Thomas, O. (1927a). A remarkable new monkey from Peru. *Ann. Mag. Natur. Hist.* **19**(9), 156—157.

Thomas, O. (1927b). The Godman—Thomas Expedition to Peru.—V. On mammals collected by Mr. R. W. Hendee in the Province of San Martin, N. Peru, mostly at Yurac Yacu. *Ann. Mag. Natur. Hist.* **19**(9), 361—375.

Thomas, O. (1927c). The Godman—Thomas Expedition to Peru.—VI. On mammals from the Upper Huallaga and neighboring highlands. *Ann. Mag. Natur. Hist.* **20**(9), 594—608.

Williams, L. (1965). "Samba and the Monkey Mind." Bodley Head, London.

Williams, L. (1967a). Breeding Humboldt's woolly monkey (*Lagothrix lagothricha*) at Murrayton Woolly Monkey Sanctuary. *Int. Zoo Yearb.* **7**, 86—89.

Williams, L. (1967b). "Man and Monkey." Andre Deutsch, London.

Williams, L. (1972). "Challenge to Survival." Andre Deutsch, London.

Williams, L. (1974). "The Woolly Monkey Book." Monkey Sanctuary Publ., Looe, Cornwall, England. (Also published as "Monkeys and the Social Instinct.")

5

Primate Conservation in Brazilian Amazonia

RUSSELL A. MITTERMEIER and ADELMAR F. COIMBRA-FILHO

I. Introduction

Brazil is the largest country in South America in both area (8,511,965 km²) and population (101,300,000); it ranks fifth in the world in area (Rebelo, 1973; Goodland and Irwin, 1975). It has within its borders more tropical rain forest (3,374,000 km²) than any other country on earth (Pires, 1974). More than 60% of the vast Amazonian hylaea, by far the largest rain forest area on earth, is found within Brazilian territory (Pires, 1974). Brazil also has a more diverse primate fauna than any other country. Forty-one species representing all sixteen genera of New World monkeys occur in Brazil, the majority of them primarily or exclusively Amazonian in distribution (Table 1). Only Colombia, with 12 genera and 22 species (Hernandez-Camacho and Cooper, 1976), Peru,

TABLE I

Primate Species Found in Brazilian Amazonia[a]

Family Callitrichidae	*Saguinus inustus*
Cebuella pygmaea	*Saguinus labiatus*
Callithrix argentata	*Saguinus midas*
Callithrix humeralifer[b]	*Saguinus mystax*
	Saguinus nigricollis
Saguinus bicolor[b]	
Saguinus fuscicollis	
Saguinus imperator	
Family Callimiconidae	
Callimico goeldii[e]	
Family Cebidae	
Aotus trivirgatus	*Saimiri sciureus*
Callicebus moloch	*Cebus albifrons*
Callicebus torquatus	*Cebus apella*
	Cebus nigrivittatus
Pithecia pithecia	*Ateles belzebuth*[f]
Pithecia monachus	*Ateles paniscus*[f]
Chiropotes albinasus[b,c]	
Chiropotes satanas	*Lagothrix lagothricha*[f]
Cacajao calvus	
(*C. calvus calvus,*[b,d]	*Alouatta belzebul*[b]
C. calvus rubicundus)[d]	*Alouatta seniculus*
Cacajao melanocephalus[e]	

[a] Except for *Ateles*, taxonomy follows Hershkovitz, 1972.
[b] Species endemic to Brazilian Amazonia.
[c] Amber sheet, vulnerable species in the IUCN "Red Data Book."
[d] Red sheet, endangered species in the IUCN "Red Data Book."
[e] Gray sheet, indeterminate species in the IUCN "Red Data Book."
[f] Species not currently listed in the IUCN "Red Data Book," but which should be.

with 12 genera and 20 species (Gardner, unpublished list), and Madagascar, with 12 genera and 19 species (Martin, 1972), can even approach Brazil's impressive primate fauna.

In spite of the great variety and abundance of Brazilian primates, remarkably little is known of their behavior, ecology, and current conservation status. The few field studies of primates of Brazilian Amazonia (e.g., Thorington, 1968a) have been short term, and most of our knowledge of these animals still stems from anecdotal reports of 19th and early 20th century naturalists (e.g., Wallace, 1852; Bates, 1863).

In this paper, we briefly review the available field data on Brazilian Amazonia's nonhuman primates, paying particular attention to conservation and threats to the survival of these animals. Most of the data on conservation status comes from two primate surveys conducted in Brazilian Amazonia in 1973. We also discuss various ways of defining Amazonia itself, and describe the major habitat types within Brazilian Amazonia.

II. Defining Amazonia

Definitions of Amazonia vary, depending largely on the interests of the authors in question. Meggers (1971, 1973) defines Amazonia by physical and climatic features, calling it:

> ... that portion of South America east of the Andes that lies below 1500 m in elevation, where rain falls on 130 or more days per year, where relative humidity normally exceeds 80% and where annual average temperature variation does not exceed 3°C. These characteristics prevail over most of the Amazon drainage, with the exception of the headwaters of the longer tributaries, and extend over the Guianas to the mouth of the Orinoco.... Vegetation consists of tropical rain forest broken by small enclaves of savanna where the soil is too porous to retain moisture during dry months (Meggers, 1973, p. 311).

Ducke and Black (1953) define Amazonia on the basis of the distribution of the genus *Hevea*, which includes the economically important rubber trees:

> The only natural limits of the hylaea are the Atlantic and the Andes; on the north and south extremes, the hylaea (rain) forest is gradually replaced by the flora of the drier neighboring countries. In such conditions, it may be convenient to take for limits of the hylaea those of the geographic area of some genus of forest trees proper to the country, well studied by botanists, and well known to everyone for its economic importance. No genus seems better suited for this purpose than *Hevea* ... (Ducke and Black, 1953, p. 2).

Cabrera and Yepes (1940) define Amazonia (*Distrito Amazónico* in their terminology) by the presence of certain characteristic mammal species, such as the marsupial *Marmosa noctivaga* and certain rodents (*Sciurillus pusillus*, *Dactilomys* sp., and *Mesomys* sp.) (Cabrera and Yepes, 1940; Fittkau, 1974). Fittkau (1969, 1974) equates Amazonia with the zoogeographical province Hylaea of the Neotropical Guiana-Brazilian subregion. He includes in Amazonia the entire tropical forest region, as well as the transitional regions between forest and other drier regions of the north and south. The area covered amounts to some 7,000,000 km², of which about 80% is covered with tropical rain forest.

All these definitions refer to roughly the same area, the main differences being in where the northwestern and southern boundaries are placed. The southern boundary is especially difficult to determine because of an extensive system of interdigitating peninsulas of savanna and forest located roughly between 48° and 62°W (Hueck, 1972). For an idea of the variation in the placement of Amazonia's boundaries, see figures in Cabrera and Yepes (1940), Ducke and Black (1953), Fittkau (1969, 1974), Meggers (1971, 1973), Hueck (1972), Haffer (1974), and Pires (1974).

For the purposes of this paper, we refer to Amazonia as the vast tropical rain forest region of northern South America, including the forests of the Amazon and Orinoco drainages and the Guianas, and bordered on the west by the Andes, on the east and northeast by the Atlantic, on the northwest by a transition zone between forest and the *llanos* (plains) of Venezuela and Colombia, and on the south by a transition zone between forest and the *cerrado* (savanna) of central Brazil. Included within the boundaries of Amazonia are a number of small *campos* (savanna-like enclaves) entirely surrounded by forest (Table II). The coastal forests of southeastern Brazil are not considered part of Amazonia. The approximate extent of Amazonia is indicated in Fig. 1 and the area it covers in Table II. When we refer to Brazilian Amazonia, we mean that part of Amazonia that occurs within the political borders of Brazil (Fig. 1; Table II).

III. Habitat Types in Amazonia

Topographically and climatically, Amazonia is a comparatively homogeneous environment (Meggers, 1971; Fittkau, 1974). The terrain is flat to moderately undulating. The mainstream of the Amazon falls only about 70 m

TABLE II

Surface Area of Various Forest Types in Brazilian Amazonia [a]

	km²
Amazonia (Brazilian and non-Brazilian)	6,000,000
Non-Brazilian Amazonia	2,300,000
Brazilian Amazonia	3,700,000
Total forested area, Brazilian Amazonia	3,374,000
Terra firme forest	3,303,000
Várzea forest	55,000
Igapó forest	15,000
Mangrove forest	1,000
Total nonforested area (including 150,000 km² of savanna, and a variety of other formations), Brazilian Amazonia	326,000

[a]Modified from Pires, 1974.

Fig. 1. Map of South America showing the approximate extent of Amazonia (modified from Hueck, 1972, p. 3). The numbers correspond to 14 different Amazonian regions recognized by Hueck (1972): 1. Delta of the Amazon. 2. Northeastern Amazonia. 3. Tocantins-Gurupí region. 4. Tapajós and middle to lower Xingu region. 5. Madeira-Purus region. 6. Eastern hylaea. 7. Rio Negro region. 8. Floodplain (*várzea*) of the Amazon and lower Madeira. 9. Acre region and Beni-Mamoré-Guaporé region. 10. Hylaea of the foothills of the Andes. 11. Caquetá—Vaupes—Guainía region. 12. Region of the right bank tributaries of the Orinoco. 13. Guianan region (Surinam, Guyana and French Guiana). 14. Delta of the Orinoco.

between the Peruvian border and the ocean (almost 3000 km) and eleven of its tributaries flow for more than 1600 km without a single fall or rapid (Meggers, 1971). Together with the equatorial location, the low elevation results in a uniform temperature. Average temperatures of the warmest and coldest months rarely differ by more than 2°C. Average annual temperature varies little, ranging from 24° to 27°C for the entire region. Average annual precipitation usually exceeds 2000 mm, is always greater than 1500 mm and may be as high as 3500 mm (Hueck, 1972).

In spite of the uniformity of temperature and the superficially homogeneous appearance of the forest, Amazonia in fact includes a wide variety of habitats. Rivers are important factors in determining habitat diversity in Amazonia. The larger rivers act as barriers to dispersal of certain plant species (Hueck, 1972). All rivers rise and fall during the year (some as much as 15 m), flooding forests along their margins. Certain kinds of rivers are rich in nutrients, while others are extremely poor. Forest composition in a particular area is affected by whether or not the forest is flooded throughout all or part of the year. Forests that are flooded are affected by the composition of the river water inundating them.

Terra firme or dry land forests are not affected by seasonal fluctuation in river level. They make up about 98% of the Amazonian forest (Table II; Meggers, 1971; Pires, 1974). Even the youngest soils of the *terra firme* date back to the Tertiary, and the soils of the Guayana and Brazilian shields, on which much of the *terra firme* forest occurs, are among the oldest geological formations on earth. The soils of the *terra firme* have, therefore, been exposed to leaching for millions of years and are generally poor (Meggers, 1971). Most of the nutrients in the *terra firme* forest are stored in the living plants, and the plants are unexcelled in their ability to capture and store nutrients from the air and from whatever decaying organic matter may be available. Plants have also evolved many chemical defenses to protect against nutrient loss to herbivorous animals in this low primary productivity habitat (Janzen, 1975).

Flooded forests are seasonally or permanently inundated by one of three major kinds of Amazonian rivers: "white water" rivers, "black water" rivers or "clear water" rivers. White water rivers, like the mainstream of the Amazon, the Japurá, the Purus, and the Madeira, originate in the mineral rich Andes and carry a heavy silt load. Their color can be described as café au lait, they have turbid opaque waters, and are slightly acidic to slightly basic (pH 6.5–8.8; Goodland and Irwin, 1975). White water rivers drain only about 12% of Amazonia, but carry 86% of the total dissolved salts and 82% of the suspended solids discharged each year (Meggers, 1971). The floodplain or *várzea* of white water rivers is very wide and may attain 100 km in some areas (Hueck, 1972). It is rejuvenated yearly by mineral-rich silts left by receding river waters, but has a lower number of tree species than surrounding areas of dry land—perhaps because of the smaller geographic area it occupies (Huber, in Hueck, 1972; Goodland and Irwin, 1975).

Black water rivers, of which the Rio Negro is the best example, are transparent, acidic (pH 3.7—5.4), usually the color of "strong tea," and carry virtually no nutrient-rich silt (Meggers, 1971; Goodland and Irwin, 1975). These rivers are characterized by low, sloping, poorly defined banks covered with *igapó*, the periodically or permanently flooded swamp forest. They originate in the very poor sands of the Guayana and Brazilian shields. As already mentioned, plants growing on such impoverished soils have developed a wide variety of secondary defensive compounds which minimize nutrient loss. Rainwater runoff into black water rivers is exceptionally rich in humic acids and probably other toxic organic compounds. This occurs because the leachate from fresh vegetation and decomposing litter on these soils is rich in phenols and other defensive chemicals and because the poor soil and the high input of phenols lead to a litter and soil community poor at degrading these secondary compounds (Janzen, 1975). Litter falling directly into the rivers from the *igapó* forests decays, resulting in oxygen-deprived waters. The biocidal nature of the plant secondary compounds, together with the low oxygen content and extremely poor nutrient content of the black water rivers, results in reduced animal communities. Biomass of invertebrates, fish, reptiles, amphibians, birds, and mammals is much lower in these rivers than in the white water areas (Roberts, 1972; Marlier, 1973; Mittermeier *et al.*, 1976, and unpublished data). Brazilian black water rivers are often referred to as *rios de fome* (rivers of hunger) because of their low productivity. Unlike the white water floodplain, which frequently includes wide expanses of treeless, savanna-like *campos* and "lakes," the inundated margins of the black water rivers are almost always covered with *igapó* forest.

Clear water rivers, like the Xingu and the Tapajós, are similar to black water rivers in the absence of suspended silt and tendency to acidity (pH 6.4—6.6). However, the banks of these rivers are high and stable and there is consequently less oxygen consuming, humic acid-rich organic matter in the water (Meggers, 1971). Animal biomass appears to be higher than in the black water rivers, but not as high as in the white water rivers.

In addition to the *várzea* and *igapó* forests and the various kinds of *terra firme* forest, several other plant formations, brought about by local soil and climatic conditions, exist in Amazonia. These include the *terra firme* savanna-like *campos* entirely surrounded by forest, the *várzea campos*, the coastal mangrove forests, the white sand *campinas*, the transitional *mata seca* (dry forest), some montane forest (e.g., on the sandstone mountains on the Brazilian-Venezuelan border), and a number of other types of limited extent (Pires, 1974; Goodland and Irwin, 1975).

In addition to these natural forest formations, slash and burn agriculture and other human activities have resulted in many secondary forest areas of variable size and development. Young secondary forests are known as *capoeira* in Brazil. Older formations are called *capoeirão*.

The different kinds of forest in Amazonia provide a variety of habitats for primates. Very little information is yet available on habitat preferences of Amazonian monkeys, but preliminary observations suggest that some species prefer or are restricted to certain kinds of forest, whereas others occupy a wide variety of habitats. Available data on this topic is discussed in Section V.

We might mention here that the dynamics of primate distribution in Amazonia (and elsewhere) is a subject that merits much further investigation. Primate distribution is affected by many factors, including the evolutionary history of the species, zoogeographical barriers, major habitat requirements, local habitat preferences, number and kind of competing primate and non-primate species per given area and, more recently, the presence of man. Some understanding of all these factors is important for primate conservationists interested in ensuring the survival of primates in some semblance of their natural habitat.

IV.   Threats to the Survival of Primates in Brazilian Amazonia

Threats to the survival of primates in Brazilian Amazonia can be divided into three major categories: habitat destruction, hunting for food and a variety of other purposes, and live capture for export to foreign markets or to serve local pet markets. The effects of these three forms of exploitation vary considerably from area to area and from species to species, depending on the extent of the forest habitat, the degree and kind of human activity in the area in question, local hunting traditions, proximity of commercial animal dealers that serve foreign markets, and size and desirability of different species as food items.

A.   Habitat Destruction

In some parts of South America, the major threat to the survival of non-human primates is habitat destruction. Southeastern Brazil is a prime example. The Atlantic coastal forests were never as extensive as those of Amazonia and the region has been settled for more than 450 years. It is the most densely populated part of Brazil, and is also the agricultural, industrial, and cattle-raising center of the country. Consequently, forest destruction has taken place for a long time and many of southeastern Brazil's unique primates are approaching extinction.

Some habitat destruction has taken place in Amazonia, especially in the *várzea* forests along the mainstream of the Amazon. This area is readily accessible and the nutrient-rich *várzea* soils are suitable for agriculture (Meggers, 1971). However, in most of Amazonia, habitat destruction has not yet become a major threat to the primate fauna, primarily because of the vast extent of the rain forest, the comparatively small human population (7,561,000—Rebelo, 1973), and the low level of technological development of most of the area. Some parts of Amazonia (e.g., the *igapó* forests of the Rio Negro) are virtually uninhabited, impossible to use for cattle-raising or agriculture, and difficult to lumber. They will probably remain the way they are for many years to come.

However, habitat destruction is on the increase in Amazonia and is certain to become a much more important factor in the near future. An enormous system of interconnecting highways [including the 5619 km *Transamazonica* (BR230), the 2323 km *Perimetral Norte* (BR210), and many others] is opening large areas of previously untouched *terra firme* forest to development (Goodland and Irwin, 1975). It is doubtful that the generally poor soils of the *terra firme* will be of much use for intensive agriculture or cattle-raising in the long term (Goodland and Irwin, 1975; N: Smith, personal communication). In spite of this, a number of commercial interests, among them Brazilian ranchers and, more recently, Arab investors (Smith, 1976), have expressed interest in (and in some cases have actually begun) clear-cutting large areas of *terra firme* for cattle-raising. These people are obviously unaware of the ecological consequences of clear-cutting (discussed in Meggers, 1971; Goodland and Irwin, 1975). In the process of finding out how useless the *terra firme* soils are, they will probably destroy the delicate ecological balance of large tracts of land in Amazonia. Before allowing large scale clear-cutting on the *terra firme*, the Brazilian government should devote considerable attention to alternate uses to which the forests might be put (e.g., utilization of native game and plant species—Smith, 1974a,; Goodland and Irwin, 1975; Mittermeier, 1975, 1976).

B. Hunting

1. Hunting for Food

At the present time, hunting of primates for food is perhaps the major threat to the survival of a number of primate species in Brazilian Amazonia. Nonetheless, the effects of hunting vary considerably from area to area, depending in large part on local hunting tradition and the availability of other protein sources (e.g., fish, turtles, and other game animals). The larger primate species are, of course, more affected by hunting for food. *Ateles, Lagothrix, Alouatta, Cebus,* and *Chiropotes* are frequently shot for food in many areas, with *Ateles* and *Lagothrix* being especially esteemed because of their size and the quality of their meat. *Callicebus, Aotus, Saimiri, Saguinus, Callithrix,* and *Cebuella* are less affected by food hunting because these small monkeys barely provide enough meat to recompense the hunter for the cost of his shotgun shell.

In some parts of Amazonia, hunting pressure is fairly heavy. For example, along the Rio Tapajós, *Ateles belzebuth marginatus, Cebus apella, Chiropotes albin-*

asus, and *Alouatta belzebul* are commonly hunted and are becoming rare (although *Saimiri sciureus*, *Callicebus moloch*, and *Callithrix argentata* are usually not persecuted and frequently occur in close proximity to human habitations). In contrast, in some parts of the upper Amazon in Brazil (e.g., between the lower Rio Japurá and the Rio Auatí-Paraná), hunting pressure is quite light. Only *Ateles* and *Alouatta* are occasionally hunted. *Cebus apella*, *Cebus albifrons*, *Cacajao calvus calvus*, and the small species are rarely or never hunted for food. Some local people even expressed disgust at the thought of eating *Cacajao* because of the human appearance of its bald red face. In this rich *várzea* area, other protein resources are quite abundant. Fish are easily caught in the rivers and nesting turtles (*Podocnemis* sp.; Mittermeier, 1975) are common during certain months of the dry season. As a result, primate hunting is of little importance.

Preliminary data from the Transamazonian highway suggest that primate hunting is not yet significant in this area either. Food hunting in 100 km² areas around three Transamazonian *agrovilas* (colonies) accounted for only three primates in a 1-year period (1974). One *Alouatta belzebul* represented 0.1% of the game intake for one community; one *Alouatta belzebul* and one *Cebus apella* made up 1.3% of the game intake for a second community; no primates were shot in the third community investigated (N. Smith, personal communication). As in the upper Amazon, alternate protein sources in the form of larger game animals (e.g., *Tapirus terrestris*, *Mazama* sp., and *Tayassu* sp.) are available. It will, however, be interesting to see if there is an increase in primate hunting as the larger game animals decrease in abundance.

In connection with this, it should be pointed out that hunting traditions in a particular area can change rapidly. One family of settlers from a region where primates are hunted can quickly change the attitudes of local people toward monkey eating, and it takes only a few hunters to wipe out the larger species of a local primate fauna, once they have concentrated on monkeys as a food item. Large cebids like *Ateles* and *Lagothrix* have a much slower reproductive rate than small species like the callitrichids (Eisenberg, 1976) and are rapidly decimated by concentrated hunting pressure.

Hunting alone can denude an area of larger primates, even in the absence of clear-cutting or other forest exploitation. In some parts of Amazonia, there are tracts of primary forest habitat in which species like *Alouatta*, *Ateles*, *Lagothrix*, and *Cebus* have been completely wiped out by heavy hunting pressure. The disappearance of species like these can, of course, also have ecological impact on the forest itself, since many monkey species act as dispersal agents for trees.

In the past, monkey hunting by aboriginal Indian populations was apparently quite heavy in some parts of Amazonia. Bates (1863), for example, reported that one tribe of about 200 Tucana Indians living near Tabatinga, Brazil, killed about 1200 *Lagothrix lagothricha* per year for food. Carvalho *et al.*

(1949) and Carvalho (1951) also mentioned heavy persecution of monkeys in the upper Rio Xingu region. We have no current data on the effects of monkey hunting by remaining Indian populations. However, the number of Indians still living a traditional lifestyle in Brazilian Amazonia is small, probably not exceeding 50,000 individuals (Goodland and Irwin, 1975). Most Amazonian Indians have been absorbed by the Brazilian population or have been severely decimated by the diseases of civilized man. When considering man's effect on primate populations in today's Brazilian Amazonia, we must primarily be concerned with the *caboclo* or "neo-Amazonian," whose ethnic origins are quite diverse and who makes up the bulk of the population.

2. Hunting for Other Purposes

Primates are sometimes hunted for purposes other than human consumption. The most important of these is for bait. Primate carcasses are used as bait for fish, turtles, and especially spotted cats. The procedure employed by *gateiros* (cat hunters) is to shoot a monkey or some other medium-sized mammal or bird, drag it through the forest for several hundred meters to leave a scent trail, and then place the freshly killed animal in a crude wooden trap. The cat, usually an ocelot (*Felis pardalis*), follows the scent of the bait, goes into the trap to retrieve it, and is captured alive. Cats trapped in this way can easily be dispatched by a shot in the head or strangled with a nylon cord, thus avoiding shot damage to the pelt and increasing its value (Smith, 1976). Since a trap line usually consists of a number of traps, each of which must be baited, many primates and other bait animals are killed. In some areas (e.g., along the Transamazonian highway—N. Smith, personal communication; in some tributaries of the Rio Negro, personal observation), hunting of monkeys for cat bait is the most serious drain on primate populations. Species as small as *Saimiri* are sometimes used.

Utilization of monkeys as fish or turtle bait is much less serious. A single primate carcass can be used to bait many fishing lines and the animals are apparently used for this purpose only if nothing else is available.

Stuffed monkeys and ornaments made of various parts of monkey skeletons (e.g., monkey skull necklaces, *Alouatta* hyoid necklaces) are sometimes sold as tourist curios in some Amazonian towns (e.g., Leticia, Colombia; Santarem, Brazil). However, this trade affects only small numbers of animals.

Chiropotes and *Pithecia* are sometimes killed to obtain their long, thickly furred tails to use as dusters. The tail is cut off at the base, the vertebrae are removed, and a stick several centimeters long is inserted in place of the vertebrae. The result is an adequate equivalent of a feather duster. Until recently, *Pithecia* tail dusters were sold in the markets of Iquitos, Peru, but they are only used very locally in Brazilian Amazonia.

C. Live Capture

Live capture of primates for export to foreign markets has never been a serious problem in Brazilian Amazonia, although large numbers have been shipped out of other parts of Amazonia (Cooper, 1968; Soini, 1972; Committee on Conservation of Nonhuman Primates, 1975.) Until recently, the major centres of exportation for Amazonian primates were Iquitos, Peru, and Leticia, Colombia. A substantial portion of the primates exported from Leticia were brought across the border from adjacent parts of Brazil, but the effects of this trade did not extend very far into Brazilian territory and were apparently largely restricted to the vicinity of the mainstream. The species most exported were *Saimiri sciureus, Saguinus fuscicollis, Saguinus mystax,* and *Saguinus nigricollis.* The Leticia trade has now been ended by the Colombian government.

Certain primate species (e.g., *Lagothrix lagothricha, Cebus apella,* and *Cebus albifrons*) are popular as pets in Amazonia and are occasionally captured to serve a local market. The usual procedure is to shoot the mother and retrieve the infant from the back of its dead mother after both have fallen to the ground. In this way, the hunter gets both meat and an infant which can be kept or sold as a pet. Unfortunately, the infant is also frequently killed by the fall or by stray shot. *Lagothrix lagothricha* is especially affected by hunting for pets. Infant *Lagothrix* make delightful pets and are much in demand. Table III is a list of primates kept as pets by people in villages and towns visited during the course of the 4-month 1973 survey. It appears that the numbers captured to serve the local pet market are small, even when one takes into account high losses during capture and from inadequate care in captivity.

Since 1967, the Brazilian Fauan Protection Law (Lei no. 5.197, January 3,

TABLE III

Primates Kept as Pets by Local People in Brazilian Amazonia [a]

Species	Number of individuals kept as pets
Callithrix argentata	1
Callicebus torquatus	1
Pithecia monachus	2
Saimiri sciureus	6
Cebus albifrons	6
Cebus apella	10
Lagothrix lagothricha	8
Ateles paniscus	1
Alouatta seniculus	1
	36

[a]Information gathered during July–November, 1973 primate survey.

1967) has prohibited commerce in native fauna. The Brazilian Endangered Species Acts (Portaria no. 303, May 29, 1968; Portaria No. 3.481, May 31, 1973) provide additional protection for the following Amazonian primates: *Cacajao calvus calvus, Cacajao calvus rubicundus, Cacajao melanocephalus, Chiropotes albinasus,* and *Callimico goeldii.* However, enforcement of these laws in the vast Amazonian region is difficult.

V. The Primates of Brazilian Amazonia

Of the 16 genera and approximately 50 species of New World monkeys, 14 genera and 30 species are represented in Brazilian Amazonia (Table 1). Only two monotypic genera, *Brachyteles* and *Leontopithecus*—both rare, endangered, and restricted to small tracts of coastal forest in southeastern Brazil—are not found in Amazonia. Fig. 2A is a species density map in which the number of

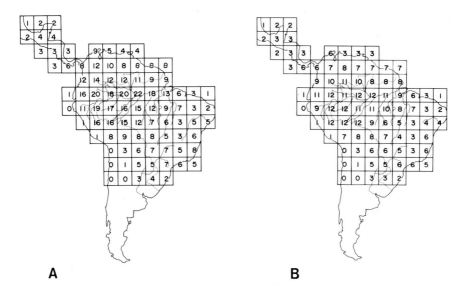

A   **B**

Fig. 2. (A) Map of South America showing the number of primate species per 500 × 500 km quadrant. Species density reaches a peak in the middle and upper Amazon regions of Brazil, Colombia, and Peru. Brazilian Amazonia has more primate species than any other region on earth. The taxonomic arrangement used in making this map follows Hershkovitz (1972), except for *Aotus, Ateles,* and *Callithrix.* We tentatively follow Brumback (1974) in recognizing three species of *Aotus,* and prefer to retain Kellogg and Goldman's (1944) arrangement of *Ateles* until further information on this genus becomes available. We also recognize five valid southern Brazilian species of *Callithrix,* where Hershkovitz recognizes only one (Coimbra-Filho and Mittermeier, 1973). (B) Map of South America showing the number of primate genera per 500 × 500 km quadrant.

primate species per 500 km × 500 km quadrant have been plotted. Figure 2B shows the number of genera per quadrant. The great diversity of primates in Brazilian Amazonia is unmatched in the world.

In this section, we briefly review the known distribution, group size, feeding behavior, and habitat preferences of primates in Brazilian Amazonia. In addition, we discuss conservation status of each genus and how its members are affected by the various threats to primate survival discussed in the previous section. The data on conservation status comes primarily from the two surveys conducted in Brazilian Amazonia in 1973. The first of these was conducted by R.A.M. and investigated 22 localities (Fig. 3) in the 4-month period from July to November, 1973. The second was carried out by R. C. Bailey (Mittermeier *et al.*, 1976; Bailey, personal communication) in the Rio Quixito in November, 1973. The Quixito is a small white water tributary of the Rio Itacuaí, which is itself a tributary of the Rio Javari (Fig. 3). None of the localities investigated were in parks or reserves and hunting pressure varied from very light to heavy. The number of primate groups encountered in each of the major habitat types during these surveys is given in Table IV. Table V is a list of vernacular names of primates in Brazilian Amazonia.

A.   Family Callitrichidae

1. *Cebuella*

a. *Distribution and Habits.*   The pygmy marmoset, *Cebuella pygmea*, is the world's smallest living monkey. It occurs in the upper Amazon region, extending into Brazil as far east as the Rio Japurá north of the Rio Solimões* and the Rio Purus south of the mainstream (Hill, 1957; de Avila-Pires, 1974). Field observations and reports of local people indicate the *Cebuella* live in small groups of two to nine individuals (Tokuda, 1968; Kinzey *et al.*, 1975; Moynihan, 1976; Ramirez *et al.*, 1976), although Hernandez-Camacho and Cooper (1976) mention that groups as large as 10−15 have been observed in Colombia. *Cebuella* feeds mainly on insects, fruit, and tree exudates, with exudates apparently making up a large part of the animal's diet (Izawa, 1975; Kinzey *et al.*, 1975; Hernandez-Camacho and Cooper, 1976; Moynihan 1976; Ramirez *et al.*, 1976). The "short-tusked" condition of the lower anterior dentition (incisors almost as long as canines) is a specialization for gouging holes in trees and inducing exudate flow, and many of the trees frequented by *Cebuella* are riddled with such holes (Kinzey *et al.*, 1975; Coimbra-Filho and Mittermeier, 1976).

Cebuella is a very adaptable animal and occurs in a variety of habitats.

---

*In Brazil, the name *Rio Solimões* is used to refer to the mainstream of the Amazon between the Colombian border and the mouth of the Rio Negro.

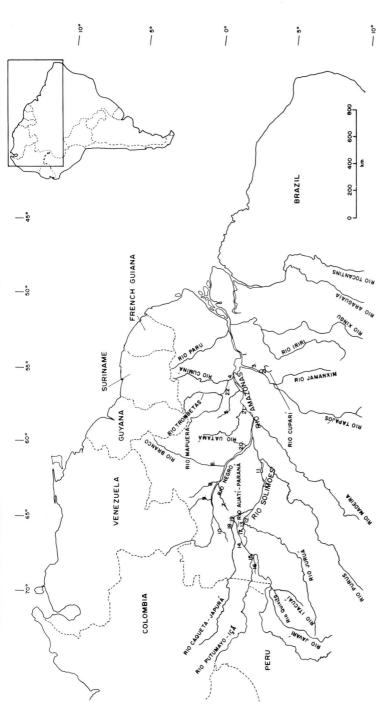

Fig. 3. Map of northern South America showing the localities covered during the two 1973 primate surveys in Brazilian Amazonia. Localities numbered 1–22 were investigated during the 4-month survey conducted by R. A. M. They are listed below. The 3-week survey by R. C. Bailey was carried out in the Rio Quixito, a tributary of the Rio Itacuaí, which is itself a tributary of the Rio Javarí. 1. Taperinha. 2. Monte Cristo, Rio Tapajós. 3. Tapaiuna, Rio Tapajós. 4. Oriximiná Rio Trombetas. 5. Lower Rio Nhamundá. 6. Boiaçú, Rio Branco. 7. Rio Cuiuni, Rio Negro. 8. Rio Araçá, Rio Negro. 9. Rio Padauiri, Rio Negro. 10. Rio Uneiuxi, Rio Negro. 11. Lago Miuá, Rio Solimões. 12. Costa da Batalha, Rio Solimões. 13. São José (= Jacaré), Rio Solimões. 14. Lago Marimari, Rio Auatí-Paraná. 15. Santo Antonio do Içá, Rio Solimões. 16. Rio Jacurapá, Rio Içá. 17. Rio Panauá, Rio Auatí–Paraná. 18. Bom Futuro, Rio Japurá. 19. Lago Maraá, Rio Japurá. 20. Rio Cuieras, Rio Negro.

TABLE IV

Number of Primate Groups Found in Each of the Major Habitat Subdivisions During the Two Brazilian Amazon Surveys[a]

	Terra firme secondary 12 km	Terra firme primary 51.1 km	Black water igapó Small rivers 237 km	Black water igapó Large rivers (276 km)	Clear water igapó 7.2 km	White water várzea Secondary 30 km	White water várzea Primary 182.4 km (160 km)	Total 519.7 km (+436 km)
Callithrix a. argentata							1	1
Callithrix a. leucippe	1	1						2
Saguinus fuscicollis fuscicollis						1	1	2
Saguinus midas midas	3	1						4
Saguinus mystax mystax							4	4
Aotus trivirgatus			1					1
Callicebus moloch		1						1
Callicebus torquatus torquatus			1			1	18	20
Cacajao calvus calvus							3	3
Cacajao calvus rubicundus			2					2
Cacajao melanocephalus			7					7
Pithecia pithecia	1							1
Pithecia monachus		1					2	3
Chiropotes albinasus					1			1
Cebus albifrons			1				1	2
Cebus apella		4	1	(+1)	1		8	14 (+1)
Saimiri sciureus		4	7	(+3)	2	3	13 (+2)	29 (+5)
Alouatta belzebul		1						1
Alouatta seniculus		7	7		2		23 (+1)	39 (+1)
	5	20	27	(+4)	6	5	74 (+3)	137 (+7)

[a]The distances covered in each habitat type are given under the respective heading. Distances in parentheses are those covered during sporadic observations. Groups in parentheses are those located during the sporadic observations. Numbers refer to groups observed from the deck of the expedition boats.

TABLE V

Vernacular Names of Primates in Brazilian Amazonia, Based on Data Collected During the 4-Month 1973 Survey

	English name	Brazilian Amazon name[a]
Cebuella pygmaea	Pygmy marmoset	Leãozinho
Callithrix argentata	Black-tailed marmoset	Macaquinho; macaquinho branco
Saguinus midas midas	Red-handed tamarin	Macaquinho; sauim; sagüi
Aotus trivirgatus	Night monkey	Macaco da noite
Callicebus torquatus	Titi; widow monkey	Zogi-zogi
Callicebus moloch	Dusky titi	Zogi-zogi
Pithecia pithecia	White-headed saki	Parauacú; paraguacú
Pithecia monachus	Monk saki	Parauacú
Chiropotes albinasus	White-nosed saki	Cuxiú; cuxiú-de-nariz-branco
Chiropotes satanas	Bearded saki	Cuxiú
Cacajao melanocephalus	Black uakari	Macaco bicó; bicó, acarí; macaco mal acabado
Cacajao calvus calvus	White uakari	Uacarí
Cacajao calvus rubicundus	Red uakari	Uacarí
Saimiri sciureus	Squirrel monkey	Macaco do cheiro; macaco amarelo
Cebus albifrons	White-fronted capuchin monkey	Caiarara
Cebus apella	Tufted capuchin monkey	Macaco prego
Ateles belzebuth	Long-haired spider monkey	Coatá; macacão
Ateles paniscus	Black spider monkey	Coatá; macaco preto
Lagothrix lagothricha	Woolly monkey	Barrigudo
Alouatta belzebul	Red-handed howler monkey	Guariba
Alouatta seniculus	Red howler monkey	Guariba

[a]The most commonly used name is underlined.

Moynihan (1976) observed it in the Putumayo region of Colombia living as a commensal of man in highly degraded patches of forest between pastures and crop fields. Between the Rio Japurá and the mainstream of the Amazon in Brazil, it occurs in both *terra firme* primary forest and in secondary forest near villages. In the Peruvian Amazon, *Cebuella* also can be found in *igapó* forest of black water rivers (Ramirez *et al.*, 1976).

b. STATUS. *Cebuella* is presently in no danger in Brazilian Amazonia. It is too small to be hunted for food or bait and adapts well to man-altered environments. It is occasionally captured as a pet by local people (Fig. 4A) and certain Indian tribes, at least in Colombia, sometimes keep pygmy marmosets to pick lice out of their hair (Hernandez-Camacho and Cooper, 1976; Bailey, personal communication). In both Colombia and Brazil, a traditional method

Fig. 4. (A) Adult *Cebuella pygmaea* being kept as a pet in Leticia, Amazonas, Colombia. (B) Family group of *Callimico goeldii*. (Photo taken by R.A.M. at the Frankfurt Zoo, Frankfurt, West Germany.)

of capturing *Cebuella* is to ring the trunk of a large tree frequented by the animal with a sticky resin. When the animal enters or leaves the tree, it gets stuck in the resin and is easily captured (Hernandez-Camacho and Cooper, 1976). However, such practices are conducted on a small scale and do not represent a threat to the survival of *Cebuella* in Brazilian Amazonia.

2. *Callithrix*

a. DISTRIBUTION AND HABITS. Two species of *Callithrix* live in Brazilian Amazonia: *Callithrix humeralifer* (including *C. h. humeralifer* and *C. h. chrysoleuca* as subspecies) and *Callithrix argentata* (including *C. a. argentata, C. a. leucippe*, and *C. a. melanura* as subspecies). Both species occur south of the mainstream. *Callithrix humeralifer* is found between the Rio Madeira and the Rio Tapajós, whereas *C. argentata* ranges between the Tapajós and the Tocantins-Araguaia, south to the Rio Tacuari and the Rio Mamoré (Hershkovitz, 1968).

No data is available on the habits of *C. humeralifer* in the wild. *Callithrix a. argentata* and *C. a. leucippe* were observed in small groups of two to four on the east bank of the Rio Tapajós during the 1973 survey. Krieg (1930) reported groups of three to eight individuals for *C. a. melanura* in the *Chaco* region to the south, and Miller (in Allen, 1916) saw one *C. a. melanura* group of about twelve individuals in Mato Grosso. Krieg (1930) found vegetable matter and insect remains in the stomachs of Chaco-dwelling *C. a. melanura*, but nothing is known of the diet of Amazonian representatives of *C. argentata*. However, diet of both Amazonian *Callithrix* spp. is probably similar to that of southern Brazilian members of the genus, which feed on insects, fruit, small vertebrates, and tree exudates (Coimbra-Filho and Mittermeier, 1976). Like *Cebuella*, *Callithrix* have the specialized "short-tusked" lower anterior dentition. Southern Brazilian species like *Callithrix jacchus* use these teeth much like *Cebuella*, but we do not know if Amazonian *Callithrix* also behave in such a manner. In the Rio Tapajós, *C. a. argentata* was observed in várzea forest and *C. a. leucippe* in terra firme forest during the 1973 survey.

b. STATUS. Amazonian *Callithrix* appear to survive well in close proximity to man and are usually not hunted for food. However, both subspecies of *C. humeralifer* and one subspecies of *C. argentata* (*C. a. leucippe*) are restricted in range. Hershkovitz (1972) considers *C. humeralifer* endangered, but we did not visit the range of this animal and have no new data on it. *Callithrix a. leucippe* is restricted to the east bank of the Rio Tapajós, between the tributaries Jamanxim and Cuparí. This area is cut by the Transamazonian highway and is being subjected to clear-cutting to make way for cattle ranches in several areas. If such habitat destruction continues, the survival of this subspecies could be threatened.

3. *Saguinus*

a. DISTRIBUTION AND HABITS.   Eight species of *Saguinus* are found in Brazilian
Amazonia: *S. bicolor* (including *S. b. bicolor, S. b. martinsi* and *S. b. ochraceous*),
S. fusicicollis (including *S. f. fuscicollis, S. f. acrensis, S. f. avilapiresi, S. f. crandalli,
S. f. cruzlimai, S. f. fuscus, S. f. melanoleucus, S. f. weddelli,* and perhaps *S. f. tripar-
titus*), *S. imperator, S. inustus, S. labiatus* (including *S. l. labiatus* and *S. l. thomasi*),
S. midas (including *S. m. midas* and *S. m. niger*), *S. mystax* (including *S. m.
mystax, S. m. pileatus,* and *S. m. pluto*), and *S. nigricollis* (Hershkovitz, 1966, 1968,
1969). *Saguinus bicolor* is found in a fairly restricted area between the lower Rio
Negro and the lower Paru de Oeste, (or Cumina) north of the Rio Amazonas
(Hershkovitz, 1966). At least eight of the thirteen recognized subspecies of *S.
fuscicollis* occur entirely or partly in Brazil, extending as far east as the Rio Japurá
north of the mainstream and the Rio Madeira to the south (Hershkovitz, 1968).
Saguinus imperator lives in the southwestern part of Brazilian Amazonia in the
vicinity of the upper Rio Purus and the upper Rio Juruá (Hershkovitz, 1968).
Saguinus inustus is found north of the mainstream, in an area between the Rio
Negro and the Rio Japurá and extending west to the Colombian border (Hill,
1957; Hershkovitz, 1969). One subspecies of *S. labiatus, S. l. labiatus,* is found
south of the Solimões between the Rio Madeira and the Rio Purus; the other,
S. l. thomasi, is known only from the type locality, the Rio Tonantins on the
north bank of the Solimões (Hershkovitz, 1968). *Saguinus m. midas* occurs from
the Rio Negro, east to the mouth of the Amazon and north to the Guianas; the
subspecies *S. m. niger* lives south of the Amazon between the Rio Xingu and
the Rio Gurupí (Hershkovitz, 1969). *Saguinus mystax* ranges south of the
Solimões as far east as the Rio Madeira (Hershkovitz, 1968), while *S. nigricollis*
extends a short way into Brazil, between the Solimões and the Rio Içá (Hershkovitz, 1966).

    Saguinus spp. live in groups of 2 to 24 individuals (Bates, 1863; Husson 1957;
Thorington, 1968a; Tokuda, 1968; Freese, 1975; Hernandez-Camacho and
Cooper, 1976; Dawson, 1976; Neyman-Werner, 1976), with an average ranging
from 3 to 8 depending on species and habitat (Thorington, 1968; Freese, 1975;
Dawson, 1976; Neyman-Warner, 1976). Fourteen groups of *s. nigricollis* counted
by Tokuda (1968) averaged an unusually high thirteen animals per group.
Saguinus spp. eat insects, fruit, small vertebrates, leaves, shoots, buds, moss, and
flowers (Bates, 1863; Enders, 1930; Fooden, 1964; Hladik and Hladik, 1969;
Moynihan, 1970; Izawa, 1975; Hernandez-Camacho and Cooper, 1976). Although they occasionally consume exudates if they happen to come across them
(Hladik and Hladik, 1969; Izawa, 1975), *Saguinus* apparently have not developed a tree-gouging/exudate-eating specialization as have *Cebuella* and at
least some *Callithrix. Saguinus* are "long-tusked," with lower incisors much
shorter than lower canines, and do not gouge holes with their lower dentition.

The *Saguinus* on which information is available are for the most part quite adaptable and capable of living in primary forest, as well as secondary formations near human habitations and plantations (da Cruz Lima, 1944; de Carvalho and Toccheton, 1969; Thorington, 1969; Hernandez-Camacho and Cooper, 1976; Dawson, 1976; Mittermeier *et al.*, 1976; Neyman-Warner, 1976).

   b. STATUS.   No data is available on the conservation status of *S. imperator*, *S. insustus*, and *S. labiatus* in Brazilian Amazonia. Hershkovitz (1972) lists *S. bicolor* as an endangered species because it is restricted in range and occurs in close proximity to rapidly growing Manaus, the second largest city in Brazilian Amazonia. We have no data on the subspecies *S. b. martinsi* and *S. b. ochraceous*, but recent observations of *S. b. bicolor* indicate that it is quite adaptable and even occurs in secondary forest within the city limits of Manaus (D. Magor, personal communication; personal observation). Nonetheless, it should be closely watched. No data is available on six of the eight races of *S. fuscicollis* that definitely occur in Brazilian Amazonia, but several (*S. f. acrensis, S. f. avilapiresi, S. f. crandalli*, and *S. f. cruzlimai*) are fairly restricted in range. *Saguinus f. fuscicollis* was one of the Brazilian species commonly brought across the border into Leticia, Colombia, but still appears to be common away from major collecting areas (Mittermeier *et al.*, 1976). Freese (personal communication) saw a captive *S. f. melanoleucus* from the Rio Tarauacá in Fonte Boa and was told by its owners that the animal was common in the Tarauacá region. *Saguinus midas midas* has a large range in Brazil and is the only callitrichid in most of the northeastern section of Brazilian Amazonia. It was the most frequently encountered callitrichid in the 4-month 1973 survey and is reportedly common in *terra firme* secondary forest (Mittermeier *et al.*, 1976; Table 4). *Saguinus midas niger* occurs in one of the most developed parts of Brazilian Amazonia, but can live in secondary forest and is even found in close proximity to Belem, the largest city in Amazonia (Bates, 1863; da Cruz Lima, 1944; de Carvalho and Toccheton, 1969; Thorington, 1969; Mittermeier *et al.*, 1976). Neither of the subspecies of *S. midas* are in any danger. No data is available on two of the subspecies of *S. mystax, S. m. pluto* and *S. m. pileatus*. The third, *S. m. mystax*, was another of the Brazilian species commonly brought across the border to Leticia, Colombia, but, like *S. f. fuscicollis*, is apparently still common (Mittermeier *et al.*, 1976; Freese, personal communication). The comparatively small range of *S. nigricollis* is centered around Leticia, but we do not know what effect the Leticia trade had on Brazilian representatives of the species.

   It is important to note that all callitrichids have a greater intrinsic rate of natural increase than the larger Cebidae. They usually have twin births and reproduce and develop more rapidly than their larger New World relatives (Eisenberg, 1976). In addition, they are rarely hunted for food because of their small size, do not appear to be popular as pets in Amazonia (Table III), and are quite capable of living in secondary situations in close proximity to man (indeed, some species may even prefer secondary forest). As a result, most Amazonian callitrichids are much less affected by man's activities than the majority of the Cebidae.

B. Family Callimiconidae

1. *Callimico*

a. DISTRIBUTION AND HABITS. *Callimico goeldii* (Fig. 4B), an unusual little monkey which shares features with both the Callitrichidae and the Cebidae, is known from a number of scattered localities in the upper Amazon region. In Brazil, it has been recorded from the Rio Yaco (Riberio, 1940); Ucuna, Rio Xapurí (state of Acre) (Thomas, 1914); Seringal Oriente, near Vila Taumaturgo, Rio Juruá (state of Acre) (Hernandez-Camacho and Barriga-Bonilla, 1966); and the Rio Mu, a tributary of the upper Juruá (Hill, 1957). Three *Callimico* groups observed by Whittemore (1972) in the Putumayo region of Colombia contained two, two, and five individuals. The diet of this animal is apparently similar to that of *Saguinus* and includes insects, fruits, and small vertebrates (Lorenz, 1971; Moynihan, personal communication). *Callimico* occurs in several kinds of habitat, including bamboo forest (Whittemore, 1972), secondary forest (Moynihan, personal communication), and terra firme primary forest (Moynihan, personal communication; Soini, personal communication).

b. STATUS. *Callimico* is a naturally rare and very sparsely distributed species about which almost nothing is known. It currently has a gray sheet in the International Union for the Conservation of Nature's (IUCN) "Red Data Book," indicating that its status is indeterminate. It would be best to give this species full protection, at least until we can learn more about its conservation status.

C. Family Cebidae

1. *Aotus*

a. DISTRIBUTION AND HABITS. The night monkey, *Aotus trivirgatus* (Fig. 5A), is widespread throughout most of Brazilian Amazonia (Hill, 1960). The world's only nocturnal monkey, it lives in small family groups of two to four individuals (Humboldt and Bonpland, 1812; Miller, in Allen, 1916; Moynihan, 1964; Tokuda, 1968; Hernandez-Camacho and Cooper, 1976) and feeds mainly on fruit, nuts, leaves, bark, flowers, gum, insects, and small vertebrates (Humboldt and Bonpland, 1812; Krieg, 1930; Hladik and Hladik, 1969; Hernandez-Camacho and Cooper, 1976). It is a very adaptable animal and occurs in a wide variety of habitats. Hernandez-Camacho and Cooper (1976) report that in

Fig. 5. (A) Juvenile *Aotus trivirgatus* being kept as a pet by a Yagua Indian woman in Peruvian Amazonia. (B) *Callicebus torquatus* from Brazilian Amazonia.

Colombia it lives up to 3200 m, is found in many forest types, and sometimes even enters coffee plantations. In Brazilian Amazonia, it is not uncommon near human habitations. *Aotus* apparently spends the day in tree holes or in accumulations of dry leaves and twigs (Bates, 1863; Hernandez-Camacho and Cooper, 1976).

b. STATUS. *Aotus* is important in biomedical research and, in some parts of its range (e.g., delta of the lower Cauca and neighboring middle Magdalena in Colombia; Hernandez—Camacho and Cooper, 1976), is trapped for export. Some Brazilian *Aotus* used to be shipped out of Leticia, but this trade has now ended. *Aotus trivirgatus* is not presently in any danger in Brazilian Amazonia. It is occasionally kept as a pet by local people (Fig. 5a), but is usually not hunted for food and, because of its nocturnal habits, rarely comes into contact with man.

2. Callicebus

a. DISTRIBUTION AND HABITS. Five of the seven recognized subspecies of *Callicebus moloch* (*C. m. cupreus*, *C. m. brunneus*, *C. m. donacophilus*, *C. m. hoffmannsi*, and *C. m. moloch*) and two of the three subspecies of *Callicebus torquatus* (*C. t. lugens*, and *C. t. torquatus*) are found in Brazil. The two species are sympatric in a number of areas. *Callicebus moloch* occurs south of the mainstream, as far east as the Rio Tocantins. *Callicebus torquatus* (Fig. 5B) is found on both banks of the Solimões, extending as far east as the Rio Purus to the south and the Rio Negro-Rio Casiquiare to the north (Hershkovitz, 1963). Both species of *Callicebus* usually live in small family groups of two to seven individuals, with an adult pair and one or two offspring being the group size most frequently encountered (Bates, 1863; Allen, 1916; Krieg, 1930; Hershkovitz, 1963; Mason, 1966, 1968; Moynihan, 1966; Tokuda, 1968; Freese, 1975; Hernandez-Camacho and Cooper, 1976). The diet includes fruit, seeds, leaves, insects, arachnids, millipedes, bird eggs, and small vertebrates (Krieg, 1930; Izawa, 1975; Hernandez-Camacho and Cooper, 1976; personal observation). Although not occurring in as wide a variety of habitats as *Aotus*, the two species of *Callicebus* are also quite adaptable. In Colombia, they occur in gallery forest, in isolated patches of woods in the middle of savannas, and in low mountainous areas (Mason, 1968; Hernandez-Camacho and Cooper, 1976). In Brazil, they can be found in *terra firme* forest, as well as flooded *várzea* and *igapó* forest (Table IV), and may exist in close proximity to human habitations (Bates, 1863; personal observation).

b. STATUS. Both *Callicebus* are occasionally hunted for food or captured as pets in Brazilian Amazonia, but do not appear to be in any danger at present. *Callicebus torquatus* was the third most frequently encountered monkey during the two 1973 surveys, being exceeded only by *Alouatta seniculus* and *Saimiri sciureus* (Table IV). Smith (personal communication) reports that it is also common in the Altamira region along the Rio Xingu.

3. Pithecia

a. DISTRIBUTION AND HABITS. Both currently recognized species of the genus *Pithecia* occur in Brazilian Amazonia. *Pithecia pithecia* is found north of the mainstream, between the Rio Negro and the Atlantic and extending north to the Guianas and Venezuela. *Pithecia monachus* (Fig. 6) ranges as far east as the Rio Tapajós on the south bank of the Amazon and to the Rio Negro on the North bank (Hill, 1960). Little data are available on *Pithecia* in the wild. Sanderson (1949) mentions groups of two to five in Surinam and two groups of *P. pithecia* observed by R.A.M. in Surinam contained two and three individuals. Eighteen *P. monachus* groups seen by Tokuda (1968) in Colombian Amazonia averaged three animals per group, and Hernandez-Camacho and Cooper (1976) also report pairs or small family groups. Freese (1975) found groups of five to eight along the Rio Nanay in Peru and groups of three in the Rio Samiria. In Brazilian Amazonia, local people reported that small groups of two to six animals were typical. The diet of *Pithecia* apparently consists primarily of fruit, nuts, berries, leaves, and insects (Sanderson, 1949; Fooden, 1964; Izawa, 1975; D. Buchanan, personal communication; personal observation). Hernandez-Camacho and Cooper (1976) indicate that in

Fig. 6. Subdadult *Pithecia monachus* from Brazilian Amazonia.

Colombia, *P. monachus* is restricted to undisturbed high rain forest. In Brazilian Amazonia, this species is found in both *terra firme* and *várzea* forest. Bates (1863) mentions it from *terra firme* on the north bank of the Amazon from the Rio Tonantins to Peru. During the two 1973 surveys, *P. monachus* was found in *terra firme* forest in the Rio Quixito and in *várzea* primary forest between the Rio Japurá, the Rio Auatí-Paraná, and the Rio Solimões. In the lower Rio Trombetas, one *P. pithecia* group was observed in *terra firme* secondary forest during the 4-month 1973 survey. On two separate occasions, solitary *P. monachus* were seen in close proximity to *Cacajao calvus calvus* groups in the Japurá-Auatí-Paraná-Solimóes region, and one of the groups of *P. pithecia* seen in Surinam was near a large group of *Chiropotes satanas chiropotes.*

b. *Status.* In Brazilian Amazonia, *Pithecia* spp. are occasionally hunted for food or to obtain their tails for use as dusters, and are sometimes kept as pets. However, they are widely (though sparsely) distributed, very quiet, shy, and quick to escape when encountered in the forest. In some parts of Colombia, *P. monachus* is known as *mico volador* (flying monkey) (Hernandez-Camacho and Cooper, 1976) and Brazilian *caboclos* and Surinam Bushnegroes consider *Pithecia* one of the most difficult monkeys to find and hunt. Neither species presently appears to be in any danger in Brazil.

4.   Chiropotes

a. *Distribution and Habits.*   Both species of *Chiropotes* are found in Brazilian Amazonia. *Chiropotes albinasus* occurs between the Rio Xingu and the Rio Madeira, at least as far south as the Rio Jiparaná, and is apparently endemic to Brazil (Hill, 1960; personal observation). *Chiropotes satanas* ranges both north and south of the mainstream. The subspecies *Chiropotes satanas chiropotes* extends from the Rio Negro to the Atlantic, north to the Guianas and Venezuela. The other subspecies, *Chiropotes satanas satanas* is found south of the Amazon, from the Xingu east perhaps as far as the coast (Hill, 1960). The eastern and southern limits of the range of this subspecies have not been determined.

Very little is known of the habits of *Chiropotes*. Brazilian *caboclos* maintain that groups range in size from 10 to 30 + animals. Miller (in Allen, 1916) mentioned a group of 15 to 20 *"Cacajao rooseveltii"* (= *Chiropotes albinasus*). In Surinam, R.A.M. observed a group of *C. s. chiropotes* that contained at least 30. *Chiropotes* spp. eat fruit, nuts, berries, and some leaves and flowers (Fooden, 1964; Grzimek, 1968; personal observation.) During the 4-month 1973 survey, *Chiropotes albinasus* was observed in *igapó* forest in the Rio Tapajós (near Monte Cristo) and was reported from *terra firme* primary forest. *Chiropotes satanas chiropotes* was reported from both *igapó* and *terra firme* primary forest in the Rio Trombetas and the Rio Negro. The group observed in Surinam was also in *terra firme* primary forest. On one occasion, *C. albinasus* was seen in close proximity to a group of *Cebus apella* in the Rio Tapajós, and *C. satanas chiropotes* is reported to occasionally associate with *Saimiri sciureus* and/or *Cebus apella* in the Rio Negro.

b. STATUS.   *Chiropotes albinasus* is currently listed as an amber sheet, vulnerable species in the IUCN's "Red Data Book." It should retain this status for the present, but may have to be upgraded to an endangered species in the future. It is heavily hunted for food in the Rio Tapajós area and has now disappeared from parts of the lower Tapajós where it was collected in the 1930's. Its range is also being cut by the Transamazonian highway and forests are being destroyed to make way for cattle pasture. Workers cutting down the forests hunt monkeys for food, and *C. albinasus* is large enough to be hunted for this purpose. It is also shot to bait cat traps and to obtain the tail as a duster.

Chiropotes satanas chiropotes is also hunted for food in the Rio Negro and Rio Trombetas, but we presently have too little information on it to be able to determine its status. *Chiropotes satanas satanas* has a relatively small range in one of the most densely inhabited parts of Brazilian Amazonia. We do not have any information on its current status, but it may well be endangered. A survey to determine its situation should be conducted as soon as possible.

5.   Cacajao

a. DISTRIBUTION AND HABITS.   The main purpose of the 4-month 1973 survey was to locate study sites for *Cacajao* and to determine the conservation status and range boundaries of the members of this genus. Two species of *Cacajao* are currently recognized: *C. melanocephalus* (Fig. 7A) and *C. calvus*, the latter including *C. calvus calvus* and *C. calvus rubicundus* (Fig. 7B) as subspecies (Hershkovitz, 1972). In Brazilian Amazonia, the black uakari (*C. melanocephalus*) is found north of the Solimões, between the Japurá, the Solimões itself, and the Negro, and also north of the Negro at least as far east as the Rio Araçá (a small north bank tributary of the Negro entering above the Rio Branco.) Several authors (Hill, 1960; Napier and Napier, 1967) give the Rio Branco as the eastern limit of the species, but local people report that it is not found on either bank of this river. *Cacajao melanocephalus* also extends north to southern Venezuela and west to Colombia (Hernandez–Camacho and Cooper, 1976). The white uakari (*C. calvus calvus*) is, as far as is presently known, restricted to the huge *várzea* "island" formed by the Solimões, the Japurá, and the Auatí-Paraná, an anastomose between the Solimões and the Japurá. During the 1973 survey, several local people told us of "white" uakaris from south of the Solimões. These reports probably originate from areas like the upper Rio Juruá, where populations of the widespread red uakari (*C. calvus rubicundus*) sometimes have whitish backs. In Brazil, the red uakari is found south of the Solimões from the Peruvian border east at least to the Rio Juruá and perhaps to the west bank of the Rio Purus. Local people report it from the Purus, but we did not investigate this river and know of no museum specimens from the area. The red uakari also occurs north of the Solimões between the Içá and Solimões. We observed it in the Rio Jacurapá, a small black water tributary of the Içá. Local people also report it from the area between the Içá and the Auatí-Paraná, but we were unable to verify this.

During the 4-month 1973 survey, *Cacajao* spp. were seen in groups ranging in

size from 10 to 30 + individuals. Two of the *C. melanocephalus* groups encountered contained more than twenty animals each, and a single *C. calvus
calvus* subgroup had eighteen. The main group from which this subgroup separated contained at least thirty animals. *Cacajao melanocephalus* is also said to live
in large groups of 30 + animals in Colombia. Reliable local informants in Brazilian
Amazonia told us that groups of 30 to 50 uakaris are not rare and that some
may be as large as 100. Olalla (1937; and in Lönnberg, 1938) also mentions *C.
calvus rubicundus* groups of 100 animals.

The diet of *Cacajao* is still poorly known. Bates (1863) and Humboldt and
Bonpland (1812) considered uakaris fruit eaters. Our field observations and
information from local informants indicate that the animals eat mainly fruit
and nuts. They are apparently capable of breaking open foods as hard as Brazil
nuts (*Bertholettia excelsa*, Lecythidaceae) with their canines (W. Thomas, personal
communication).

Cacajao spp. are most restricted in habitat than most other Amazonian primates. During the 1973 survey, *C. melanocephalus* was only observed in *igapó*
forests of small, black water rivers and "lakes." Local people said it occasionally
moves into *terra firme* forest at certain times of the year, but we have not yet
been able to confirm this. The entire known range of *C. calvus calvus* is *várzea*
forest of the white water Solimões and Japurá. The two groups of *C. calvus
rubicundus* encountered during the 1973 survey were in *igapó* forest of a small,
black water tributary. All uakaris appear to avoid the margins of the larger
rivers.

Uakaris sometimes associate with other species. *Cacajao calvus rubicundus* was
seen feeding in the same tree as a group of *Saimiri sciureus*, and *C. calvus calvus*
was, on two occasions, found in close proximity to *Pithecia monachus*. Caboclos
in the Rio Negro region report that *C. melanocephalus* often forms mixed groups
with *Saimiri sciureas* and/or *Cebus apella*, or with *Cebus albifrons*.

b. STATUS.   *Cacajao calvus rubicundus* and *C. calvus calvus* are listed as red sheet,
endangered species in the IUCN's "Red Data Book." *Cacajao melanocephalus* was
also listed as a red sheet species, but was recently changed to a gray sheet, indeterminate species. *Cacajao melanocephalus* is one of the more abundant primates along the small, black water rivers between the Negro and the Japurá.
Seven of nineteen primate groups encountered in this area during the 1973
survey were *C. melanocephalus*. However, the animal is apparently quite rare
everywhere else that it occurs. Nothing is known of its current status in southern
Venezuela, but it was uncommon there even in Humboldt's time. Hernandez-
Camacho and Cooper (1976) consider its position in Colombia as 'very precarious" because it is rare, heavily hunted tor food, and restricted in choice of
habitat. In Brazilian Amazonia, the animal is usually not hunted for food, but is
occasionally shot for fish, turtle, or cat bait (Fig. 7A). *Cacajao melanocephalus* is

Fig. 7. (A) Two adult female *Cacajao melanocephalus* shot for fish bait in the Rio Cuiuni, a tributary of the Rio Negro, Brazilian Amazonia. (B) Juvenile *Cacajao calvus rubicundus*.

best considered a vulnerable species, but is not presently endangered with extinction because of the remoteness of much of its Brazilian range.

Cacajao calvus calvus is very restricted in range and, as pointed out by Bates (1863) over 110 years ago, "is rare even in the limited district which it inhabits." Fortunately, it occurs in an area where monkey hunting is uncommon. It is sometimes shot for fish bait or for human consumption, but several people living within its range expressed great disgust at the thought of eating an animal that looked so human. However, as mentioned in Section IV, attitudes toward primate hunting can change fairly quickly with the influx of new settlers or with decreasing abundance of alternate protein resources. Should the attitude toward hunting white uakaris suddenly change, the entire population could be severely decimated or wiped out within several years. We therefore consider it best to retain this animal as an endangered species, and strongly recommend that a reserve be created within its range.

Bates (1863) noted that white uákaris were in great demand as presents for influential people living down river. The animals were caught by Indian hunters, who shot them with blowguns and arrows dipped in diluted *urarí* poison. When the wounded monkey fell out of the tree, a pinch of salt was placed in its mouth as an antidote and it revived. As far as we were able to gather, this monkey is no longer in demand in the local pet trade.

We do not presently have sufficient data on *C. calvus rubicundus* to be able to comment on its status in Brazil.

6. *Saimiri*

a. DISTRIBUTION AND HABITS.   The squirrel monkey, *Saimiri sciureus* (Fig. 8), is widespread in Brazilian Amazonia. It is found in the entire region north of the mainstream and extends as far east as the Rio Tocantins on the south bank. The southern limits of the range extend beyond the borders of Amazonia (Krieg, 1930; Hill, 1960).

Squirrel monkeys live in larger groups than any other New World monkey. Baldwin and Baldwin (1971) reported a group of 200 to 300 + animals from the upper Amazon region of Brazil, near the Colombian border, and three others ranging in size from 120 to 300+ from upper Amazonian Peru. Sanderson (1957) mentions a group of 550 from Guyana. Other group counts and estimates range from 11 to 50+ animals (Thorington, 1967, 1968b; Tokuda, 1968; Baldwin and Baldwin, 1972a; Bailey *et al.*, 1974; Sponsel *et al.*, 1974; Freeze, 1975; Hernandez-Camacho and Cooper, 1976; Klein and Klein, 1976; Bailey, personal communication). During the two 1973 Amazon surveys, we encountered a total of 29 *Saimiri* groups that ranged in size from 20 to 70 animals (Mittermeier *et al.*, 1976; Bailey, personal communication).

Fig. 8. Male *Saimiri sciureus*. (Photo taken by R.A.M. at the Monkey Jungle, Goulds, Florida.)

Squirrel monkeys eat a wide variety of foods, including fruit, berries, nuts, flowers, buds, seeds, leaves, gum, insects, arachnids, and small vertebrates (Sanderson, 1949; Fooden, 1964; Thorington, 1967, 1968; Hladik and Hladik, 1969; Baldwin and Baldwin, 1972a; Izawa, 1975; Hernandez-Camacho and Cooper, 1976; Bailey, personal communication). They can live in many different kinds of forest habitat. In Colombia, they may be found in gallery forest, low canopy sclerophyllous and hillside forests, palm forests, and *terra firme, igapó,* and *várzea* rain forests (Hernandez-Camacho and Cooper, 1976). In French Guiana, *Saimiri* sometimes lives in coastal mangrove forests (M. Condamin, personal communication) and in the southern reaches of its range occurs in the *Chaco* of Paraguay (Krieg, 1930). In Brazilian Amazonia, we encountered these monkeys in *terra firme* primary and secondary forest, *várzea* forest, and *igapó* forest of black and clear water rivers. However, they appear to prefer river margins, perhaps because insects are abundant in such ecotone areas. Only four of the 29 groups we saw in 1973 were in *terra firme* forest (Table IV). In Surinam, Sanderson (1949) also found *Saimiri* to be partial to river margins.

Saimiri is more frequently seen in mixed species associations than any other Amazonian primate. During the two 1973 surveys, we recorded it four times with *Cebus apella*, once with *Cebus albifrons*, and once with *Cacajao calvus rubi-*

cundus. Local people reported that it sometimes associates with *Cacajao melano-cephalus* and *Chiropotes satanas.* Many of the mixed species associations observed were probably nothing more than temporary gatherings of highly mobile primate species at a common food source (e.g., a fruiting tree). However, the association between *Cebus apella* and *Saimiri* appears to be more complex. It has been noted a number of times in other parts of South America (Thorington, 1967, 1968b; Klein and Klein, 1973; Hernandez-Camacho and Cooper, 1976) and may be long term in some areas.

b. STATUS. *Saimiri* is perhaps the numerically most abundant and least threatened species in Brazilian Amazonia. It can live in a variety of habitats, is frequently seen close to human habitations, and is rarely persecuted. Occasional specimens are captured as pets or shot for bait or human consumption, but such exploitation does not even make a dent in the population. Only in the area immediately adjacent to Leticia, Colombia have Brazilian *Saimiri* been decimated by large-scale live trapping for export (Baldwin and Baldwin, 1971; Hernandez-Camacho and Cooper, 1976). However, given the large numbers of *Saimiri* in Brazilian Amazonia, even such exploitation has nothing more than a local effect.

During the two 1973 surveys, *Saimiri sciureus* was the second most frequently encountered primate (29 groups; Table IV). Only *Alouatta seniculus* was seen more often (39 groups; Table IV), and this species may exceed *Saimiri* in total biomass. However, *Saimiri* groups average much larger than those of *A. seniculus* and the number of individual *Saimiri* seen far outnumbered not only *Alouatta* but all other primate species combined.

7. *Cebus*

a. DISTRIBUTION AND HABITS. Three species of *Cebus* are found in Brazilian Amazonia. *Cebus albifrons* (Figs. 9A and 10A) extends into Brazil as far east as the Rio Negro—Rio Branco north of the Amazon and the Rio Tapajós south of the Amazon (Hershkovitz, 1949). *Cebus apella* is widespread throughout most of Brazilian Amazonia, both north and south of the mainstream (Hill, 1960). *Cebus nigrivittatus* occurs north of the mainstream, between the Rio Negro—Rio Branco—Rio Catrimani and the Rio Paru, north to the Guianas and Venezuela (Hill, 1960). *Cebus nigrivittatus* and *C. albifrons* are probably allopatric [although maps in Hill (1960) indicate a possible zone of sympatry between the Rio Branco and the Rio Catrimani]. *Cebus apella* is sympatric with *C. albifrons* or *C. nigrivittatus* in many parts of Amazonia.

Group size in *Cebus* monkeys varies from 2 to 40 animals (Bates, 1863; Hensel, 1867; Chapman, 1937; Kühlhorn, 1939; Krieg, 1948; Husson, 1957; Oppenheimer, 1968; Tokuda, 1968; Oppenheimer and Oppenheimer, 1973; Freese, 1975; Baldwin and Baldwin, 1976; Durham, 1976a; Hernandez-Camacho and

Fig. 9. Juvenile *Cebus* monkeys being kept as pets in *Amazonia*. (A) *Cebus albifrons.* (B) *Cebus apella.*

Fig. 10. (A) Two juvenile *Cebus albifrons* being kept as pets by young girl at Miuá, near Codajás, Rio Solimões, Brazilian Amazonia. (B) Adult male *Cebus apella.*

Cooper, 1976; Klein and Klein, 1976; Milton and Mittermeier, 1977). Small groups of two to five individuals are often seen in areas of heavy hunting pressure or in isolated forest patches or second growth (Hernandez-Camacho and Cooper, 1976; personal observation). Ten groups of *C. apella* counted by Tokuda (1968) in the Colombian Amazon averaged twelve animals per group, whereas four groups of *C. albifrons* averaged seven per group.

Cebus spp. have a very eclectic diet, eating a greater variety of foods than any other New World monkey. Foods eaten include fruit, nuts, berries, seeds, leaves, buds, flowers, shoots, bark, gum, insects, arachnids, eggs, small vertebrates, and even certain kinds of marine life in coastal areas (e.g., oysters, crabs) (Kühlhorn, 1954; Hill, 1960; Fooden, 1964; Thorington, 1967, 1968; Oppenheimer, 1968; Hladik and Hladik, 1969; Izawa, 1975; Hernandez-Camacho and Cooper, 1976; Milton and Mittermeier, 1977).

All *Cebus* are quite flexible in choice of habitat. *Cebus apella* is especially adaptable and has a larger range than any other species of New World monkey. The range of the genus *Cebus* is the second largest in the New World, being exceeded only by *Alouatta*. In Colombia, *C. apella* occurs "in virtually every type of humid forest from gallery, to palm, to rain forest and including both seasonally flooded and nonflooded forest. It is found in broken or isolated forest and in second growth; it also crosses open ground in passing from one forest segment to another" (Hernandez-Camacho and Cooper, 1976). *Cebus albifrons* is also capable of living in secondary forest and has even been collected in brackish water mangroves (Hernandez-Camacho and Cooper, 1976). *Cebus nigrivittatus* lives in dry forest in the *llanos* of Venezuela, as well as mature tropical rain forest in the Guianas (Oppenheimer and Oppenheimer, 1973; personal observation). *Cebus apella* is also found in the Paraguayan *Chaco* and in dry forest areas of eastern Brazil (Krieg, 1930; personal observation). In Brazilian Amazonia, *C. apella* was observed in *terra firme* primary forest, *várzea* forest and *igapó* forests of black and clear water rivers during the 1973 surveys. *Cebus albifrons* was seen in black water *igapó* forest and reported from *várzea* and *terra firme* primary forest. *Cebus apella* was much more common (14 groups; Table 4) than *C. albirrons* (2 groups; Table 4). Izawa (1975) also reported *C. albifrons* as being less abundant than *C. apella* in Colombian Amazonia. Reports from local people indicate that, in areas where the two species are sympatric, *C. apella* usually occupies one kind of habitat and *C. albifrons* another. *Cebus nigrivittatus* was not encountered during the 1973 surveys, but only a small part of its range was investigated.

Cebus apella is frequently seen in mixed species associations with *Saimiri* (see section on *Saimiri*) and was once observed in close proximity to *Chiropotes albinasus* during the 4-month 1973 survey. It is reported to occasionally associate with *Cacajao melanocephalus* as well. *Cebus albifrons* forms mixed groups with *Saimiri* (but less frequently than *C. apella*) and is also reported to associate with *Cacajao melanocephalus*.

b. STATUS. *Cebus apella* and *C. albifrons* are not presently endangered in Brazilian Amazonia, but appear to be diminishing in areas where hunting pressure is heavy. *Cebus* spp. are not as sought after as *Alouatta, Lagothrix,* and *Ateles,* but are good to eat and large enough to recompense the hunter for his efforts. Like most other medium to large monkeys, they are sometimes used to bait cat traps.

Cebus infants are popular as pets because they are active, intelligent, and very hardy in captivity (Figs. 9A, B, and 10A). They are not as popular as *Lagothrix* and do not command as high a price, but are more abundant and consequently easier to obtain. In Amazonia, they are kept as pets more frequently than any other primates (Table III).

Hernandez-Camacho and Cooper (1976) report that *C. apella* and *C. albifrons* are hunted as agricultural pests in Colombian Amazonia and are strongly affected by habitat destruction in the Columbian llanos. In Brazilian Amazonia, habitat destruction has not yet become a major problem for *Cebus* populations and they do not appear to be persecuted as agricultural pests.

Capture for export is not a threat to *Cebus* in Brazilian Amazonia. Brazilian *Cebus* were occasionally exported from Leticia, Colombia, but usually in small numbers (Hernandez-Camacho and Cooper, 1976).

Cebus apella was the fourth most frequently encountered species during the two 1973 surveys, exceeded only by *Alouatta seniculus, Saimiri sciureus,* and *Callicebus torquatus* (Table IV).

8. *Ateles*

a. DISTRIBUTION AND HABITS. Kellogg and Goldman (1944) recognize four species of *Ateles*. However, a number of recent authors (e.g., Moynihan, 1967, 1970; Hershkovitz, 1972; Hernandez-Camacho and Cooper, 1976) have tended to lump all spider monkeys into a single species, for which the name *Ateles paniscus* is used. The taxonomic status of the *A. geoffroyi* group of Mexico and Central America versus South American *Ateles* is not clear. A zone of hybridization or sympatry exists between *A. geoffroyi panamensis* and *A. fusciceps rufiventris* in eastern Panama and should be investigated. Recent data on karyology also raises questions about relationships among the South American *Ateles* (Heltne and Kunkel, 1975). Until further data becomes available, we prefer to follow the taxonomic arrangement of Kellogg and Goldman (1944)—although it seems likely that all *Ateles* in Brazilian Amazonia are indeed conspecific.

Using Kellogg and Goldman's taxonomic arrangement, there are two species of *Ateles* in Brazilian Amazonia. *Ateles paniscus* is divided into two subspecies: *A. p. paniscus,* which occurs between the Rio Negro–Rio Branco and the Atlantic, north to Venezuela and the Guianas, and *A. p. chamek* (Fig. 11A), which occurs south of the mainstream as far east as the Rio Juruá and south to western

Fig. 11 (A) Juvenile *Ateles paniscus chamek.* (B) Juvenile *Lagothrix lagothricha lagothricha* at ease in a monkey-sized hammock. Juvenile woolly monkeys make delightful pets and are much persecuted for this reason and for their excellent meat.

Matto Grosso (Kellogg and Goldman, 1944). *Ateles belzebuth* is divided into three subspecies, two of which occur in Brazilian Amazonia. *Ateles b. belzebuth* enters the north-western corner north of the Rio Negro, while *A. b. marginatus* (Fig. 12A) ranges between the Rio Tapajós and the Rio Tocantins on the south bank of the Amazon.

Spider monkeys live in groups of 2 to 40 animals (Bartlett, 1871; Allen, 1916; Carpenter, 1935; Geijskes, 1954; Wagner, 1956; Eisenberg and Kuehn, 1966; Durham, 1971, 1975; Freese, 1975; Hernandez-Camacho and Cooper, 1976; Klein and Klein, 1976; personal observation). Group size varies considerably with habitat and altitude. Durham (1971), for instance, found that it decreased with altitude from a mean of 18.5 at 275 m to 11 at 576 m, 7 at 889 m, and 4.5 at 1424 m. Spider monkey groups also tend to break up into small subgroups, rather than moving together as a single cohesive unit (Carpenter, 1935; Eisenberg and Kuehn, 1966; Klein and Klein, 1976).

The diet of *Ateles* spp. consists largely of fruit, supplemented by nuts, berries, seeds, buds, flowers, leaves, insects, arachnids, and eggs (Bartlett, 1871; Carpenter, 1935; Wagner, 1956; Hladik and Hladik, 1969; Richard, 1970; Durham, 1971; Izawa, 1975; Hernandez-Camacho and Cooper, 1976).

Fig. 12. (A) Adult *Ateles belzebuth marginatus* shot as cat bait by a *gateiro* working along the Cuiabá-Santarem highway. (Photo courtesy of N. Smith.) (B) Female and juvenile *Alouatta seniculus* feeding. This monkey was the most frequently encountered primate during the two 1973 surveys in Brazilian Amazonia. (Photo by R.A.M. in the Monkey Jungle, Goulds, Florida.)

In Brazilian Amazonia, *Ateles* tends to prefer *terra firme* primary forest. Hernandéz-Camacho and Cooper (1976) note that Colombian *Ateles* may be absent from apparently suitable habitats because of heavy hunting pressure. This may also be true in Brazil and could account for the rarity or absence of spider monkeys along the banks of major rivers and in *terra firme* areas easily accessible to man.

b. STATUS. *Ateles* is one of the two most vulnerable primates in Brazilian Amazonia. It is large and very good to eat, and has been subjected to heavy hunting pressure throughout most of its range—even in areas where other primates are usually not molested. Its size and noisy habits make it easy to locate, follow, and hunt. *Ateles* is also persecuted by cat hunters who use it as bait (Fig. 12A) and infant *Ateles* are captured for sale as pets (although *Ateles* are by no means as popular as *Lagothrix* and *Cebus*). *Ateles* should be given complete protection in all countries in which it occurs and should be placed in the IUCN "Red Data Book" as an amber sheet, vulnerable species. In Brazilian Amazonia, a thorough survey is needed to learn more about the status of this monkey and to find sites for the establishment of reserves.

9. *Lagothrix*

a. DISTRIBUTION AND HABITS. The genus *Lagothrix* is represented by one species, *L. lagothricha*, and three subspecies in Brazilian Amazonia. *Lagothrix lagothricha lagothricha* (Fig. 11B) occurs on the north bank of the Solimões, as far east as the Rio Negro. *Lagothrix lagothricha poeppigii* and *L. lagothricha cana* range south of the mainstream, with the former extending as far east as the Rio Juruá and the latter occurring between the Juruá and the Tapajós (Fooden, 1963).

Recent reports indicate that woolly monkeys travel in groups of 4 to 35 individuals (Tokuda, 1968; Durham, 1975; Kavanagh and Dresdale, 1975; Hernandez-Camacho and Cooper, 1976; Mittermeier *et al.*, this volume, Chapter IV). Durham (1975) found an average of 13 at 600 m and 7 at 1500 m. Their diet consists primarily of fruit, supplemented by leaves, seeds, berries, and some insects (Miller, in Allen, 1916; Izawa, 1975; Hernandez-Camacho and Cooper, 1976; Mittermeier *et al.*, this volume, Chapter 4). Little data is available on the habitat preferences of these monkeys in Brazilian Amazonia, since they were not encountered during the 1973 surveys. Miller (in Allen, 1916) reported them "in the great lake-swamp on the Solimoens (= Solimões), feeding in the low berry brushes." In Colombia, *Lagothrix* lives in gallery forests, palm forests, flooded and nonflooded primary rain forest, and cloud forest, but not in secondary forest (Hernandez-Camacho and Cooper, 1976). Colombian *L. lagothricha lugens* range as high as 3000 m (Hernandez-Camacho and Cooper, 1976).

b. STATUS. Like *Ateles*, *Lagothrix* is a vulnerable genus. Woolly monkeys have long been heavily hunted for their meat, which is said to be excellent (Bates,

1863; Bartlett, 1871; Allen, 1916; Hernandez-Camacho and Cooper, 1976), and to obtain infants for sale as pets. Infant *Lagothrix* make very attractive pets (Fig. 11B), are much in demand, and command higher prices than any other Amazonian primate. At least in Colombia (Izawa, 1975), and undoubtedly also in Brazil, *Lagothrix* is used as bait for cat traps. Small numbers of Brazilian *Lagothrix* were exported from Leticia each year, but this drain on the population has now ended. Hunting pressure and the demand for infants, together with the woolly monkey's slow reproductive rate and preference for undisturbed forest habitat, have placed this genus in a precarious position throughout its range. As with *Ateles*, we feel that *Lagothrix lagothricha* should be given complete protection wherever it occurs, that it should be placed in the IUCN "Red Data Book" as an amber sheet, vulnerable species, and that a thorough survey should be conducted to learn more of its status and to find suitable sites for reserves.

10. *Alouatta*

a. DISTRIBUTION AND HABITS. Two species of *Alouatta* are found in Brazilian Amazonia. *Alouatta seniculus* (Fig. 12B) is widespread, being found as far east as the Rio Madeira south of the mainstream and in all of Brazilian Amazonia north of the mainstream. *Alouatta belzebul* is only found south of the Amazon, from the Rio Madeira east to the state of Maranhão (Hill, 1962).

Alouatta is the best studied and most wide-ranging of all Neotropical non-human primate genera. Central American *Alouatta palliata* have been studied by a number of workers (e.g., Carpenter, 1934, 1953, 1962, 1965; Collias and Southwick, 1952; Altmann, 1959; Bernstein, 1964; Chivers, 1969; Richard, 1970; Baldwin and Baldwin, 1972b, 1973, 1976; Mittermeier, 1973; Milton, 1975; and others) and *Alouatta caraya* from northern Argentina, southern Brazil, Paraguay, and Bolivia has been briefly investigated (Krieg, 1928; Pope, 1966, 1968). *Alouatta seniculus* has the largest range of the genus and has been studied in Venezuela and Trinidad (Neville, 1972a, 1976) and in Colombia (Klein and Klein, 1976), Aside from Sanderson's (1949) report of groups of 20 to 40 animals in Surinam and Humboldt's mention of up to 40 animals in a single tree in Venezuela (Humboldt and Bonpland, 1812), most field workers have observed small groups of 2 to 16 for this species (Bates, 1863; Osgood, 1912; Racenis, 1952; Tokuda, 1968; Neville, 1972a,b; Freese, 1975; Izawa, 1975; Hernandez-Camacho and Cooper, 1976; Klein and Klein, 1976). The 35 groups observed by Neville (1972a) ranged in size from four to sixteen and averaged approximately nine animals per group. Toduka (1968) counted twelve groups that averaged four animals per group. Freese (1975) saw groups of two to seven, with an average of approximately five per group. Hernandez-Camacho and Cooper (1976) mention groups of two to fifteen, with an average of six to eight. During the two 1973 surveys in Brazilian Amazonia, a total of 39 *A. seniculus* groups were encountered. They were all fairly small, ranging in size from two to eight animals. Four groups were accurately counted and averaged 4.25 animals per group (Table VI).

The small group sizes reported for *A. seniculus* are similar to those of *A. caraya* (Miller, in Allen, 1916: two for seven animals per group; Krieg, 1928: nine groups, range three to ten animals per group, average 5.9; Pope, 1968: seventeen groups, range four to fourteen animals per group, average 7.9), but differ from the larger groups of *A. palliata*, which range from 2 to 35 and average from 8.2 to 18.9 animals per group (Carpenter, 1934, 1953, 1963; Collias and Southwick, 1952; Altmann, 1959; Bernstein, 1964; Chivers, 1969; Baldwin and Baldwin, 1972b; Mittermeier, 1973).

The other Amazonian species, *A. belzebul*, has never been studied in the wild, nor has *A. pigra* from southern Mexico and *A. fusca* from southern Brazil.

Alouatta spp. eat more leaves than any other New World monkeys. In addition, they eat fruit, seeds, nuts, flowers, buds, bark, pith, clay, gum and, on rare occasions, insects (Rengger, 1830; Carpenter, 1934; Sanderson, 1949). Kühlhorn, 1954; Altmann, 1959; Fooden, 1964; Hladik and Hladik, 1969; Richard, 1970; Neville, 1972a; Mittermeier, 1973; Izawa, 1975; Hernandez-Camacho and Cooper, 1976).

Alouatta seniculus occupies a wide variety of habitats. In Colombia, it is found in mangrove swamps of the Caribbean coast, gallery forests in the *llanos* and other relatively dry areas, deciduous tropical forests, cloud forests (including oak forests) up to 3200 m, flooded and unflooded rain forests, and extremely small, isolated forest patches and second growth (Hernandez-Camacho and Cooper, 1976). The Venezuelan animals studied by Neville (1972a, 1976) lived in gallery forest in the *llanos*; those in Trinidad lived in swamp forest. Osgood (1912) reported *A. seniculus* from mangrove thickets in Venezuela. In Brazilian

TABLE VI

Composition of Four *Alouatta seniculus* Groups Counted During the 4-month 1973 Survey in Brazilian Amazonia

Group	Adult males	Adult females	Juveniles	Infants	Total
Vicinity of Oriximiná	1	3		1	5
Vicinity of Oriximiná	1	1		1	3
Rio Jacurapá	1	2	2	2	7
Rio Panauá	1	1			2
Total	4	7	2	4	17
Average	1.0	1.75	0.5	1.0	4.25

Amazonia, *A. seniculus* also lives in many different habitats, being found in *terra firme* primary forest, *várzea* forest, and *igapó* forests of the black and clear water rivers (Table 4). It is the only large primate that can still be seen fairly frequently along the mainstream of the Amazon in Brazil. The two groups of *A. belzebul* encountered during the 4-month 1973 survey were in *terra firme* primary forest.

b. STATUS. *Alouatta seniculus* is a remarkably resistant primate. It is large and good to eat (although not as sought after as *Ateles* and *Lagothrix*, Hernandez-Camacho and Cooper, 1976) and is hunted in much of Brazilian Amazonia. It is frequently persecuted even in areas where most other primates are not hunted, and is used to bait cat traps. The animal is easy to locate because of its incredibly loud vocalizations (the specialized hyoid apparatus is larger in *A. seniculus* than in any other *Alouatta*; Hershkovitz, 1949), and its slow-moving habits make it relatively easy to hunt. Nonetheless, it continues to be abundant in Brazilian Amazonia, can often be found at high densities in close proximity to human habitations (even in areas of moderate hunting pressure, e.g. along the lower Rio Nhamundá), and was the most frequently encountered primate during the two 1973 surveys (39 of 137 groups—Table IV). It is not presently endangered in Brazilian Amazonia, but should be carefully watched. In areas where it has been subjected to heavy hunting pressure for a long time (e.g., vicinity of Iquitos, Peru, personal observation; Magdalena and Cauca valleys in Colombia, Hernandez-Camacho and Cooper, 1976), it has become rare or disappeared entirely.

Alouatta belzebul is hunted along the Rio Tapajós. However, along the main part of the Transamazonian highway, it is apparently still common and rarely hunted by colonists (N. Smith, personal communication).

Hernandez-Camacho and Cooper (1976) note that *Alouatta* skins (and also those of *Lagothrix* and *Aotus*) are sometimes used locally in Colombia as ornamentation for brow bands of horse bridles. The same authors mention that the hyoid apparatus of *A. seniculus* is used as a drinking vessel in parts of Colombia because it is believed to have therapeutic properties for curing goiters. In some parts of Brazilian Amazonia, drinking from an *Alouatta* hyoid is thought to ease labor pains during birth and in Surinam it is thought to cure stuttering. Another unusual *Alouatta* medication, from the Peruvian Amazon, is used for a cough: put ground howler hair in boiling water, let cool, and drink (C. Freese, personal communication).

VI.   The Future of Primate Conservation in Brazilian Amazonia

The foregoing account indicates that the majority of Brazilian Amazonia's nonhuman primates are not in immediate danger of extinction. Nonetheless, it is necessary to realize that Amazonia is a rapidly changing land. The remote-

ness, inaccessibility, and sheer magnitude of the region has in the past protected much of the wildlife from excessive human exploitation. Today, however, Amazonia can no longer be considered remote and the elaborate 14,000 km
pioneer highway system presently under construction (Goodland and Irwin,
1975) is ending the area's inaccessibility. The enormous size of the Amazonian
forest still serves to buffer civilization's effect on the wildlife. However, as man
slowly but surely penetrates all corners of the vast hylaea, the animals' remaining refuges are beginning to disappear.

Brazil has an obligation to ensure the continued survival of a representative
sample of the world's most varied primate fauna—not to mention the rest of
Amazonia's unparalleled animal and plant life. The best way to accomplish
this is, of course, to create a system of national parks and reserves in the area
(Coimbra-Filho, 1972). At the present time, Araguaia National Park, located
in the central-western portion of the state of Goias, is Brazil's only Amazonian
national park.* Its proposed size is 2,000,000 hectares, but as of 1972 the effective size was only 3000 hectares (Jorge Padua *et al.*, 1974). Once fully implemented, Araguaia will be a highly significant park and will protect many
important Amazonian species. However, since it is located in the transition zone
between Amazonian forest and the savanna-like *cerrado*, it cannot be considered
entirely representative of Amazonia.

Many new parks are needed in Brazilian Amazonia and they should be established as quickly as possible (i.e., as soon as carefully conducted studies have
determined which areas are most suitable for park or reserve status, Coimbra-
Filho, 1972). The longer the government waits, the more difficult it will become
to find large tracts of undisturbed public land. Once land becomes attractive to
commercial interests and falls into private hands, the problems involved in
acquisition of land for parks and reserves increase manifold.

The Brazilian government is presently considering approximately twenty
areas as possible sites for national parks and reserves, and also has plans for a
number of "ecological stations" dedicated to research (P. Nogueira-Neto, personal communication). The government has also discussed the possibility of
creating international frontier parks with neighboring countries like Colombia,
Peru, and Ecuador. If such areas are to have any long-term significance, they
must be both well-protected and large. "Paper parks," created thousands of
kilometers away in Brasilia, are useless if local Amazonians are not aware of
their existence and if protective legislation is not enforced. In addtion, the size
factor is of critical importance. Small parks may be able to preserve a portion of

*Reference is often made to Xingu National Park and Tumacumaque National Park, both
located within Brazilian Amazonia (e.g., von Puttkamer, 1975; Goodland and Irwin, 1975).
However, these protected areas have large Amerindian populations and are, in effect, Indian
reserves rather than true national parks.

the fauna for a certain period of time (and may be the only answer for saving particular endemic species with very restricted ranges), but in the long run will not ensure the survival of a truly representative sample of Amazonia's wildlife. Terborgh (1974, 1975) has determined, by analogy with Caribbean land bridge islands, that natural extinction rates are much slower in large parks and reserves than in relatively small ones. He points out that "reduction of extinction rates to acceptable levels (less than 1% of the initial species complement per century) requires reserves of substantial size, on the order of 1,000 mi^2" (Terborgh, 1975, p. 378). International frontier parks would be one way of making parks as large as possible.

Brazil's primate fauna is unsurpassed in the world. Nonetheless, few Brazilians realize what a significant national heritage it really is. Along with the establishment of parks and reserves, another essential step is to initiate a carefully structured primate conservation program, hopefully with international assistance. Such a program would have among its goals assessing the current conservation status of primates (especially species thought to be vulnerable or endangered), monitoring primate populations on a regular long-term basis to ensure their continued survival, and educating Brazilians in the value of Brazil's unique wildlife heritage and the skills needed to save it from destruction.

We can talk about primate conservation in Brazilian Amazonia (and elsewhere) for many years to come. However, if we are to accomplish anything, we must have a clear picture of the status of the animals we are trying to protect and, most important of all, we must have strong support from the Brazilians themselves. The former is necessary to determine what species and what areas are most in need of protection, the latter is essential if protection is to become a political reality.

Acknowledgments

We would like to thank N. J. H. Smith and R. C. Bailey for allowing us to use unpublished data from their work in Amazonia. Thanks also to the New York Zoological Society for partly funding the 4-month 1973 survey and to the Fundação de Amparo à Pesquisa do Estado de São Paulo (FAPESP) for making available the University of São Paulo's EPA boats in which the survey was conducted. P. E. Vanzolini and the EPA provided information on important monkey localities in Amazonia. Other field work by R.A.M. in Amazonia was partly funded by an NSF Graduate Fellowship. A.F.C.-F.'s work in Brazil was conducted while a fellow of the Brazilian National Research Council. We are grateful to the following people for reading and providing very helpful comments on early drafts of this paper: R. C. Bailey, J. Cassidy, R. W. Cooper, J. Fleagle, C. Freese, M. Marcus, N. J. H. Smith, P. Soini, and R. E . Thorington, Jr. Illustrations were prepared by G. Carter.

References

Allen, J. A. (1916). Mammals collected on the Roosevelt Brazilian expedition, with field notes by Leo. E. Miller. *Bull. Amer. Mus. Natur. Hist.* **35**, 559—610.

Altmann, S. A. (1959). Field observations on a howling monkey society. *J. Mammal.* **40**, 317—330.

Avila-Pires, F. D. (1974). Caracterização zoogeográfica da Provincia Amazonica. II. A familia Callithricidae e a zoogeográfia Amazonica. *An. Acad. Brasil. Cienc.* **46**(1), 159—185.

Bailey, R. C., Baker, R. S., Brown, D. S., von Hildebrand, P., Mittermeier, R. A., Sponsel, L. E., and Wolf, K. E. (1974). Progress of a breeding project for non-human primates in Colombia. *Nature* (*London*) **248**, 453—455.

Baldwin, J. D. and Baldwin, J. I. (1971). Squirrel monkeys (*Saimiri*) in natural habitats in Panama, Colombia, Brazil and Peru. *Primates* **12**(1), 45—61.

Baldwin, J. D. and Baldwin, J. I. (1972a). The ecology and behavior of squirrel monkeys (*Saimiri oerstedi*) in a natural forest in western Panama. *Folia Primatol.* **18**, 161—184.

Baldwin, J. D. and Baldwin, J. I. (1972b). Population density and use of space in howling monkeys (*Alouatta villosa*) in southeastern Panama. *Primates* **13**(4), 371—379.

Baldwin, J. D. and Baldwin, J. I. (1973). Interactions between adult female and infant howling monkeys (*Alouatta palliata*). *Folia Primatol.* **20**, 27—71.

Baldwin, J. D. and Baldwin, J. I. (1976). Primate populations in Chiriqui, Panama. *In* "Neotropical Primates: Field Studies and Conservation" (R. W. Thorington, Jr., and P. G. Heltne, eds.), pp. 20—31. National Academy of Sciences, Washington, D. C.

Bartlett, E. (1871). Notes on the monkeys of eastern Peru. *Proc. Zool. Soc. London*, 217—220.

Bates, H. W. (1963). "The Naturalist on the River Amazons." John Murray, London.

Bernstein, I. S. (1964). A field study of the activities of howler monkeys. *Anim. Behav.* **12**, 92—97.

Brumback, R. A. (1974). A third species of the owl monkey (*Aotus*). *J. Hered.* **65**, 321—323.

Cabrera, A. and Yepes, J. (1940). "Mamiferos Sud Americanos," Vol. I. Historia Natural Ediar.

Carpenter, C. R. (1934). A field study of the behavior and social relations of howling monkeys. *Comp. Psychol. Monogr.* **10**(2), 1—168.

Carpenter, C. R. (1935). Behaviour of the red spider monkey (*Ateles geoffroyi*) in Panama. *J. Mammal.* **16**, 171—180.

Carpenter, C. R. (1953). Grouping behavior of howling monkeys. *Arch. Neerl. Zool.* **10**, 45—50.

Carpenter, C. R. (1962). Field studies of a primate population. *In* "Roots of Behavior" (E. L. Bliss, ed.), pp. 286—294. Harper and Row, New York.

Carpenter, C. R. (1965). The howlers of Barro Colorado Island. *In* "Primate Behavior" (I. DeVore, ed.), pp. 250—291. Holt, Rinehart and Winston, New York.

Carvalho, C. T. and Toccheton, A. J. (1969). Mamíferos do nordeste do Pará, Brasil. *Rev. Biol. Trop.* **15**(2), 215—226.

Carvalho, J. C. M. (1951). Relações entre os indios do alto Xingu e a fauna regional. *Publ. Avulsas Mus. Nac. (Rio de Janeiro)* **7**, 1—32.

Carvalho, J. C. M. (1952). Notas de viagem ao Rio Negro. *Publ. Avulsas Mus. Nac. (Rio de Janeiro)* **9**, 1—92.

Carvalho, J. C. M., de Lima, P. E., and Galvão, E. (1949). Observações zoológicas e antropológicas na região dos formadores do Xingu. *Publ. Avulsas Mus. Nac. (Rio de Janeiro)* **5**, 1—48.

Chapman, F. M. (1937). My monkey neighbors on Barro Colorado. *Natur. Hist.* **40**, 471—479.

Chivers, D. J. (1969). On the daily behavior and spacing of howling monkey groups. *Folia Primatol.* **10**, 48—102.

Coimbra-Filho, A. F. (1972). Mamíferos ameaçados de extinção no Brasil. *In* "Espécies da Fauna Brasileira Ameaçadas de Extinção," pp. 13—98. Academia Brasileira de Ciencias, Rio de Janeiro.

Coimbra-Filho, A. F. and Mittermeier, R. A. (1973). New data on the taxonomy of the Brazilian marmosets of the genus *Callithrix* Erxleben, 1777. *Folia Primatol.* **20**, 241—264.

Coimbra-Filho, A. F. and Mittermeier, R. A. (1976). Tree-gouging, exudate-eating and the "short-tusked" condition in *Callithrix* and *Cebuella*. *In* "Biology and Conservation of the Callitrichidae," in press. Conference held at Front Royal, Virginia, August 18—20, 1975.

Collias, N. E. and Southwick, C. H. (1952). A field study of population density and social organization in howling monkeys. *Proc. Amer. Phil. Soc.* **96**, 143—156.

Committee on Conservation of Nonhuman Primates (1975). "Nonhuman Primates—Usage and Availability for Biomedical Programs." National Academy of Sciences, Washington, D. C.

Cooper, R. W. (1968). Squirrel monkey taxonomy and supply. *In* "The Squirrel Monkey" (L. A. Rosenblum and R. W. Cooper, eds.), pp. 1—29. Academic Press, New York.

da Cruz Lima, E. (1944). "Mamíferos da Amazonia. Introdução Geral e Primatas." Museu Paraense Emílio Goeldi de Historia Natural e Etnografía, Belem do Pará.

Dawson, G. (1976). Troop size, composition and instability in the Panamanian tamarin (*Saguinus oedipus geoffroyi*): ecological and behavioral implications. *In* "Biology and Conservation of the Callitrichidae," in press. Conference held at Front Royal, Virginia, August 18—20, 1975.

Ducke, A. and Black, G. A. (1953). Phytogeographical notes on the Brazilian Amazon. *An. Acad. Brasil. Cienc.* **25**(1), 1—46.

Durham, N. M. (1971). Effects of altitude differences on group organization of wild black spider monkeys (*Ateles paniscus*). *Proc. 3rd. Int. Congr. Primatol. (Zürich* 32—40.

Durham, N. M. (1975b). Some ecological, distributional, and group behavioral features of Atelinae in southern Peru, with comments on interspecific relations. *In* "Socioecology and Psychology of Primates." (R. A. Tuttle, ed.). Aldine, Chicago.

Eisenberg, J. F. (1976). Comparative ecology and reproduction of New World monkeys. *In* "Biology and Conservation of the Callitrichidae," in press. Conference held at Front Royal, Virginia, August 18—20, 1975.

Eisenberg, J. F. and Kuehn, R. E. (1966). The behavior of *Ateles geoffroyi* and related species. *Smithson. Misc. Coll.* **151**(8), 1–63.

Enders, R. K. (1930). Notes on some mammals from Barro Colorado Island, Canal zone. *J. Mammal.* **11**, 280–292.

Fittkau, E. J. (1969). The fauna of South America. In "Biogeography and Ecology in South America" (E. J. Fittkau, J. Illies, H. Klinge, G. H. Schwabe, and H. Sioli, eds.), Vol. 2 pp. 624–658. (Monogr. Biol. 19.) Dr. W. Junk N. V., the Hague.

Fittkau, E. J. (1974). Zur ökologischen Gliederung Amazoniens. I. Die erdgeschichtliche Entwicklung Amazoniens. *Amazoniana* **5**(1), 77–134.

Fooden, J. (1963). A revision of the woolly monkeys (genus *Lagothrix*). *J. Mammal.* **44**, 213–247.

Fooden, J. (1964). Stomach contents and gastro-intestinal proportions in wild-shot Guianan monkeys. *Amer. J. Phys. Anthropol.* **22**, 227–232.

Freese, C. (1975). A census of non-human primates in Peru. In "Primate Censusing Studies in Peru and Colombia," Report to the National Academy of Sciences on the activities of Project AMRO-0719, Pan American Health Organization, WHO.

Geijskes, D. C. (1954). Het dierlijk voedsel van de Bosnegers aan de Marowijne. *Vox Guyanae* **1**(2), 61–83.

Goodland, R. J. A. and Irwin, H. S. (1975). Amazon jungle: green hell to red desert? *Develop. Landscape Manage. Urban Planning* **1**, 155 pp.

Grzimek, B. (1968). "Grzimek's Tierleben," Mammals I, Vol. 10. Kindler Verlag, A. G., Zürich.

Haffer, J. (1974). Avian speciation in tropical South America. *Publ. Nuttall Ornithol. Club* (*Cambridge*) **14**, viii + 390 pp.

Heltne, P. G. and Kunkel, L. M. (1975). Taxonomic notes on the pelage of *Ateles paniscus paniscus*, *A. p. chamek* (sensu Kellogg and Goldman, 1944) and *A. fusciceps ruffiventris* (= *A. f. robustus*, Kellogg and Goldman, 1944). *J. Med. Primatol.* **4**, 83–102.

Hensel, R. (1867). Beiträge zue Kenntniss der Thierwelt Brasiliens. *Zool. Gart. (Frankfurt)* **8**, 361–374.

Hernandez-Camacho, J. and Barriga-Bonilla, E. (1966). Hallazgo del género *Callimico* (Mammalia: Primates) en Colombia. *Caldasia* **9**(44). 365–377.

Hernandez-Camacho, J. and Cooper, R. W. (1976). The nonhuman primates of Colombia. In "Neotropical Primates: Field Studies and Conservation" (R. W. Thorington, Jr., and P. G. Heltne, eds.), pp. 35–69. National Academy of Sciences, Washington, D. C.

Hershkovitz, P. (1949). Mammals of northern Colombia. Preliminary report no. 4. Monkeys (Primates), with taxonomic revisions of some forms. *Proc. U.S. Nat. Mus.* **98**, 323–427.

Hershkovitz, P. (1963). A systematic and zoogeographic account of the monkeys of the genus *Callicebus* (Cebidae) of the Amazonas and Orinoco river basins. *Mammalia* **27**, 1–80.

Hershkovitz, P. (1966). Taxonomic notes on tamarins, genus *Saguinus* (Callithricidae, Primates), with descriptions of four new forms. *Folia Primatol.* **4**, 381–395.

Hershkovitz, P. (1968). Metachromism or the principle of evolutionary change in mammalian tegumentary colors. *Evolution* **22**, 556–575.

Hershkovitz, P. (1969). The evolution of mammals on southern continents. VI. The recent mammals of the neotropical region: zoogeographical and ecological review. *Quart. Rev. Biol.* **44**(1), 1—70.

Hershkovitz, P. (1972). Notes on New World monkeys. *Int. Zoo Yearb.* **12**, 3—12.

Hill, W. C. O. (1957). "Primates. Comparative Anatomy and Taxonomy," Vol. III. Hapalidae. Univ. Press, Edinburgh.

Hill, W. C. O. (1960). "Primates. Comparative Anatomy and Taxonomy," Vol. IV. Cebidae, Part A. Univ. Press, Edinburgh.

Hill, W. C. O. (1962). "Primates. Comparative Anatomy and Taxonomy," Vol. V. Cebidae, Part B. Univ. Press, Edinburgh.

Hladik, A. and Hladik, C. M. (1969). Rapports trophiques entre végétation et primates dans la forêt de Barro Colorado (Panama). *Terre Vie* **23**, 25—117.

Hueck, K. (1972). "As Florestas da América do Sul." Editôra Polígono, S. A., São Paulo.

Humboldt, A. and Bonpland, A. (1812). "Recueil d'observations de Zoologie et d'Anatomie Comparée," Vol. I. F. Schoell and G. Dufour, Paris.

Husson, A. M. (1957). Notes on the primates of Suriname. *Studies on the Fauna of Suriname and other Guyanas* **1**, 13—40.

Izawa, K. (1975). Foods and feeding behavior of monkeys in the upper Amazon Basin. *Primates* **16**(3), 295—316.

Janzen, D. (1975). Tropical black-water rivers, animals, and mast fruiting by the Dipterocarpaceae. *Biotropica* **6**(2), 69—103.

Jorge Padua, M. T., Magnanini, A., and Mittermeier, R. A. (1974). Brazil's national parks. *Oryx* **12**(4), 452—464.

Kavanagh, M. and Dresdale, L. (1975). Observations on the woolly monkey (*Lagothrix lagothricha*). *Primates* **16**(3), 285—294.

Kellogg, R. and Goldman, E. A. (1944). Review of the spider monkeys. *Proc. U.S. Nat. Mus.* **96**, 1—45.

Kinzey, W., Rosenberger, A. L., and Ramirez, M. (1975). Vertical clinging and leaping in a neotropical anthropoid. *Nature (London)* **255**, 327—328.

Klein, L. L. and Klein, D. J. (1973). Observations on two types of neotropical primate intertaxa associations. *Amer. J. Phys. Anthropol.* **38**(2), 649—653.

Klein, L. L. and Klein, D. J. (1976). Neotropical primates: aspects of habitat usage, population density, and regional distribution in La Macarena, Colombia. *In* "Neotropical Primates: Field Studies and Conservation" (R. W. Thorington, Jr., and P. G. Heltne, eds.), pp. 70—78. National Academy of Sciences, Washington, D.C.

Krieg, H. (1928). Schwarze Brüllaffen (*Alouatta caraya* Humboldt). *Z. Säugetierk.* **2**, 119—132.

Krieg, H. (1930). Biologische Reisestudien in Südamerika. XVI. Die Affen des Gran Chaco und seiner Grenzgebiete. *Z. Morphol. Ökol. Tiere* **18**, 760—785.

Krieg, H. (1948). "Zwischen Anden und Atlantik: Reisen eines Biologen in Südamerika." C. Hanser Verlag, München.

Kühlhorn, F. (1939). Beobachtungen über das Verhalten von Kapuzineraffen in freier Wildbahn. *Z. Tierpsychol.* **3**, 147—151.

Kühlhorn, F. (1954). Gefügesetzliche Untersuchungen an Neuweltaffen (*Cebus apella* L. und *Alouatta caraya* Humboldt). *Z. Säugetierk.* **20**, 13—36.

Lönnberg, E. (1938). Remarks on some members of the genera *Pithecia* and *Cacajao* from Brazil. *Ark. Zool.* **30A**(18), 1—25.

Lorenz, R. (1971). Goeldi's Monkey *Callimico goeldii* Thomas 1904 preying on snakes. *Folia Primatol.* **15**, 133—142.

Marlier, G. (1973). Limnology of the Congo and Amazon rivers. *In* "Tropical Forest Ecosystems in Africa and South America: A Comparative Review" (B. J. Meggers, E. S. Ayensu, and W. D. Duckworth, eds.), pp. 223—238. Smithsonian Inst. Press, Washington, D.C.

Martin, R. D. (1972). Adaptive radiation and behaviour of the Malagasy lemurs. *Phil. Trans. Royal Soc. London* **264**(862), 295—352.

Mason, W. (1966). Social organization of the South American monkey, *Callicebus moloch*. A preliminary report. *Tulane Stud. Zool.* **13**, 13—28.

Mason, W. (1968). Use of space by *Callicebus* groups. *In* "Primates, Studies in Adaptation and Variability" (P. Jay, ed.), pp. 200—216. Holt, Rinehart and Winston, New York.

Meggers, B. J. (1971). "Amazonia, Man and Culture in a Counterfeit Paradise." Aldine-Atherton, Chicago.

Meggers, B. J. (1973). Some problems of cultural adaptation in Amazonia, with emphasis on the pre-European period. *In* "Tropical Forest Ecosystems in Africa and South America: A Comparative Review" (B. J. Meggers, E. S. Ayensu, and W. D. Duckworth, eds.), pp. 311—320. Smithsonian Inst. Press, Washington, D.C.

Milton, K. (1975). Urine-rubbing behavior in the mantled howler monkey *Alouatta palliata*. *Folla Primatol.* **23**, 105—112.

Milton, K. and Mittermeier, R. A. (1977). A brief survey of the primates of Coiba Island, Panama. *Primates*, in press.

Mittermeier, R. A. (1973). Group activity and population dynamics of the howler monkey on Barro Colorado Island. *Primates* **14**(1), 1—19.

Mittermeier, R. A. (1975). A turtle in every pot. *Anim. Kingdom* **78**(2), 9—14.

Mittermeier, R. A. (1977). The *Podocnemis* turtles of Amazonia—river turtles as a major protein resource for man. *Oryx*, in press.

Mittermeier, R. A., Bailey, R. C., and Coimbra-Filho, A. F. (1976). Conservation status of the Callitrichidae in the Brazilian Amazon, Surinam and French Guiana. *In* "Biology and Conservation of the Callitrichidae," in press. Conference held at Front Royal, Virginia, August 18—20, 1975.

Moynihan, M. (1964). Some behavior patterns of platyrrhine monkeys. I. The night monkey (*Aotus trivirgatus*). *Smithson. Misc. Coll.* **146**(5), 1—84.

Moynihan, M. (1966). Communication in the titi monkey, *Callicebus*. *J. Zool.* **150**, 77—127.

Moynihan, M. (1967). Comparative aspects of communication in New World primates. *In* "Primate Ethology" (D. Morris, ed), pp. 236—266. Weidenfeld and Nicholson, London.

Moynihan, M. (1970). Some behavior patterns of platyrrhine monkeys. II. *Saguinus geoffroyi* and some other tamarins. *Smithson. Contrib. Zool.* **28**, 1—77.

Moynihan, M. (1976). Notes on the ecology and behavior of the pygmy marmoset, *Cebuella pygmaea*, in Amazonian Colombia. *In* "Neotropical Primates: Field Studies and Conservation" (R. W. Thorington, Jr., and P. G. Heltne, eds.), pp. 79—84. National Academy of Sciences, Washington, D.C. August, 1972, Battelle Conference Center, Seattle, Washington.

Napier, J. R. and Napier, P. H. (1967). "A Handbook of Living Primates." Academic Press, New York.

Neville, M. K. (1972a). The population structure of red howler monkeys (Alouatta seniculus) in Trinidad and Venezuela. Folia Primatol. 17, 56—86.

Neville, M. K. (1972b). Social relations within troops of red howler monkeys (Alouatta seniculus). Folia Primatol. 18, 47—77.

Neville, M. K. (1976). The population and conservation of howler monkeys in Venezuela and Trinidad. In "Neotropical Primates: Field Studies and Conservation," (R. W. Thorington, Jr., and P. G. Heltne, eds.) pp. 101—109. National Academy of Sciences, Washington, D.C.

Neyman-Warner, P. (1976). The ecology and social behavior of the cotton-top tamarin in Colombia. In "Biology and Conservation of the Callitrichidae," in press. Conference held at Front Royal, Virginia, August 18—20, 1975.

Olalla, A. M. (1937). I. Un viaje a pesquizas zoológicas hacia el Rio Juruá, Estado del Amazonas, Brasil—1936. II. Notas de campo, observaciones biológicas. Rev. Mus. Paulista (Sao Paulo) 23, 233—298.

Oppenheimer, J. R. (1968). Behavior and ecology of the white-faced monkey, Cebus capucinus, on Barro Colorado Island, C. Z. Unpubl. doctoral dissertation, Univ. Illinois, Chicago, Illinois.

Oppenheimer, J. R. and Oppenheimer, E. C. (1973). Preliminary observations of Cebus nigrivittatus (Primates: Cebidae) on the Venezuelan llanos. Folia Primatol. 19, 409—36.

Osgood, W. H. (1912). Mammals from western Venezuela and eastern Colombia. Field Mus. Natur. Hist. (Zool. Ser.) 10(5), 29—66.

Pires, J. M. (1974). Tipos de vegetação da Amazonia. Brasil Florestal 5(17), 48—58.

Pope, B. L. (1966). Population characteristics of howler monkeys (Alouatta caraya) in northern Argentina. Amer. J. Phys. Anthropol. 24, 361—370.

Pope, B. L. (1968). Population characteristics. In "Biology of the Howler Monkey" (M. R. Malinow, ed.), pp. 13—20. S. Karger, Basel.

von Puttkamer, J. (1975). Brazil's Kreen-Akarores, requiem for a tribe? Nat. Geog. 147(2), 254—269.

Racenis, J. (1952). Some observations on the red howling monkey (Alouatta seniculus) in Venezuela. J. Mammal. 33, 114—115.

Ramirez, M., Freese, C., and Revilla, J. (1976). Notes on the ecology of the pygmy marmoset, Cebuella pygmaea, in northeastern Peru. In "Biology and Conservation of the Callitrichidae," Conference held at Front Royal, Virginia, August 18—20, 1975. in press.

Rebelo, D. C. (1973). "Transamazônica: integração em marcha," 243 pp. Ministério de Transportes, Rio de Janeiro.

Rengger, J. R. (1830). "Naturgeschichte der Säugethiere von Paraguay." Schweigerhauserschen, Basel.

Ribeiro, A. de M. (1940). Commentaries on South American primates. Mem. Inst. Oswaldo Cruz 35, 779—851.

Richard, A. (1970). A comparative study of the activity patterns and behavior of *Alouatta villosa* and *Ateles geoffroyi*. *Folia Primatol.* **12**, 241–263.

Roberts, T. R. (1972). Ecology of fishes in the Amazon and Congo basins. *Bull. Mus. Comp. Zool. Harvard Univ.* **143**(2), 117–147.

Sanderson, I. T. (1949). A brief review of the mammals of Suriname (Dutch Guiana), based upon a collection made in 1938. *Proc. Zool. Soc. London* **119**, 755–789.

Sanderson, I. T. (1957). "The Monkey Kingdom." Doubleday, Garden City, New York.

Smith, N. (1974a). Agouti and babassu. *Oryx* **12**(5), 581–582.

Smith, N. (1974b). Destructive exploitation of the South American river turtle. *Yearb. Assoc. Pac. Coast Geog.* **36**, 85–102.

Smith, N. (1976). The cat skin trade in the Brazilian Amazon. *Oryx*, **13**(4), 362–371.

Soini, P. (1972). The capture and commerce of live monkeys in the Amazonian region of Peru. *Int. Zoo. Yearb.* **12**, 26–36.

Sponsel, L. E., Brown, D. S., Bailey, R. C., and Mittermeier, R. A. (1974). Evaluation of squirrel monkey (*Saimiri sciureus*) ranching on Santa Sofia Island, Amazonas, Colombia, *Int. Zoo. Yearb.* **14**, 233–240.

Terborgh, J. (1974). Preservation of natural diversity: the problem of extinction prone species. *Bioscience* **24**(12), 715–722.

Terborgh, J. (1975). Faunal equilibria and the design of wildlife preserves. *In* "Tropical Ecological Systems—Trends in Terrestrial and Aquatic Research" (F. B. Golley and E. Medina, eds.), Ecological Studies 11, pp. 369–380. Springer Verlag, New York.

Thomas, O. (1914). On various South American mammals. *Ann. Mag. Nat. Hist.* **13**(8), 345–363.

Thorington, R. W., Jr. (1967). Feeding and activity of *Cebus* and *Saimiri* in a Colombian forest. *In* "Neue Ergebnisse der Primatologie" (D. Starck, R. Schneider, and H.-J. Kuhn, eds.), pp. 180–184. Gustav Fischer Verlag, Stuttgart.

Thorington, R. W., Jr. (1968a). Observations of the tamarin *Saguinus midas*. *Folia Primatol.* **9**, 95–98.

Thorington, R. W., Jr. (1968b). Observations of squirrel monkeys in a Colombian forest. *In* "The Squirrel Monkey" (L. A. Rosenblum and R. W. Cooper, eds.), pp. 69–85. Academic Press, New York.

Thorington, R. W., Jr. (1969). The study and conservation of New World monkeys. *An. Acad. Brasil. Cienc. Suppl.* **41**, 253–260.

Tokuda, K. (1968). Group size and vertical distribution of New World monkeys in the basin of the Rio Putumayo, the Upper Amazon. *Proc. 8th Congr. Anthropol. Sci.*, Vol. I, Anthropology, pp. 260–261. Science Council of Japan, Ueno Park, Tokyo.

Wagner, H. O. (1956). Freilandbeobachtungen an Klammeraffen. *Z. Tierpsychol.* **13**, 302–313.

Wallace, A. R. (1852). On the monkeys of the Amazon. *Proc. Zool. Soc. London*, pp. 107–110.

Whittemore, T. C. (1972). Observations of *Callimico goeldii* and *Saguinus mystax*. Unpubl. manuscript, Harvard Univ. Dept. of Anthropology, Cambridge, Massachusetts.

6

The Red Ouakari in a
Seminatural Environment:
Potentials for Propagation
and Study

ROY FONTAINE and FRANK V. DuMOND

I. Introduction

Captive propagation efforts have played a vital role in the conservation of several rare and endangered mammals. Continued world-wide environmental depredation suggests that these programs will play an even greater role for an increasing number of species. Captive propagation should not replace the preservation of wild populations in protected natural habitats. Instead, the establishment of captive colonies capable of maintaining their numbers without input from wild stocks should serve to reduce the exploitation of wild populations by providing a continuous supply of nonhuman primates for display, education, research, and other applications.

This report describes a program designed to establish a self-perpetuating population of red ouakaris (*Cacajao calvus rubicundus = Cacajao rubicundus*) in a seminatural environment at Monkey Jungle in Goulds, Florida. Discussion will include the acclimation of *Cacajao calvus rubicundus* to a seminatural environment, the management of this species under these conditions, a broad overview of the behavior of red ouakaris, and other information of general interest to institutions maintaining ouakaris. The first author's intensive observations of red ouakaris at this site during 1970—1971 and 1973—1975 provide the basis for most of the statements included in this paper. The more than 2000 hours of observation obtained over the course of this project are supplemented by the observations of the second author who established the group in 1961–1962. In addition, the results of a questionaire and survey distributed to institutions holding *Cacajao* complement the present review of captive red ouakaris.

II. The Status of the Red Ouakari

A. The Status of the Red Ouakari in the Wild

The International Union of the Conservation of Nature and Natural Resources (IUCN) regards the red ouakari as an endangered animal. Two factors appear largely responsible for the red ouakari's endangered status: (1) the small number of specimens exported by major animal dealers, and (2) the lack of sufficient data describing the red ouakari's range and numbers in the upper Amazon basin (Harrisson, 1971).

Substantial current data describing the geographic range and numbers of wild red ouakari populations are not available. R. A. Mittermeier's (personal communication, 1975) recent investigation of the zoogeography of *Cacajao* dealt primarily with the wild status of the white ouakari (*Cacajao calvus calvus*) and the black-headed ouakari (*Cacajao melanocephalus*). According to Mittermeier, white ouakaris inhabit the fluvial "island" formed by the Rio Solimões, the Rio Jupará, and the Rio Auati-Paraná (an anastomosis of the former two rivers). Populations of *C. c. calvus* apparently intergrade with populations of *C. c. rubicundus* west of the Rio Auati-Paraná. Mittermeier (personal communication, 1975) found red ouakaris on a small south bank tributary of the Rio Içá. This observation agrees with Soini's (1972) recent assessment of the red ouakari's largely Peruvian distribution—the Rios Putumayo and Napo interfluvial basin and the area between the Rios Yavari and Ucayali. Although little detailed information is available describing the range and status of wild *C. c. rubicundus* populations, general treatments of this topic (e.g., Hill, 1960; Hershkovitz, 1972; Soini, 1972) attribute to *C. c. rubicundus* a much larger range than that described by Mittermeier for *C. c. calvus*. In spite of the ambiguous quality of available estimates of the red ouakari's natural distribution, Hershkovitz (1972) stated that wild red ouakari populations are in no immediate danger of depletion. Of course, a heavy influx of settlers into this animal's range, with its attendant hunting and habitat destruction, would certainly undermine the security of these wild populations.

Hunting and live collecting pressures have been considered by some to be the chief threats to the survival of wild red ouakari populations (Prince Philip and Fisher, 1970). Live collecting certainly dates back to the nineteenth century

(Bates, 1863) while hunting probably dates to pre-Columbian times. Prior to the general prohibition of the primate trade, red ouakaris unintentionally trapped in pursuit of legally captured species were either sold locally as pets or killed for food. The extent of such local exploitation remains unassessed. However, Miami's past commerce in live animals with Leticia, Colombia, and Iquitos, Peru provides an index of recent live collecting pressures on *Cacajao calvus rubicundus*. Shipments of *Cacajao* from Iquitos numbered only 137 out of a total primate shipment of 87,749 during the period from January 1962 to October, 1964. The recent primate purchases of the Tarpon Zoo dealership (Leticia's major primate dealer) suggest extensive depletion of wild ouakari populations in the vast collecting grounds serving Leticia. This organization purchased only fifteen ouakaris during a 6-year period beginning in 1962 (Cooper, 1968). More recently, Thorington reported a total of 200 ouakaris imported into the United States in 1968 and 1969. Although these import—export figures are for the genus *Cacajao*, two conditions justify the assumption that specimens of *C. c. rubicundus* constituted most of the trade in ouakaris. First, red ouakaris account for over 80% of the current captive stock of *Cacajao*. Second, the major centers for the export of upper Amazonian fauna lie within the range of *C. c. rubicundus* but not within those of *C. c. calvus* and *C. melanocephalus*. Since the legal trade that provided the basis for these figures has ceased, and since the development of an illicit export trade in *Cacajao* remains unlikely because dealers derive their chief profits from volume trade, live collecting probably represents a marginal threat to wild populations in 1975. Instead, hunting for food probably presents the greatest current threat to the survival of red ouakaris while habitat destruction looms large as a future threat.

The establishment of an export ban by the governments of Brazil, Ecuador, and Peru is the only protection from continued persecution enjoyed by ouakaris in their native habitat. The most important conservation measure, the establishment of protected forest preserves, remains a goal for the future.

B. Review of Captive Groups

Specimens of *C. c. rubicundus* constitute most of the captive ouakari population. The International Zoo Yearbook's (1974) rare animal census reported only one specimen of *C. melanocephalus* and seven specimens of *C. c. calvus* distributed among four collections. In contrast, it lists many more specimens of *C. c. rubicundus* in current captive collections.

In order to offer a more comprehensive review of the captive red ouakari, we mailed a written questionaire to all institutions listed as holders of these animals in the International Zoo Yearbook's rare animal census. In addition to basic census data, this inventory sought information dealing with individual life histories, social and sexual behavior, details of husbandry, etc.

TABLE I

Census of Captive Red Uakaris in 1973–1975

Institution	Males	Females	Present total	Total viable births in colony history
Brownsville, Texas [a] (Gladys Porter Zoo)	2	4	6	0
Cologne, Germany [b] (Artiengesellschaft zoologischer garten)	1	1	2	0
Colorado Springs, Colorado [a] (Cheyenne Mt. Zoo)	1	3	4	2
Frankfurt, Germany [a] (Zoologischer Garten)	3	2	5	5
Goulds, Florida [a] Monkey Jungle	2	4	6	9
Inuyana, Japan [b] (JMC Primate Zoo)	1	1	2	0
Miami, Florida [a] (Crandon Park)	1	2	3	0
Milwaukee, Wisconson [a] (Milwaukee Zoological Park)	3	1	4	0
Oklahoma City, Oklahoma [a] (Oklahoma City Zoo)	1	3	4	0
San Diego, California [a,c] (San Diego Zoological Garden)	0	1	1	1
San Leopoldo, Brazil [b] (Parque Zoologico do Rio Grande do Sul)	0	1	1	0
Twycross, Great Britain [b] (Twycross Zoo)	0	1	1	0
	15	24	39	17

[a]The authors' survey of 1975.
[b]The International Zoo Yearbook (1974) representing institutional holdings in 1973.
[c]Napier and Napier (1967).

Table I indicates that most institutions holding red ouakaris completed the survey. In the absence of a response to the questionaire, we entered the 1973 figures obtained from the International Zoo Yearbook (1974) into Table I. The current captive population of 39 red ouakaris in twelve collections generally agrees with Thorington's (1972) count of 36 captive specimens in 1970.

The red ouakari's scarlet face and chestnut-red hair make this animal a desirable item for exhibition; consequently, zoos hold all the red ouakaris accounted for in Table I. Laboratory studies rarely employ ouakaris as subjects.

The pet trade apparently absorbed many of the ouakaris imported into the United States prior to *Cacajao's* protection under the United States Endangered Species Act. Retail pet stores and disillusioned pet lovers supplied some of the ouakaris currently living in North American Zoo collections. These former pets probably lack significance for captive propagation efforts.

Table I shows that the captive red ouakari population currently consists of twelve small groups. The typical size of wild red ouakari troops remains unknown. However, R. A. Mittermeier (personal communication, 1975) recently observed groups of 5 to 30 *C. c. calvus* and 15 to 25 *C. melanocephalus*. Mittermeier believes that the group of five white ouakaris represented a temporary subunit split off from a larger group. Thus, if group size in *C. c. rubicundus* roughly parallels that characteristic of *C. c. calvus* and *C. melanocephalus*, then the largest captive red ouakari groups represent what would be exceptionally small naturally occurring population units. The presumed atypical social environment of existing captive red ouakari groups has probably been a factor in this animal's poor captive breeding record.

C.   Potentials for Propagation and Study

There has not been interinstitutional effort to breed ouakaris; consequently, the current distribution of red ouakari holdings fails to maximize the propagation potential of the small captive population. Only 15 of the 39 red ouakaris counted in our survey live in proved breeding groups. Eleven live either alone or in units where reproduction remains unlikely. The remaining thirteen specimens live in three groups (the Oklahoma City Zoo, the Gladys Porter Zoo, and Crandon Park) that hold some promise for breeding but which have not, as yet, produced any young. The failure of the Oklahoma City Zoo's ouakaris to produce young is probably due to the relative youth of their sexually active male who was born at the Frankfurt Zoo in 1970. Since the Gladys Porter Zoo also reports the possession of a male that displays apparently adequate copulatory behavior, the situation at this institution seems favorable for breeding. However, since this institution lacks precise age data, it is possible that this sexually competent male is not yet sufficiently mature to impregnate. Crandon Park maintains a sexually competent male that has sired a single stillborn offspring.

Few institutions have successfully maintained ouakaris for more than a fraction of their potential lifespan. Nevertheless, considerable progress has been made toward fulfilling the first requirement of a captive breeding program. the establishment of adequate husbandry procedures designed to keep these monkeys alive for extended periods (Crandal, 1964; Hill, 1965). Diet no longer seems to be a limiting factor in the successful maintenance of red ouakaris. A diet emphasizing fruit, vegetables, raw peanuts, commercial monkey pellets, hard-boiled eggs, and nonfat milk fortified with a vitamin and mineral supplement (including vitamin D_3) appears adequate to maintain general good health, growth, and lactation in captive red ouakaris.

The second step of a captive propagation effort, the promotion of mating behavior, poses a more difficult problem. Our survey did not reveal a specific sociological formula that would ensure red ouakari matings. Respondents to our survey reported the occurence of copulations among ouakaris maintained in pairs, multimale groups, and unimale groups. Nevertheless, our survey did suggest, though the sample is small, that copulations occur most frequently in groups where the ratio of adult males to adult females is at least 1:2. Furthermore, red ouakaris maintained as pairs seems much less prone to copulation than uakaris living in larger groups. Copulations were not reported among the ouakaris living in the Milwaukee Zoological Park where three adult males live with a single adult female.

Extensive taming and/or life as a pet during early development seems to impair a ouakari's capacity for normal adult sexual behavior. The psychological damage produced by excessive human orientation seems to affect reproductive behavior in the male more severely than in the female. Male ouakaris with pet experience have never been observed to engage in any form of mating behavior. At Crandon Park, a single female with 1 year of experience as a pet frequently copulated with an apparently normal male and eventually produced a stillborn infant.

Only four institutions have been successful in the production of viable red ouakaris. Three groups (Colorado Springs, San Diego, and Frankfurt) had a single wild-born female that produced all the viable births. The Monkey Jungle colony is exceptional in having had four parous females (including a captive-born one) in its history.

The goal of producing multiple generations of captive *C. c. rubicundus* remains beyond the grasp of most current breeding programs. Yet the attainment of this goal provides the chief rationale for undertaking the expense of a captive propagation effort. At the present time, Monkey Jungle remains the only institution to have successfully bred captive-born ouakaris. However, since most of the eight captive-born ouakaris currently living in zoo collections have not reached sexual maturity, a final assessment of their reproductive potential remins a task for the future.

The research potential of the current captive population of *C. c. rubicundus* remains unrealized. A cooperative interinstitutional study of the reproductive and developmental biology of the red ouakari would increase the effectiveness of current propagation efforts. Furthermore, since red ouakaris are the most easily maintained and the most numerous of the captive *Pitheciinae*, the study of *C. c. rubicundus* as a representative pitheciine would advance current knowledge of cebid biology.

Monkey Jungle's seminatural environment provides a setting that approaches the natural habitat of the red ouakari. Semifree-ranging conditions facilitate broad behavioral studies; studies of long-term population dynamics; and other topics that do not demand capture, restraint, and handling. This report includes a substantial portion of the available data describing the behavior of the Monkey Jungle's red ouakari colony.

Ouakaris maintained in relatively small cages can provide valuable material for detailed studies of the reproductive and developmental biology of *C. c. rubicundus*. For example, the collection of data on the menstrual cycle represents a practical adjunct to captive breeding programs. Similarly, reproducing caged colonies of red ouakaris could provide basic data describing physical growth and dental eruption sequences.

III. History of the Monkey Jungle Group

A. The Colony in 1975

The Monkey Jungle ouakari colony consists entirely of representatives of those extreme western populations of *Cacajao calvus rubicundus* that were formerly classified by Hill (1960) as *Cacajao rubicundus ucayali*. All the red ouakaris living at Monkey Jungle display the pelage features diagnostic of *C. r. ucayali*: a nape concolorous with the back and a relatively sharp demarcation between the red of the nape and the grizzled grey hairs of the forepart of the crown.

The current troop of *C. c. rubicundus* at Monkey Jungle consists of six animals: an adult male (CM3), three adult females (WF3, WF4, and CF3), a juvenile male (CM5), and an infant (CF4). [In the code used throughout this report, the first letter indicates whether the animal is captive-born (C) or wild-caught (W); the second letter designates the animal's sex; and the number indicates the animal's order of birth or introduction into the colony.] The red ouakaris of Monkey Jungle share a 1.6 hectare simulated tropical American "wet forest" with groups of *Alouatta seniculus*, *Saguinus fuscicollis*, and *Saimiri sciureus*.

B. Introduction and Development of the Group

The establishment of the red ouakari colony at Monkey Jungle began approximately 8 months after the successful introduction of 37 squirrel monkeys into the "Rainforest." Thirteen red ouakaris of vague provenience, but all representative of western populations of *C. c. rubicundus*, served as the foundation stock. Animal dealers supplied all of these founders; consequently, several of these animals had experience as pets which weighed against their integration into ouakari society.

The introduction of wild-caught ouakaris occurred during the warmer months of 1961 and 1962. In order to ensure new colony members a few months of experience in the "Rainforest" prior to the onset of winter, no introductions occurred after September.

Table II summarizes the topic of this section: the chronology of introduction and adaptation of the red ouakari colony.

On April 8, 1961 two male infants (WM1 and WM2) were released into the

TABLE II

Chronology of Introduction and Adaptation of Monkey Jungle's Uakaris

Date	Change	Total	Event description
April 8, 1961	+2	2	Two infants (WM1 and WM2) released
April 29, 1961	−1	1	WM1 dies (intestinal *Trichomonas* suspected)
May 16, 1961	+1	2	Juvenile (WF1) released
July 3, 1961	+2	4	A pair of juveniles (WM3 and WM4 released)
July 16, 1961	+2	6	Two adults (WF2 and WM5) introduced
July 22, 1961	−1	5	WM5 dies (suspected intestinal parasites)
August 20, 1961	−1	4	WM2 dies (hawk predation suspected)
September 1, 1961	+1	5	Adult male (WM 6) released
September 3, 1961	−1	4	WM6 dies (impaled on broken limb)
March 9, 1962	+1	5	An adult male released (WM7)
September 22, 1962	+4	9	Three large juveniles (WF3, WF4, and WF5) and a subadult (WM8) released
June 23, 1963	+1	10	WF2 gives birth to CM1
Late 1963	−2	8	WM8 and WF5 disappear
Early 1964	−2	6	WM3 and WM4 removed from the "Rainforest"
July 1964	+2	8	WF3 gives birth to CF1; WF4 gives birth to CM2
July 1966	−1	7	WM7 dies (intestinal infection)
Mid 1966	−2	5	CM2 and CF1 disappear
August 29, 1966	+1	6	CM3 born to WF2
August 30, 1966	+1	7	CF2 born to WF3
January 28, 1968	−1	6	CF2 falls and dies
October 24, 1969	+1	7	CF3 born to WF4
October, 1969	−1	6	CM1 dies (secondary infection following extraction of canine teeth)
November, 1972	−1	5	WF2 disappears
May 2, 1973	+1	6	WF3 gives birth to CM4
May 31, 1973	+1	7	CF3 gives birth to CM5
March 14, 1974	−1	6	WF1 disappears
July 28, 1974	+1	7	CF4 born to WF3
March 18, 1975	−1	6	CM4 dies (atypical pneumonia)

"Rainforest." WM2 appeared to be about one-quarter grown while WM1 seemed slightly larger. WM1 died shortly after introduction while WM2 survived for several months. The short captive life of these infants suggested that older ouakaris might make a healthier adjustment to the seminatural environment than the then more readily available infants.

WF1, a juvenile female purchased from a local dealer, joined WM2 on May 16, 1961. The same dealer supplied Monkey Jungle with two more juveniles (WM3 and WM4) 2 months later.

Monkey Jungle acquired two adult specimens, WF2 and WM5, on July 16, 1961. Since WM5 survived for only a few days, another adult male (WM6) was acquired on September 1, 1961. This monkey died only 2 days after introduction. Finally, an adult breeding male (WM7) was successfully introduced on March 9, 1962. The last introduction of wild-born ouakaris occurred on September 22, 1961 with the release of three large juvenile females (WF3, WF4, and WF5) and a subadult male (WM8). WF5 and WM8 died the following year whereas WF4 and WF3 remain in the present colony.

The social structure of the ouakari colony began to stabilize in early 1963. However, a series of attacks directed to tourists (including one resulting in substantial injury) and the direction of severe aggression against howler monkeys led to the extraction of the canines and lower incisors; and eventually, the permanent removal of WM3 and WM4 from the "Rainforest." The removal of these human oriented, fearlessly aggressive, and poorly integrated specimens from the colony produced the largely harmonious society that has persisted to the present. The ouakari colony gradually peripheralized WF1, a former pet, over the course of the next 10 years. Apparently, overt aggression directed against WF1 never played an important role in this animal's peripheralization which resulted in the failure of the rest of the group to groom her, answer her vocalizations, or otherwise respond to her misplaced social signals.

During the period of *C. c. rubicundus'* establishment in the "Rainforest," other primate species were adapting to this environment as well. Interactions between ouakaris and other monkey species generally ranged from indifferent to friendly. The habit of WM3 and WM4 of attacking red howlers (*Alouatta seniculus*) stood out as the chief exception to the general tenor of the red ouakari's relationships with other *Ceboidea* in the early years of the colony.

C. Births, Geniology, and Multigenerational Breeding

The first birth in the "Rainforest" occurred on June 23, 1963 when WF2 gave birth to CM1. In the meantime, WF3 and WF4 reached maturity; in July of 1964, both of these animals produced viable young. The only mature male in the colony during the corresponding mating periods, WM7, sired CM1, CM2, and CF1. We ruled out the remaining pair of males (WM3 and WM4) as candidates for paternity because of their relative youth and their failure to display normal social behavior.

WF2 gave birth to CM3 on August 29, 1966, nearly 3 years after her first captive birth. The next day, WF3 produced an infant (CF2). WM7, who died in July, 1966, sired both infants.

As a result of WM7's death, the colony lacked an adult male for several years. Finally, CM1 reached maturity in early 1969 and WF4 gave birth to CF3 on October 24 of that year. Unfortunately, CM1 died and once again the colony lacked a reproducing male.

CM3 reached reproductive condition in the fall of 1972; thus far, he has sired three young. In May of 1973, WF3 gave birth to CM4 and CF3 gave birth to CM5. The next year, WF3 gave birth to CF4 in July while CF3 did not produce any offspring that year. On April 26, 1974, CF3 manipulated and carried about an object very much resembling an aborted fetus for several minutes while displaying behavior indicative of high general arousal. Regrettably, we could not recover this object for closer examination after CF3 dropped it.

Only two of the wild-born founders remain in the current "Rainforest" colony. The rest of this group was born in the seminatural environment.

The breeding record indicates the occurrence of wild-born male by wild-born female crosses (WM7 × WF2, WM7 × WF3, WM7 × WF4), captive-born male by wild-born female crosses (CM1 × WF4, CM3 × WF3), and captive-born female by captive-born male crosses (CM3 × CF3). All of these matings produced viable young that survived for at least 18 months.

The mating of two captive-born individuals (CM3 and CF3) to produce CM5 provides the chief basis for an optimistic assessment of the ouakari's reproductive future in seminatural environments. Since this particular mating represents the reproduction of second- and third-generation captive-born animals, it demonstrates that rearing in the seminatural environment can supply all the necessary inputs for reproductive success in both sexes.

A tendency toward inbreeding exists in all small closed populations such as the Monkey Jungle ouakari colony. At the present time, CM5 is the only inbred animal currently living in the group. CM3's parents, WF2 and WM7, are the same as CF3's grandparents on the paternal (CM1) side. Thus, CM5, the product of a CM3 × CF3 mating, may be more homozygous than a randomly selected wild ouakari. In the future, trades with other breeding colonies may be undertaken in order to introduce new genetic material and to restrict the phenotypical expression of deleterious recessives.

D. Losses, Longevity, and Infant Mortality

It was often impossible to determine the cause of death among animals lost in the seminatural environment. Indeed, many ouakaris were merely presumed dead when they ceased to be seen in the "Rainforest." The seminatural environment is large enough to hide a cadaver and deliberate searches for them have frequently failed or led to the discovery of decomposed remains. During periods when the ouakaris were not subject to quantitative observation, deaths often went unnoticed for several weeks; consequently, the information in this section is not as complete as we might wish it to be.

The early losses in the colony probably resulted from a continuation of the traumas associated with separation from the wild and introduction into an alien social and physical environment.

The first loss occurred on April 29, 1961 when a small infant (WM1) died. His death was attributed to intestinal parasites probably acquired prior to release.

The next introductory death occurred on July 22, 1961 when WM5 died of suspected intestinal parasites only a few days after his introduction into the "Rainforest."

On August 20, 1961, the smallest member of the original pair of infants (WM2) disappeared. This infant may have fallen victim to hawks (*Buteo jamaicensis*) that were preying upon marmosets while ouakaris were being established.

WM6 died on September 3, 1961 after only 2 days in the "Rainforest." He died by impaling himself on a broken limb. The intense agonistic activity associated with the introduction of WM6 in the colony may have been a factor in this accident.

No additional losses occurred for about 2 years. Then, in late 1963, WM8 and WF5 suddenly disappeared.

During July, 1966 an intestinal infection killed WM7. A pair of captive-born ouakaris, CM2 and CF1, also disappeared about this time. A series of inquiries concerning the care and feeding of ouakaris immediately after the disappearance of one of these infants suggested that this animal had been filched by one of the tourists. Unlike older ouakaris who regard humans with considerable suspicion, young ouakaris often solicit human attention; as a result, the theft of one is not difficult to imagine. In fact, in the spring of 1971, workmen at Monkey Jungle successfully thwarted an attempted theft of an extremely friendly juvenile female (CF3) of the same general age as the one stolen in 1966. The circumstances surrounding the loss of the other animal of this pair remain unexplained.

CF2, the next captive-born fatality, lived only 17 months in the "Rainforest." On January 28, 1968 she fell from a tree, struck her head on the rocky jungle floor, and suffered injuries including a fractured skull and loss of unilateral coordination that proved fatal after a few days. When CM1 reached maturity, he continued to interact with the public in a friendly manner. For example, he once held hands and bipedally "danced" with a young girl. Fearing a repetition of the previous episode involving WM3 and WM4, the descision was made to extract this animal's large canines. Since WM3 and WM4 rapidly recovered from their tooth extractions and were introduced immediately without ill effect, CM1 was returned to the "Rainforest" after a brief absence in order to minimize social damage. Unfortunately, food apparently impacted in the lower extraction wound and infection set in. By the time CM1 could be caught and treated, the infection had advanced to a generalized septicemia which failed to respond to antibiotic therapy, Thus, CM1 died in October, 1969.

A pair of wild-born female ouakaris died of unknown causes in recent years. WF2 died during November, 1972 after over 11 years in the "Rainforest." Since she appeared to be middle aged when acquired in 1961, old age may have been a factor in her death. WF1 died on March 14, 1974 after nearly 13 years of living in the seminatural environment. WF1 received treatment for pneumonia and *Strongyloides* parasitism during February, 1973 and a recurrence of these ailments may have contributed to her death.

CM4 became very ill during his 22nd month. He received treatment for pneumonia and *Strongyloides* parasitism in February, 1975. After an apparent recovery, his illness recurred during March and he died of an atypical pneumonia on March 18, 1975. This animal's health rapidly declined following the unusually early birth of a sibling (CF4) which led to his abrupt and apparently stressful weaning.

Five of the ouakaris that were born at Monkey Jungle have died. A single captive-born adult died as a result of secondary infection following a tooth extraction. The remaining four captive-born ouakaris all died during the latter part of their second year. Other institutions that report the successful breeding of ouakaris similarly note high mortality in the second year. Perhaps, the intense social development and increasing independence from the mother that characterize the ouakari in his second year produces stress leading to a lowering of resistance to disease. The case of CM4 and CM5 illustrates the possible relationship between weaning and susceptibility to a variety of medical hazards. CM4 and CM5 were at this critical age in early 1975 when they both received treatment for pneumonia and *Stongloides*. In addition, CM5 received treatment for dental abscesses (left maxillary PM2 and PM3) that represented a recurrence of a previous maxillofacial abscess. CM5 recovered but CM4 did not. It is of interest that CM4's general activity level and apparent health gradually declined after he experienced very early and abrupt weaning due to CF4's birth. CM5 did not suffer through any of these events and his weaning occurred over a much greater time span allowing easier adjustment to independence. Accidental losses of ouakaris in their second year may well relate to increasing independence from protective maternal responses without the full development of patterns of behavior enabling the young ouakari to cope with environmental hazards.

The observation of ouakaris from birth to senescence will ultimately provide precise estimates of the life expectency of *Cacajao*. The present colony includes two females (WF3 and WF4) that have lived in the "Rainforest" since September 22, 1962. The comparison of the external appearance of these animals on receipt with familiar captive-born specimens suggests that these animals were born in 1959. WF3's production of live births in 1972 and 1973 demonstrates a lack of fatigue of the reproductive apparatus after 15 years and several births. Neither WF3 nor WF4 currently display impaired movement or other signs of senility. At introduction into the "Rainforest," WF2 displayed facial aging apparently equivalent to the facial aging displayed by WF4 at 12 years. Nevertheless, WF2 lived in the "Rainforest" for over 11 years and delivered two viable off-spring while in captivity. Thus, the longevity of the better adjusted ouakaris at Monkey Jungle suggests a life expectancy of 15 to 20 + years for appropriately maintained captive red ouakaris.

IV. Description of the Monkey Jungle "Rainforest"

A. Monkey Jungle

Monkey Jungle is a commercial zoological park devoted to the naturalistic display of nonhuman primates in social units comparable to those encountered in the field. The "Rainforest" is one of Monkey Jungle's major facilities. At one time or another the "Rainforest" has accommodated social groups of the following sympatric platyrrhine monkeys: *Alouatta seniculus, Aotus trivirgatus, Cacajao c. rubicundus, Callimico goeldii, Pithecia monachus, Saguinus fuscicollis*, and *Saimiri sciureus* (Iquitos).

B. Physical Characteristics

Monkey Jungle lies at approximately 25.4°N latitude and 80.2°W longitude or 35.4 km south of Miami, Florida.

Subtropical climate characterizes the environs of Monkey Jungle. Department of Commerce (1968) figures reveal an annual rainfall ranging from 142.2 to 152.4 cm. About 70% of this precipitation falls in heavy summer rains. These relatively short tropical storms typically occur in the afternoon. Individual storms, particularly easterly waves and hurricanes moving north from the Caribbean, may add up to 38 cm of precipitation to any given annual total (Raisz, 1964).

The annual mean relative humidity and the annual mean daily maximum relative humidity are 72 and 92%, respectively. The relative humidity rarely drops below 30% and its rises to 90% + on all but a very few nights each year.

A mean annual high 29.2°C and a mean annual low of 17°C reflect the area's mean annual temperature of 23.1°C (U.S. Dept. of Commerce, 1968). Freezes occur in less than one-quarter of the years (U.S. Department of Commerce, 1968). These rare freezing spells occur in conjunction with equally rare relative humidity readings below 30%.

Table II reveals that the occurrence of deaths bears no obvious relationship to season. Red ouakaris apparently tolerate occasional cold spells without serious difficulty. Although CM4's death occurred in temporal association with unusually cold weather, this animal's general health declined well before the onset of the cold spell.

C.   Natural Ecological Characteristics of the "Rainforest"

The "Rainforest" consists of a 1.6 hectare parcel of a larger hammock. A hammock is a tract of subtropical hardwood jungle typically growing on land a few meters higher above sea level than the surrounding area. In most instances, these hammocks represent the northern limit of the distribution of the broad-leaf trees found in them whose southern distribution extends into South and Central America. In southern Florida, high pinelands, prairies, and flatwood pinelands generally surround hammocks.

Hammock growth in the "Rainforest" consists of several subecologies (Fig. 1) including a tall central subclimax area, a central area of secondary growth where the species structure suggests the development of a climax community, a low scrub area and gallery adjacent to the periphery, and a flatwood pineland-hammock ecotone in the western portion of the seminatural environment.

The following description of the seminatural environment's flora will concentrate on those species with the greatest apparent biomass. It could easily be lengthened to include the discussion of the many species represented by relatively few specimens and/or low apparent biomass.

Figure 1 schematically represents the apparent relative abundance of the floristic dominants of the seminatural environment. Whenever appropriate, subsequent discussion of the flora of the "Rainforest" will treat each subecology in terms of two divisions: a layer of tall trees forming a canopy with occasional emergent crowns (upper case letters in Fig. 1) and an understory of smaller trees, shrubs, aroids, vines, etc. (lower case letters in Fig. 1). These divisions are by no means absolute and several species (notably *Nectandra coriacea* and *Coccoloba diversifolia*) straddle both divisions.

The hardwood species *Lysiloma bahamensis* and *Quercus virginiana* dominate the canopy in the central climax area of the "Rainforest." Other indigenous trees well represented at this level in the central area include: *Nectandra coriacea, Bursera simaruba, Ficus aurea, Simarouba glauca, Sideroxylon foetidissimum, Coccoloba diversifolia, Rapanea guianensis, Roystonea elata, Exothea paniculata,* and *Prunus myrtifolia.* The tallest trees (individual specimens of *F. aurea*) stand as high as

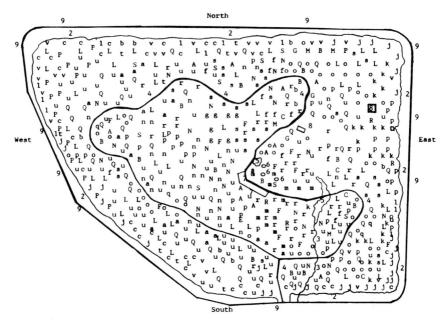

Fig. 1. Schematic representation of the seminatural environment. Upper-case letters represent relative densities of dominant trees of the canopy. Lower-case letters represent relative densities of dominant plant forms in the understory. The arabic numbers designate man-made structural features of the "Rainforest." L, *Lysiloma bahamensis*; Q, *Quercus virginiana*; N, *Nectandra coriacea*; B, *Bursera simaruba*; F, *Ficus aurea*; S, *Simarouba glauca*; A, *Aiphanes spp*; P, *Pinus elliottii*; M, *Sideroxylon foetidissimum*; R, *Roystonea elata*; C, *Coccoloba diversifolia*; G, *Rapanea guianensis*; E, *Exothea paniculata*; u, *Psychotria undata*; o, *Philodendron oxycardium*; c, *Chiococca alba*; r, *Rhaphidophora aurea*; j, *Jasminum dichotomum*; s, *Syngonium podophyllum*; a, *Ardisia solanacea*; k, *Croton furgusonii*; l, *Solanum verbascifolium*; v, *Vitrus rotundafolia*; p, *Philodendron spp*; f, *Philodendron fenzlii*; g, *Philodendron sellonum*; t, *Tetrazygia bicolor*; n, *Coccoloba diversifolia*; q, *Philodendron radiatum*; b, *Ardisia escallonioides*; m, *Malvaviscus arboreus*; 2, grassy perimeter; 4, walking trail; 5, pond. 6, feeding station; 7, viewing area; 8, shed or small structure; 9, fence; 3, stream.

25 m. Hurricanes periodically blow away much of the canopy and many of the emergent crowns; consequently, the modal height of large trees in the central area lies around 15 m. These larger trees grade into an understory of smaller trees and shrubs. *Coccoloba diversifolia*, *Psychotria undata*, and *Ardisia solanacea* predominate at this level.

Nectandra coriacea dominates the central area of secondary growth which is in the process of recovery from damage suffered in a severe 1960 hurricane. This region supports large populations of *Psychotria undata*, *Ardisia solanacea*, *Coccoloba diversifolia*, *Lysiloma bahamensis*, and *Bursera simaruba*.

The native flora of the seminatural environment's peripheral areas presents a marked contrast with the central area's native flora. The western periphery consists of a flatwood pineland-hammock ecotone. In this region, *Lysiloma bahamensis* and *Pinus elliottii* form mixed stands dominating the upper forest layer while *Chiococca alba*, *Solanum verbascifolium*, *Tetrazygia bicolor*, *Jasminum dicotomum*, and *Vitrus rotundafolia* predominate in the understory. The species composition of the low scrub and gallery areas around the "Rainforest" periphery greatly resembles the understory of the flatwood pineland-hammock ecotone. In addition to the previously listed indigenous forms, one frequently encounters *Psychotria undata*, *Ardisia escallonioides*, and *Trema micrantha* in the region.

The indigenous faunal community consists primarily of Florida Everglades wildlife: raccoons (*Procyon lotor*), opossums (*Didelphis marsupialis*), rabbits (*Sylvilagus p. palustris*), squirrels (*Sciurus c. carolinensis*), rats (*Rattus r. frugivorous*), skunks (*Mephitis mephitis*), mice (*Mus musculus*), several species of snakes (the constrictor *Elaphe quadrivittata* predominates), hawks especially *Buteo jamaicensis*), vultures (*Calhartes aura*), several species of small birds, tree frogs, small lizards, and a great variety of invertebrates.

The intentional elimination of native fauna did not accompany the establishment of primates species in the "Rainforest." However, the great expansion of the squirrel monkey population in the "Rainforest" has led to the elimination of tree frogs and small lizards and an absence of nesting birds.

D.   Modifications of the Environment

In order to create the visual effect of a "living museum" for the public, the natural hammock flora was enriched by the addition of more than 100 species of exotic forest palms, 20 species of *Philodendron*, 15 species of surface aroids, and 50 species of hardwoods including heavy plantings of *Ficus* sp.

The effect of these introductions was largely limited to the understory in the central areas of the "Rainforest." The occasional occurrence of *Philodendron oxycardium* and *Philodendron* spp. in the forest periphery and flatwood pineland-hammock ecotone represents the chief modification of the flora of these areas.

Most of the exotic trees failed to compete with the well-established native hammock varieties which remain dominant in the canopy of the central area. Sporadic examples of *Aiphanes* sp., *Ficus* sp., *Chrysalidocarpus* sp.,*Persea americana*, *Exiobotryia japonica*, *Citrus paradisi*, *Ptychosperma alexanderi*, *Cococus plumosa*, and *Tabebuia heterophylla* account for most of the large exotic trees of the central areas of the "Rainforest."

Exotic aroids, in contrast, have become plentiful in the understory of the central areas of the "Rainforest." Indeed the great abundance of these aroids provides the chief ecological contrast between the central areas and the essentially unaltered peripheral areas of the seminatural environment. The most abundant aroids are *Philodendron oxycardium, P. fenzlii, P. sellonum, P. radiatum, Rhaphidophora aurea,* and *Syngonium podophyllum.*

Most of these introductions thrive in climates characterized by annual rainfall in the order of 450 cm. In order to promote the survival of these rainforest plants in Florida's climate, an irrigation system ensuring a yearly total of 450 cm of moisture was constructed. This system is turned on from 6:00 to 8:30 AM on alternate days or more frequently during dry weather.

A grassy perimeter and a 2.4-m chicken wire fence were built around the "Rainforest" in order to confine the monkeys within the seminatural environment. A covering of sheet metal inclined inward at an angle of 20° overlies the upper 75 cm of the fence. This covering makes escape by scaling the fence a difficult motor task. The grassy perimeter varying in width from 5 to 12 m hinders escape by producing a break in the otherwise continuous foliage. Escape would be an easy matter if this perimeter were absent; however, with this barrier present, an animal intent on escaping must complete a leap involving at least a 10 m horizontal component or learn a way to negotiate the fence.

These barriers proved effective in confining the ouakaris to the "Rainforest" during the critical early stages of introduction. WM3, WM4, WF2, and CM5 did learn to negotiate the fence; however, these monkeys always returned to the "Rainforest" after short journeys to surrounding areas.

A number of modifications of the seminatural environment were undertaken in order to accommodate the 300,000 visitors who pass through the "Rainforest" each year. These include a 1.5-m walking trail that winds through the "Rainforest", a stream that ends in a small pond stocked with piranha, and finally, a viewing area where the public can sit and observe the monkeys as they feed at the provisioning site. Provisioning ensures visitors the opportunity to observe monkeys in the "Rainforest." Without this practice, most of the animals would disappear into the thick foliage of the "Rainforest" where they would be overlooked by the public.

V.   Ecological Utilization of the Environment

A.   Locomotion and Routing

The red ouakari's practice of favoring identifiable routes in travel between distant points in the "Rainforest" is an important aspect of their tendency to restrict activities to only a portion of the available space. In travel between two distant points, ouakaris favor the shortest arboreal route consisting of a continuous network of solid supports. Travel over these preferred routes usually involves quadrupedal pronograde walking and running. Leaping, brachiation and other apparently more strenuous activities are used to traverse gaps in the continuity of the solid support network. The presence of a section in a preferred route requiring anything other than simple quadrupedal pronograde locomotion usually indicates that considerable distance would have to be added to the route in order to maintain continued travel over a solid support network.

The constancy of a pathway's constituent structural elements and the corresponding behavioral constancy observed in travel over it reflects the frequency of use of that pathway. An intensively used route, such as the one connecting the provisioning area with the southeast quarter of the "Rainforest," is remarkably constant with respect to its support sequences and the locomotor behaviors used in traveling over them.

The intensive use of preferred routes in travel between distant points is largely an adult characteristic. Immature ouakaris typically vary the support sequences used in travel between two points. In addition, young ouakaris often interrupt progression to engage in play and other activities that disrupt the continuity of travel over established routes.

Several alternative preferred routes often connect distant points in the "Rainforest." The individual's age, sex, general health, and a variety of motivational features influence the selection of a particular route on any given occasion. The following example illustrates the complexity of the factors effecting route selection in travel between distant points in the "Rainforest." When WF3 was not carrying an infant, she consistently used the same route as the other adult females in travel between the provisioning area and the southeast quarter of the "Rainforest." The final segment of this route passes over the public walkway; consequently, it brought WF3 into closer spatial proximity with the public than she normally tolerated. When WF3 was carrying an infant, she used this route in most of her travel from the southeast to the provisioning area when her hunger motivation allowed her to better tolerate proximity to people. However, once WF3 had finished eating, she would not use this route in return to the southeast when people stood under the critical crossing. Instead, she would carry her infant over a much longer but nevertheless remarkably consistent route in traveling from the provisioning area to the southeast.

B. Stratification

One can view any forest environment as a collection of horizontal strata that provide sites for various activities. The simplest classification of "Rainforest" strata divides the environment into the following major levels: a region of tall trees (10 + m) and emergent crowns often forming a continuous canopy; a lower level (3−10 m) including an often continuous network of thick branches arising from the major trunks of the dominant hardwood species; a low-lying layer of shrubs and young trees that typically grades into a physiognomically contrasting layer of vines and shrubs on the forest periphery; and finally, the "Rainforest" floor.

Ouakaris use all levels of the "Rainforest." However, behavioral differences correlated with age, sex, and life experiences affect the distribution of individual ouakaris among the various "Rainforest" strata. In addition, weather and diurnal periodicity affect the distribution of ouakaris among the strata of the "Rainforest."

Ouakaris use the highest stratum of the "Rainforest" primarily for sleeping and intensive foraging during the early morning and late afternoon. Adult males who rarely ascend to the tops of emergent crowns and who travel less extensively about the "Rainforest" seem to use this level much less comprehensively than representatives of other age—sex classes.

On occasion, ouakaris have been observed to ascend to the tops of the highest trees in order to attend to disturbing extragroup stimuli originating beyond the boundaries of the "Rainforest."

Weather influences the amount of time spent in the highest stratum of the "Rainforest." During cold weather (0°–10°C), the ouakaris seek exposure to the sun by increasing the amount of time spent in the upper stratum. During the summer ouakaris tend to avoid this stratum during the midday heat, but they ascend to this level after heavy summer rains.

Ouakaris focus diurnal activity in the shaded middle stratum of the "Rainforest." This middle stratum, which includes a largely continuous succession of solid supports arising from the main trunks of the dominant hardwood species, provides the setting for much of the ouakari's travel about the seminatural environment. The solid supporting structures of this level permit minimized interindividual distances and, consequently, this forest stratum provides the setting for much of the ouakari group's social life. It provides the physical support facilitating such activities as grooming and copulation. Ouakaris also seek the protection from solar radiation offered at this stratum for the lengthy midday resting periods that are a conspicuous aspect of their diurnal activity cycle. Although the intermediate forest level provides relatively little natural food, the ouakaris frequently move to food-bearing terminal branches at this level to feed for short bouts interspersed between longer bouts of rest.

A zone of immature trees, shrubs, vines, etc., occurs below the middle forest level. Juvenile ouakaris make extensive use of this zone. The adults, in contrast, only occasionally descend to feed on the seasonally abundant fruit available at this level. Adult females, in particular, appear uneasy while foraging in the low stratum. They have frequently been observed carrying small fruit-bearing branches from this level to a higher perch for leisurely consumption.

Juveniles also use the low stratum for vigorous locomotor play and rough and tumble play. However, terrestrial activity is an even more strongly juvenile characteristic than the above. Juvenile ouakaris frequently descend to the ground to forage, play, and manipulate objects lying on the ground. These activities tend to occur during the midday period when the remaining ouakaris rest in the middle stratum. Adult terrestrial activity, in contrast, is extremely limited. It occurs primarily in association with drinking and the acquisition of provisioned food. Adult males, in keeping with their general preference for lower strata, descend to the forest floor far more frequently than adult females.

A few generalizations emerge from the observation of the use of "Rainforest" strata by ouakaris. First, stratification follows a diurnal cycle involving activity at higher levels followed by a late morning shift to lower strata. This shift reverses itself toward the end of the day when the ouakaris return to higher strata. Second, within the framework of the diurnal cycle, they prefer low strata when air temperatures are high and high strata when air temperature is low. Third, adult females apparently prefer higher strata than either adult males or juveniles. Finally, adult males express much less variability in their use of forest strata when compared with the activity of females and immature animals.

C. Daily Activity Rhythm

The behavior of red ouakaris in the seminatural environment conforms to a strictly diurnal activity cycle; activity begins at sunrise and ends at sunset.

The red ouakari's day begins with the first light of dawn, just prior to the time when color becomes apparent to a human observer. Urination, defecation, vigorous scratching, and autogrooming generally occupy the first 5−15 minutes prior to travel away from the sleeping trees. Adult females frequently groom and nurse dependent young for several minutes before moving off to begin foraging.

The ouakaris soon move over well-established routes from the sleeping trees to distant food sources. The ouakaris typically devote the next 2 hours to intensive foraging at high levels. During this period, the group tends to remain dispersed over an area of at least 1.0 hectares.

After approximately 1.5−2.0 hours of intensive foraging, the formerly dispersed ouakaris coalesce in the high to middle strata of the south central area of the "Rainforest." This gathering around 08:30 generally coincides with the arrival of the caretaker staff and it may represent a response to the increased human activity occurring at this time of day.

Several interrelated behavioral trends characterize the period from about 08:30 to the first provisioning at 10:40. These trends include: a general decrease in the rate of foraging, a replacement of foraging by play among the immature ouakaris, a decrease in interindividual distance, and finally, an increase in the rate of social interaction.

Grooming and other forms of social interaction appear especially intense during the 30 minutes period prior to the 10:40 feeding. As the time of this feeding approaches, the adult females and young become quite agitated; social grooming increases to high rates and the probability of inter- and intraspecific aggression reaches its daily peak. Once food is offered, the ouakaris lose no time in crowding in with the squirrel monkeys for their share of the 10:40 feeding.

The ouakaris normally leave the provisioning area around 10:55. A common sequence of events following this departure involves travel to the southeast quarter over established routes, drinking from the stream, and finally travel to the south central area for rest and participation in stable grooming relationships. Less commonly, they follow an extended route about the "Rainforest" that includes periodic stops for foraging and other activities. This circuit of the "Rainforest" eventually returns to the provisioning area.

The general agitation preceding the first provisioning does not recur in connection with the 11:45 feeding; instead, the ouakaris travel directly to the feeding stations from relatively dispersed points with a minimum of preliminaries. The ouakaris stay longer at this feeding than any other. Indeed, adult males and juveniles frequently spend 30 minutes or more at this feeding. The general sequence of events following this feeding resembles that following the first provisioning; however, the animals now seem much less inclined to take long trips about the "Rainforest."

General activity declines following the completion of the second trip away from the feeding stations. Midday travel consists largely of movement between the provisioning site and preferred resting spots in the southeast and south-central portions of the "Rainforest." During this period, stable grooming relationships are reinforced and a low level of general activity prevails. During the midday, the ouakaris often ignore offerings of food and trips to the feeding stations are often skipped at 13:55, 14:55, and 15:55. The young ouakaris frequently engage in play or otherwise maintain relatively high levels of activity while the adults rest.

Seasonal factors effect midday activity. During the short winter day, ouakaris tend to forage at a moderately active pace and they regularly attend midday provisionings. These adjustments probably reflect the combined effects of the increased activity typically correlated with relatively low temperatures and the reduction of the foraging period that follows the last feeding during the winter months. The State of Florida's adherence to Daylight Savings Time serves to reinforce the latter factor by concentrating the seasonal variation in day length to the period after the last provisioning.

Ouakaris generally become more active with the passing of the midday temperature peak. Thus, more ouakaris attend the last feeding than the previous ones. In accordance with previously stated seasonal factors, the period of increased activity after the last feeding is most conspicuous in the summer.

Events after the last provisioning follow a seasonally dictated course. During the winter, darkness follows soon after the last provisioning; therefore, the ouakaris move to the sleeping sites soon after this feeding. In the summer, when several hours separate the last feeding and dusk, the ouakaris ascend to high "Rainforest" strata to forage. Foraging near the close of day seems most characteristic of adults; juveniles and larger infants often spend much of this period engaged in vigorous play.

The ouakari group gradually disperses over the course of the foraging period following the last feeding. Their travels during this period eventually lead to the flatwood pineland-hammock ecotone where preferred sleeping trees are located.

A number of specific sleeping sites that the ouakaris have used every night for as long as 5 years are found in the western fringe of the seminatural environment. The most consistently used sleeping places are certain multiple crotches in the crowns of tall specimens of *Pinus elliottii* where mothers sleep in contact with dependent young. The rest of the ouakari colony sleeps widely dispersed in a crescent-shaped strip of forest along the "Rainforest" periphery extending from the northwest corner to the south central border. Here, individual ouakaris sleep in high trees on single limbs away from the main trunk or on forks within secondary branches.

Ouakaris arrive at the sleeping trees about 30 minutes prior to sunset. Before assuming the characteristic sleeping posture in which all four limbs are flexed under the body, ouakaris sit, sprawl, autogroom, etc., while their responsiveness to external stimuli slowly wanes. Adult females with infants often groom and nurse them during this period.

The ouakaris typically assume the flexed sleeping posture, cease all movement, and apparently fall asleep by the time perception of color by a human observer ceases.

Once settled down for the night, the ouakaris do not normally move until dawn.

D.   Foraging and Drinking

Foraging, the activities involved in the acquisition and consumption of food growing in the "Rainforest," represents one of the principle activities of the ouakaris. The high rate of foraging engaged in by the red ouakaris in the seminatural environment provides one of the chief behavioral contrasts between life there and life in the typical zoo setting. The dispersal and social disengagement of ouakaris during foraging in the "Rainforest" provides them with periods when they are spared the constant and intense social stimulation dictated by the forced proximity of the typical caged environment.

Intensive foraging occurs during two daily peaks (06:00–08:30 and 17:00–19:00) with the latter one severely curtailed in the winter. Sporadic foraging occurs throughout the day and ouakaris occasionally overlook provisioned food in favor of items growing in the "Rainforest."

Ouakaris feed in a relatively active fashion, typically moving a few meters from one feeding place to another within 60 seconds. They employ a number of common foraging techniques. In one common method, the animal sits on a terminal branch, bends the extreme food-bearing end to the mouth, and either bites the food off without manipulating it or picks it off with the free hand. In another common technique, the ouakari breaks off a food-bearing terminal and carries it to a solid support where the food may be either directly bitten off or picked off with the hands. In these techniques, the ouakari typically consumes buds and leaves by directly biting them off the stem whereas fruits and lysiloma seed pods are manipulated prior to eating.

In addition to sitting, the ouakaris display a wide variety of postures in the acquisition of food. In order to accommodate to the location of desired food items, ouakaris will stand bipedally; suspend by the hindlimbs; suspend by a pair of ipsilateral limbs; and suspend by both hindlimbs in combination with a single forelimb.

The ouakaris have incorporated much of the vegetation available in the "Rainforest" into their diet. The products of native Florida trees constitute the bulk of the ouakaris' foraged diet. Exotic hardwoods and palms provide a modest supplement to the native items, while the abundant aroids serve more as ornamentation than as dietary items.

Table III provides a broad overview of the foraged plant foods of the red ouakari. The extent of use of a particular dietary item reflects the abundance of the species producing the item, the duration of the phenological change associated with the occurrence of the item, the location of the item with respect to favored travel routes and strata, preference for the item and, finally, the occurrence of competition for that item with other species, particularly *Saimiri sciureus*.

Lysiloma bahamensis, the principle food species of the red ouakaris and the most abundant large hardwood, provides the best illustration of the factors affecting the incorporation of items growing in the "Rainforest" into the diet of the red ouakaris. Ouakaris eat the flowers, leaves, green twigs, cambium, and especially the seeds of this leguminous tree. Indeed, the seeds of the Bahama lysiloma are the most important food item growing in the "Rainforest." The ouakaris intensively feed on them when the seed pods begin to appear in the spring. The persistence of lysiloma seeds amplifies their value as a food item and the ouakaris consume large numbers of them for several months until they deplete the supply. Thus, lysiloma seeds constitute a more important element of the diet than such highly preferred items as the relatively ephemeral fruits of *Sideroxylon foetidissimum* and *Simarouba glauca*. The occurrence of large num-

TABLE III

Plant Species Used for Food by the Red Uakaris in the Seminatural Environment [a]

Species		Leaves	Buds	Flowers	Fruits [b]	Seeds [b]
Aiphanes acanthophylla	E			R	R	
Ardisia escallonioides	N		M		M	
Ardisia solanacea	N	M	M	R	M	
Bursera simaruba	N	F	F	R	F	
Carica papaya	E			R	R	
Chiococca alba	N			R	F	
Chrysalidocarpus sp.	E			R	R	
Chrysophyllum oliviforme	N	R	R	R	R	
Cococus plumosa	E			R	R	
Coccoloba diversifolia	N	F	F	R	R	
Exiobotryia japonica	E			R	R	
Exothea paniculata	N	M	M	R	F	
Ficus aurea	N	M	F	R	M	
Hamelia patens	N			R		
Lysiloma bahamensis	N	M	M	M		F
Nectandra coriacea	N	M		R	F	
Philodendron oxycardium	E		R			
Philodendron spp.	E		R			
Pinus elliottii	N					R
Prunus myrtifolia	N	R		R	R	
Psidium gaujava	E			R	R	
Psychotria sulzneri	N	R	R	R	R	
Psychotria undata	N	M	F	M	F	
Ptychosperma alexanderi	E			R	R	
Quercus virginiana	N	F	F	R		M
Rapanea guianensis	N	R	M	M	F	
Roystonea elata	N			R	R	
Schefflera actinophylla	E	R	R			
Schinus terebinthefolius	E	R	R		R	
Sideroxylon foetidissimum	N	R	R		M	
Simarouba glauca	N	F	M	R	F	
Solanum verbascifolium	N				F	
Tabebuia heterophylla	E	R	R			
Tetrazygia bicolor	N	R	R	R	M	
Trema lamarckiana	N				R	
Trema micrantha	N				R	
Vitrus rotundafolia	N				M	

[a] Key to table: N, native species; E, exotic species; R, rare use of the item; M, moderate use of the item; F, frequent use of the item.

[b] Note: The ouakaris consume both the flesh and seeds of items listed under the heading "fruit" while they eat only the seeds of those items listed under the heading "seeds."

bers of lysiloma pods in the high strata where intensive foraging occurs and the minimal competition between *Cacajao* and *Saimiri* for this resource greatly increases the importance of lysiloma seeds as a staple food item.

Table III demonstrates that ouakaris hardly depend on any one species for food; a number of items are heavily utilized. The diet of the red ouakari in the "Rainforest" emphasizes immature, green fruit; buds; and new growth leaves. *Cacajao* also consume a wide variety of flowers when they become available for brief periods as well as the cambium of several hardwoods (notably *L. bahamensis* and *Bursera simaruba*). *Cacajao's* consumption of new growth reflects a positive preference for younger foliage over older or mature leaves. Consumption of green fruit, on the other hand, usually reflects competition with the large *Saimiri* colony for the limited supply of favored fruits.

The ouakaris' habit of breaking off branches in combination with a population density that surely exceeds natural conditions is changing the floral characteristics of the "Rainforest." For example, lysiloma trees in the "Rainforest" are stripped of their seed pods by October when trees elsewhere bear abundant seed. This may have a negative effect on the reproductive success of these trees. Furthermore, ouakaris damage lysilomas by girdling some of the larger limbs in their quest for cambium and by breaking branches during feeding and aggressive encounters. Nevertheless, the fast growing lysiloma seems capable of tolerating these depredations. *Bursera simaruba* and *Coccoloba diversifolia* seem much less robust to *Cacajao's* destructive habits. Excessive use of the young foliage of these species has damaged most of the specimens in the seminatural environment. Indeed, these species may prove unable to withstand the destructive foraging habits of *Cacajao* indefinitely. Thus, the destructive feeding habits of red ouakaris may ultimately reduce the number of forage species available in the seminatural environment.

Contrasts among individuals with respect to the use of strata relate to contrasts in the foraged diet that serve to reduce competition within the group for food. For example, young ouakaris frequently feed above the ground in the low scrub zone of the forest periphery. Products of *Psychotria undata, Solanum verbascifolium,* and *Chiococca alba* provide a substantial portion of the foraged diet of young ouakaris. The adults, although attracted to these food sources, rarely exploit them because of their reluctance to descend close to the ground where these species occur. When feeding on items near the ground, the adults frequently maintain some contact with mid-level forest zones (frequently by hindlimb suspension). When adults descend near the ground to feed, they typically stay only long enough to break off a food-laden branch, which they carry to a higher level for consumption. Juveniles and older infants feed near the ground in the same confident manner displayed when foraging in the high strata. They, therefore make efficient use of the resources available in the low scrub strata.

Ouakaris occasionally supplement their mainly vegetarian diet with small quantities of invertebrate prey. With the exception of small infants, ouakaris search for grubs by tearing the bark off dead limbs and extracting the grubs from the rotten wood. Adult females occasionally snatch at flying insects as they pass within reach and subsequently eat them. On only one occasion has an adult female ever been observed to actively hunt and stalk mobile insect prey. Young ouakaris consume the larvae of wrapper moths that live in *Psychotria undata* leaves as well as small terrestrial snails. These items are absent from the adult diet because they do not occur in the strata where the adults typically forage and travel.

Cacajao's diet accounts for much of this animal's water intake. However, ouakaris also obtain water in the "Rainforest" by licking it from leaves, by obtaining it from hollows in trees where rainwater collects, and finally, by descending to the ground to drink from the artificial stream. The latter source is currently the chief water source, but ouakaris did not exploit it until 1970.

The acquisition of this habit of descending to the ground to drink from the stream was observed in some detail for several members of the current uakari group. WF3 and WF4 acquired this habit by observing their offspring, CM4 and CF3, respectively, drinking from the stream. CF3 passed the habit on to her infant CM5 and WF3 passed it on to CF4 after she acquired it from CM4.

Ouakaris occasionally visit tree holes containing water even though they more frequently descend to the stream to drink. They obtain water from these holes by plunging the hand in and licking and sucking the water off the hand, or if the individual is small enough, by sticking the head in the hole and drinking.

E.   Provisionization

Monkeys in the "Rainforest" receive provisions in order to facilitate their observation by tourists, to discourage the monkeys from soliciting food from the visiting public, and to provide a stable food source that compensates for seasonal fluctuations in the abundance of naturally occurring foods.

Provisioning occurs at three feeding stations consisting of small framelike structures about 7 m apart. An attendant places food in aluminum pans and then raises them about 4 m via a chain and pulley system to the hungry monkeys at the feeding station. Food is offered in this fashion at 10:40, 11:45, 12:55, 13:55, 14:55, 15:55, and 16:55. The 10:40 provisioning consists of a vitamin-enriched mixture of whole wheat bread soaked in milk and eggs. The remaining feedings include: Wayne Monkey Diet (Allied Mills; Chicago, Illinois), bananas, apples, peanuts, grapes, sunflower seeds, oranges, carrots, celery, string beans, and other fruits in season. A short lecture by a guide involving a description of the "Rainforest" and its inhabitants precedes each feeding. This speech, in turn, is preceded by an announcement by the guide directing the public to the site of provisioning. Ouakaris over 18 months of age clearly associate this announcement with the start of provisioning. If sufficiently motivated, they will run toward the provisioning site before the attendant finishes the first few words of this announcement.

The behavior of the ouakaris at the feeding stations does not reflect a dominance hierarchy or competition for food. The order of feeding fails to follow a regular pattern associated with dominance in other situations. In addition, animals of widely varied status feed side by side without any agonistic exchanges.

The behavior of the ouakaris at the feeding stations seems to feature social disengagment. Vocal exchanges and other forms of social communication decline to low levels when the ouakaris partake of provisioned food. Adult ouakaris tend to avoid eye contact with conspecifics by turning away from them when feeding in close quarters at the provisioning site. However, this behavior does not reflect dominance and it is practiced by all ouakaris in spite of apparent social dominance. Adult males and juveniles feed in close quarters at the feeding stations far more frequently than adult females. The latter often carry their share of the food several meters above and away from feeding stations for consumption. However, since they do this without regard for the presence or absence of conspecifics, this habit reflects not an orientation to a dominance hierarchy but the previously noted general preference for higher strata. This practice does, of course, result in disperal during eating, and it reinforces social disengagement during feeding.

An outstanding aspect of the adaptation of *Cacajao* to the "Rainforest" was the rapidity with which even freshly caught specimens adjusted to provisioning. Indeed, provisioned food was often eaten by fresh-caught wild adults within the first hour of introduction into the "Rainforest." This represents a much greater ease in adjusting to the provisioning practice than that displayed by *Saimiri* and *Alouatta*. Perhaps the behavior patterns facilitating social disengagement during provisioning played a role in the rapid adaptation of the red ouakari to the practice of provisioning in the seminatural environment.

VI. Specific Characteristics of the Red Ouakari

A. Physical Description

This section will attempt to provide a broad qualitative account of the external characteristics of the red ouakari in order to provide colony managers and field workers with a general conception of the bearing of healthy animals. Available general descriptions of the external characteristics of these animals (Hill,

1960; Napier and Napier, 1967; and Sanderson, 1957) and descriptive information gathered in the seminatural environment do not always agree. Discrepancies probably arise from the necessity of basing descriptions of *Cacajao* on a few observations of preserved specimens of unknown age or on the short-term observation of a few caged specimens. Since age–sex variation significantly effects many of the external characteristics discussed here, it will be treated under each heading whenever appropriate.

The discussion of the red ouakari's external morphological characteristics will emphasize discrepancies between previous reports and current observations, information that supplements previous descriptive accounts, and previously overlooked aspects of this animal's external characteristics. [See Hill (1960) for a detailed account of other external features of *Cacajao*.] Management policies that weigh against the regular capture and restraint of specimens for tissue sampling, palpation, etc., in combination with the small number of specimens used to demonstrate developmental phenomena limit the precision of the information reported here.

1. General body Pelage

Chestnut-red general body pelage distinguishes *C. c. rubicundus* from *C. c. calvus*. *Cacajao c. rubicundus* does not possess a natal coat; the neonate is sparsely covered with hair of the adult color. The newborn ouakari presents a blackish cast, particularly on the extremities, that gradually disappears as the general body pelage becomes more dense during the first 3 months. During the first 2 years the immature ouakari's coat appears to maintain a relatively uniform length over the most densely haired regions of the body: the dorsal surface of the trunk, the distal surface of the limbs, and the tail. However, beginning in the third year, the hair of the arms, back, and shoulders gradually lengthens to form a mantle. This feature becomes especially prominent in adult males which also develop a bushy tail in adulthood. The full attainment of the male's mantle during either the sixth (CM2) or the seventh (CM3) year exaggerates the adult male's robust appearance and it serves to emphasize the contrast between adult males and other ouakaris.

2. Alopecia of the Scalp

The development of pattern baldness in both sexes of *C. calvus* is one of the external characteristics distinguishing this species from *C. melanocephalus*.

Uno *et al.* (1969) correlated the histological properties and enzyme activities of the hair follicles with the degree of baldness in one adult female and two juvenile female specimens of *C. rubicundus* (= *C. c. rubicundus*). Since they did not report exact ages for their subjects, we can not directly relate this report of the chronology of development of baldness to their study. Thus, the present descrip-

Fig. 2. A series of photographs illustrating the development of pattern baldness in CF3. The juvenile CF3 at 18 months displays a full scalp (a) that is lost by the time she is 34 months of age (b). At 4 years of age, CF3 carries her infant 2 in the typical dorsal carriage position. CF3's baldness has progressed very little beyond the condition seen at 34 months. Her infant 2's scalp has not yet reached its peak apparent thickness (c). (Photos by Roy Fontaine.)

Fig. 2c.

tion serves primarily to provide field workers and colony managers with material for estimating·age in red ouakaris.

The loss of hair in the frontal and parietal regions of the scalp occurs over much of the red ouakari's life. The hair that the red ouakari loses tends to be shorter than the general body pelage. It also contrasts in color being primarily gray but becoming rufous laterally and posteriorly where it grades into the general body pelage.

The neonate is born with a gray scalp composed of extremely short hair that is well differentiated from the forehead. As the young infant's general body pelage fills out during the first 2 months, it grades into the gray hair of the scalp causing an increase in apparent hair thickness.

This increase in apparent thickness is reinforced by continued growth of the gray hair of the crown. This relatively thick hair persists during the first 2 years of life (Fig. 2a,c).

A major change occurs in the third year with the result that the 3-year-old scalp approaches the adult condition (Fig. 2b) while most of the remaining characteristics of the animal retain juvenile form. Once the ouakari is 3 or 4 years old, its degree of baldness apparently stabilizes and the animal seems to maintain the same degree of baldness until it is about 10 to 15 years old. Then, the process accelerates so that one can distinguish older and younger adults (Fig. 3a,b).

Fig. 3. Two pictures illustrating the continued development of pattern baldness in WF1. WF1 at approximately 11 years of age (a: photo by Patricia Snyder) has more hair in the frontal and parietal regions than she does at 14 years of age (b: photo by Roy Fontaine).

3. Facial Coloration

Descriptions of *C. calvus* typically cite this monkey's scarlet or pink face and forehead as a critical feature separating this species from *C. melanocephalus*. Some debate exists concerning the degree of saturation of red facial coloration in this species. Bates (1863) observed wild specimens and described them as scarlet-faced. Sanderson (1957), on the other hand, observed captive individuals and described them as pink-faced. Hill (1965), noting that most captive specimens are pink-faced, demonstrated that exposure to sunlight helped to maintain a bright scarlet facial coloration in captive adult ouakaris.

In the seminatural environment, the facial coloration of red ouakaris over 3 years of age has always been a bright scarlet hue while younger animals have always been some lighter color as Bates (1863) noted. Thus, exposure to sunlight can only maintain red facial coloration in the older animals capable of sustaining it in the first place.

During the first 2 months, the face of the infant is a very unsaturated pink in the region of facial prognathism; the forehead is basically gray with only the slightest trace of pink showing through the darker color. As the infant matures, the grayish cast of the face is gradually replaced by a pink color. Nevertheless, the pink face remains quite unsaturated throughout the first year and the grayish cast persists through the first 6 months.

The young ouakari's face darkens throughout the second and third year. The major change, from pink to scarlet, however, occurs during the fourth year. In CF3 and CM3, the two individual's whose development has been most thoroughly studied, this color change followed a few months after the establishment of adult pattern baldness. The development of scarlet facial coloration in CF3 occurred in conjunction with her first pregnancy early in her fourth year. The change from pink to scarlet facial coloration was accompanied by major behavioral changes (such as an increase in the use of adult locomotor patterns and a great decline in the frequency of play). These events occurred so rapidly that caretakers reported CF3 missing when she suddenly acquired these adult characteristics. Once a red ouakari acquires a scarlet face, changes in facial coloration occur very slowly throughout adult life. In some individuals (WF1, WF4), facial coloration changed from scarlet to crimson with increasing age.

Ouakaris at Monkey Jungle have never displayed the sudden facial flush from pink to red that is commonly cited as an emotional response in this species (Sanderson, 1957; Hill, 1960; Forbes, 1896). The reverse case, however, the changing of facial color from pink to white occurred when one individual went into psychic shock during handling necessitated by medical treatment. This observation, in combination with the observation of a desaturation of facial color at death in CM4, indicates the importance of circulation in the maintenance of facial coloration. However, pigmentation may also play a role in this process,

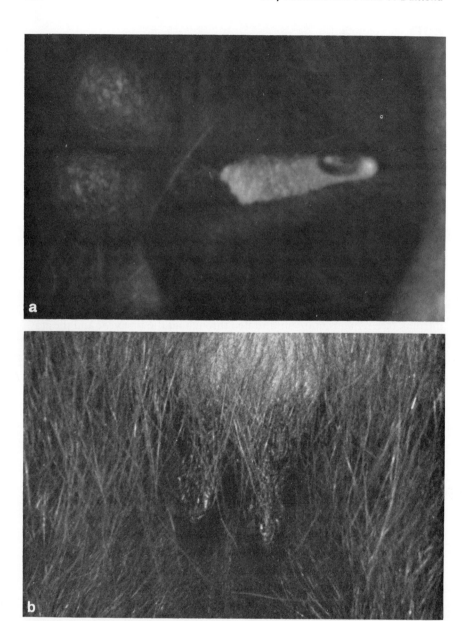

Fig. 4. Ventral view of the external genitalia of an adult male, CM3, at 9 years of age (a: photo by Roy Fontaine). A ventral view of the labia of an adult female, CF3, at 5 years of age (b: photo by Roy Fontaine). CF3 at 18 months rolls on her back in play exposing the undifferentiated genitalia of the juvenile (c: photo by Patricia Snyder).

Fig. 4c.

since CM4 retained a pink facial hue for several weeks after his death. Bates (1963) reported similar events for a wild-caught specimen of *C. c. calvus.*

4.   External Genitalia

The present section supplements Hill's (1960) previous account of *Cacajao's* external genitalia and it features age-dependent contrasts in the gross appearances of the red ouakari's external genitalia.

Examination of the external genitalia of *Cacajao* allows easy sexual identification of adults while providing uncertain clues as to the sexual identity of younger animals. The most probable basis for this condition is the occurrence in the male of a developmental stage of several years duration in which the external genitalia and other characteristics of the male approach the adult female in form. Female red ouakaris lack this developmental stage which begins in the male's fourth year. Their development proceeds directly from the juvenile condition to adulthood in the fourth year. Thus, the external genitalia of infants and juveniles appear similar in both sexes because these young ouakaris are in the process of developing the apparently similar external genitalia of adult females and subadult males.

In the adult male the testes are fully descended. The scrotum is a black, pendulous, flaccid, asymmetrically bilobed sac (Fig. 4a). It is entirely hairless as described by Kinzey (1971).

The penis of the adult male appears large in proportion to other perineal structures. The prepuce encloses the proximal half of the body while tiny spicules cover the exposed distal portion of the body. The glans is poorly differentiated from the body and, as mentioned by Hill (1960), the penis is laterally compressed along its length. With the exception of a ring of black pigmentation around the external urethral meatus, the distal portion of the penis is entirely unpigmented.

The external genitalia of the subadult male red ouakari presents a marked contrast with the adult male condition. The subadult male scrotum consists of a pair of rugose, black, hairless, symmetrical, parapenial scrotal alae that superficially resemble the adult female labia. Roughly triangular, and broadly attached to the pubic region, each pendulous scrotal lobe is distinct from its opposite member, but connected by a bridge of apparently similar tissue postpenially. The testes appear undescended at this level of development. The prepuce covers the entire length of the subadult's glans penis which only becomes visible during erection.

Hill (1960) provides a detailed account of the external genitalia of *C. rubicundus ucayali* (= *C. c. rubicundus*). In comparison with males observed in the "Rainforest," his description conforms to a transitional stage between the adult and the subadult.

The sexual identification of juvenile and infant ouakaris on the basis of field characteristics is problematical. Male and female external genitalia do not have obvious differentiating characteristics during the first 2 years of life (Fig. 4c). In these animals, the external genitalia consist of three hillocks in the pubic region that are more darkly pigmented than the surrounding regions. During vigorous locomotor play in the male, erection of the penis occurs allowing the unpigmented penis to protrude beyond the prepetual anlage. The absence of visible erection characterizes the young female.

The external genitalia of the adult female resemble those of the subadult male. The glans clitoridis and preputium clitoridis of adult females in the seminatural environment form a dusky hillock anterior to the rima pudendi. To each side of the rima pudendi lie black, rugose, pendulous labia (Fig. 4b). These structures superficially resemble the scrotum of the subadult male; however, their relatively uniform lateral thickness, their relatively gracile character, and their tendency to form an acute angle anteriorly allow discrimination of the adult female labia and the subadult male scrotum. A constriction of the labia about halfway along their ventral extent which may correspond to a demarcation between the labia minora and the labia majora provides another characteristic facilitating the identification of adult females.

Hill's (1960) description of the female genitalia of a specimen of *C. rubicundus* (= *C. c. rubicundus*) probably represents a female of about 3 years of age that was in the process of acquiring adult characteristics. The pendulous character of the labia was not developed in this specimen. Hill's (1960) account suggests that the pendulous labia of the adult female ouakari may represent a fusion of the labia minora and the labia majora with the portion homologous with the labia major extending most ventrally.

5.   Sternal Glands

Red ouakaris possess a centrally located, darkly pigmented rugose glandular field on the anterior pectoral region overlying the body of the sternum. These sternal glands which occur in both sexes became apparent in the third year in CM5 and CF3.

The function of these glands remains unclear. However, the frequent practice of adult males of washing this general region with urine, aromatic substances (citrus fruits), and irritants (*Begonia* sp., *Rhaphidophora aurea*) in the context of agonistic display as well as the occasional rubbing of this region on tree limbs by adult females suggests that these glands may be involved in olfactory signaling. Perhaps these structures are homologous with the sternal glands of *Ateles* (Hill, 1962) and *Callicebus* (Moynihan, 1966).

B.   Sexual Dimorphism

Available quantitative studies of sexual dimorphism in *C. c. rubicundus* deal primarily with osteometrics. Napier and Napier (1967) summarized available metrical data (Thomas, 1928; Lonnberg, 1938) with the statement that the ratio of female to male head and body length and the ratio of female to male tail length are, respectively, 90 and 92%.

We have not undertaken metrical studies of red ouakaris living in the "Rainforest." Instead, the present description of sexual dimorphism is based on the superficial examination of living specimens of *C. c. rubicundus*.

Sexual dimorphism becomes conspicuous with adulthood. Since the external characteristics of subadult males generally resemble those of adult females, sexually dimorphic characteristics in the adult male correspond to developments away from the adult female-subadult male condition.

Adult males appear large and robust. Their hairy mantle and bushy tail exaggerates their size making them appear about 20% larger than adult females and subadult males. Conspicuous contrasts between the locomotor behavior of adult males and other age—sex classes suggest that the apparent dimorphic contrasts in body build may have functional significance. The locomotor behavior of adult males involves much less agile movement than that of the adult females and subadult males. The locomotion of the adult male ouakari strongly emphasizes quadrupedal pronograde modes of function. In contrast with other age-sex classes, adult males in the "Rainforest" brachiate and perform suspensory postures infrequently.

Perhaps the most striking contrast between the adult male ouakari and the subadult male is the presence of paired muscular masses overlying the frontal and parietal bones of the former (Fig. 5a,b). These structures which probably correspond to the temporalis muscles lend a square appearance to the adult male's face. In CM3, these muscular structures developed in the seventh year in conjunction with the rapid development of typical adult male external genitalia, pelage characteristics, body size, and general body proportions.

C.   The Use of External Characteristics to Assess Age—Sex Class

Table IV summarizes the preceding information in the form of a field guide to the age—sex classification of *C. c. rubicundus.*

The classification of individuals as adult females, adult males, and subadult males applies previously reported descriptions of external morphological characteristics. Additional, largely behavioral, data are needed to facilitate the identification of the younger animals who are classified under the following headings: infant 1, infant 2, juvenile 1, and juvenile 2.

A grayish-faced ouakari who remains in continual contact with his mother and who rides laterally in the flexure of his mother's thigh is classified as an infant 1. The sparsely haired infant 1's lateral carriage may have a thermostatic function.

An infant 2 is an older ouakari that is variably dependent on maternal transport. At the beginning of this period, the infant 2 is totally dependent on his mother for travel; but at the end of this period the infant 2 has achieved full locomotor independence. The mother typically carries the infant 2 about on her back. Lateral carriage in the flexure of her thigh greatly declines in frequency in the third and fourth month as the infant 2's general body pelage fills out.

A relatively unsaturated pink face characterizes the juvenile 1. A ouakari enters this period partially dependent upon maternal nourishment, but it leaves this period fully weaned. A high rate of motor development leading to a complex locomotor repertoire that often finds expression in play marks this period.

The juvenile 2 red ouakari has achieved basic independence from its mother. However, the presence of the maternal bond is revealed by the mother's high rate of grooming of her juvenile 2 and her display of protective responses when her juvenile 2 becomes embroiled in an intraspecific squabble or is faced with an external threat.

The reader should regard the age—sex categories outlined in this report as tentative. Observations on a larger sample are needed in order to assess the variability of diagnostic characteristics. The extent to which wild populations adhere to the pattern of development observed in the seminatural environment and other captive situations remains unknown.

Fig. 5. The subadult CM3 at 5 years and 7 months lacks the paired muscular masses overlying the frontal and parietal bone (a: photo by Patricia Snyder) that distinguish the same animal as an adult of 9 years (b: photo by Roy Fontaine).

TABLE IV

A Field Guide to the Age–Sex Classification of *C.c. rubicundus*

	Infant 1	Infant 2	Juvenile 1	Juvenile 2	Subadult male	Adult male	Adult female
Age	1–2 months	3–11 months	12–24 months	25–36 months	3–5 or 6 years	5 or 6 years	3 years
Apparent % of adult female size	10–25	25–60	60–85	85–100	100	120	100
General body pelage	Sparse	Thick, uniform length	Thick, uniform length	Developing mantle	Mantled	Exaggerated mantle, bushy tail	Mantled
Facial coloration	Gray cast, pink trace	Unsaturated pink	Unsaturated pink	Saturated pink	Scarlet	Scarlet	Scarlet
Alopecia of scalp	Very short, thick hair	Hairy scalp	Hairy scalp	Stage of transition	Bald	Bald	Bald
Frontal-parietal muscle mass	Absent	Absent	Absent	Absent	Absent	Present	Absent
Sternal glands	Absent	Absent	Absent	Faint gray	Present black	Present black	Present black
Maternal dependence	Helpless	Dependent	Dependent to independent	Independent	Independent	Independent	Independent
Locomotor behavior	Lateral carriage	Dorsal carriage	Autonomous complexly variable	Autonomous complexly variable	Variable, nimble	Emphasizes quadrupedalism	Variable, nimble
External genitalia	Undifferentiated	Undifferentiated	Undifferentiated	Transition stage	Labiaform scrotum	Pendulous scrotum	Pendulous labia

VII. Reproductive Patterns

A. Sexual Behavior

Observations of the sexual behavior of *Cacajao* at Monkey Jungle have been largely confined to sexual relationships between juvenile males and adult females. The adult male's involvement in defensive responses to human intruders interferes with his conduct of sexual relations in the presence of observers. Enough observations of adult male sexual behavior have been collected, however, to allow description of the major elements of *Cacajao*'s sexual behavior. The recognition of sexual behavior in *Cacajao* is facilitated by obvious contrasts between sexual patterns and other aspects of the social behavior of ouakaris. Most behavior patterns serving a sexual function fail to occur in other contexts.

Behavioral observations suggest that chemical stimuli transmit information concerning female sexual attractiveness. Sniffing of the female genitalia invariably precedes copulation in both adults and juveniles. Less frequently, licking of the female genitalia by the male accompanies this precoital sniffing. Nonreceptive females indicate intolerance of these sexual preliminaries by escape and/or aggressive behavior.

Visual stimuli play a secondary role in the initiation of sexual activity in *Cacajao*. Receptive females at Monkey Jungle present their external genitalia to the male by standing quadrupedally with the tail raised and with the hindquarters oriented toward the male. Females at the Oklahoma City Zoo vary this pattern slightly; they lower the forequarters during genital presentations. Although genital presentation signals female receptivity to genital sniffing and/or copulation, females also receive these acts without actively soliciting them via genital display.

Behavioral observation of ouakaris in the seminatural environment yields only an unrefined conception of female sexual cyclicity. Adult female ouakaris seem to have a lunar cycle with a midcycle period of about a week during which they are sexually attractive to the males. A shorter 3- or 4-day period of sexual receptivity occurs within this 1-week period. During the remaining days of female sexual attractiveness, the females either reject male sexual advances or escape from them.

Genital sniffing by the male alternated with rapid movement of the female away from the male in the most thoroughly studied instances of adult female–juvenile male sexual interaction in the "Rainforest." These activities were interpreted as an attempt by the female to lead the juvenile male away from human observers rather than as an attempt to avoid conspecifics, since copulations sometimes occurred within a few meters of other ouakaris including the adult male.

Copulations in the "Rainforest" have always occurred on solid horizontal supports that the ouakaris seem to seek out for this purpose (Fig. 6). The adult female displays some postural variability during copulation. She may lie prone with all four limbs flexed under her body or she may allow her limbs to dangle freely beneath her. Less frequently, the female lowers her hindquarters and maintains them in flexion during copulation while raising the forequarters. During copulation, the male sits behind the female with the thighs abducted and flexed. In juvenile copulations, the males were observed to flex their hind-limbs at the knees allowing the feet to maintain a firm grip on the small of the female's back (Fig. 6). The latter aspect of juvenile copulation may have been overlooked in adult male copulation. However, since this feature of juvenile copulation serves to "lock" the male into a secure position of dorsal-ventral contact with the female, it may prove to be a typical aspect of copulation in *Cacajao*.

Copulation in *Cacajao* requires several minutes. The typical copulation con-sists of a series of intromissions each accompanied by multiple pelvic thrusts by the male partner occasionally aided by complementary movements by the female partner. Periods of withdrawal alternate with periods of intromission. During these withdrawal periods behavior is quite variable ranging from general body contact to rapid travel to another location to continue copula-tion.

Fig. 6. Copulation involving a juvenile 1 male (CM4) seated behind his partner an adult female (WF4). (Photo by Roy Fontaine.)

Postcoital behavior in the adult male usually includes smelling of the female genitalia followed by a bout of mutual social grooming. The juvenile male, in contrast, usually leaves the female after a few cursory genital sniffs.

B. The Annual Reproductive Cycle

Reproductive data collected on animals removed from their natural habitat must be regarded with caution. An unequivocal account of the red ouakari's annual reproductive cycle awaits the collection of comparative data derived from the observation of wild populations.

In their review of the annual reproductive cycle in monkeys and apes, Lancaster and Lee (1965) stressed the importance of two kinds of data in the evaluation of reproductive cyclicity: the seasonal distribution of copulations and the seasonal distribution of births.

Table V summarizes the seasonal distribution of births in the three colonies that report the occurrence of viable captive births. These data suggest that colony managers in the North Temperate Zone should expect captive red ouakari colonies to display a birth season of about 6 months duration extending from May to October. Since widely different environmental conditions confront the ouakaris living in each institution listed in Table V, it is not possible to identify the external stimuli triggering the regular periodicity of the birth season.

The collection of data suggesting the seasonal occurrence of sexual behavior in *Cacajao* is aided by the absence in captive ouakaris of behavior serving both sexual and general communicative functions. In comparison with *Macaca*, the assessment of sexual behavior is easy since *Cacajao* does not include patterns such as genital presenting, pelvic thrusting, etc. within its repertoire of non-sexual social signals. On the other hand, the ouakari's long mantled fur hinders the observation of intromission and ejaculation and the detection of copulation plugs.

Table V includes a frequency count of the copulations of CM4 and CM5 with parous adult females and a similar count of the genital sniffing of all male ouakaris living in the "Rainforest" during a recent 1 year period. The lack of quantitative data describing the incidence of copulation by CM3 during the indicated period reflects this animal's involvement in the threatening of human intruders, an activity that interferes with copulation. During this period, CM3 displayed interest in the females by sniffing and licking their genitalia; at other times, he has copulated with them. For example, CM3 engaged in copulations in December, 1972, January, 1973, and February, 1973.

In the history of the Monkey Jungle red ouakari colony other adult males have engaged in copulation during December (CM3), March (CM1, WM7), and April (WM7). The combined data describing sexual behavior in the "Rainforest" suggests the occurrence of a mating season there that minimally overlaps with the birth season.

TABLE V

The Annual Reproductive Cycle

	January	February	March	April	May	June	July	August	September	October	November	December
Birth occurrence												
Monkey Jungle												
Goulds, Florida	0	0	0	0	2	1	3	2	0	1	0	0
Cheyenne Mountain Zoo												
Colorado Springs, Colorado	0	0	0	0	1	0	0	0	0	1	0	0
Frankfurt Zoo												
West Germany	0	0	0	0	0	1	1	2	1	0	0	0
Sexual behavior												
Frequency of observed copulation												
Monkey Jungle												
July 1974–1975	12	4	3	1	1	0	0	0	0	37	4	13
Gladys Porter												
Brownsville, Texas 1973–1974	4	1	5	2	0	2	4	2	1	1	1	0
Frequency of genital sniffing												
Monkey Jungle	12	11	4	11	0	0	0	0	0	20	1	4

The occurrence of seasonal sexuality in other captive situations is more difficult to evaluate because of the absence of comparable quantitative data. The Cheyenne Mt. Zoo and the Oklahoma City Zoo both report seasonal sexuality but quantitative supporting data are not available. The Gladys Porter Zoo, on the other hand, reports the sporadic occurrence of copulation throughout the year.

The significance of the failure of ouakaris at the Gladys Porter Zoo to display seasonal incidence of copulation remains unclear. It is of interest to note, however, that all females at this institution that have been described as participating in copulation lived in isolation for a substantial portion of their lives prior to their introduction into the Gladys Porter Zoo collection. Thus, the social environment may play a role in the maintenance and establishment of seasonal sexuality in *Cacajao*. Perhaps the absence of seasonal sexuality relates to the failure of the Gladys Porter Zoo ouakaris to enjoy any reproductive success.

C. Age of Reproductive Capability and Maturity

Only the most limited data are available describing the age at which red-ouakaris become capable of species-typical sexual behavior, the age at which females first become capable of giving birth, and the age at which males become capable of impregnating.

Male ouakaris apparently copulate and display sexual interest in receptive females long before they are capable of fertilizing them. Copulations that appeared to involve intromission were first observed in CM4 at the age of 18 months and CM5 at the age of 17 months. CM3, an individual who has sired several offspring, was not observed to engage in copulation at a comparable age because the ouakaris were not under careful study during CM3's early development. Nevertheless, CM3 displayed sexual interest in and received genital presents from colony females as a 4-year-old subadult, at least 30 months prior to proved breeding success. An individual born at the Frankfurt Zoo on July 25, 1970 received genital presents and copulated with adult females in November, 1973, and perhaps earlier, without fathering offspring.

Two male ouakaris whose development from birth to adulthood has been observed in the "Rainforest" became proved breeders at about 6 years of age. The combined observations suggest that reproductive capability in the male comes with the development of the full complement of adult male characteristics listed in Table IV and that the performance of species-typical sexual behavior in the male does not always reflect reproductive maturity.

CF3 is the only female whose development from birth to adulthood has been observed in the "Rainforest." CF3 gave birth to her first infant at the age of 3 years and 7 months. In contrast with the frequent observation of sexual activity in the juvenile male, the performance of sexual behavior by CF3 was not noted prior to adulthood. CF3 became preganant a few months after acquiring the typical external characteristics of the adult female ouakari (Table IV). Thus, given the limitations imposed by $N = 1$, the information gathered in the seminatural environment suggests that the female ouakari, in contrast with the male, comes into breeding condition at the relatively early age of 3 years.

D. Spacing of Pregnancies

Knowledge of the expected interbirth interval aids the planning and evaluation of captive propagation efforts.

The data collected from institutions that maintain multiparous females indicate that breeding females typically display an interbirth interval of about 2 years duration. For example, the five interbirth intervals of the Cheyenne Mt. Zoo and the Frankfurt Zoo ranged from 23 to 28 months. The occurrence of two lengthy periods (July, 1966—January, 1968; October, 1969—October, 1972) during which the red ouakari colony at Monkey Jungle lacked an adult male has been a barrier to the collection of data describing the interbirth interval in this group. However, when proved breeding males have been continuously present over the duration of interbirth intervals, these intervals have usually been of about 2 years duration. In the single exceptional case, where the interbirth interval was substantially shorter than 2 years (15 months), the atypically abrupt weaning of the displaced infant seemed to initiate a general decline in health that eventually proved fatal at 23 months.

VIII.   Some Aspects of Social Behavior in *Cacajao c. rubicundus*

The present discussion of the social behavior of red ouakaris will focus on those aspects of the social milieu that are assumed to reflect conditions favorable for breeding.

A. Allogrooming

A high rate of allogrooming distinguishes *Cacajao* from the other primates currently sharing the "Rainforest" with them: *Alouatta seniculus, Saguinus fuscicollis,* and *Saimiri sciureus.*

The frequent practice of allogrooming in a red ouakari group attests to the presence of the harmonious social relations that foster breeding. Institutions that report little or no allogrooming also report the absence of breeding and high levels of intragroup aggression. Indeed, a cohesive group of ouakaris can be conceived of as a network of dyadic grooming relationships. In this report, the term "grooming bout" refers to a sequence of grooming motions that is separated from similar motions by a time interval greater than that separating any of the grooming motions within the bout. The term "grooming relationship" refers to the long-term summation of grooming bouts within particular dyads.

Certain aspects of the motor behavior involved in grooming represent generalizations of the ouakari's normal pattern of manipulation while other aspects emerge as unique to the context of grooming. The strong pronation of the forearms, for example, suggests the derivation of grooming from the act of fine prehension as described by Bishop (1964). On the other hand, the tendency of both hands to participate in grooming, the tendency of each hand to perform a different grooming motion, and the tendency of the grooming activities of each hand to alternate during the course of the typical grooming event represent manual activities seldom seen outside of the context of grooming.

The manual activities associated with grooming fall into four broad categories: holding, resting, parting, and picking. Holding occurs during the grooming of an appendage such as the tail. It differs in no essential way from ordinary sustained manipulation. Since ouakaris often limit manual grooming activity to a single hand, a state of inactivity often characterizes the resting hand which remains ready to participate in grooming. Frequent changes from active manual activity to resting occur over the course of the typical grooming bout. The continued pronation of the resting forearm facilitates these role changes. A common method of parting the hair in grooming involves a raking motion of the abducted digits. Ouakaris direct long series of raking motions to a given portion of the pelage in a systematic fashion ensuring the thorough examination of that region of the groomee's coat. Another method of parting the hair in grooming bouts resembles the prehension of intermediate size objects (see Bishop, 1964). In this situation, the groomer flexes all five digits with the result that the hair of the groomee becomes firmly flexed between the distal phalanges and the proximal palmar surface. The manual and oral removal of objects from the fur of the groomee punctuates the course of the typical grooming bout. In most cases, the groomer removes objects by performing picking motions resembling ordinary fine prehension in which the groomer opposes d1 and the enlarged thenar pad. Less frequently, the groomer removes particles by grasping them between d2 and d3.

Dyadic allogrooming bouts vary along several dimensions. The chief variables include the duration, the rate of occurrence, and the reciprocity of grooming bouts.

The dyadic character of most grooming bouts in *Cacajao* provides the background for their reciprocity. Stable triangular grooming bouts typically consist of two adult females grooming a nonreciprocating adult male or infant 2. Triangular grooming events among adults are volatile; they usually break up when one of the triad leaves to make the bout dyadic.

The role of groomer and groomee alternates throughout the course of a reciprocal grooming bout. In the "Rainforest," the reciprocal grooming of adult females appears to be the chief focus of positive social interactions facilitating group cohesion.

Ouakaris mark changes in grooming roles by a variety of cues. Thus, in one common pattern, the groomer stops grooming and solicits allogrooming from the partner by means of a display in which the solicitor sits with the vertebral column erect while looking up and away from the grooming partner. Alternately, a ouakari might invite grooming by lying prone with the limbs flexed underneath the body. Vocalization frequently accompanies these changes in grooming roles. Solicitations of allogrooming do not always produce the desired effect. If the partner receiving the grooming invitation either fails to initiate grooming or also displays a posture inviting grooming, the grooming bout terminates and the two partners move away from each other. Reciprocal allogrooming bouts usually end with at least one partner either vocalizing or displaying postures to solicit further grooming. The groomee rarely terminates a grooming bout.

Dyadic grooming bouts between adult females seem inherently unstable. Individual reciprocal grooming bouts always seem on the verge of breaking up via the failure of a groomee to accept an invitation to continue the grooming bout in the role of the groomer. This is true even when the individuals involved include a mother and her adult offspring or when the grooming partners have maintained a stable grooming relationship of at least 5 years duration. Since reciprocal allogrooming in the red ouakari demands a high level of responsiveness to the many vocal and postural cues emitted and exchanged over the course of a typical bout, the occurrence of stable reciprocal grooming relationships among ouakaris indicates the presence of a well-integrated social group whose members have developed considerable sensitivity to the individual meaning of emitted social signals. An adult female with a history as a pet (WF1) failed to develop reciprocal allogrooming relationships with other adult females after over 12 years in the "Rainforest." WF1 failed to respond to social signals emitted by other group members. She ignored postural cues soliciting active grooming and she often failed to answer social signals emitted by other group members. Her failure to attend to emitted social signals may have contributed to her lifelong nulliparity.

Reciprocity in allogrooming bouts varies with the age—sex class of the participants. As previously noted, adult females in stable grooming relationships usually reciprocate allogrooming. Adult males, in contrast, rarely reciprocate. When they do, their active grooming consists of a few cursory manual grooming motions followed by the immediate assumption of a posture soliciting grooming.

However, unlike the situation observed among adult females, the failure of an adult male to continue a grooming bout in the active mode does not lead to the termination of the grooming bout or to the weakening of the dyadic grooming relationship.

Reciprocity in allogrooming evolves over the course of the individual's life. During infancy, all grooming is undirectional. The infant 1 receives frequent and exclusive grooming from its mother. The infant 2 is groomed by other adult females as well. In order to counter the mother's tendency to carry her infant 2 away in response to the presence of conspecifics, adult females usually groom the mother prior to turning attention to her infant 2. Active participation in allogrooming increases in frequency throughout the juvenile 2 period in both sexes. In the male, reciprocity in allogrooming continues over the course of the subadult developmental period. However, when the male passes to adulthood, his tendency to reciprocate allogrooming declines such that the adult male almost always displays the passive mode of participation in grooming bouts.

The rate of occurrence of specific dyadic grooming bouts provides a further means of assessing the social structure of a red ouakari group. The most common grooming bout is the grooming of an infant by its mother. Mothers groom their juvenile offspring less frequently than they groom their infants but more frequently than they groom other adults. A mother does not display a greater tendency to groom adult or subadult offspring than she displays toward unrelated adult members of the group.

Observations in the seminatural environment indicate that the most common dyadic grooming bouts, the maternal grooming of dependent offspring and the unidirectional grooming of adult males by adult females, are nonreciprocal. On the other hand, the rate of occurrence of dyadic grooming bouts among juvenile 2's, subadult males, and adult females seems correlated with its reciprocity. The consistent failure of individuals in the latter age–sex classes to reciprocate in grooming bouts leads to a weakening of grooming relationships that in its extreme form produces peripheralization of the nonreciprocating ouakari.

The duration of individual grooming bouts, the third major axis of variation, appears positively correlated with their rate of occurrence. The relationship between duration and reciprocity, on the other hand, seems more complex. Reciprocal grooming bouts among adult females are usually of intermediate duration (5–10 minutes) while nonreciprocating grooming bouts among adult females typically break up after less than 1 minute. Unidirectional adult female–adult male grooming bouts tend to be lengthy (20 minutes) while the duration of the unidirectional maternal grooming of infants and juveniles is highly variable. Thus, although a mother may groom her dependent offspring for brief periods during nursing, pauses in play, etc., she may also groom her dependent offspring for an extended time period during those parts of the daily cycle characterized by low levels of general activity.

B.   Spatial Relations

The tendency of red ouakaris to abstain from general body contact with conspecifics and to maintain an interindividual distance of several meters during foraging, resting, autogrooming, etc., is one of the most conspicuous features of the social behavior of these animals. Grooming is the chief instance when one observes ouakaris in contact with each other. Yet, the groomer frequently marks brief pauses in continuing grooming bouts by moving a few meters away from the groomee. Although adult females with associated infants and juveniles often rest in contact with them, this tendency begins to decline in infant 2's who often rest a meter or so from their mothers.

During most of the day, the ouakaris disperse over about 0.4 hectares where they maintain an interindividual distance of about 3 to 15 meters for such activities as coordinated travel, foraging, and resting. The maintenance of this interindividual distance involves minimal aggression, although noisy squabbles may break out over possible violations of individual space. These squabbles usually involve juveniles who sometimes attempt to establish general bodily contact with adult females other than their mothers and who receive some form of mild punishment for their efforts. Adult males, in contrast, frequently tolerate juvenile violations of individual space (Fig. 7).

Fig. 7. CM3 tolerates the sudden establishment of body contact with him by two infant 2's (CM4, CM5). Adult males are the only adults other than the mother that tolerate general body contact with infants. (Photo by Roy Fontaine.)

The tendency of ouakaris to maintain an interindividual distance of several meters did not interfere with their adaptation to the forced proximity and contact occasioned by provisioning. Although they occasionally squabble (never seriously) in anticipation of the presentation of food at the feeding stations, the probability of aggression declines with the availability of food. Apparently, each ouakari concentrates on grabbing handfuls of food while ceasing to respond to the spatial location of conspecifics. This social disengagement during feeding is a conspicuous aspect of the behavior of red ouakaris in the "Rainforest." Although ouakaris often carry food away from the feeding stations and consume it several meters from conspecifics, this practice does not reflect the consistent use of coercive behaviors in maintaining interindividual distances because any individual can feed within reach of any other individual at the feeding station without provoking an agonistic interaction.

Although mutual avoidance seems to maintain interindividual spacing among red ouakaris, mechanisms facilitating proximity also operate because ouakaris neither randomly disperse about the "Rainforest" nor do they maximize interindividual distance. Antiphonal vocalization probably provides the chief means by which individuals maintain contact with the group at acceptable interindividual distances. In addition, the vocalizations associated with inter- and intraspecific fighting lead to the aggregation of the colony with references to the disturbance. In these stressful situations, juveniles and infant 2's often succeed in establishing temporary contact with normally intolerant conspecifics.

C. Mother—Offspring Interactions

An understanding of the general character of interactions between a mother and her offspring provides background information for the sound management of *Cacajao* in captivity.

All births in the seminatural environment have occurred at night. A mother with a neonate tends to maintain an increased interindividual distance during her infant's first month. The mother carries her thinly haired infant 1 ventrolaterally in the flexure of the hip. During the first month, the infant 1 remains quite inactive, but his rate of activity increases throughout his second month. The mother ouakari typically displays a passive attitude toward her infant 1; she allows it to move about her body unhindered and she permits ad libitum assess to the nipple. Although the mother is quick to return the infant 1 to the customary ventrolateral carrying position in the mother's hip flexure when he signals distress by vocalization, the infant seems to assume most of the initiative in maintaining appropriate contact with the mother.

Two correlated features of mother—infant 2 interactions during the third and fourth months stand out: an increase in the amount of time the infant spends away from the mother and a replacement of ventrolateral by dorsal carriage.

During the latter part of the infant 2 period, independent locomotion gradually replaces dorsal carriage. Mothers typically encourage locomotor independence by moving away from their infant's approach. This causes the infant to follow with the result that the mother can effectively lead her infant a few hundred meters in this fashion.

During the infant 2 period, the frequency of the mother's social interactions with other members of the group returns to postnatal levels. This allows the infant to become involved with other members of the group. The infant first receives grooming from other adult females in the context of reciprocal grooming events involving the mother and another adult female.

The infant 2 becomes adapted to solid food over the course of this developmental period. He probably obtains his first solid foods by stealing them from his mother. The high rate of infant 2—mother contact in combination with the infant 2's tendency to manipulate and orally explore objects within its reach eventually leads it to take food items held in his mother's hands or feet. At first, the mother tolerates such theft. Later on, she resists it by turning away from the infant. However, once the infant 2 gains sufficient skill at stealing food to enjoy some success, the maternal prevention of this practice involves more active modes of restraint, such as, gently pushing the infant away. Since the mother does not severely punish the infant for food theft, the infant probably stops stealing when it learns that the independent acquisition of food results in faster satisfaction of hunger than food theft.

The mother displays protective behavior toward her infant 2 in response to any disturbance. Mutual approach is the most common consequence of disturbance. However, the mother actively retrieves the infant from situations, objects, and animals that simultaneously attract the infant 2 and upset the mother. The friendly approach of infants to humans and the maternal interference with such approach is the prime example of this. Mothers occasionally physically restrain infant 2's to prevent their approach to humans and other attractive but disapproved stimuli. In spite of these protective maternal responses, the infant red ouakari seems to play the major role in determining the rate of mother—infant contact.

The young ouakari enters the juvenile 1 period rather dependent upon his mother's milk and he leaves this period fully weaned. The termination of nursing in the juvenile has occurred at various times in the seminatural environment. CM4 was abruptly weaned at 16 months on the day of a sibling's birth. The termination of nursing in CM5 occurred at 22 months as a result of an artificial separation from the group necessitated by CM5's illness. Apparently, lactation ceased during the 1-month period of CM5's absence. CF3 nursed as late as 20 months when detailed observations on the group were temporarily suspended. Extrapolation from available data suggests that weaning is completed early in the third year in the absence of a sibling's birth.

In comparison with *Macaca*, weaning in *Cacajao* involves minimal maternal punishment. Mothers frequently terminate or prevent nursing by turning their backs to the infant, thus effectively blocking nipple contact. Mothers occasionally push away persistent juveniles and, in the most extreme cases, they may take one of the juvenile's hands and give it a gentle bite. Although these bites are highly ritualized (the teeth apparently do not occlude), the juvenile responds to them with loud vocalizations, tail wagging, and other signs of extreme emotionality. Since extensive nonnutritive nipple contact does not characterize the behavior of captive ouakaris, the decline in the juvenile 1's rate of nursing is probably a direct reflection of declining milk production. During the juvenile 1 period, the mother does not display a constant response to nipple contact. On any given day, she may allow her juvenile to nurse while performing behaviors which seem to encourage nursing such as grooming, erect posture exposing the nipple, and manual support of the juvenile 1 during nursing. Later on during the same day, on the other hand, the mother may avoid or deliver mild punishment to her juvenile when it attempts to nurse. The simultaneous display of these dual maternal response sets occurs through-out the juvenile 1 period as the rate of nursing declines.

The red ouakari gains basic independence during the juvenile 2 period. Nevertheless, a mother displays stronger defensive responses to events threatening her juvenile 2 than she displays to stresses facing unrelated members of the group. The mother grooms her juvenile 2 at a high rate in comparison with her older offspring and members of the group at large. In addition, the failure of the juvenile 2 to reciprocate does not lead to the termination of the grooming event.

During the fourth year the social bond between a mother and her offspring continues to weaken. As a result, the mother does not display obvious differential responses to her subadult and adult offspring.

D. Play

Ouakaris from the infant 2 to the juvenile 2 stage in the seminatural environment spend several hours of each day engaged in play. Infant 1's play much less than this while adults and subadults play only rarely. Since the estimated 3 hours per day devoted to play by *Saimiri sciureus* in the "Rainforest" may represent an artifact of seminatural conditions (Baldwin and Baldwin, 1974), it is also possible that the frequency of play observed in *Cacajao* may reflect some aspect of seminatural conditions such as an alteration of the space-time pattern produced by provisioning. Nevertheless, the frequent occurrence of play provides an index of the general health of young, captive ouakaris, since a sudden decline in the rate of play occurs when the health of the young ouakari declines below optimal levels.

Most of the play in the seminatural environment involves the following activities: exploration, intense, nondirectional locomotor activity, and social play.

Infant 1's of only a few weeks of age engage in exploratory play. They manipulate twigs, leaves, and other objects in the environment while maintaining contact with the mother. Perhaps the first object of the infant 1's manual exploration is the mother who seems to tolerate some rather rough handling by her infant. The oral exploration of various objects soon follows their manual exploration. The expanded activities of the infant 2 and the juvenile 1 expose them to a wide variety of objects; consequently, all manner of items, including potentially hazardous ones are manipulated and chewed. These activities decline through the juvenile 2 period and become rare in older ouakaris. Subadults and adults restrict manual and oral exploration to presumed novel or unfamiliar objects, such as, snake skins, books, and cameras.

Once the infant gains sufficient locomotor competence to move away from his mother, intense, nondirectional locomotion becomes an important activity. The circularity of this play mode is a most striking feature of it. Once circular locomotor play becomes incorporated into the behavioral repertoire of the infant, it remains a predominant feature of the young ouakari's behavior until the time of transition to adult or subadult status.

Circular locomotor play consists of the repetitive performance of a definite sequential arrangement of specific elements of the locomotor repertoire with reference to specific substrate elements in the environment. During circular locomotor play, the ouakari slowly transforms one circular pattern into another by the successive elimination and addition of motor elements into the circular play pattern. The infants tend to maintain the integrity of the circular patterns for a longer period of time than the juveniles who readily transform the motor sequences. The juveniles often create more complex circular patterns than the infants. In addition, they also seem to display a greater fluidity in moving from nondirected circular locomotor play to other forms of play such as manual and oral exploration and social play.

Red ouakaris of all age—sex description beyond the infant 1 stage engage in social play characterized by distinctive facial expressions, vocalizations, postures, and motor activity.

Infant 2's and juveniles perform most of the social play observed in the "Rainforest." The most conspicuous components of social play include wrestling, mock biting, and grappling. The occurrence of a cackling vocalization and a play face allow an observer to distinguish play from other behavior.

Although young ouakaris often play on the ground, adults limit play to arboreal settings. Arboreal play typically involves suspension by either the hindlimbs, the hindlimbs and a single forelimb, or two ipsilateral limbs while the hands grapple and poke at the play partner. Mock bites are delivered and each individual seems to try to dislodge his opponent. Suspensory posture is often critical to the occurrence of play in adults. Large juveniles sometimes induce social play with adults by rapidly charging them and knocking them off balance into a suspensory posture. If the juvenile fails to knock the adult off balance, however, a mild rebuff of the juvenile usually results. At times a seated adult will briefly grapple or playfully poke at a juvenile suspended overhead; however, most instances of adult and subadult play are suspensory.

Terrestrial social play in the "Rainforest" is limited to red ouakaris from the infant 2 to the juvenile 2 stage. Terrestrial play, in contrast with arboreal play, involves much chasing, charging, lunging, and other noncontact play modes. Wrestling on the ground involves ventral–ventral contact and dorsal–ventral contact in combination with mock biting.

Much of the social play of juvenile red ouakaris living in the "Rainforest" is interspecific. Juvenile red ouakaris actively pursue play with all primate species found in the seminatural environment. They modify their social play to accommodate the interspecific play partner. Thus, play with *Saimiri sciureus* occurs principally on the ground while play with *Alouatta seniculus* is almost exclusively arboreal. Similarly, play between red ouakaris and *Saguinus fuscicollis* consists largely of noncontact modes while play with *Saimiri* and *Alouatta* involves considerable physical contact.

E. Social Roles and Social Dominance

The term "role" refers to a consistent response pattern performed by certain individuals in a society. Since certain roles of the adult females and immature ouakaris have been previously discussed, the present section will deal primarily with the role of the adult male red ouakari. Since the construct "dominance" is often conceptually linked with the social roles of adult males, it merits consideration under the present heading.

A linear dominance hierarchy maintained by agonistic interactions is reported as a prominent characteristic of caged ouakari groups. In the seminatural environment, in contrast, a status hierarchy only becomes apparent to a human observer after many hours of observation. Spatial displacements and priority of access to favored foods fail to provide a reliable measure of dominance among the ouakaris of the seminatural environment. Instead, ouakaris express dominance in noisy fighting which occurs periodically in a variety of settings. The consistent winning of these fights by particular individuals who slap, bite, and drive away the subordinate animals defines dominance in the red ouakari. Although these noisy disputes are decisive, they have never been observed to result in visible injury. The biting of conspecifics by red ouakaris appears highly ritualized in both the seminatural environment and in caged settings; the use of the well-developed canines against conspecifics has not been reported in any captive group. Most of the fighting in the "Rainforest" since 1970 has involved adult females. In the current colony, the adult male (CM3) has asserted dominance over the most dominant female (WF4) on less than ten occasions in over 1000 hours of observation.

The chief role that the dominant animals in the seminatural environment display with respect to subordinates is to intervene in noisy fights among them. Since the adult male is dominant over the largest number of conspecifics, he intervenes in the largest number of disputes. Although the dominant intervener occasionally sides with one of the antagonists, it is more common for the intervener to move between the opponents in the dispute and to perform aggressive gestures with orientation away from the adversaries. Dominant ouakaris frequently orient toward humans in the context of intervention in noisy disputes in the seminatural environment. In addition to direct intervention, dominant ouakaris often approach and/or direct aggressive gestures toward the vicinity of the noisy fights of subordinates.

The performance of various aggressive gestures with reference to stimuli external to the group stands out as a conspicuous role of the adult male red ouakari under all conditions of captivity. The aggressive gestures include such diverse activities as branch shaking, branch rocking, branch breaking, carrying and dropping branches, hind leg extension, hind leg rubbing, rapid flexion of the hindlimbs causing the hindquarters to bounce, strutting about with hyperextended limbs, and a stereotyped display involving the washing of the chest with urine. Adult male ouakaris in the seminatural environment perform various combinations and sequences of these gestures with orientation to such extragroup stimuli as humans, howler monkeys, motor vehicles, aircraft, and potential predators. The current adult male in the "Rainforest" (CM3) devotes at least 1–2 hours per day to activity of this type. Comparable levels have been reported for caged adult males. CM3's current high rate of aggressive display with reference to extragroup stimuli was absent during his subadult period; it emerged in conjunction with the development of the typical adult male morphological features of *C. c. rubicundus*. Although CM3 directs many aggressive displays to red howlers (the chief nonhuman primate antagonists of *Cacajao* in the "Rainforest") and he directs aggressive displays to the rarely occurring birds of prey (usually *Buteo* sp.) seen in the "Rainforest," it is the adult females who participate most actively in intense agonistic interactions with these species. Thus, they rush ahead of the displaying adult male to bite and slap the howlers or to drive off perched hawks. Previous adult males ouakaris, especially the partially tame ones (WM3 and WM4), did not share CM3's tendency to limit aggression against howlers to gestures; they displayed the same type of direct aggression against howlers that is currently limited to the adult females. During CM3's subadult period the colony lacked an adult male ouakari; consequently, WF2, the most dominant adult female, performed most of the aggressive responses to extragroup stimuli.

Adult males and females play complementary roles in the coordination of travel about the "Rainforest." The movement of the older adult females into adjacent areas of the "Rainforest" often precipitates following by the younger adult females and by the immature ouakaris. The adult males typically take up the rear in group progression. The participation of the adult male in group progression is important because the group eventually returns to the vicinity of the adult male if he fails to follow the travel initiated by adult females.

Adult males seem to play a role in the socialization of the young. With the exception of the mother, the adult male is the first group member whom the infant 2 approaches from a distance. The adult male is especially attractive to the infant 2's and juvenile 1's who frequently initiate contact with him. The response of the adult male to such contact ranges from restrained ritualized aggression to tolerance. In contrast, adult females (other than the mother) consistently reject general body contact. The abruptness with which the youngster initiates contact with the adult male governs the extent of tolerance displayed by him. Thus, precipitous contact tends to elicit ritualized aggression while gradual approach leading to contact with the adult male normally results in tolerance of contact. The general circumstances surrounding the young ouakari's establishment of contact with the adult male also modify the male's tolerance of contact. For example, the adult male generally tolerates abrupt contact initiated by infants who have been frightened by a loud noise or other strong stimuli (Fig. 7).

The attraction of the adult males to infants has varied among the different adult males that have lived in the "Rainforest." For example, CM3 seemed to ignore infants until they initiated contact with him. WM7, on the other hand, displayed a great deal of interest in infants. WM7 groomed neonates in contact with their mothers and he chased other ouakaris away from the vicinity of mothers with neonates—thus maintaining the relative isolation of the mother—infant dyad.

F. Interspecific Relations

Ouakaris in the "Rainforest" generally display low response rates to non-primate species. They typically fail to respond to small vertebrates such as rats, mice, tree frogs, toads, and song birds. Occasionally, a sudden movement by one of these in close quarters may produce a short-lived startle response. Ouakaris actively avoid large specimens of the constrictor *Elaphe quadrivittata*, which preys on squirrel monkeys. They also chase off alighted hawks with piloerection and loud vocalizations. The overhead flight of hawks and other large, rarely occurring birds elicits rapid ascension and vocalization toward the passing bird. In contrast, the turkey vulture, the raptorial bird most commonly observed over the "Rainforest," fails to elicit an alarm response even though it presents a flight silhouette similar to that of the buteonine hawks that elicit strong defensive responses. The ouakaris even ignore the great aggregations of 50 to 100 turkey vultures that periodically pass over the "Rainforest" in the dry season, sometimes at heights less than 10 m over the treetops.

Interactions between red ouakaris and other primate species frequently occur in the seminatural environment. In general, infant 2 to juvenile 2 ouakaris tend to play with other primates while adults either ignore them or act in a hostile fashion toward them. Most of the playful interactions of the young involve *Saimiri sciureus*. Infant 2's and juveniles may spend about an hour each day engaged in play with the young squirrel monkeys. Play with howlers and marmosets is less frequent. Young ouakaris attempt to play with adult squirrel monkeys on occasion, but they are consistently rejected by them.

The frequent occurrence of play between *Cacajao* and *Saimiri* in the "Rainforest" may be a factor in the sensitivity that captive-born ouakaris of all ages display to the behavior of squirrel monkeys. Although adult ouakaris occasionally slap at squirrel monkeys or push them away from provisioned food sources, competition between *Cacajao* and *Saimiri* is minimal and positive interactions between them often occur. Ouakaris respond to the alarm calls of squirrel monkeys with either orientation to the source of disturbance or ascent if the ouakaris are on or near the ground. Captive-born ouakaris also respond with threats to aggression directed by humans toward squirrel monkeys. In addition, CM3, an adult male ouakari, occasionally orients aggressive gestures to noisy fights among the squirrel monkeys. Although positive responses to squirrel monkeys are more common among the captive-born ouakaris, they are not limited to them. WF1, an adult female that the other ouakaris usually ignored because of behavioral abnormalities, frequently solicited and received grooming from juvenile squirrel monkeys.

Relations between adult ouakaris and adult red howlers are often contentious. Fights often occur. Recent disputes have usually begun as a result of active interference by an adult ouakari in the playful wrestling of young ouakaris and howlers. This action precipitates defensive responses by the adult howlers leading to a major interspecific dispute involving most individuals of both species.

The relative dominance of ouakaris and howlers has wavered throughout the history of the "Rainforest." At the present time, the red ouakaris and red howlers have apparently partitioned the "Rainforest" into two somewhat separate ranges. The howlers concentrate their activity in the northern portion of the "Rainforest," while the ouakaris live primarily in the southern section. At the provisioning site the ouakaris favor the southern feeding station while the howlers feed almost exclusively at the northwest feeding station.

Cacajao's responsiveness to man is complex. The chief variables effecting a semifreeranging red ouakari's responsiveness to humans include the individual's age and sex, the monkey's early experience as either a pet or a member of a ouakari group, and finally, the animal's status as wild- or captive-born.

Infant 2 to juvenile 2 ouakaris tame easily and generally respond positively to humans in the seminatural environment. Captive-born infant 2's first display responsiveness to humans by approaching them with high-pitched vocalizations, scalp retraction, and other signs of agitation. These approaches by the infant 2's usually precipitate maternal retrieval when they first appear. Eventually, the mother learns to tolerate her infant's approaches to people. Meanwhile, the emotional responses that accompany friendly approach diminish in intensity; the frequency of approach increases; and approach to the point of contact with humans becomes common. The initiation of friendly interactions with people is a conspicuous behavioral characteristic of all juvenile ouakaris that have lived in the "Rainforest."

The juvenile's friendly orientation to people changes with the acquisition of adult and subadult morphological features. The effects of pet experience and captive versus feral-born status are especially conspicuous in adult males. Both nonpet, wild-born adult males, and mother-reared, captive-born adult males tend to respond to humans with hostility while maintaining a distance of several meters between themselves and humans. The rate of occurrence of hostile gestures seems greater among captive-born adult males. Adult males with an early history as a pet are less consistently aggressive. Periods of playfulness alternate with periods of aggression in an unpredictable fashion. The failure of these former pets to develop the tendency to maintain a distance between themselves and humans makes these adult males very dangerous to people.

In comparison with adult males, adult females direct low levels of aggression to humans. Instead, they generally either ignore humans or solicit food from them. Wild-born, nonpet adult females direct more gestural aggression to humans than adult females with other backgrounds. A wild-born, pet-reared adult female (WF1) displayed the least gestural aggression to humans; she even directed friendly gestures to them. CF3, a captive-born, mother-reared adult female displayed relatively little aggression to people until her juvenile son was captured for medical treatment. From that point, CF3 directed aggression to humans at levels comparable to those observed in wild-born, nonpet adult females. Adult females with highly varied individual histories have learned to solicit food from the public. Only two adult females in the history of the "Rainforest" never learned to solicit food. Both were wild-born, nonpet specimens. One was imported as an adult; the other was a juvenile 2 in transition to adult status.

Adult ouakaris are generally quite aware of being watched. They frequently avert their eyes from a human observer only to glance back intermittently as if to check on his movements. Direct observation by humans occasionally stimulates aggressive gestures in adult ouakaris.

IX. Suggestions for a Captive Propagation Effort

The "Rainforest" facility represents an ideal in the captive management of *Cacajao* that may be impossible or impractical to duplicate in the limited space of the typical urban zoo. Since some breeding has occurred in caged groups, conditions in the seminatural environment may exceed the lower bounds for the establishment of a reproducing population.

The management procedures suggested in this report are based on fragmentary information derived from observations in the "Rainforest" and from our survey of other captive situations. Definitive procedures for the care and management of *Cacajao calvus* in captivity await the development or a more thorough understanding of the reproductive physiology, social behavior, nutritional needs, and diseases of this endangered species.

A. Optimal Physical Conditions

Definitive information on optimal and minimal cage size for *Cacajao* is not available. The limited data provided by institutions breeding *Cacajao* suggest tentative space requirements for ouakaris in captivity. The Frankfurt Zoo maintains a reproducing group of red ouakaris (including an adult female and two adult males) in two connecting cages (3 × 4 × 3 m; 6 × 4 × 6 m). The Cheyenne Mt. Zoo reports a pair of viable births in a group of red ouakaris maintained in a 1.83 × 2.44 × 2.44 m cage. The condition of parity in only one of the two adult females in this group in spite of breeding in the nulliparous female suggests that lack of available space may have contributed to this individual's reproductive failure. Crandon Park maintained a pair of ouakaris in a cage of about the same size as that housing the Cheyenne Mt. group before moving them to larger quarters. In the old cage, the adult male of this pair frequently and inadvertently collided with the female in the context of aggressive displays directed to the public. These collisions stimulated the female to respond with intense defensive threats which, in turn, stimulated contact aggression by the male. A similar incident may have been a factor in inducing abortion because miscarriage occurred on a holiday when abnormally large crowds increased the frequency and intensity of the male's aggressive displays.

The design of housing for a group of five to ten *C. c. rubicundus* might profit from application of the observation that ouakaris in the seminatural environment tend to maintain a minimal interindividual distance of 3 m for most activities. A cage with linear dimensions twice this value (6 m) would allow ouakaris to maintain interindividual distances approaching minimal values maintained in the "Rainforest." A further reduction in stress may be achieved without a corresponding increase in linear cage dimensions by the construction of opaque partitions dividing the cage into several sections remaining normally accessible to all animals. This practice would permit the voluntary isolation of individuals, dyads, etc, within the ouakari group and would facilitate the temporary isolation of individuals for manipulation by management. The construction of indoor-outdoor cages is a good practice to follow since sunlight maintains red facial coloration in *C. calvus* (Hill, 1965). However, the indoor (winter) quarters should still provide at least 216 m³ of living space. Institutions wishing to maximize the efficiency of limited resources might find it worthwhile to house a small troop of *Saimiri sciureus* in the same cage as their ouakaris. This practice would necessitate a corresponding increase in cage volume; of course, it would have to be done without knowledge of the extent to which these two species serve as antagonistic reservoirs of infectious microorganisms. The installation of perches also increases the effective volume of the cage. If perches are installed, metal pipes are recommended because they are easier to clean than platforms. Also, *Cacajao* seems adapted to locomotion on curved surfaces. Several zoos equip ouakari cages with rubber ropes, plastic chains, and other structures that seem to facilitate the performance of varied locomotor behaviors and to encourage ouakaris to engage in suspensory play behavior. Zoos displaying *Cacajao* should construct at least one set of perches or ropes between the remaining equipment and the viewing area in order to allow the adult male to vent his aggression to the public without inadvertently orienting it to conspecifics.

The captive adult male ouakari's habit of spending a substantial amount of time orienting aggressive gestures to humans is a widely reported feature of this animal's behavior. Since these vigorous displays are incompatible with copulation, housing for ouakaris should be located in an area where the public may be excluded during periods when adult females appear sexually attractive and receptive.

The recommended practices for increasing the efficiency of living space impede the maintenance of proper hygiene. A clean environment is especially important for red ouakaris because their coat seems to suffer more from suboptimal sanitation than that of other primates. Contamination and deterioration of the coat of ouakaris transferred from the "Rainforest" to cages with solid floors became apparent in only a few days. The construction of a floor of 5 × 5 cm 6-gauge electroweld wire elevated about 50 cm above the concrete cage floor would greatly alleviate pelage contamination in *Cacajao*.

Ouakaris seem capable of adapting to temperature variation that greatly exceeds that characteristic of their native habitat. In the "Rainforest" they have survived brief periods of temperature extremes ranging from about −2° to 38°C without difficulty. Most of the time, temperature in the seminatural environment varies between 17° and 30°C. As noted previously, the occurrence of deaths and illnesses in the "Rainforest" bears no obvious relationship to temperature extremes. The Frankfurt Zoo maintains indoor quarters at 20°–26°C while plastic flaps permit ad libitum access to the outside even in January when average surface temperatures range from 0° to 10°C. The Oklahoma City Zoo prevents access to the outside quarters when the temperature falls below 5°C. Although red ouakaris seem capable of tolerating greater temperature extremes, a temperature range of 15° to 25°C, or its wind-chill equivalent, should be maintained in the interior quarters.

Sanderson (1957) has argued that ouakaris develop respiratory disorders when exposed to a desiccated atmosphere. However, zoo colonies in semiarid climates (Gladys Porter Zoo and Cheyenne Mt. Zoo) do not report the occurrence of respiratory disorders. Nevertheless, a relative humidity value similar to that characteristic of a tropical wet climate (around 80%) should be maintained in the indoor quarters.

Ventilation in indoor quarters is recommended. An adequate air exchange rate without the creation of drafts should be maintained. Indoor quarters should provide access to direct sunlight and outdoor quarters should include adequate shelter from rain and direct solar radiation.

B.   Optimal Social Conditions

Current restrictions on the importation and exportation of *Cacajao* inhibit the establishment of the captive population with the greatest potential for sustained reproduction: a wild group captured, transported, quarantined, and established in captivity as an intergrated social unit. At the present time, and probably in the future, zoos will have to settle for much less.

Definitive field data concerning group size, socionomic sex ratio (ratio of adult females to adult males), etc., is currently unavailable. Thus, provisional guidelines for the encouragement of a captive social environment that will favor breeding must rely on observations of ouakaris in captivity.

Observations of ouakaris in the "Rainforest" and other captive situations suggest that the presence of several adult females in a group does not inherently limit the production of viable offspring or the occurrence of copulation in any given female and that most interactions between adult females are friendly. Thus, the housing of more than one adult female in a single facility can be recommended with some degree of confidence. The general character of interaction patterns among adult males seems much less clear. The Frankfurt Zoo's group with two adult males and a single adult female has produced five young since 1967. The fact that copulation is limited to only one of the adult males suggests that intermale competition may limit reproduction to a single adult male in a multimale captive group. At the Milwaukee Zoological Park, where a single adult female lives with three adult males, considerable aggression has been observed among the adult males. The high level of aggression reported among the adult males in this group may be related to the failure of copulation to occur in this colony. Observations in the "Rainforest" do not point to a consistent pattern of interaction between adult male red ouakaris. For example, when WM8 was introduced into the "Rainforest" along with three females the established adult male of the group, WM7, directed much aggression to him although WM7 did not direct aggression to the introduced females. On the other hand, WM7 did not display intolerance to those adult males (WM3 and WM4) who matured to adult status after 2 years in the "Rainforest" and who had lived as immatures in the "Rainforest" prior to WM7's introduction.

Ideally, a group reflecting natural sociological conditions ought to be established in captivity (probably multi-adult female, possibly multi-adult male). However, in light of the current shortage of available captive ouakaris, a group composed of an adult male and three or four adult females would probably offer the best current propagation prospects. The use of a single adult breeding male is not without risks since many of the currently available adult males have lived as pets and are not properly socialized. The "social rehabilitation" of former pet ouakaris has never followed their release into the "Rainforest" and similar experiences with pets have been reported by other zoos. For these reasons, the behavior of the adult male should be observed over an extended period in order to verify the adult male's capacity for normal social and sexual behavior. The failure of the adult male to show normal adult sexual and social behavior patterns during the first year of group living probably indicates that the adult male in question will never become an effective breeder and that he ought to be exchanged for another adult male who might succeed in the breeding situation. (Alternately, if the proper facilities are available, artificial insemination to competent females might be tried.)

The very limited available information suggests that the best prospects for captive breeding include (1) captive-born individuals raised in social groups (2) adult specimens confiscated at port of entry by USDI officials or similar government personnel and (3) if import and export permits can be acquired, fresh, wild-caught adults.

C. Nutrition and Diet

The precise nutritional requirements for *C. c. rubicundus* remain unknown. However, many zoos have maintained ouakaris in apparent good health for long periods on a diet essentially similar to that suggested by Crandal (1964) consisting of greens and fruits, commercial monkey pellets, milk, hard-boiled eggs, and optionally, small quantities of meat. The addition of vitamin D_3 to the diet of most captive red ouakaris seems to represent the most important supplement to Crandal's recommendations. This practice is strongly recommended because vitamin D_3 is more active physiologically than vitamin D_2 in several species of New World monkeys (Lang, 1968).

The chief contrast between the diet of ouakaris living in the seminatural environment and those in other captive situations lies in the food foraged from the "Rainforest." No attempt has been made to measure the nutritional composition of this portion of their diet. However, lysiloma seeds and rapidly growing new leaves and buds may contribute considerable protein. At the very least, the high cellulose content of the foraged dietary items guarantees adequate roughage. Thus, the stool of healthy ouakaris in the "Rainforest,' in contrast with that usually seen in other captive situations, is in the form of relatively hard pellets. Several zoos currently follow the practice of providing caged ouakaris with browse (foliage, twigs, etc.) This practice is recommended because it provides beneficial roughage.

The constant availability of food in the seminatural environment provides another contrast between living conditions in the "Rainforest" and life in the typical zoo. As noted previously, ouakaris in the "Rainforest" are offered food seven times each day. Ouakaris are probably adapted to continuous feeding; consequently, the high frequency of provisioned feedings in combination with the constant availability of naturally growing food allows food intake to approach presumed natural conditions.

The content of the provisioned diet offered in the seminatural environment is fundamentally similar to that offered to most ouakaris living in zoos. The first provisioning consists of a soft mash of whole wheat bread, eggs, non fat milk, and multiple vitamins including vitamin D_3. The eggs and non fat milk provide an important protein source. The ouakaris consume several handfuls of this mash since it is offered to them when their hunger is maximal. The soft consistency of this item makes it suitable for infants. The other provisionings include apples, bananas, carrots, celery, grapes, melon, pears, raisins, raw peanuts, string beans, sunflower seeds, Wayne Monkey Diet (Allied Mills; Chicago, Illinois), and other fruits in season.

The practice of feeding the nutritionally more important items first is recommended. Each zoo maintaining ouakaris will have to determine food preferences experimentally in order to devise a feeding schedule that will ensure the intake of the less palatable and the nutritionally more important items. Since ouakaris may satiate on favored foods of low nutritional value, public feeding should be prohibited. Commercial monkey pellets, a staple protein source of ouakaris in many zoo collections, seem to be a low preference food item in the "Rainforest" and other captive situations. Zoos depending on the high nutritional value of this item should take extra care to determine whether or not certain individuals are consistently rejecting the commercial monkey pellets in favor of preferred food items. In evaluating the consumption of monkey pellets, observers should be cautioned against confusing feeding behavior with oral and manipulative behaviors. Ouakaris often manipulate, chew, and otherwise treat as food, objects that they ultimately drop uneaten. Several primate species find commercial monkey pellets more palatable after they are softened by soaking in water. This practice is especially recommended for infants who might be entirely unable to chew dry commercial monkey pellets.

Water should be available to captive red ouakaris on an ad libitum basis.

D. Diseases and Health

Current knowledge of the diseases effecting *Cacajao* is indeed fragmentary. This section gathers together the information obtained in the survey of the captive red ouakari population and the data available in the current literature on platyrrhine parasitism.

The delicacy of these animals makes the close monitoring of all aspects of husbandry a major part of the maintenance of ouakaris. The establishment of a diet regimen guaranteeing adequate nutrition, scrupulous sanitation, the use of nontoxic paints and building materials in the construction of living quarters, the provision of adequate space and devices to promote exercise and the maintenance of muscle tone, and protection from temperature extremes provide the best protection against disease. *Cacajao* will not survive on the minimal care that is often adequate for the more hardy laboratory primates.

In the survey of the ouakari in captivity, only a single case of probable dietary deficiency could be identified. In this case, an adult female with the clinical symptoms of rickets (see Ruch, 1959) was observed in an inspection of a laboratory colony of four red ouakaris. Oranges constituted the bulk of this individual's diet because she apparently failed to adapt to the hard monkey pellets that were also provided. Apparently, a combination of dietary deficiency and poor general husbandry led to the extinction of this colony within 18 months.

Oral disorders are occasionally reported in red ouakaris. Robinson (1974) described a case of multiple canine fistulae in a female red ouakari and provided details of its successful treatment. Abscesses of the gums have been reported as an occasional problem at the Frankfurt Zoo.

At Monkey Jungle, dental abscesses (left maxillary PM2 and PM3) that were probably derived from a previous maxillofacial abscess afflicted CM5. This case is the most thoroughly documented case that we surveyed in which a moribund individual was brought back to health; consequently, it will be treated in some detail. CM5 was removed from the "Rainforest" on March 3, 1975 and a maxillofacial abscess complicated by severe secondary pneumonia was diagnosed.

A fistula leading from the maxillofacial region to the oral cavity was enlarged in order to facilitate pus discharge. Then, CM5 was treated with 50 mg/kg chloramphenicol (Chloromycetin, Parke-Davis, Detroit, Michigan) every 8 hours for a period of 14 days. During this period, a humidifier kept the air space in his cage moist. When the secondary pneumonia was cured and the abscess was under control, treatment with chloramphenicol was replaced by 10 days of oral administration of 100 mg/kg tetracycline hydrochloride (Veterinary Panmycin Aquadrops; Upjohn, Kalamazoo, Michigan) every 8 hours. CM5 was returned in apparent good health to the "Rainforest" on April 7, 1975. On May 8, 1975 CM5's left maxillofacial region was severely swollen and the left eye was closed. CM5 was immediately captured and abscesses of the left maxillary PM2 and PM3 (representing a probable continuation of CM5's previous maxillofacial abscess) were diagnosed. These teeth were extracted and the corresponding alveoli were flushed with saline solution. CM5 received treatment for the next 32 days with 50 mg/kg tetracycline hydrochloride administered orally every 8 hours. During this treatment period, the veterinarian cleaned the healing extraction wounds of impacted food, etc., every 3 or 4 days. CM5's diet during this period was restricted to soft foods in order to limit impacting. Treatment with tetracycline hydrochloride was discontinued once the extraction wounds had finally healed. CM5 was returned to the "Rainforest" on June 14, 1975 after a period of observation to assess the effects of antibiotic withdrawal. CM5, who was about 2 years old during this experience, did not seem to suffer major social damage as a result of his separation from the rest of the ouakari colony.

Gastrointestinal diseases usually indicated by diarrhea have been a problem in several captive ouakari groups. Nonfatal infections of *Shigella* sp. and *Salmonella* spp. have occurred at the Cheyenne Mt. Zoo. Intestinal protozoa have also caused illness in *Cacajao*. Several ouakaris at the Oklahoma City Zoo have been successfully treated for infections of *Balantidium coli*, *Trichomonas* sp., and *Giardia* sp. *Trichomonas* sp. was a suspected cause of the introductory death of WM1 who lived for 3 weeks in the "Rainforest." However, as noted by Ruch (1959), the occurrence of a large infection of *Trichomonas* usually reflects prior poor health. An important intestinal parasite of *Cacajao* appears to be nematodes of the genus *Strongyloides*. Serious infections of *Strongyloides* sp. have occurred in three individuals in the "Rainforest:" CM4, CM5, and WF1. WF1 and CM5 responded to treatment with orally administered thiabendazole, MSD (Mintezol; Merck, Sharp, & Dohme; West Point, Pennsylvania). CM4 died of atypical (viral?) pneumonia that may have been secondary to the strongyloidiasis. "Hookworms" caused the death of a 20-month-/old red ouakari born at the Cheyenne Mt. Zoo. "Tapeworms" were treated successfully with paromomycin sulfate at the Oklahoma City Zoo.

Respiratory infections are an important malady of captive red ouakaris. Several individuals in the "Rainforest" have experienced pneumonia. In addition to the case of atypical pneumonia noted above, four additional cases of pneumonia have been treated in the "Rainforest." In two of these cases, the pneumonia was believed to be a secondary infection. In the other two cases, the animals (WF1 and WM3) were treated successfully with peroral tetracycline hydrochloride. The susceptivity of *Cacajao* to *Myobacterium tuberculosis* remains unknown although the Oklahoma City Zoo did report the occurrence of a false positive to initial tuberculin testing that failed confirmation in subsequent roentgenography.

Two deaths at the Gladys Porter Zoo were attributed to infectious hepatitis. Impure water was the suggested contaminant. Crandal (1964) cited hepatitis as the cause of death in a red ouakari kept for over 8 years at the Philadelphia Zoological Park.

Toxoplasma apparently infects natural populations of *Cacajao* (Kuntz and Myers, 1972). Ruch (1959) reported a fatal case of toxoplasmosis in a red ouakari housed at the Philadephia Zoo.

The occurrence of blood parasites in *Cacajao* has been noted several times. Although infections of *Plasmodium brasilianum* have been reported several times for wild specimens of *C. calvus* (Deane, 1972; Kuntz and Myers, 1972; Ruch, 1959), *Plasmodium* infections have not been reported in captive ouakaris. Infection by the hemoflagellate, *Trypanosoma cruzi*, an important human pathogen causing Chagas' disease, has been reported as a natural disease of *Cacajao* (Dunn, 1968). A fatal case of Chagas' myocarditis and heart failure in a captive red ouakari and its treatment is described in detail by Larsy and Sheridan (1965).

X.   Comparative Perspectives

It seems appropriate at this point to consider some broad speculative contrasts between red ouakaris and other cebids. (Discussion will emphasize contrasts between *Cacajao* and the other cebids currently living in the "Rainforest:" *Alouatta* and *Saimiri*.)

Ouakaris in the seminatural environment maintain intermediate activity levels in comparison with the more active squirrel monkeys and the relatively lethargic howlers.

The ouakari's repertoire of expressive behaviors compares with that of the more advanced cebids such as *Cebus, Lagothrix,* and *Ateles*. A well developed repertoire of vocalizations, postural displays, and facial expressions is available for intraspecific communication. Subtle variations of the basic units of communication as well as variation in their combination and sequencing yield extensive possibilities for communication.

In contrast with *Saimiri* and *Alouatta*, and perhaps other cebid genera, *Cacajao* engages in frequent allogrooming. The tension often apparent in reciprocal allogrooming and the variation in the form and frequency of the social signals temporally associated with allogrooming suggest that allogrooming at high rates in the red ouakari represents a relatively recent evolutionary acquisition.

Dominance hierarchies are infrequently expressed and group organization and direction rely on more subtle interactions. Social roles in group defense, coordination of travel, etc. appear more highly evolved than those expressed by *Saimiri*; indeed, the level of complexity of social organization and role differentiation in *Cacajao* seems reminiscent of that seen in *Cebus*.

The passive expression of maternal care in ouakaris reflects the typical cebid condition while the relatively long period of infant dependency suggests affinities with the more advanced cebids such as *Ateles* and *Cebus*. The observation that ouakari mothers display high rates of protective care when infant 2's and juvenile 1's signal stress or engage in prohibited activities during periods of mother-offspring spatial sepration suggests that the passivity of maternal behavior during mother—offspring contact reflects a highly evolved level of mother—offspring behavioral coordination during contact that masks a latent capacity for more active modes of maternal behavior. Thus, these active modes only become conspicuous when infant 2's develop considerable locomotor independence. For example, WF3 did not begin to use her body as a bridge for CF4's passage over wide foliage gaps until CF4 had matured to the point where separation of WF3 and CF4 by a gap exceeding the latter's locomotor capacities had become a possibility.

In comparison with *Alouatta* and *Saimiri*, *Cacajao* displays a high level of behavioral plasticity. *Cacajao*'s practice of varying behavior to suit interspecific play partners is the best illustration of this. Investigative and manipulatory behaviors seem more highly developed in *Cacajao* than in *Alouatta* and *Saimiri*, although ouakaris do not approach the level of sophistication displayed by *Cebus* with respect to these dimensions. Sticks, stones, and other objects are extensively manipulated by juveniles who sometimes respond to these objects as though they were social play partners. Even adult ouakaris attend to and manipulate novel objects. *Cacajao*'s orientation toward inanimate objects may be related to their tendency to incorporate them into behavior patterns such as the breaking and dropping of branches in agonistic contexts.

Acknowledgments

Mr. Fontaine's investigation of the behavior of *Cacajao* in 1970—1971 was supported by NSF Grant GB 8348 to D. K. Candland. His research was supported by housing from Monkey Jungle, Inc. for the duration of the study reported in this paper.

The authors wish to thank Dr. Robert W. Cooper, Dr. Patricia Scollay, Mr. Robert C.

Baily, and Ms. Betsy Fine for their careful reading and their comments and criticisms of the manuscript.

The authors wish to thank the following individuals who provided details of their experience with *Cacajao* in captivity: Ms. Carol Brubaker (Oklahoma City Zoo,) Dr. Lawrence Curtis (Oklahoma City Zoo), Dr. R. Faust (Frankfurt Zoo), Mr. Robert Bullerman (Milwaukee Zoological Park), Dr. Gordon Hubbell (Crandon Park), Mr. Mark Rich (San Diego Zoo), Mr. Greeley Stones (Gladys Porter Zoo), Mr. John Snelling (Oklahoma City Zoo), and an anonymous contributor from the Cheyenne Mt. Zoo.

References

Baldwin, J. D. and Baldwin, J. I. (1974) Exploration and social play in squirrel monkeys (*Saimiri*). *Amer. Zool.* **14**, 303–315.

Bates, H. W. (1863). "The Naturalist on the River Amazons." E. P. Dutton, New York.

Bishop, A. (1964). Use of the hand in lower primates. *In* "Evolutionary and Genetic Biology of the Primates" (J. Buettner-Janusch, ed.), Vol. II, pp. 133–225. Academic Press, New York.

Cooper, R. W. (1968). Squirrel monkey taxonomy and supply. *In* "The Squirrel Monkey" (L. A. Rosenblum and R. W. Cooper, eds.), pp. 1–30. Academic Press, New York.

Crandal. L. S. (1964). "Management of Wild Animals in Captivity." Univ. of Chicago Press, Chicago, Illinois.

Deane, L. M. (1972). Plasmodia of monkeys and malaria eradication in Brazil. *"Int. Zoo Yearb."* **12**, 26–60.

Dunn, F. L. (1968). The parasites of *Saimiri*: in the context of platyrrhine parasitism. *In* "The Squirrel Monkey" (L. A. Rosenblum and R. W. Cooper, eds.), pp. 31–68. Academic Press, New York.

Forbes, W. A. (1896). "A Handbook to Primates." Edward Stuart, London.

Harrisson, B. (1971). "Conservation of Nonhuman Primates in 1970." S. Karger, Basel.

Hershkovitz, P. (1972). Notes on New World monkeys. *"Int. Zoo Yearb."* **12**, 3–12.

Hill, C. A. (1965). Maintenance of red facial coloration in the red uakari. *"Int. Zoo Yearb."* **5**, 140–141.

Hill, W. C. O. (1960). "Primates, Comparative Anatomy and Taxonomy," Vol. IV. Edinburgh Univ. Press, Edinburgh.

Hill, W. C. O. (1962). "Primates, Comparative Anatomy and Taxonomy," Vol V. Edinburgh Univ. Press, Edinburgh.

Kinzey, W. G. (1971). Male reproductive system and spermatogenesis. *In* "Comparative Reproduction of Nonhuman Primates" (E. S. E. Hafez, ed.), pp. 85–114. Thomas, Springfield, Illinois.

Kuntz, R. E. and Myers, B. J. (1972). Parasites of South American primates. *'Int. Zoo Yearb.'* **12**, 61–68.

Lancaster, J. B. and Lee, R. B. (1965). The annual reproductive cycle in monkeys and apes. *In* "Primate Behavior: Field Studies of Monkeys and Apes" (I. DeVore, ed.), pp. 486–513. Holt, Rinehart, and Winston, New York.

Lang, C. M. (1968). The laboratory care and clinical management of *Saimiri* (squirrel monkeys). *In* "The Squirrel Monkey" (L. A. Rosenblum and R. W. Cooper, eds.) pp. 394—416. Academic Press, New York.

Lasry, J. E. and Sheridan, B. W. (1965). Chagas' myocarditis and heart failure in the red uakari. *Int. Zoo Yearb.* **5**, 182—184.

Lonnberg, E. (1938). Remarks on some members of the genera *Pithecia* and *Cacajao* from Brazil. *Ark. Zool.* **30A** (18), 1—25.

Moynihan, M. (1966). Communication in the Titi monkey. *J. Zool.* **150**, 77—127.

Napier, J. R. and Napier, P. H. (1967). A Handbook of Living Primates. Academic Press, New York.

Prince Philip, H. R. H. Duke of Edinburgh and Fisher, J. (1970). "Wildlife Crisis." Cowles, New York.

Raisz, E. (1964). "Atlas of Florida." Univ. of Florida Press, Gainesville, Florida.

Ruch, T. C. (1959). "Diseases of Laboratory Primates." W. B. Saunders, Philadelphia.

Robinson, P. T. (1974). Multiple canine teeth fistulae in a red uakari monkey. *Vet. Med. Small Anim. Clin.* **69**, 699.

Sanderson, I. T. (1957). "The Monkey Kingdom." Chilton Books, New York.

Soini, P. (1972). The capture and commerce of live monkeys in the amazonian region of Peru. *Int. Zoo Yearb.* **12**, 26—36.

Thomas, O. (1928). The Godman-Thomas expedition to Peru. The mammals of the Rio Ucayali. *Ann. Mag. Nat. Hist.* **10**, 249—265.

Thorington, R. W. (1972). Importation, breeding and mortality of New World primates in the United States. *Int. Zoo Yearb.* **12**, 18—23.

Uno, H., Aldachi, K., and Montagna, W. (1969). Baldness of the red uakari (*Cacajao rubicundus*); Histological properties and enzyme activities of hair follicles. *J. Gerontol.* **24**, 23—28.

U.S. Department of Commerce (1968). "Climatic Atlas of the U.S." Weather Bureau, Washington, D.C.

7

The Socioecological Basis for the Conservation of the Toque Monkey (*Macaca sinica*) of Sri Lanka (Ceylon)

WOLFGANG P. J. DITTUS

I.   Introduction

In a recent outline of nature conservation in Sri Lanka (Ceylon), Crusz (1973) stressed the need for a sound ecological basis to any conservation efforts. Broadly, such information concerns the factors that govern a species' existence, such as its resource requirements, population dynamics, social behavior and organization, and its relationship with other species including man.

My aim here is to contribute to the understanding of the relationships between such factors, particularly as they pertain to the conservation of the toque macaque (*Macaca sinica* Linnaeus 1771) of Sri Lanka, which I studied in its natural habitats over a period of 7 years.

II.   General Information

A.   The Geography and Climatic Zones of Sri Lanka

The island of Sri Lanka is a geographically related fragment of the Indian subcontinent and is separated from it by only a shallow strait. Sri Lanka is 435 km long, 225 km wide, and encloses an area of 65,000 km². About one-fifth of the island is mountainous reaching its highest point at 2527 m. The mountains are confined to the south and central regions of the island, and are surrounded by the lowland plains which make up the remaining four-fifths of the island.

The climate of Sri Lanka is equatorial and is subject to two monsoon winds which blow in opposition to one another. The moisture laden southwest monsoon (May to September) originates over the warm south Indian Ocean and sheds most of its rain on the southwest corner of the island and on the mountain facing it. On the leeward side of the mountains, it constitutes a strong desiccating wind over the vast northern and eastern lowland plains. The north-east monsoon (November to February) originates over cool north Asian landmasses and brings some rain to the northern and eastern plains. Intermonsoonal cyclones and convectional rains bring additional water to all parts of the island, mainly in April and October.

The interaction between climate and varied topography has given rise to several climatic zones (Fig. 1) each with its own vegetational features. Following Mueller-Dombois and Sirisena (1967), the Wet Zone (D) is subdivided into the lowland rain forest (Zone D_1) below 1000 m, the hilly midland rain forest (Zone D_2) above 1000 m, and the montane rain forest (Zone D_3) above 1500 m. The Dry Zone (B) extends over most of the lowland plains in the north and east, and is separated from the Wet Zone by an Intermediate Zone (C). In the northwest and southeast, the Dry Zone grades into the Arid Zones, A_1 and A_2, respectively. The demarcation of the latter zones is most arbitrary; Fernando (1968) confines them much closer to the coasts.

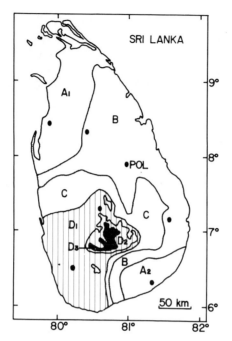

Fig. 1. Sketch map of Sri Lanka indicating the distribution of the three subspecies of *M. sinica* in relation to the climatic zones (see text, Section II, A). Light shade indicates *M. s. sinica* (Zones A_1, A_2, B, and C). Lines indicate *M. s. aurifrons* (Zones D_1 and D_2). Dark shade indicates *M. s. opisthomelas* (Zone D_3). Study sites (●) within the Wilpattu and Ruhunu National Parks are in Zones A_1 and A_2, respectively. Anuradhapura and the Gal Oya National Park are in Zone B, northwest and southeast of Polonnaruwa (POL), respectively. The Sinharaje Forest Reserve and the Udawattekelle Sanctuary are in Zone D_1 southwest and north of the mountains (Zones D_2 and D_3), respectively. The Horton Plains are in Zone D_3.

Since vegetational form largely determines animal distributions, each of these zones has its typical fauna (see Henry, 1970; Kirtisinghe, 1957; Phillips, 1935; and Eisenberg and McKay, 1970).

B. Taxonomy and Morphology

Macaca sinica is endemic to Sri Lanka and belongs to the subfamily of Old World monkeys, Cercopithecinae, which also includes other macaques, baboons (*Papio, Theropithecus*), vervets or guenons (*Cercopithecus*), drills (*Mandrillus*), and others. Toque macaques share the subgenus *zati* with their nearest relatives, the bonnet macaques (*M. radiata*) of southern India, and possibly with *Macaca assamensis* (Hill and Bernstein, 1969) of northeastern India and Assam.

The toque macaque is long tailed and agile both on the ground and in the trees. It is about one-half the size of the commonly known rhesus macaques (*M. mulatta*); the average weight of wild adults is 5.7 kg for males, and 3.6 kg for females. Its common name refers to a well formed caplike whorl of hair (toque) radiating symmetrically outward from the crown of its head (Figs. 2 and 3). Hill (1932) and Pocock (1932) describe the morphological differences between *M. sinica* and *M. radiata*. Basically, *M. sinica* is somewhat smaller, possesses a brighter pelage with orange and reddish hair, and instead of flesh color, its ears, lower lip, and borders of the eyelids are pigmented black. In distribution, *M. sinica* is confined to forested regions. No groups have been observed living in towns, as is often reported for *M. mulatta* of India (e.g., Southwick *et al.*, 1961).

Fig. 2.   A juvenile male (3 ½ years old) of the *M. s. sinica* subspecies.

Three races have been recognized. Their distributions, as described by Eisenberg and McKay (1970) and Hill (1974), are delimited according to major climatic or phytogeographic zones (Fig. 1). The most widely distributed subspecies is the forma typica, *M. s. sinica* (Linnaeus 1771), which occurs throughout the Dry and Arid Zone plains (Zones B, A_1, and A_2) and the Intermediate Zone (C). In animals of this subspecies, the hair of the toque and pelage are generally short and appear golden-light brown; the skin is white or very pale blue and is often patchy (Fig. 2).

Fig. 3.   An adult female of the *M. s. aurifrons* subspecies. The toque (head hair) of these rain forest inhabitants is in a long umbrellalike arrangement compared to that of the Dry Zone race (see Fig. 2).

Macaca s. aurifrons (Pocock, 1932) is confined to the lowland and midland Wet Zones (D_1 and D_2). It differs from the dry zone race mainly by possessing much longer toque hair, a conspicuous golden-orange or reddish pelage, and a lesser degree of reddening in the facial skin of females (Fig. 3). The montane subspecies, *M. s. opisthomelas* (Hill, 1942) occurs on the highest mountain plateau (Zone D_3) above 1500 m. Specimens of this subspecies have shorter limbs, darker skin, possess a longer, denser, and duskier pelage than the other subspecies, and the females lack red faces.

C.   The Study Populations

Information was collected from all subspecies in their various habitats, specifically from the Wilpattu and Ruhunu National Parks in the northwest (A$_1$) and southeast (A$_2$) Arid Zones, respectively; the Sinharaje Forest Reserve in the lowland rain forest (D$_1$), Udawattekelle Sanctuary, Kandy, in the midland rain forest (D$_2$), the Ohiya and Horton Plains regions in the highland forests (D$_3$), and from Anuradhapura in the Dry Zone plains (B).

In order to gain a greater understanding of toque macaque society and its relation to the environment, particularly as socioecological relationships influence the population, my efforts were concentrated on one population in the Archaelogical Sanctuary at Polonnaruwa in the Dry Zone plains (B) (Fig. 1).

The habitat of the Polonnaruwa study site consists of natural semi-evergreen forest. In areas immediately surrounding archeological ruins, the shrub layer and some trees have been cleared. A quantitative appraisal of the forest structure was undertaken to aid ecological interpretations (Dittus, 1977a).

The forest can be conveniently categorized into a predominantly nonwoody herb layer, a shrub layer of woody vegetation up to 5 m in height, and a tree layer above 5 m. The tree layer consists of a closed canopy between 15 and 18 m high and a discontinuous layer of tall emergent trees up to 25 m.

The number of tree species characteristic of the Polonnaruwa study area was 46, representing 36 genera and 21 families. As in other forests of the Dry Zone, *Drypetes sepiara* (Euphorbiaceae) was the most common species constituting 21.3% of forest trees. Most species had few individuals represented. Emergents were defoliated for part of the year, while most trees in the canopy layer retained their foliage. This feature, and measures of species diversity, placed the semievergreen forest as an intermediate type between tropical evergreen rain forest and monsoon or seasonally defoliated forest (Walter, 1971).

The study site was a peninsula of forest, bordered by a lake, stream, or irrigation channel, and active and abandoned cultivation. It was continuous with more extensive forests along a narrow section. Sri Lanka's other primate species occurred in the study site; this included the gray langur (*Presbytis entellus*), purple-faced langur (*P. senex*), and slender loris (*Loris tardigradus*). Other mammals typical of the region were also present, and included in the axis deer (*Axis axis*), muntjac (*Muntiacus muntjac*), mouse deer (*Tragulus meminna*), jackal (*Canis aureus*), fishing cat (*Felis viverrina*), jungle cat (*F. chaus*), mongooses (*Herpestes edwardsi* and *H. smithi*), and many others. However, larger mammals, the elephant (*Elephas maximus*), leopard (*Panthera pardus*), sloth bear (*Melursus ursinus*) wild pig (*Sus scrofa*), and sambar deer (*Cervus unicolor*) were absent. Domestic cattle and water buffalo (*Bubalus bubalis*) occurred in open places within the study area.

[Nomenclature follows that of Eisenberg and McKay (1970).]

D.   Research Schedule and Methods

The study was initiated in September 1968 and was continuous over $3\frac{1}{2}$ years until early 1972. A 2-month field trip in March and April of 1975 served for collection of certain demographic and social information. Observations between 1972 and 1975 were sporadic and were maintained by my assistant, Mr. S. M. S. Farook.

The Polonnaruwa study population consists of nearly 450 macaques from eighteen troops whose home ranges are contiguous and which together constitute a deme (local population). General ecological and demographic records and certain selected types of information were maintained on all eighteen troops. Five troops were chosen for more intensive study. Two troops of 18 and 36 individuals, respectively, served as the focal study troops for detailed quantitative behavioral and ecological information.

Nearly all animals in the study population have been individually identified and photographic records are available for most of them. Identification was based on variations in facial coloration, structure of the head hair or toque, color of pelage, size, and sex. Facial features were particularly individualistic. Differences in the distribution and size of dark or red pigment spots and scars, frequently on the ears, were most helpful. Only females had red faces. Observations were made with the aid of binoculars. Troops were never provisioned or intentionally interfered with. The main study troops were well habituated by the end of 2 years; some animals approached the investigator to within 1 m or less. For detailed quantitative information, individuals were followed from dawn to dusk, and often for several days in succession.

III.   The Ecology and Behavior of *Macaca sinica*
       at Polonnaruwa

In an attempt to give a brief overview of social and ecological relationships, I have omitted much technical information. The reader may refer to reports that have been or will be published elsewhere.

A.   The Structure and Demography of the Population

Toque macaques live in social groups or troops that average 20.6 (standard error of the mean = 2.5) animals excluding first year infants, and 24.7 (standard error of the mean = 3.0) including infants as neonates. The size of troops ranged from 8 to 43 in Polonnaruwa and the frequency distribution of their size is skewed, the modal class being less than the average (Dittus, 1974).

The composition of the population by age and sex class is given in Table I.

TABLE I

The Age and Sex Structure of the Population and Rates of Mortality[a]

		Males		Females	
Age class	Age (years)	Number	Percentage mortality[b]	Number	Percentage mortality[b]
Infant	0–1	68	39.5	59	52.6
Juvenile	1–5	83	13.5	54	20.0
Subadult	5–7	23	32.8	—	—
Adult male	7–30	48	8.2	—	—
Adult female	5–30	—	—	111	5.2
Total		222		224	

[a]Adapted after Dittus (1975).
[b]The average percentage dying annually.

The total number of males almost exactly equaled that of females. There were 2.3 times as many adult females as adult males, but among the juveniles and infants there were more males than females. The sex ratio at birth was 1:1, and the ages of juveniles were known or estimated according to the morphology of juveniles with known ages. The differences in the sex ratios among the juveniles and adults are therefore not attributable to sex differences in the rates of maturation; instead, they reflect differences in sex-specific rates of mortality (Table I), as outlined by Dittus (1975). The causation and significance of these facts will be examined later in relation to population regulation (Section III, F).

B. The Forest as a Resource Area

The life sustaining requirements of the toque macaque may be regarded as food, water, and refuges for resting and sleeping, and for the avoidance of predators, the direct sun and rain. An enumeration of the specific resources, and an examination of the manner in which they are utilized follows below.

1. Refuges from Predators, the Sun and Rain

Anatomically the toque macaque is structured for arboreal life (Grand, 1972), but travels with ease on the ground. When alarmed or under threat of predation, animals rapidly fled to neighboring trees. Although animals were more concealed in the shrub layer, individual shrubs rarely served as refuges under intense threat of danger. Extensive open places were either circumvented or crossed rapidly, with caution and in unison as a compact group. Adult males were at the head and rear of such traversing groups. Ground versus arboreal foraging was a direct function of the location of food sources. In behaviors, like resting or grooming, where animals had a freedom of choice of stratum use, a marked arboreal preference was evident (Table II).

Daytime resting and grooming occurred on the large lower branches of trees possessing a dense canopy that provided shade. Direct sunlight was generally avoided, when possible, even during foraging. Sunning occurred only when the fur was wet from rain or early morning dew.

Rain induced animals to sit hunched on the lower branches of trees—usually several animals huddling together. Some animals definitely chose shelter beneath large overhanging branches, or nestled close to protective concavities along the tree trunk. Others sat exposed in the rain. Very heavy rains always induced resting in the trees.

2. The Selection of Sleeping Trees

In marked contrast to daytime siestas beneath the tree canopy, at night the macaques slept in the tops of canopy trees, preferably in emergents. A general criterion of sleeping trees was that the trunk itself was of a width, smoothness, and lack of branches along its lower reaches so that the macaques themselves could not climb it. Entrance to the sleeping tree was gained by way of smaller adjacent "access" trees or a series of them. Ideally the thin flexible branches of the access trees barely reached the branches of the sleeping tree. Inaccessibility was such that otherwise independent infants were, and often had to be, carried by adults. Generally, sleeping loci were at the terminal forks toward the outer third of large branches that extended far from the tree trunk (Table III). During strong winds, sturdier, more central branches were used. Sleeping near the trunk was rare. Much shifting and testing of branches occurred before animals settled for the night.

At sunrise, I would frequently find sleeping parties in the same location and with the same ordering of their members as the night before. Shifting during the night did occur, but rarely.

TABLE II

Total Time[a] of Activity per Forest Stratum

	Number of minutes engaged in activity		
	Foraging	Resting and grooming	Daytime sleeping
Ground	1099	1039	0
Arboreal	2149	4136	441

[a]Total time was 148 hours over 15 days.

TABLE III

The Distribution of Sleeping Parties Along the Length of Large Canopy Branches[a]

Location	Number of sleeping parties located
Near the trunk	2
Less than half way outward from trunk along branch	1
More than half way outward from trunk along branch	7
At the terminal forks of a large branch	52

[a]Data are from four troops taken over 25 days; sleeping parties consisted of one to eight animals.

All emergent species were used at some time, but only two species (*Adina cordifolia* and *Schleichera oleosa*) were used more than expected according to their availability in the forest (Dittus, 1974). Together these two species constituted 76% of 196 trees of twelve species used for sleeping. In one troop they represented 97% of all trees used.

Each troop had several preferred sleeping trees. For example, one troop (of eighteen animals) used 88 different sleeping trees on 190 nights spread over 3 years; 47% of the sleeping trees were used only once, whereas 10% of the trees were used 83% of the time (Dittus, 1974). On any one night two or more adjacent trees were generally used. Such groves were located in various parts of the home range, but the most frequently used were central. Compared to small troops, large troops used a greater number and diversity of trees on any one night (Dittus, 1974).

The use of sleeping trees on successive nights was generally staggered, as only 15% of trees were used again on a successive night. In part, this reflects day to day variations in foraging routes. Often, however, daily routes terminated in the same area, and different sleeping groves and trees were nevertheless chosen. This suggests a strategy to counter predation.

Because of the great number of trees in the forest and the fact that each troop had sleeping trees in all parts of the home range, one might assume that sleeping trees are not a limiting commodity. However, several facts combine to suggest the opposite: (1) Troops were very selective as to the type of tree used for sleeping; (2) The troop is a spatially cohesive unit and dispersal over great distances is not favored. A troop settling for the night, therefore, was not confronted with a choice of sleeping trees in its entire home range; rather, its choice was limited by the specific locality where it terminated its movement for the day. (3) Animals often displaced one another from sleeping sites, suggesting that they competed for them. As a consequence, some animals, particularly of large troops, often utilized trees and sleeping sites that were subjectively judged to be

unsafe because they could be easily reached from the ground. (4) Troops occasionally displaced one another from sleeping trees. (5) If an anti-predator strategy involving the staggered use of sleeping trees were operative, then it would
further limit the number of available sleeping places.

If sleeping trees are limiting, the limitation would be more severe for a large
troop than a small one which can be accommodated by relatively fewer sites per
amount of space.

3. Water Resources

During the rainy season macaques seldom drank from sources of accumulated
water. Apparently the moisture content of foods plus that obtained by licking
the wet fur and wet leaves provided an adequate amount of water. During the
dry season (May to September) water was obtained from the following sources:
(1) water holes, (2) the lake, (3) irrigation channel which bordered parts of the
study area, (4) ancient wells, and (5) water accumulated in tree cavities.

The nature of the water source varied for each troop. Water in tree cavities
dried out early in the dry season, and even though water holes occurred throughout the study area, few retained water permanently. The distribution of water
during the drought had a profound effect on daily movements of troops. Many
troops visited a permanent source of water at least once daily. In some instances
parts of the home range were utilized only during the drought for access to
water, and resulted in use of corridors into the home range of neighboring troops.

Judging from the frequent treks to water during the drought, water was an
essential requirement for the macaque.

4. Selectivity in Food Resources

Of the 46 species of trees considered typical of the semi-evergreen forest at
Polonnaruwa, the macaque exploited 41 (89%) for some vegetal part. A list of
these species and the relative frequency with which various parts of each species
were consumed has been given elsewhere (Dittus, 1974). In summary, 11 species
were used frequently, 6 occasionally, and 24 species were used rarely. Frequency
of use is, of course, a function of availability. Some vegetal parts might be considered "unavailable" because of their toxicity; most mature leaves and the fruits
of some species, even though abundant, were never eaten. The fruits of some of
these avoided species are known to be poisonous or unpalatable to the local
human residents. However, some of these same fruits were eaten by the langurs
of Polonnaruwa. Hence, if toxicity was a factor, it was specific only to the digestive system of the macaque. Amerasinghe *et al.* (1971) have shown that the anatomical and histological structures in the digestive systems of these primate
species differ significantly and correlate with diet.

Considering palatable foods, a definite preference for the fruits of figs (*Ficus*
sp.) was evident. Food items among the remaining plant species were chosen

more or less according to their availabilities (Dittus, 1974). In the one troop analyzed, fig fruits made up at least 51% of the total diet by dry weight, and 44% was
from one species, *F. amplissima*. Overall, fruits constituted over 70% of the diet
by dry weight. The remainder was of leaf shoots, leaves, flowers, seeds, petioles,
mushrooms, fungi, herbs, grasses, roots, tubers, the pith of succulent twig growths,
and resin.

Insects and various invertebrates contributed only 1−2% of the total diet by
dry weight but, on the average, approximately 20% of daily foraging time was
involved in the search for these creatures. Typically, insects were sought after
a major meal on fruit and during more relaxed foraging for other items of food.
Insects were generally sought on the underside of leaves of most tree species. In
certain seasons, for brief periods, insects were major food items. Termites, during
their nuptial flights, constituted the major dietary item for several days. Flushes
of caterpillars or weaver ants (*Oecophylla* sp.) were similarly exploited.

Carnivory was opportunistic, noncooperative, and included lizards (*Calotes*
sp.), some birds, their nestlings and eggs, the nestlings of tree mice (*Vandeleuria
oleracea rubida*), adult and young palm squirrels (*Funambulus palmarum kelaarti*),
and a small unidentified fish species.

5. Differential Feeding in Vegetation Layers

With the exception of two emergent species (*Ficus bengalensis* and *Schleichera oleosa*), the tree species used for food were characteristic of the canopy.
In addition, trees were exploited somewhat differently than shrubs. A comparison of parts used regardless of species identity or weights consumed
(Table IV) indicated that among trees the macaques preferred fruit (chi-
square = 7.9, p <0.01) over all other parts combined, but among shrubs
there was no significant preference between the number of shrub species
exploited for fruit versus those exploited for leaves (chi-square = 3.6, p >0.05).
This difference is related to the fact that shrub fruits were very small compared
to most tree fruits, and afforded less mass per hand picking action than a terminal bunch of leaves. If the total weights of shrub fruit versus shrub leaves con-

TABLE IV

**Comparison between Trees and Shrubs of the Number of Species that are
Exploited for Different Vegetal Parts**

	Vegetal part			
	Fruit	Flower	Leaf	Other
Number of tree species exploited	36	10	9	2
Number of shrub species exploited	17	9	10	3

sumed had been incorporated into the estimate, a preference for leaf shoots might have been indicated.

The marked seasonality at Polonnaruwa directly affected plant phenology (Rudran, 1970; Muckenhirn, 1972) and, hence, determined the types and quantities of foods available in the different layers at different times. During March and again toward the end of the dry season (September), the fruits from trees were least available (Muckenhirn, 1972). At such times the herb and shrub layers were exploited more intensely.

Foraging in the herb layer during the drought was not rewarding because most of it was parched. The thickest herb growth was in open exposed areas that were avoided by macaques for reasons of temperature control, difficulty in keeping together, and exposure to predators. Food items in the herb layer were generally small, or access to larger morsels, such as roots and tubers, involved much energy output. While foraging on herbs, the length of daily travel routes increased, and resting decreased as most daylight hours were devoted to the search for food.

During fruit scarcity in March, many shrub species produced new leaf shoots which constituted an important alternate food source. The shrub layer was again exploited toward the end of the drought. The species of shrubs that were of greatest food value to the macaque belong to the *Glycosmis* and *Randia-Carissa* associations that make up the climax shrub growth throughout most of the Dry Zone (Dittus, 1974). Typically, they occur under the closed canopy of trees and lack almost any herbaceous growth. Presumably because of their protection in the shade of trees, these shrubs retained sufficient moisture to carry some new leaves and flowers. The macaques consumed these succulent parts as well as the mature leaves of some of the shrub species.

In summary, the mainstay of the macaques' diet was provided by the fruits of mainly canopy species, especially figs, but during periods of fruit shortage, the herb and especially shrub layers provided important sources of alternate foods.

C. The Relationships between Home Range Size, Troop Size, and Resource Availability

The home ranges of fifteen troops were assessed by charting troop movements on maps drafted from large-scale (1:2,380) aerial photographs. A troop's home range was considered as that area of land which it had occupied at some time during the period of observation. Home range size was an asymptotic function of observation time (Dittus, 1974), but after 150 hours of observation, home range did not increase much for an average-sized troop. Hence, all troops were observed for at least 150 hours, although most were observed for more than this. Observations were taken throughout the year because of seasonal differences in home range use.

Adjoining home ranges overlapped extensively; areas of exclusive use were small or nonexistent. The size of home ranges varied between 17 hectares for a small troop of 8 animals, to 115 hectares for a large one of 37 animals. In addition, there was a positive correlation ($r = 0.65, p < 0.02$) between the number of animals in the troop (N), and the maximum extent of the home range (HR) measured in hectares. The relationship was defined by the equation $HR = 9 + 1.5\ N$. The regression coefficient of HR on N was significant at $p < 0.01$ (t test).

Inherent in the concept of home range is that its size should be a function of the energy requirements of its inhabitant(s). Large-sized mammals have greater energy requirements and therefore exhibit larger home ranges than small-sized ones (McNab, 1963). It seems logical, therefore, that the increase in home range size with troop size is a reflection of a greater need for resources in large troops than in small troops. However, the more resources a home range harbors, the greater its carrying capacity. Hence, troops living in habitats rich in resources might be expected to have relatively smaller home ranges than those of comparable troop size that do not inhabit resource-rich areas.

Considering water resources, all troops had access to a permanent source during the drought, some achieving this only through using corridors or extensions of their home ranges at such times.

An attempt was made to assess availability of plant resources per home range. In vegetation studies, the concept of minimal area refers to the smallest area of land that provides sufficient space and combinations of habitat conditions for a particular plant community to develop its characteristic combination of plant species and structure (Cain and Castro, 1959). From a vegetation analysis, minimal area for the Polonnaruwa forest trees was approximately one-quarter hectare (Dittus, 1977a). All macaque home ranges were sufficiently large to have incorporated representatives of all tree species in proportions typical of the Dry Zone forests of Polonnaruwa.

There were floristic variations between home ranges, but most occurred in the shrub and herb layers. This was partly the result of past and present human interference. Considerable local variation also occurred in the availability and distribution of emergent species, as is typical of Dry Zone forests (Andrews, 1961; Oudshoorn, 1961a,b; McCormack and Pillai, 1961a–e. Since most emergent species were of no direct food value to the macaque, this variation was probably inconsequential to their feeding ecology.

The variation in the shrub layer was of particular interest because of the importance of shrub species in the *Glycosmis* and *Randia-Carissa* associations as alternate food sources during shortages of fruit in the tree layer. A survey of the number of these shrubs per home range indicated an inverse correlation ($r = -0.61, p < 0.01$) between home range size (HR) in hectares and the number of shrubs (S) per individual (I) in the troop. The relationship was approximated by the equation $HR = 57 - 0.94\ S/I$. The regression coefficient of HR on S/I was significant at $p < 0.05$ (t test). That is, home ranges having many of these shrub species were smaller than those with few, in view of the number of animals each home range supported. The tree density in these shrub associations was 1.3 times greater than in the remainder of the forest, and it is possible that this may also have contributed to the relative richness of home ranges.

Two troops not included in the above analysis fed extensively from a rice mill that bordered the study area, and another troop fed from the municipal garbage dump outside the study area. Compared to the natural food sources of the other troops, their food sources were spatially concentrated and constantly available. The home ranges of these three troops were considerably smaller than predicted by their troop size.

These data suggested that troops adjusted their home ranges in accordance with the availability of critical resources per individual in the troop.

D. Interaction with Other Species

Of the three other species of primate in Polonnaruwa, the loris is a solitary, nocturnal insectivore (Petter and Hladik, 1970), and the langurs are diurnal group-living leaf eaters. *Presbytis senex* utilizes mostly the canopy and emergent tree species, and rarely comes to the ground. *Presbytis entellus* uses canopy trees and the ground layers of vegetation, and spends much time on the ground (Muckenhirn, 1972). Fruit contributed a much smaller fraction of the total dietary intake of both langur species than in the macaque (Hladik and Hladik, 1972). Macaques invariably displaced both species of langur from any contested resources.

Sri Lanka's only large predator, *Panthera pardus*, is absent from Polonnaruwa; instead feral dogs preyed on macaques. The macaque's response to dogs was one of alarm, vigilance, and avoidance. Large birds of prey and snakes elicited similar responses, although snakes, even pythons (*Python molurus pimbura*), were often approached with curiosity. Most other animals were either ignored, chased, or investigated by young macaques.

E. The Relation between Macaque and Man

Because of the Buddhist and Hindu ethics prevailing in Sri Lanka, the macaque, sometimes mistaken as the Hindu langur god, Hanuman, enjoys a certain degree of protection. This is quickly lost, however, wherever cultivation borders on macaque habitat, since macaques eat man's crops. Consequently, cultivators poison, shoot, trap, mutilate, or otherwise harass them.

A conspicuous feature of the Polonnaruwa population was that macaques fed on cultivated plants or on garbage predominantly during periods of scarcity of natural foods. Even then, such feeding was confined mainly to those troops whose home ranges had been largely cleared of the shrub layer and where domestic cattle and water buffalo grazed the herb layer. Troops whose home ranges encompassed predominantly intact forest and which also bordered cultivation were rarely observed feeding on crops, and not all troop members ventured onto cultivated land. No troop relied on crops as a major source of food, and such feeding invariably ceased with the onset of fruiting periods.

The avoidance of cultivated areas was partly a function of the risks of harm from dogs and humans, but may also reflect a true preference for forest foods.

In many other areas macaques survived in small patches of remnant forest or scrub, or within wooded parkland as in the Sacred Area of Anuradhapura. Crops and garbage were important food sources in such areas. However, a striking characteristic of these animals was their poor state of health; many had missing or broken limbs, and on the average they weighed significantly less than animals of forest-living troops (Dittus, unpublished data).

Toque macaques possess a strong unpleasant odor, and to my knowledge are not hunted for food. Although young animals are sometimes captured as prospective pets, their highly active and exploratory (i.e., destructive) nature leads to their demise or to a life chained to a post.

F.   Population Regulation

1.   Net Rate of Population Growth

The net rate of growth of a population (R_0) is the outcome of rates of birth and immigration that increase population size, and rates of death and emigration that decrease it. I have shown elsewhere (Dittus, 1975) that the population of macaques at Polonnaruwa was at equilibrium ($R_0 = 1$). Fluctuations in population size did occur, but corresponded to the annual birth season and subsequent mortality and emigration (Fig. 4). On an annual and superannual basis, the population maintained its same level. This state of balance in a wild population starkly contrasts the rapid increase in population size ($R_0 > 1$) exhibited by the managed colonies of Japanese macaques (*M. fuscata*) at Takasukiyama (Itani *et al.*, 1963; Itani, 1975) and of rhesus macaques at Cayo Santiago (Koford, 1965) and at La Paraguera (Drickamer, 1974). These colonies are provisioned with a superabundance of food.

Birth rates of 0.688 infants born per adult female per year were sufficient to theoretically permit rapid population growth in the wild toque macaque, and the equilibrium state was achieved primarily through high rates of mortality. Of all males born, 89% died prior to adulthood, which was reached at approximately 7 years; likewise, 85% of all females born died prior to adulthood, attained at 5 years (the average age of first pregnancy). Once adulthood was

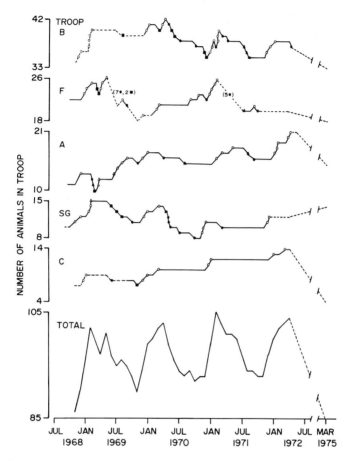

Fig. 4. The fluctuations in troop size are shown for five different troops (A, B, C, F, and SG) and for the total of these troops. Birth, ○; death, ●; immigration, △ ; emigration, ▲; loss to unknown cause, ■; discontinuous observation period, - - - -. Figures within parentheses represent surmised events of uncertain date but within the time span indicated.

reached, rates of mortality remained very low for both sexes, but were somewhat higher among adult males than among adult females (Table I).

Although predation, mainly from feral dogs, occurred, it was not the major cause of mortality.

2. Mortality in Relation to Differential Access to Vital Resources

During the nonbreeding season approximately 82% of all threats in a troop of 37 animals occurred during foraging. Because of the fundamental importance of food and water to the macaque's well being, threat behavior that occurred only during foraging was examined. The frequencies of threat between individuals of different age and sex were summarized and compared to theoretical or expected frequencies. These were calculated according to the number of individuals per age and sex class, and to differences in spatial association between them. An age—sex class was considered dominant if its members threatened those of another class more than expected. It was found that adult males threatened each other and all other age—sex classes more than expected; adult females were dominant to juveniles and infants; and juveniles dominated infants. Among the juveniles and infants, the males dominated the females. Overall, female juveniles and infants were threatened about twice as often as the males of these age classes. Subadult males were peripheral to the troop and interacted minimally with the remainder of the troop.

The rank order between age—sex classes is listed in Table V according to a summary of threats given and received per class during foraging. Two different sized troops were analyzed (Dittus, 1977b).

A threat during foraging stopped the threatened individual from feeding or prevented it from approaching the feeding area of the dominant individual. In 36% of all threats during foraging, the dominant displaced the subordinate and consumed the food that the subordinate had found. Hence, the ratio of threats received to those given (Table V) roughly reflect differential access to food and water resources between age and sex classes; adult males had the

TABLE V

Ratio of Threats Received to Those Given per Age—Sex Class During Foraging[a]

	Adult male	Sub-adult male	Adult female	Juvenile Male	Juvenile Female	Infant Male	Infant Female	Total
Number of animals in troop B	4	3	8	9	9	3	1	37
Frequency of threats								
Given (g)	344	28	355	355	333	52	13	1480
Received (r)	33	22	299	350	563	126	87	1480
Ratio r/g[b]	0.1	0.8	0.8	1.0	1.7	2.4[c]	6.7	
Number of animals in troop A	2	3	4	3	3	1	2	18
Ratio r/g[b]	0.1	0.4	0.9	1.3	2.4	0.0[c]	3.7	

[a]Adapted after Dittus (1977b).

[b]The magnitude of the ratio correlates inversely with social rank.

[c]The infant male of Troop A was a neonate, mostly carried by the mother and foraging very little.

greatest relative access and infant females the least. Indeed, the usurpation of food sources, or exploitation down the dominance hierarchy, had the effect of imposing lower foraging success on subordinate classes. With decreasing rank, duration of foraging increased, feeding rates decreased, and the proportion of low quality foods in the total diet increased (Dittus, 1977). Among adult males and juveniles, subordinates weighed less than dominants (Dittus, 1977b).

These data suggest that mortality is socially imposed through direct competition for vital resources, which is most acute for animals prior to adulthood. The greater mortality manifest by the youngest animals in general, and by the females relative to males among the juveniles and infants, was probably a direct result of the much greater frequency that they were prevented access to resources.

3. Mortality in Relation to Breeding and Migration

During the mating season, rates of threatening increased twofold, but were confined primarily to adult males, estrous females, and subadult and older juvenile male "followers" that were directly involved in mating activity. Most threats issued from adult males. The incidence of wounding increased for these animals during the mating season.

Adult males recently deposed from high status in the dominance hierarchy and subadult males wandered between troops during the breeding season. Rates of migration were highest in subadult males and, on the average, all males left their home troop before reaching adulthood. Migration was a stressful period because the subadult immigrants were relegated to low priority of access to resources by the new troop, and were frequently wounded from fights with established males. The peak in mortality among subadult males (Table I) coincided with the peak in migration in males, which suggests that the rigors of migration underlie the observed mortality. Adult males migrated at lesser rates than the subadults and had better chances of access to resources. Their rates of mortality were less than those of subadults, but higher than those of adult females which never shifted between troops (Dittus, 1975).

Insofar as mortality is socially imposed in direct reference to food resources, I proposed (Dittus, 1977b) that the density and age–sex composition of the population is regulated in accordance with the carrying capacity of the environment. Such regulation appears to act predominantly within troops (fig. 4) and some evidence indicates that occasionally troops expand at the expense of others (Dittus, 1977b). The limiting resources appear to be mainly the availabilities of food and water. Subordinate animals have least priority to favored sleeping places and similar refugia; under conditions where such might be in short supply, as in large troops, the availability of such resources may also influence population growth.

IV.   The Ecology of the Subspecies of
Macaca Sinica

The habitats of the three subspecies differ greatly; therefore, the ecology of each subspecies will be consdered separately.

A.   *Macaca sinica sinica*

The Dry Zone race is the most widely distributed. Its forest habitat grades from the relatively wet Intermediate Zone to the Arid Zones (Fig. 1). Throughout these areas macaques have been sighted only in areas of natural forest and in protected parklands. The Polonnaruwa study site is close to a lake, and probably exemplifies relatively moist habitats in the Dry Zone.

In the Arid Zones macaques investigated in the Wilpattu and Ruhunu National Parks, where they were restricted to riverine forests, and to areas immediately surrounding permanent bodies of water, such as tanks or villus. Average annual rainfall in the Arid Zone is between 950 and 1500 mm, and is highly seasonal. There are 4—5 months of drought.

The forests over most of these arid areas are low stature scrub or seasonally dry forest, with a Low diversity of tree species. In some areas almost pure stands of one tree species (*Drypetes sepiara*) prevail (Oudshoorn, 1961b). Compared to the moister forests, as in Polonnaruwa, fig species are rare, and phenological activity is more synchronized between tree species. Seasons of fruit shortage are, therefore, more accentuated in these arid regions (Muckenhirn, 1972). Studies of forest productivity, measured by the amount of litter (leaves, flowers, fruits, and small twigs) falling from trees and shrubs indicated at total dry weight of 2 metric tons/hectare/annum in the seasonally dry forests, and 5 tons/hectare/annum from forests subject to the same climate, but located near permanent water. The latter compared favorably to the 4.5 tons/hectare/annum in the Polonnaruwa forest (Hladik and Hladik, 1972).

The macaques' absence from the extensive dry regions of the Arid Zone forests, therefore, is very likely attributable to a lack of a perennial water supply and the resulting lack of an adequate year-round food supply.

During the dry season most mammals, including leopards, concentrated their activities near permanent water holes (Beck and Tuttle, 1972; Eisenberg and Lockhart, 1972). A zone of unwooded grassland, generally 50—150 m wide encircled these water holes, such that access to water involved a risk of being preyed upon. However, macaques were not represented among 51 records of leopard predation in the Wilpattu National Park (Muckenhirn and Eisenberg, 1973). There were no dogs in the park. It is likely, therefore, that the macaque population in these forests was held in balance primarily according to resource availability and through the same social mechanisms that were operative in the Polonnaruwa population.

The carrying capacity of the Polonnaruwa forest is approximately 100 maca-ques per km², or a biomass of 300 kg/km² (Dittus, 1975). Judging from the pro-ductivity, species diversity, and structure of riverine and related forests, which were comparable to those at Polonnaruwa, the ecological density of macaques in these restricted habitats was probably of the same order of magnitude as in the Polonnaruwa forests. Estimates for arid scrub forests as a whole, which incor-porate areas not inhabited by the macaque, indicated a 300 times lesser density, or 0.3 macaques per km² (Eisenberg *et al.*, 1972).

B. *Macaca sinica aurifrons*

The habitats of the rain forest race receive an average of over 3000 mm rain per year. Seasonal differences are slight, and most areas receive an excess of rain-fall at all times of the year. The natural forests here are high in species diversity and are dominated by the family Dipterocarpaceae (Merritt and Ranatunga, 1969; McCormack, 1961). The herb and shrub layers are sparse and most fruits occur in the canopy layers. The canopy is multilayered; the uppermost layer averages 24—27 m in height. Emergent species may reach up to 40 m.

Macaques were observed only briefly in the Sinharaje Forest Reserve; all macaques were in the canopy. Productivity in this forest was measured as 6.4 metric tons/hectare/annum dry weight of litter fall (Hladik and Hladik, 1972). No direct estimate of macaque density is available but, if the measure of forest productivity bears any relation to its carrying capacity for macaques, one might predict a macaque density at least equal to that in the semi-evergreen forest at Polonnaruwa. This prediction is supported by my subjective impression of a relatively high macaque density in the Udawatakelle Sanctuary of Kandy loca-ted near the midland rain forest, where macaques were more easily observed. Most macaques were in the canopy; one troop which fed on refuse was observed on the ground.

C. *Macaca sinica opisthomelas*

The montane rain forest habitat of this highland race receives in excess of 2000 mm rainfall annually. Seasonal differences are slight and no drought occurs. The average annual temperature is 15° to 16°C or 10°C lower than in most of the lowlands.

The forest canopy consists of one layer of trees, generally no higher than 10 m, with few or no emergents. Mosses, lichens, and epiphytes grow abundantly over the trunks and branches of trees, as much moisture is brought by clouds.

The shrub layer is extremely, dense, often posing an impenetrable tangle, and the herb layer is sparse.

Most trees are evergreens and exhibit a slight peak in fruiting in the period of least rainfall, between May and September, although fruit occurred throughout the year and no shortage was evident (Rudran, 1970). No figs grow at these higher elevations.

Much of the highland vegetation occurs as open *sholas*, or isolated patches of forest in grassland. The type locality of this race was given as Horton Plains (Hill, 1942), a dissected plateau of *shola* forests at 2300 m elevation. Macaques here were found only in very large patches and in continuous forest (R. Rudran, personal communication). Extensive open grasslands separated the smaller patches of forest. Perhaps the danger posed by leopards in traversing such areas prevented the macaques from using small isolated patches of forest.

Estimates of productivity in this type of forest at Horton Plains, yielded a dry weight of 5 metric tons/hectare/annum (Hladik and Hladik, 1972). This implies a carrying capacity for a macaque population comparable to that at Polonnaruwa. However, the macaque's distribution, limited to large tracts of forest, suggests a somewhat lesser overall density, and previous estimates indicated approximately thirteen macaques per km^2 (Eisenberg *et al.*, 1972). This density is likely to be considerably less toward the ridges of mountains where the forest trees are in poor form and dwarfed.

V. Estimation of Available Habitat and Population Size per Subspecies

Given that macaques are strictly forest dwellers, the density data may be extrapolated to estimate the total population size of each subspecies, based on an estimate of the area and quality of their respective forest habitats.

An inventory of the forest of Sri Lanka was carried out between 1956 and 1959 by the Hunting Survey Corporation Limited, under the auspices of the Ceylon-Canada Colombo Plan (e.g., Andrews, 1961). Its purpose was purely economic. Through the use of aerial photographs and field surveys, the extent of forests in each climatic zone was measured and classified according to merchantable yields of timber: nonproductive, low, medium, and high yield. Such differences refer mainly to the quantity of trees with large boles or trunks. Since tree productivity, in terms of fruit, flowers, and leaves (i.e., primate foods) is unrelated to trunk size, differences in yield, per se, are inconsequential to macaque feeding ecology. Notwithstanding, yield correlates with certain site characteristics, such as species diversity, density, size, and distribution of trees per species. Such data have been presented quantitatively and have been supplemented by extensive descriptions and photographs of forest structure, shrub species, moisture, and other ecological conditions. These data serve the present purposes well because the classification of forests by climatic zone adheres closely to the geographic distribution of the subsepcies of macaques, and the descriptive and quantitative data, that accompany the foresters' classification by merchantable yield, are relevant to macaque ecology and permit an appraisal of the suitability of the habitat for macaques.

I established three categories of macaque habitat: optimal, variable, and marginal. In adapting the forester's yield classification to my own, I used the following criteria.

Forests are considered as "optimal" habitat if they have an equal or greater species diversity, comparable or more complex forest structure, and an equal or greater amount and seasonal distribution of rainfall than the semi-evergreen forest of Polonnaruwa. They include the following foresters' yield classes: low, medium, and high yieldforests of the Wet Zone, low and medium yield forests of the Intermediate Zone, and the medium yield forests of the Arid and Dry Zones. There are no high yield forests in the Arid, Dry, and Intermediate Zones. Medium yield forests of the Dry and Arid Zones are found in areas of moisture as in riverine forests (after Andrews, 1961; McCormack and Pillai, 1961a,b,c,d,e; Oudshoorn, 1961a, b; Merritt and Ranatunga, 1959; McCormack, 1961).

Habitats considered as "marginal" incorporate predominantly the scrub and low stature—low species diversity stands that are characteristic of the arid and disturbed areas that typically are not inhabited by macaques, except in the locally restricted richer forest near places of permanent water. Such areas have been classified as nonproductive in the Arid and Dry Zones (Andrews, 1961; Oudshoorn, 1961 a, b; McCormack and Pillai, 1961 a–e).

Forests considered as "variable" habitat for the macaque, correspond to the low yield forests of the Dry and Arid Zones. They are the most extensive and somewhat variable. Some have a species diversity less than at Polonnaruwa and are subject to longer drought periods. They occupy extensive flat and rolling areas, mostly away from permanent water. The nonproductive forests of the Wet and Intermediate Zones are also considered as variable habitat.

The amount of macaque habitat available was computed according to the areas of forests of different yield and type given by Andrews (1961). However, these forest inventories were taken almost two decades ago, during which time much forest has been cleared. Calculations according to figures given by Wijesinghe (1972) indicate that since 1956 there has been an annual rate of deforestation of 1.06%. The amount of habitat available in 1976 was calculated, therefore, by subtracting 18% of the estimate for 1959 (Table VI).

Since optimal habitat was judged according to the forest conditions at Polonnaruwa, the density of 100 macaques per km^2, as determined from the Polonnaruwa population, was applied to these forests. Marginal habitat is equivalent to that in the Wilpattu National Park, and the density estimate of 0.3 macaques per

TABLE VI

Preliminary Estimates of Available Habitat and Population Size per Subspecies of *Macaca sinica*

Habitat quality	Area of available forest habitat (km²)		Numerical density macaques/ km²	Population estimate for 1976	Estimate of biomass (metric tons)
	1959	1976			
M. s. sinica					
Optimal	1,522	1,248	100	124,800	374.4
Variable	12,667	10,387	30	311,610	934.8
Maginal	10,890	8,930	0.3	2,679	8.1
Total	25,079	20,565		439,089	1,317.3
M. s. Aurifrons					
Optimal	1,639	1,344	100	134,400	403.2
Variable	639	524	30	15,720	47.2
Total	2,278	1,868		150,120	450.4
M. s. opisthomelas					
Variable	138	113	13	1,469	4.4
Total for species	27,495	22,546		590,678	1,772.1

km² as determined from studies in that park (Eisenberg *et al.*, 1972) was used. The density estimate for variable habitat was derived by averaging three estimates from the Dry Zone forests; that from Polonnaruwa, Wilpattu, and a third, of 1.1 macaques per km² (after McKay, 1973), from the Gal Oya National Park in the eastern Dry Zone. This average is approximately 30 macaques per km². The density estimate derived from the Horton Plains region was applied to the montane rain forest in general.

The estimates of available habitat and the corresponding population size per subspecies is given in Table VI. It must be emphasized that these extrapolations are only very rough approximations, since virtually nothing is known directly of the density and ecology of the macaque in nearly all of the areas.

VI.   Conservation Prospects and Suggestions

The size of macaque populations is in close balance with the carrying capacity of their habitats through behavioral processes which curtail or increase population size in direct relation to resource availability. The macaque relies heavily on fruit and insects for food, and plays an important role in forest maintenance through seed dispersal and insect control. Because of these relationships, the cropping of macaque populations is both unnecessary and undesirable as a means of habitat protection from overexploitation of food resources. Such overexploitation and resulting habitat degradation appears to be confined primarily to natural populations of large herbivores, such as the elephant, that have been prevented access to their normal ranges, or to introduced populations of domestic cattle (McKay, 1973). Preservation of the toque macaque requires conservation of its natural habitats.

The greatest threat to natural habitat is the rapidly expanding human population and its concommitant demands for land for agriculture, plantation, and forest products. For example, the human population of Sri Lanka has expanded exponentially from 3.6 million in 1900 to 13 million by 1972, representing an average annual growth rate of approximately 3.6% (after Crusz, 1973). Most of this human population is concentrated in the agriculturally suitable Wet Zone which comprises about one-third of the island, but now has less than 8% of its forests. The island-wide average rate of deforestation during the past two decades is approximately 1.06% annually (after Wijesinghe, 1972) and will increase under present pressures for greater timber exploitation in the last remnants of rain forest and for agricultural expasion in the Dry Zone. At present rates the estimated 22,546 km^2 of remaining forests will vanish within one human life-span.

The most threatened of the toque macaque subspecies is *M. s. opisthomelas* of the montane rain forests, with an estimated population of 1500 animals. In the past much of these montane rain forests have been cleared for tea plantations and through slash and burn agriculture. Three Wildlife Reserves exist in these areas but they are not secure (Hoffmann, 1968a). Potato planting has been established in the Horton Plains Reserve—one of the most unique tropical montane areas.

The subspecies of langur (*P. senex monticola*) that shares these montane habitats with the macaque is under legal protection. The macaque occurs at only one-third the density of this langur and should also be given legal protection. In addition to securing the survival of *M. s. opisthomelas*, the preservation and protection of its habitat is vital for the conservation of 25% of Sri Lanka's endemic animal species (Crusz, 1973) and a large number of endemic and unique subspecies.

Macaca s. aurifrons occurs in greater numbers than its montane relative, but its lowland and midland rain forest habitat is in the greatest danger of destruction through shifting agriculture and timber exploitation, even in the only legally protected remnant of the relict rain forest, the Sinharaje Forest Reserve. The problems facing such preservation in Sri Lanka have been detailed by Gopalasingham (1959).

The economic importance of maintaining Sri Lanka's rain forest for the conservation of water and soil, and as future timber reserves, has been pointed out by de Rosayro (1959). To this end, areas of protected and completely undisturbed rain forest are invaluable for the scientific management of neighboring areas insofar as they provide ecological standards with which to compare the effects of man's exploitation on the soil, vegetation, animals, and rainfall. Such areas also provide a reservoir of plants and animals for the recolonization of regenerating forests (Struhsaker, 1975). Tourism in the Luquillo rain forest of Puerto Rico attracts up to 4000 Sunday visitors a day (Odum, 1970). The superior economic potential of tourism over timber extraction in undisturbed rain forest is clear, even on a short-term basis. This seems especially applicable to Sri Lanka's economy, which rests partly on an expanding tourist industry.

Crusz (1973) lists 3850 km² which have been set aside as National Parks or Sanctuaries. The two largest, Wilpattu (1095 km²) and Ruhunu (1090 km²), lie in the Arid Zones and offer only marginal habitat for the macaque because they live only in gallery or other moist forests. The total macaque populations in these parks is estimated to be less than 400 and 350, respectively. The third largest, Gal Oya (259 km²), lies in the eastern Dry Zone. McKay (1973) estimated that the total macaque population there and in the surrounding regions (459 km²) is between 200 and 250 animals. My observations suggest this may be an underestimation. Together these three major parks comprise 2637 km², or 68% of National Wildlife Reserves, but protect merely 1000 to 1500 of the *M. s. sinica* subspecies. These parks alone are obviously inadequate to safeguard this subspecies. Extensive areas of forest, and much of it optimal habitat for the macaque, are still intact outside these parks. However, the establishment of National Wildlife Reserves enclosing optimal habitat for this subspecies is essential in view of current rates of deforestation.

The establishment of any Sanctuary inhabited by macaques by necessity should allow a bordering zone of 200 to 300 m of treeless grassland, which macaques infrequently cross, to ensure the peaceful interface between man and macaque. Such zones might provide grazing areas for herbivorous species.

Small town sanctuaries are an attractive solution to the preservation of local floras and faunas and are of special value to meet the present need for the biological training of students and school children (Atapattu and Wickremasinghe, 1974). The protection of small local reservoirs poses a problem, however, in highly populated areas. Large areas of high quality Dry Zone forests are not only vauable for conservation, but, like protected rain forests, may promise greater economic returns and pose fewer problems in the enforcement of conservation laws. Large reserves may be indicated on maps of Sri Lanka, but until their legal status is defined, and the enforcement of conservation laws is taken seriously, such areas will soon be cultivated (Hoffmann, 1968a, b).

The toque monkey is only one organism in a rich biome that will survive with the preservation of its habitats. Often this monkey is a source of our amusement, yet in its own society and economy it adheres to a wisdom more ancient than that of human philosophers and statesmen: it has come to terms with the quantity of its kind, assuring a quality of life to its future generations.

The quality of our own lives may be enhanced if only the words of Gautama Buddha were heeded more seriously: "The forest is a peculiar organism of unlimited kindness and benevolence that makes no demands for its sustenance and extends generously the products of its life activity; it affords protection to all beings, offering shade even to the axeman who destroys it."

Acknowledgments

Research was supported, in part, by National Institute of Mental Health Grant RO1MH15673-01 and-02; Research Grant 686 from the National Geographic Society; Smithsonian Institution Foreign Currency Program grant SFC-7004, awarded to Professor J. F. Eisenberg and Dr. Suzanne Ripley; and Research Grant 1442 from the National Geographic Society awarded to the author. Financial support during the phases of data analysis and writing was provided through Professor J. F. Eisenberg, National Zoological Park, Smithsonian Institution, and by a Smithsonian Institution Postdoctoral Fellowship awarded the author. I am especially grateful to Professor J. F. Eisenberg whose unfailing support, advice, and encouragement made this research a reality. I am indebted to Professor H. Crusz of the University of Sri Lanka for logistic support, and to the Government of Sri Lanka for permission to work in their country. I thank especially Messrs. S. M. S. Farook, S. Waas, and G. DeSilva for their devoted assistance in the field, and Miss L. Andradi and Ms. Wy Holden for secretarial help. I thank professor H. Crusz, Ms. A. Baker-Dittus, Drs. E. Morton, N. A. Muckenhirn, and J. Seidensticker for criticisms of the manuscript.

References

Amerasinghe, F. P., Van Cuylenberg, B. W. B., and Hladik, C. M. (1971). Comparative histology of the alimentary tract of Ceylon primates in correlation with the diet. *Ceylon J. Sci. (Bio. Sci.)* **9**(2), 75–87.

Andrews, J. R. T. (1961). "A Forest Inventory of Ceylon." Government Press, Colombo.

Atapattu, S. and Wickremeasinghe, C. S. (1974). Sri Lanka's Gal Oya National Park: Aspects and prospects. *Biol. Conserv.* **6**(3), 219–222.

Beck, B. B. and Tuttle, R. H. (1972). The behavior of gray langurs at a Ceylonese waterhole. *In* "The Functional and Evolutionary Biology of Primates" (R. Tuttle, ed.), pp. 351–377. Aldine Atherton, Chicago, Illinois.

Cain, S. A. and Castro, G. M. de O. (1959). "Manual of Vegetation Analysis." Harper, New York.

Crusz, H. (1973). Nature conservation in Sri Lanka (Ceylon). *Biol. Conserv.* **5**(3), 199–208.

De Rosayro, R. A. (1959). Editorial notes: The place of forestry in the wet zone. *Ceylon Forester* **4**,(2), 99–102.

Dittus, W. P. J. (1974). The ecology and behavior of the toque monkey, *Macaca sinica*. Ph. D. Dissertation, Univ. of Maryland, College Park, Maryland.

Dittus, W. P. J. (1975). Population dynamics of the toque monkey, *Macaca sinica. In* Socioecology and Psychology of Primates" (R. H. Tuttle, ed.), pp. 125—152. Mouton Publ., The Hague.

Dittus, W. P. J. (1977a). The ecology of a semi-evergreen forest community in Sri Lanka. *Biotropica,* in press.

Dittus, W. P. J. (1977b). Behavioral regulation of density and composition in a wild population of the toque monkey, *Macaca s. sinica. Behaviour,* in press.

Drickamer, L. C. (1974). A ten-year summary of reproductive data for free-ranging *Macaca mulatta. Folia Primatol.* **21,** 61—80.

Eisenberg, J. F. and Lockhart, M. (1972). An ecological reconnaissance of Wilpattu National Park, Ceylon. *Smithson. Contrib. Zool.* **101,** 1—118.

Eisenberg, J. F. and McKay, G. M. (1970). An annotated checklist of recent mammals of Ceylon with keys to the species. *Ceylon J. Sci. (Bio. Sci.)* **8**(2), 69—99.

Eisenberg, J. F. Muckenhirn, N. A., and Rudran, R. (1972). The relation between ecology and social structure in primates. *Science* **176,** 863—874.

Ferando, S. N. U. (1968). "The Natural Vegetation of Ceylon." Lake House, Colombo.

Gopalasingham, E. (1959). The role of protection in forestry. *Ceylon Forest.* **4**(2), 201—206.

Grand, T. I. (1972). A mechanical interpretation of terminal branch feeding. *J. Mammal.* **53**(1), 198—201.

Henry, G. M. (1971). "A Guide to the Birds of Ceylon," 2nd ed. Oxford Univ. Press, New York.

Hill, W. C. Osman. (1932). The external characters of the bonnet monkeys of India and Ceylon. *Ceylon J. Sci.* (B) **16,** 311—322.

Hill, W. C. Osman. (1942). The highland macaque of Ceylon. *J. Bombay Natur. Hist. Soc.* **43**(3), 402—406.

Hill, W. C. Osman (1974). "Primates, Volume VII, Cynopithecinae." Wiley, New York.

Hill, W. C. Osman and Bernstein, I. S. (1969). On the morphology, behavior and systematic status of the Assam macaque (*Macaca assamensis* McClelland 1839). *Primates* **10,** 1—17.

Hladik, C. M. and Hladik, A. (1972). Disponibilities alimentaires et domaines vitaux des primates a Ceylan. *Terre Vie,* pp. 149—215.

Hoffmann, T. W. (1968a). A plea for orderly land use in Ceylon. *Loris* **11**(4), 176—180.

Hoffmann, T. W. (1968b). A National Reserve is far more than a Sanctuary. *Loris* **11**(4), 199—200.

Itani, J. (1975). Twenty years with Mount Takasaki monkeys. *In* "Primate Utilization and Conservation" (G. Bermant and D. G. Lindburg, eds.), pp. 103—125. Interscience, New York.

Itani, J., Tokuda, K., Furuya, Y., Kano, K., and Shin, Y. (1963). The social construction of natural troops of Japanese monkeys in Takasakiyama. *Primates* **4**(3), 1—42.

Kirtisinghe, P. (1957). "The Amphibia of Ceylon." Published by the author, Colombo.

Koford, C. B. (1965). Population dynamics of rhesus monkeys on Cayo Santiago. *In* Primate Behavior" (I. DeVore, ed.), pp. 160—174. Holt, Rinehart and Winston, New York.

McCormack, R. J. (1961). "A Management Inventory of the Morapitiya-Runakanda-Delgoda Forest Area, Ceylon." Government Press, Colombo.

McCormack, R. J. and Pillai, A. K. (1961a). "A Forest Inventory of the Teravil-Oddussuddan Forest Reserve, Ceylon." Government Press, Colombo.

McCormack, R. J. and Pillai, A. K. (1961b). "A Forest Inventory of the Nuwaragala and Forest Reserves and the Ratkarawwa Proposed Reserve, Ceylon." Government Press, Colombo.

McCormack, R. J. and Pillai, A. K. (1961c). "A Forest Inventory of the Kumbukkan Forest Reserve, Ceylon." Government Press, Colombo.

McCormack, R. J. and Pillai, A. K. (1961d). "A Forest Inventory of the Panama Proposed Reserve, Ceylon." Government Press, Colombo.

McCormack, R. J. and Pillai, A. K. (1961e). "A Forest Inventory of the Kantalai Forest Reserve, Ceylon." Government Press, Colombo.

McKay, G. M. (1973). Behavior and ecology of the Asiatic elephant in southeastern Ceylon. *Smithson. Contrib. Zool.* **125**, 1–113.

McNab, B. K. (1963). Bioenergetics and the determination of home range size. *Amer. Natur.* **97**(894), 133–139.

Merritt, V. G. and Ranatunga, M. S. (1969). Aerial photography survey of Sinharaja forest. *Ceylon Forest.* **4**(2). 103–156.

Muckenhirn, N. A. (1972). Leaf-eaters and their predator in Ceylon: Ecological roles of gray langurs, *Presbytis entellus*, and leopards. Ph.D. Dissertation, Univ. of Maryland, College Park, Maryland.

Muckenhirn, N. A. and Eisenberg, J. F. (1973). Home ranges and predation of the Ceylon leopard. *In* "The World's Cats," Vol. 1, Ecology and Conservation (R. L. Eaton, ed.), pp. 142–175. World Wildlife Safari, Winston, Oregon.

Mueller-Dombois, D. and Sirisena, V. A. (1967). "Climate Map of Ceylon." Ceylon Survey Department, Colombo.

Odum, H. T. (1970). Summary: An emerging view of the ecological system at El Verde. *In* "A Tropical Rain Forest: A Study of Irradiation and Ecology at El Verde, Puerto Rico" (H. T. Odum, ed.). Division of Technical Information, U.S. Atomic Energy Commission, Oak Ridge, Tennesee. Section I. pp. 191–289.

Oudshoorn, W. (1961a). "A Forest Inventory of the Attavillu Forest Reserve and Proposed Reserve, Ceylon." Government Press, Colombo.

Oudshoorn, W. (1961b). "A Forest Inventory of the Madhu Forest Area, Ceylon." Government Press, Colombo.

Petter, J. J., and Hladik, C. M. (1970). Observations sur le domaine vital et la densite de population de *Loris tardigradus* dans les forets de Ceylan. *Mammalia* **34**(3), 394–409.

Phillips, W. W. A. (1935). "Mammals of Ceylon." Dulau and Company, London.

Pocock, R. I. (1932). The long-tailed macaque monkeys (*Macaca radiata* and *M. sinica*) of southern India and Ceylon. *Bombay J. Natur. Hist. Soc.* **35**, 267–288.

Rudran, R. (1970). Aspects of ecology of two subspecies of purple-faced langurs (*Presbytis senex*). M.Sc. Dissertation, Univ. of Sri Lanka, Colombo, Ceylon.

Southwick, C. H., Beg, M. A., and Siddiqi, M. R. (1961). A population survey of rhesus monkeys in villages, towns and temples of northern India. *Ecology* **42**(3), 538–547.

Struhsaker, T. T. (1975). "The Red Colobus Monkey." University Press, Chicago, Illinois.

Walter, H. (1971). "Ecology of Tropical and Subtropical Vegetation." Van Nostrand, New York.

Wijesinghe, L. C. A. de S. (1972). The role of forestry in the development of Ceylon's land resources. *Proc. 27th Annu. Ceylon Assoc. Advan. Sci. Sess. Part 2*, pp. 179–198.

8

The Status of the Barbary Macaque *Macaca sylvanus* in Captivity and Factors Influencing Its Distribution in the Wild

JOHN M. DEAG

I. Introduction

The Convention on International Trade in Endangered Species of Wild Fauna and Flora signed in Washington, D.C. in 1973 lists *Macaca sylvanus* as a species which may become threatened with extinction if trade is not regulated to avoid overexploitation (King, 1974). This article summarizes the information on the distribution of Barbary macaques and attempts to assess the danger from habitat destruction and trapping. In the past, most international trade in Barbary macaques involved animals culled from the Gibraltan population but recently large numbers have been exported from Morocco. This species has never been used in quantity for laboratory research but future trapping of large numbers of animals for this purpose always remains a possibility. Since the species has now been recognized as endangered, it is timely to assess the number of animals in captivity and their reproductive success. My own observations recorded here were made in Morocco during 1968 and 1969 when I was based at the University of Bristol.

II. Distribution and Habitat

The Barbary macaque has a discontinuous distribution with isolated populations scattered in Morocco, Algeria, and Gibraltar. Zeuner (1952) and Napier and Napier (1967) record it in Tunisia. At present this is incorrect but it probably lived there in historic times (Joleaud, 1931a). Forbes (1897) and Lydekker (1894) recorded the species from Gibraltar and some distance into Spain. An exciting new development is reported by P. Ballesteros (personal communication); some Barbary macaques have recently escaped from captivity in Spain and are living and reproducing in the wild.

Only brief comments will be made on distribution, abundance, and habitat, since, in 1974, D. M. Taub made a detailed survey of the animals in North Africa, the results of which will be published shortly. Taub estimated that there were between 9,500 and 21,500 Barbary macaques in North Africa; 77% in Morocco, and 23% in Algeria (Taub, 1975).

A. Morocco*

The species is found in the Middle Atlas, High Atlas and Rif (Joleaud, 1931a; Heim de Balsac, 1936; Panouse, 1957; Deag and Crook, 1971; Whiten and Rumsey, 1973; Deag, 1974). It is most abundant in the central areas of the

---

*It has not been possible to include a map showing locations. The interested reader is referred to Deag and Crook (1971); "The Times Atlas of the World," Vol. 4, Times Publishing Co., London (1956); and The Michelin Map of Morocco, Sheet 169, Paris (no date).

Middle Atlas—Taub (1975) estimated that up to 98% of the Moroccan monkeys live there. The most dense populations, variously estimated to be 60—70/km² (Deag, 1974) or 43/km² (Taub, 1975) occur in the mixed cedar (*Cedrus libanotica atlantica*) and holm oak (*Quercus ilex*) forests found at an altitude of over 1600 m. There is little permanent habitation in these forests and no cultivation, but sheep, cattle, and some goats are grazed in the forests during the summer and autumn. Commercial exploitation includes felling cedar for timber and oak for firewood and charcoal production. In spite of these activities, the monkeys are generally less disturbed than at lower altitudes, unless they happen to be in an area subject to felling. The social behavior and ecology of monkeys in cedar forests have been studied by Deag (1970, 1973, 1974, 1977; Deag and Crook, 1971) and Taub (1975).

At lower altitudes, the mixed cedar and oak forests give way to almost pure stands of holm oak (Deag, 1974). These forests are also grazed and, being closer to villages and cultivated land, the monkeys are more disturbed. Near Azrou, for example, they are reputed to raid crops and rubbish dumps. I suspect that the monkeys living in the poorer forests (principally of oak) in the eastern and southern Middle Atlas (e.g., those near Beni Mellal and El Ksiba) are under a similar pressure from man. Their density is certainly much lower than in the central Middle Atlas (Taub, 1975). In the High Atlas, the habitat consists of holm oak and juniper (*Juniperus phoenicia*) scrub on steep mountain slopes. This habitat represents the southwestern limit of the species' range and the monkeys are limited to one or two isolated populations. Above some villages the slopes are terraced for cultivation and the monkeys are reputed to raid crops (Deag and Crook, 1971).

In the Rif, the monkeys live in precariously small populations in the more isolated areas. In these mixed-wood forests (Deag, 1974; Taub, 1975) they are clearly under pressure from man both through habitat destruction (Taub, 1975) and hunting (see Section IV).

B. Algeria

Historical accounts are provided by Joleaud (1931a) and Heim de Balsac (1936). The following account is based on Taub's recent survey (Taub, 1975). Barbary macaques are found in seven localities in three main regions: the Chiffa Gorges (South-southwest of Alger and Blida); the Grand Kabylie (three sites from the coast at Bejaia westward to Djurduraa; and the Petit Kabylie (three sites from Guerrouch westward to Kerrata). In all areas the animals are less abundant than in the central Middle Atlas of Morocco. There are three main habitat types: cedar/cedar—holm oak mixes; Portuguese oak (*Q. faginea*)/ cork oak (*Q. suber*) mixes; and scrub vegetation in gorges. Apart from Taub's survey there has been little attempt to study Algerian monkeys. Some brief observations were made by Kearton (1924).

C. Gibraltar

Two groups of monkeys are currently found on Gibraltar and in the period 1950–1970 these included 28–45 animals (Burton and Sawchuk, 1974). It is unknown whether the original Gibraltan monkeys were endemic or introduced. There were certainly monkeys there when the British gained control in 1704 and since that date numerous introductions have occurred, the earliest in 1740 (Kenyon, 1938). For further details on the history and status of this population see Lydekker (1894), Forbes (1897), Hooton (1942), Zeuner (1952), Morris and Morris (1967) and Hill (1974). The monkeys are cared for and partially fed by the British Army who permit only one adult (or near adult) male per group (Burton, 1972)–a highly unnatural group structure (Deag and Crook, 1971). Behavioral and ecological observations have been made by MacRoberts and MacRoberts (1966), MacRoberts (1970), and Burton (1972).

III. Factors Influencing Distribution

Basic habitat requirements are a suitable supply of food throughout the year, trees or cliffs for sleeping and escaping from predators, and access to a permanent supply of water for use during dry years.

A. Prehistoric Factors

Zoogeographically this species presents some interesting problems (Tappen, 1960). It is unique in several ways, since it is the only macaque living outside Asia, macaque in Africa, nonhuman primate living north of the Sahara, primate other than man living in Europe today. Only one nonhuman primate (the Japanese macaque *M. fuscata*) lives in a more northerly position. To explain the discontinuous distribution of macaques one must recourse to fossils, comparison with other mammals that show Afro-Asiatic connections, and speculations on past climatology and land bridges. This is too specialized a topic to be considered here at length and the interested reader is referred to Lydekker (1894), Trouessart (1900), Joleaud (1931a), Cabrera (1932), Heim de Balsac (1936), Zeuner (1952), Cooke (1963), Kurten (1968), and Napier (1970).

Fossil animals that have been called *M. sylvanus* were distributed over much of Europe during the middle Pleistocene interglacials (Zeuner 1952; Kurten, 1968). Older European fossils are usually given different specific names but this is probably invalid and our scant knowledge of these fossils makes it difficult to pinpoint the time at which the separation of *M. sylvanus* and the Asiatic macaques occurred (Delson, personal communication). Jolly (1965) considered that the Pleistocene glaciations emptied Europe of its macaques and that sea barriers prevented recolonization. The most severe conditions in southern Europe were associated with the last (Wurm) glacial, 10,000–30,000 years ago. At that time ice covered Europe down to a line approximately from London to Prague to Dnepropetrousk (south Ukraine). There were also subsidiary ice fields in the Alps. South of the ice sheets an area of permafrost must have made the climate unsuitable for macaques over most of Europe. It is possible, however, that the Iberian Peninsula, the south of France, and Italy still provided suitable macaque habitats.

Why are macaques able to survive today in northerly latitudes with harsh climates? I believe that a crucial factor is their ability to fully exploit a wide variety of food resources. They obtain food (often using considerable manipulative skill) from all parts of their habitat—in the trees, on and under the ground, and under rocks (Deag, 1974). Note also the ability of Japanese monkeys to adopt new skills for handling novel foods (Frisch, 1959). Because of their universal feeding habits, macaques can obtain food throughout the year in spite of seasonal changes in its availability in different parts of the habitat. A temperate primate feeding on tree leaves and fruits could not find a sufficient variety of food throughout the year. The short tail and long winter coat of *M. sylvanus* and *M. fuscata* are adaptations to cold winters. At Ain Kahla, my Middle Atlas study site (Deag and Crook, 1971), winters are wild, wet, and snowy; temperatures as low as −18°C have been recorded (Deag, 1974).

B.   More Recent Factors

There is no evidence to support Sanderson's suggestion (quoted by Tappen, 1960) that *M. sylvanus* was introduced into North Africa during historic times. Joleaud (1931a) and Zeuner (1952) reviewed the factors that may be responsible for the species' present distribution. They concluded that deforestation by man led to the loss of the species from Europe (or its restriction to Gibraltar) and to its discontinuous distribution in the more inaccessible areas of North Africa. The monkeys have become extinct or less common in areas (e.g., the E. Kabylie) where they were used as food. Elsewhere they have been protected by local customs (Joleaud, 1931a, b).

A systematic comparison of the distribution in 1974 with earlier records from the literature and personal contacts enabled Taub (1975) to confirm Joleaud's and Zeuner's conclusions. There has been a contraction of the species' range in North Africa over the last century, a process that has continued during the last decade in both Morocco and Algeria. The main factor is undoubtedly deforestation, although the situation is also being aggravated by hunting (see Section IV).

IV.   Hunting, Trapping, Internal Use, and Export of Monkeys

It is difficult to discover how many monkeys are caught or killed each year and the uses to which they are put. Some problems are real enough: Who knows how many are casually shot, caught for sale in markets, or used as household pets or in fairs? On the other hand, large-scale trapping operations could be closely monitored by government departments and the data made public. The destruction of large carnivores probably means that losses due to man have mainly replaced other forms of predation.

The inclusion of *M. sylvanus* in the Convention on Trade in Endangered Species of Wild Fauna and Flora means that, in the future, export permits are required from the country of origin before import into another country is permitted (King, 1974). There is, of course, no international control over the issuing of export permits.

A.   Morocco

During our stay in the cedar forests we were told of monkeys being shot throughout the Middle Atlas (apparently during 1955) because of the damage caused by them stripping cedar bark (Deag, 1974). Another report indicated that about 50 monkeys may have been caught at Ain Kahla during 1967 for the Institut Pasteur, Casablanca. Several hundred other monkeys were reputedly caught for the same organization elsewhere in Morocco before 1968. The Instituts Pasteur in Casablanca and Tangier did not reply to my enquiries (see Section VI) about the number of animals they have used from 1964 through 1974.

During my study, at least ten monkeys (mostly young) were caught or shot in the vicinity of Ain Kahla. The people responsible included soldiers, shepherds, and foresters. The monkeys were taken as pets, sold to foreign tourists or reputedly to the Institut Pasteur. Some quickly died from inadequate care. Captive animals were seen in several villages, being used for display, entertainment, or simply as household pets. In the High Atlas monkeys are known to be shot for raiding crops. Shooting is prevalent in the Rif and has contributed to the destruction of the monkeys and their restriction to small unstable groups in the most isolated places (Whiten, personal communication).

A recent and most important development has been the capture of many macaques for export to European wildlife parks and other establishments. The undoubtedly incomplete figures I have obtained are summarized in Table I. Altogether, over 425 monkeys (418 from Table I plus those recorded above) have been removed from the forests since 1968. More than 95% have been exported chiefly to wildlife parks and zoos. I have reason to believe that there may have been other exportations not included in the above totals.

TABLE I

Details of the Barbary Macaques Known to Have Been Caught in Morocco
Since 1969

Year	Number and sex (M/F/?)	Mortality during capture	Destination
1969	52 (16/36)[a]	20	La Montagne des Singes, France
1971	150[b]	?	La Montagne des Singes, France
1972	2[a]	?	Museum National d'Histoire naturelle, Brunoy, France
1974	62[b]	< 5%	Parc de Rocamadour, France
1974	15 (8/7)	?	Faculty de Medecine, Madrid, Spain
1974	5	?	Military Hospital Rabat (1), Institut Pasteur, (4), Maroc
1974	> 48 (13/31/4)	?	Tierpark Rheine, W. Germany[c]
1974	1 (1/0)	?	Zoorama de Chize, France
1974	10 (2/8)	?	Zoo Safari, Quexigal, Spain
1974 or 1975	50 (ca)	?	Bellewaerde, Belgium

Total > 395 plus at least 23 killed during capture[d]

[a]Caught near Azrou.
[b]Caught near Azrou, Beni-Mellal, El Ksiba or Mrirt.
[c]Total caught unknown. Many of these animals were sent to other zoos from the Tierpark Rheine.
[d]Some other animals caught in 1968–1969 are mentioned in the text.

The monkeys have been protected by law for some time. An official permit for capturing animals (from the Ministère de l'Agriculture et de la Réforme Agraire) has been theoretically required, with the annual maximum catch apparently limited to 100 animals. In my experience these regulations have not been strictly enforced. It is worth noting that, according to Taub (1975), the Moroccan Eaux et Forêt currently consider the species to be common. In the past, trapping and export were apparently arranged by the persons wanting the monkeys. More recently (1974), the situation seems to have been regularized with only the Parc Zoologique National, Temera, Rabat and the Parc Zoologique Municipal, Casablanca, being permitted to export the animals. Most are caught in oak forests; the tall cedars allow the monkeys a good opportunity for escape. They are darted or caught in traps near watering points, housed temporarily in Morocco, and then exported. With experience, losses during the whole process including transportation abroad (but not necessarily deaths after arrival, see Section VI) can be kept to less than 5% (Turckheim, personal communication).

There is, therefore, no trapping at present on the scale seen in India (Southwick *et al.*, 1970) but this might easily occur if *M. mulatta* becomes more difficult to obtain and there is insufficient control of exports. The absolute number of animals captured (at least 425 over 7 years) may seem small but it should be viewed against Taub's (1975) *maximum* estimate of 17,000 monkeys in the whole of Morocco. To understand the impact of this trapping one also needs to know the location so that numbers caught can be related to population density. The lack of information on the use of *M. sylvanus* in Moroccan laboratories and the large exports to Europe are a cause for concern and should be watched carefully. The Convention on International Trade in Endangered Species should provide a stimulus for monitoring traffic in this animal and a more open attitude among officials.

B.   Algeria

I am grateful to Dr. J. P. Thomas and Dr. A. R. Dupuy for providing most of the following information. The monkeys are officially protected (under the control of the Ministère de l'Agriculture et de la Révolution Agraire) and apparently may not be killed, captured, or exported. In some places there is, however, little actual control and there are exceptions to this protection for the purpose of supplying medical laboratories and for protecting crops. Damage to crops has to be proved, but, on the whole, damage to trees and crops is not considered an important problem. The number of animals caught per annum is fixed and has perhaps been in the order of several dozen per annum since 1964. There has been no large-scale trapping and exporting as in Morocco, but there may be a small unofficial trade involving tourists. There are several National Parks which include Barbary macaques but Taub (1975) considered most of them to be too small for long-term conservation.

The Institut Pasteur d'Algérie and the Institut des Sciences Médical: Service de Physiologie et de Biologie use Algerian caught Barbary macaques supplied to them by the Direction des Eaux et Forêts. The former institute expects to take animals from the wild in the near future (Director of the Institut Pasteur, Alger, personal communication). It is probable that in the recent past monkeys have been exported from Algeria to l'Institut Pasteur in neighboring Tunisia.

C.   Gibraltar

The demography of the Gibraltan animals has been studied by Burton and Sawchuk (1974) from whom the following details are taken. The size of the monkey population is regulated and surplus animals sent to zoos. During the period 1950–1970 the colony was expanding, unlike the previous two decades when animals were imported from North Africa to supplement the dwindling numbers. Forty-four animals born between 1950 and 1970 inclusive were exported, representing 44% of the total loss from the population. There was no significant difference in the number of males and females exported but more males than females were exported as juveniles. At least six of the zoos returning questionnaires (see Section VI) had received monkeys from Gibraltar at one time or another. Given that there is limited space for the monkeys on Gibraltar and animals have to be culled, the colony is providing a useful reservoir of animals for zoos without depleting the natural populations in North Africa.

V.   Monkeys and the Regeneration of the Moroccan Cedar Forests

Since the monkeys feed on the cedars (Deag, 1974) it is important to comment on any damage that they incur and to assess its significance in relation to other demands on the forest.

The loss of seeds, germinating seedlings, and male strobiles (all of which are eaten) probably has little effect on regeneration since they are produced in huge quantities. The eating of cedar leaves is not thought to significantly stunt tree growth since the monkeys did not concentrate on the terminal shoots. The removal of bark was the most obvious and the most destructive of their cedar-eating habits. The loss of side branches and the occasional removal of bark from the main stem of younger trees must result in a loss of timber. At Ain Kahla, bark was only stripped during 2 months of the year and since monkeys repeatedly returned to places where stripping had been started this localized the damage. Even in that brief season, cedar was only one component of a very varied diet (Deag, 1974). Compared with the other losses of timber from the forest (official felling, large losses of immature timber caused by felling operations, illegal felling by shepherds) the losses due to monkeys are probably extremely small. They may, however, assume economic "importance" simply because man, in overexploiting the timber, feels that other losses must be kept to a minimum.

It is my opinion (admittedly based on only a short period in the forest) that the regeneration of cedar at Ain Kahla is not keeping up with exploitation— a conclusion shared by Taub (1975). The factors controlling regeneration are extremely diverse and depend very much upon a suitable balance of temperature, humidity, minerals, and fungal mycorrhiza, etc. (Marion, 1953—1955; Lepoutre, 1957—1961, 1962—1963; Pujos, 1964). Reduced regeneration may be a direct consequence of the exploitation rate and methods. The removal of significant numbers of mature cedars and oaks may upset the temperature and humidity balance at ground level. The cedar seedlings are very susceptible to drying out in calcerous soils and in some areas this will be exaggerated by loss of shade. Immature trees, already established in felling areas, are often destroyed during felling operations and during the carting of timber. The grazing of sheep and goats in the forest also severely limits the establishment of new trees but this appears to be the least studied factor influencing regeneration.

The monkeys did not browse on the seedlings once the cotyledons had opened. In their extremely selective feeding these were not touched. This was in marked contrast to the sheep (and the few goats) that grazed in the forest. During 1969, the sheep were excluded from the forest until June. In only 1 or 2 weeks after the sheep arrived there was a drastic change in the vegetation. It was beaten flat in wide paths created by the passage of flocks of about 200 sheep. They were not selective feeders; it seemed as if a moving flock ate more or less anything in its path. The sheep browsed on the lower cedar branches. At the end of the summer, when pasture was poor, the shepherds lopped off side branches or even felled tall trees to feed them. It is therefore unlikely that young seedlings escaped being eaten. The adverse influence of grazing animals and the minimal damage caused by seed eaters on the regeneration of forests is recognized for other trees (Mellanby, 1968). The shepherds realized that they were overexploiting the pastures; many of them moved on after a month or so, leaving a smaller density of sheep in the valley. The grazing of herbaceous plants in summer probably caused little long-term damage; most annuals had dropped their seeds before grazing commenced and perennials had subterranean bulbs and tubers. Grazing must nevertheless reduce the food available to the monkeys during the summer. The only other major herbivores in the forest were wild pigs, hares and small rodents. These must compete for some food items but, compared with the huge biomass of sheep, cattle, and goats, they probably have little effect on the herbaceous vegetation.

In the Assaka-n-Ouam cedar forest (northeast of Bekrit) the damage by monkeys is said to be more extensive than at Ain Kahla, with very young trees being heavily damaged by bark stripping (Arnaud, personal communication). This forest is a small (ca. 21 km²) isolated one and it would be interesting to compare it with Ain Kahla to determine the reason for this difference. It may well be related to population density and the availability of other food. It is essential that extensive data on population density, damage to trees, etc., is collected before any attempt is made to control the number of monkeys in the Moroccan forests. Most monkeys for export are caught in oak forests where they do no damage and their eating of caterpillars may be beneficial. From the forester's point of view it might have been wiser to catch these monkeys in places like Assak-n-Ouam where damage to cedars is said to be important.

Since the forests have been traditional grazing grounds for generations (Métro, 1958) it is difficult to control grazing. Most families previously herded their own sheep but now shepherds are often employed. This may be one factor contributing to the high densities of sheep. Perhaps regeneration could be promoted by restricting grazing in some parts of the forest for many years at a time and by restricting the number of sheep in the entire area.

The monkeys have been in the forests for thousands of years and it is to be hoped that man, in his search for an instant cure to his economic problems, will not see fit to destroy them.

VI.   Barbary Macaques in Captivity

To find out about Barbary macaques in captivity, a questionnaire was prepared and circulated. Twenty-two replies were received from organizations that use or have used this species. The questionnaire requested information on the following:

(a) For the years 1964—1974 inclusive: colony size, births, other additions to the colony; the number of losses due to deaths, animals killed for experimental purposes, and animals transferred to other establishments. Animals were classified by sex and approximate age: mature (= assumed to be capable of breeding); immature or born during the year.

(b) The year the species was first kept or used.

(c) The source of the current (i.e., post-1964) animals and the original source if the species was kept before 1964.

(d) The names of organizations to which animals have been transferred.

(e) Whether any animals are expected to be taken from the wild in the near future.

(f) The uses to which the animals are put (display to public, breeding, laboratory use with research area specified).

(g) The names of other organizations keeping or using the species.

While interpreting the results of this exercise it should be noted that replies were not obtained from some organizations reputed to keep the species, and that some organizations were not included in the circulation of the questionaire. In most cases only incomplete information was available. Births recorded as stillbirths are not recorded. Information from other sources (e.g., Int. Zoo Yearb., published by the Zoological Society of London) has been included where appropriate.

A.   Laboratories

In spite of Galen's precedent in the second century (Temkin and Straus, 1946), *M. sylvanus* has been little used for medical research. With the exception of a new colony in Madrid, little information has been obtained on colony size and the number killed per annum.

In Algeria during the 1950's and early 1960's, Barbary macaques were extensively used in the Faculté de Medecine, Alger for neurohistological and physiological studies (Girod *et al.*, 1955; Barry and Lefranc, 1961; Girod, 1961, 1962; Bianchi, 1962). At present, locally caught monkeys (Section IV) and some bred in the laboratory are used in the Institut Pasteur, Alger (for medical experiments and cell cultures) and in the Institut des Sciences Médicale: Service de Physiologie et de Biologie (for physiological and medical experiments including endocrinology and neurophysiology). Females are used in preference to males. There are plans to establish a primatology center around 1980 (Institut Pasteur d'Algerie, J. P. Thomas, F. Bourlière, personal communications). In Morocco, the Institut Pasteur, Casablanca kept monkeys in the 1930's and 1940's and used them in studies of typhus (Blanc and Woodward, 1945) and beriberi (Leblond and Chaulin-Servinière, 1942). The latter article implies that the colony was established in 1935 and that in 1940 it held a minimum of 48 animals. Locally caught monkeys are still used in Morocco (for an unknown purpose) at the Instituts Pasteur at Casablanca and Tangier and in the Royal Military Hospital, Rabat (Y. Raymond, personal communication). In Tunisia the Institut Pasteur does not keep monkeys (A. Chadli, personal communication) but apparently they were used there in the past.

In Europe a colony of fifteen wild-caught Moroccan animals was recently established at the Faculty de Medicine, Universidad Autonoma, Spain. They are being used for studies of brain and behavior, and social behavior (J. M. R. Delgado, personal communication). In the United States Barbary macaques are rarely used for medical research; only two cases involving single animals have come to my attention (Lindsay and Chaikoff, 1966; J. C. Hixon, personal communication).

There is no need for concern about the number of animals used in laboratories outside North Africa. Inside North Africa the situation may be very different. I have received information that hundreds of monkeys may have been supplied to laboratories in Morocco (Section IV). The precarious nature of the Algerian monkey populations (Taub, 1975) makes the capture of monkeys especially undesirable. The accurate figures needed to properly assess the situation are unobtainable. For the moment, laboratories must be considered important consumers, producing none or few laboratory-bred monkeys for their own use.

B. Zoos and Wildlife Parks

A list is given in Appendixes 1 and 2 of 56 zoos and wildlife parks known (or reputed) to keep Barbary macaques in 1964–1974. Because of the incomplete information at my disposal the following analysis involves certain assumptions and deductions. While the results must therefore be approximate, I believe they help to reveal the status of the species in captivity. Gibraltar is largely omitted from this discussion. As recorded above (Section IV) the colony is an important source of animals for zoos: 146 young (excluding still births) were born during 1950—1970 and 44 animals exported (Burton and Sawchuk, 1974).

Most zoos and wildlife parks reported that their animals were kept for display to the public and for breeding. Several colonies have been used for studies of behavior and for education (Zoological Society of San Diego, California; Metropolitan Toronto Zoological Society, Toronto; La Montagne des Singes France; National Zoological Park, Washington, D.C. (Lahiri and Southwick, 1966); Kingdoms Three, Atlanta, Georgia).

1. The Number of Barbary Macaques in Zoos and Wildlife Parks

In 1974 (or thereabouts) the colonies listed in Appendix 1 held 568 animals (Table II). Three parks held 418 of these: 368 in France at La Montagne des Singes (established 1969) and the Parc de Rocamadour (established 1974), and 50 in Belgium at Bellewaerde. Other colonies also exist (Appendix 2) and if we assume four animals per colony this gives a minimum of 700 in zoos and parks during 1974. To this should be added an additional 30 or so for Gibraltar. Although the organizations listed in Appendix 1 held 568 animals in 1974, they handled a minimum of 744 during 1964—1974 (Table II).

2. Births in Zoos and Wildlife Parks

During 1964—1974, thirty-three zoos and wildlife parks bred Barbary macaques (Appendixes 1 and 2) producing 321 young (Table III). Of these 166

TABLE II

The Minimum Number of Barbary Macaques in Zoos and Wildlife Parks, 1964—1974[a]

	Number and sex when known (M/F/unsexed)	
Number of individuals present during 1974		
Zoos (eighteen questionnaires + three other zoos)	200	(55/91/54)
La Montagne des Singes, Kintzheim	306	
Parc de Rocamadour (excludes animals transferred from	62	
Kintzheim in 1974)	568	
Number of individuals present during 1964—1974		
Zoos (eighteen questionnaires + three other zoos)	314	(113/139/62)
La Montagne des Singes (1969—1974)	368	(103/153/50)
Parc de Rocamadour (as above)	62	
	744	

[a]Full details are given in Appendix 1.

TABLE III

The Minimum Number of Barbary Macaques Born in Zoos and Wildlife Parks, 1964–1974[a]

	Number and sex when known (M/F/unsexed)	
Number of live births during 1974		
Zoos (nineteen questionnaires + one other zoo)	12	(4/2/6)
La Montagne des Singes, Kintzheim	70	
	82	
Number of live births during 1964–1974 inclusive		
Zoos (nineteen questionnaires + twenty-one other zoos)	155	(60/59/36)
(of the 155 at least 24 died in their first year)		
La Montagne des Singes (1969–1974)	166	(35/33/166)
	321	

[a] From the *Int. Zoo Yearb.* (up to, but not including, 1974) and other sources.

were produced by La Montagne des Singes (1969–1974) and 50 by the National Zoological Park, Washington, D.C. (1964–1974; 78 births in 1950–1974). A minimum of 105 were produced by the remaining thirty-one organizations. In 1974, La Montagne des Singes produced 70 young; the remaining zoos produced 12. The latter is an underestimate since the *Int. Zoo Yearb.* has not published the 1974 figures. The other figures given above overestimate colony productivity since they do not allow for infants deaths. Some indication of these is given in Table III, but this must itself be an underestimate.

3. Mortality in Zoos and Wildlife Parks

The records of deaths are too patchy for analysis beyond a superficial level. Some zoos (for example Paington, U.K., Chester, U.K., and the National Zoological Park, Washington, D.C.) have a creditable record with self-supporting colonies, few or no animals being added from outside and births balancing deaths or supplying animals for transfer elsewhere. The two most productive colonies were the National Zoological Park, Washington, D.C. (with 14 deaths and 50 births in 1964 to 1974) and La Montagne des Singes (with 77 deaths and 166 births in 1969 to 1974). If one considers the twelve organizations listed in Appendix 1 which recorded deaths (other than La Montagne des Singes and the National Zoological Park Washington) then the overall picture is poor: 53 deaths to 42 births, coupled with 73 animals being added to the colonies from outside.

A recent report (from an impeccable source) shows just how easily a complete colony can succumb to disease. About 50 recently imported Barbary macaques died from salmonella at Bellewaerde. This is made even more disturbing by the stated intention of replacing the animals with further imports.

4. Status in Captivity—Conclusions

Although the evidence is fragmentary it is obvious that there are sufficient Barbary macaques in captivity to form adequate breeding stocks. With good management both large and small colonies can be self-supporting and can produce an excess of births over deaths. Some zoos are, however, essentially consumers dependent for their stock on Gibraltar, more successful colonies, or wild-caught animals. The establishment of satisfactory trading arrangements between organizations would make further exports from North Africa completely unnecessary.

I must admit to being initially apprehensive about the scale of the trapping organized by La Montagne des Singes. The evidence presented above shows their colony to be extremely productive. If the excess animals are made available to others, so negating any further demand for exports, then the initial exports may have been justified.

VII.   Conclusions and Recommendations

The Barbary macaque is under considerable danger from habitat destruction, trapping, and hunting. Human disturbance is most prevalent in the peripheral parts of the animal's range where isolated populations continue to decline and become extinct. In addition, the monkeys living in the more central areas of Morocco are being exploited and several hundred have recently been caught. This disturbance is coupled with the long-term threat of habitat destruction through the overexploitation of the cedar and oak forests. Quite apart from any moral, ecological, scientific, or medical arguments (Southwick et al., 1970) that can be put forward for their protection, these monkeys represent an important natural resource. With an increase in leisure time available to the Moroccan and Algerian people and the increasing number of tourists visiting their countries, the monkeys may become appreciated as an important asset.

There are now sufficient animals in captivity to make further exports from North Africa unnecessary. The animals breed satisfactorily in captivity and both small and large colonies can be self-supporting.

Recommendations

1. In Morocco a proper system of protection should be introduced and adequately supervised parks established. Taub (1975) has recently put forward

specific proposals concerning parks in both Morocco and Algeria, and also for the training of foresters in conservation.

2. The commercial exploitation of the North African forests should be reviewed to ensure their long-term stability. Devastated areas should be reforested with native species (see also Taub, 1975).

3. The trapping of monkeys should be strictly controlled. It must be supervised by people who are aware of the relative abundance of the animals in different areas and of the importance of not reducing populations below a level critical for their survival. Records should be kept of the number, age, and sex of animals caught, the number killed during capture, the date, and locality.

4. Exports from Morocco should be prohibited. If this is not done they should be strictly controlled and made only to organizations competent at keeping and breeding animals and who guarantee not to immediately sell off small numbers to third parties. Records should be kept of all exports showing the trapping location and the age and sex of the animals. In the present survey, information of this type came largely from the importers.

5. The use of monkeys for research purposes in North Africa should be strictly controlled and laboratories only permitted to take animals from authorized organizations. Records should be kept of the number, age, and sex of animals and what they are used for. Laboratories in North Africa are obviously in a very favorable position for setting up breeding colonies which would make it unnecessary to trap wild animals.

6. Consideration should be given to establishing a Register of Endangered Species in Captivity. Details of colonies and the transfer of animals could be collected biennially. The Register need not be a stud book. In the present survey many organizations could only supply incomplete information. An effort should be made to keep good records of all endangered animals.

7. Zoos which currently consume more animals than they produce should be under an obligation to improve their breeding success. Both they and new zoos should no longer take possession of additional wild-caught animals.

VIII.   Appendix 1. Zoos and Wildlife Parks that Replied to the Questionnaire on Barbary Macaques in Their Collection 1964—1974

For each organization the person who completed the questionnaire is noted in parentheses followed by five numbers in parentheses separated by slashes which represent the following data:

First number: Colony size in 1974.
Second number: Number of live births in 1974.

Third number: Mean colony size to nearest integer, 1964–1974 inclusive, years without any monkeys excluded.

Fourth number: Total number of individuals that lived in colony, 1964–1974.

Fifth number: Number of live births in 1964–1974 inclusive.

Zoo addresses are given in the *Int. Zoo Yearb.* **14**, 1974. Most of these organizations did not keep Barbary Macaques for the entire period 1964–1974. Some no longer keep the species.

North of England Zoological Society, Chester, *United Kingdom* (Mr. G. S. Mottershed and Mr. W. H. Timmis) (8/0/7/13/8)

Paignton Zoological and Botanical Gardens, Paignton, *United Kingdom* (Mr. W. E. Francis) (4/1/4/11/8)

Zoological Society of London, *United Kingdom* (Dr. M. R. Brambell) (0/0/2/4/1)

Belle Vue Zoological Gardens, Manchester, *United Kingdom* (The Zoo Manager) (7/1/4/14/4)

National Zoological Park, Washington, D. C., *United States* (Mr. M. S. Roberts) (7/0/24/61/50)

Zoological Society of San Diego, San Diego, California, *United States* (Mr. M. Rich) (7/1/5/13/7)

Lincoln Park Zoological Gardens, Chicago, Illinois, *United States* (Mr. S. L. Kitchener) (2/0/2/6/1)

Kansas City Zoological Garden, Kansas City, Kansas, *United States* (Ms. J. C. Hixon) (7/0/7/13/1)

New York Zoological Society, New York, *United States* (Mr. M. MacNamara) (15/2/11/22/11)

La Montagne des Singes, Kintzheim, *France* (Mr. G. deTurckheim) (306/70/168/368/166)

Parc de Rocamadour, *France* (Mr. G. deTurkheim) (62 + 45 transferred from Kintzheim during 1974/?/107?/107?/?)

Zoorama Européen de la Forêt de Chizé, Beauvouir-sur-Niort, *France* (The Director) (2/0/2/2/0)

Menageries du Jardin des Plantes, Paris, *France* (Dr. J. Rinjard and Mme. K. Kurdi) (6/0/7/17/5)

Calgary Zoological Society, Calgary, *Canada* (Mr. D. R. Banks) (2/0/2/4/0)

Metropolitan Toronto Zoological Society, Toronto, *Canada* (Mr. C. Rabey) (12/3/12/13/4)

Zoo Safari "Quexigal," Cebreros, *Spain* (Mr. P. Ballesteros) (10/0/10/10/0)

Tierpark Rheine, Rheine, *West Germany*, (Dr. W. Salzert) (30/4/30/30/4)

Tiergarten, Nürnberg, Nürnberg, *West Germany* (Dr. M. Kraus) (9/0/9/9/0)

Vivarium Darmstadt, Darmstadt, *West Germany* (Dr. H. Ackerman) (4/0/4/4/0)

Westfalischer Zoologischer Garten, Münster, *West Germany* (Dr. H. Reichling) (5/0/5/5/0)

Information on colony size is also available for the following:

Zoo de Vigo, *Spain* (P. Ballesteros, personal communication) (2 females, 1974)

Kingdoms Three, Stockbridge, Atlanta, Georgia, *United States* (E. O. Smith, personal communication) (11 animals, 1975; The *Int. Zoo Yearb.* records two births for 1973)

Bellewaerde, Zillebeke, *Belgium* (50 animals, 1975)

IX. Appendix 2. Other Zoos Known (or Reputed) to Keep Barbary Macaques, 1964–1974

Zoos known to have bred the species are marked by an asterisk (*); those that are reputed to keep this species by a question mark (?). Addresses for most zoos will be found in the *Int. Zoo Yearb.* **14**, 1974.

* Chessington Zoo, Surrey, United Kingdom.
* Jersey Wildlife Preservation Trust, Jersey, Channel Islands.

* Zoological Society of Glasgow, Calder Park, Glasgow, United Kingdom.
* Flamingo Park Zoo, Malton, United Kingdom.
* Norfolk Wildlife Park and Pheasant Trust, Great Witchingham, United Kingdom.
  Sherwood Zoological Park, Hucknall, United Kingdom.
  Exmouth Zoo, Exmouth, United Kingdom.
  Ravensden Zoo, Ravensden, United Kingdom.
  Zoo de Thoiry, Thoiry, near Paris, France.
* Parque Zoologico de Jerez de la Frontera, Spain.
? Barcelona Zoo, Barcelona, Spain.
* Jardin Zoologico de Aclimacao em Portugal, Lisbon, Portugal.
  Dierenpark Wassenaar, Wassenaar, Netherlands.
* Städt Tiergarten Landau, Landau, West Germany.
  Tierpark, Berlin, E. Berlin, East Germany.
* Zoologicka Zahrada, Bonjnice, Czechoslovakia.
* Slaski Ogrod Zoologiczny, Chargow-Katowice, Katowice, Poland.
  Tiergarten Schönbrunn, Austria.
* Parc Zoologique National, Temara, Rabat, Morocco.
* Les Jardins Exotique, Rabat, Morocco.
  Parc Zoologique Municipal, Casablanca, Morocco.
  Parc Zoologique du Hamma, Alger, Algeria.
  Parc Animalier d'Annaba, Annaba, Algeria.
  Parc de Loisirs, Alger, Algeria.
* Parc Zoologique de la Ville de Tunis, Tunis, Tunisia.
* Lincoln Park Zoo, Oklahoma City, Oklahoma, United States.
* Milwaukee Country Zoological Park, Milwaukee, Wisconsin, United States.
* Highland Park Zoological Gardens, Pittsburgh, Pennsylvania, United States.
  Overton Park Zoo, Memphis, Tennessee, United States.
* Philadelphia Zoological Garden, Philadelphia, Pennsylvania, United States.
* Rare Feline Breeding Compound, Centre Hill, Florida, United States.
  Monkeys' Paradise, Rimouski, Quebec, Canada.
* Japan Monkey Centre, Inuyama, Japan.

A new park containing *M. sylvanus* was opened during 1975 at Affenberg Salem, Mendleshavsen, West Germany (Director, Dr. K. Werner). It is stocked with animals from La Montagne des Singes and Parc de Rocamadour.

Acknowledgments

My thanks are due to the persons listed in Appendix 1 who kindly completed my questionnaire and to those whose unpublished information in noted in the text. Additional assistance was received from Dr. L. Barbier, Professor F. Bourlière, Dr. F. D. Burton, Rosemary Deag, Dr. A. Dufour, Dr. Y. Raymond, Dr. J. Remfry, D. M. Taub, and G. de Turckheim. During my fieldwork invaluable assistance was provided in Morocco by L'Administration des Eaux et Forêts et de la Conservation des Sols. My field research from the Department of Psychology, University of Bristol was financed by Leverhulme Research Awards, The Wenner-Gren Foundation, and the University of Bristol.

References

Barry, J. and Lefranc, G. (1961). Étude neurohistologique de l'eminence médiane chez *Macacus sylvanus*. *C. R. Soc. Biol.* **155**, 1037—1040.

Bianchi, M. (1962). Noradrénaline et adrénaline dans la surrénales du singe et du lapin. *C. R. Soc. Biol.* **156**, 1992—1995.

Blanc, G. and Woodward, T. E. (1945). The infection of *Pedicinus albidus* Rudow, the maggot's louse on typhus carrying monkeys (*Macacus sylvanus*). *Amer. J. Trop. Med.* **25**, 33—34.

Burton, F. D. (1972). The integration of biology and behavior in the socialization of *Macaca sylvana* of Gibraltar. *In* "Primate Socialization" (F. E. Poirier, ed.), pp. 29—62. Random House, New York.

Burton, F. D. and Sawchuk, L. A. (1974). Demography of *Macaca sylvanus* of Gibraltar *Primates* **15**, 271—278.

Cabrera, A. (1932). Los mamiferos de marruecos. *Trab. Mus. Nac. Cienc. Nat. Madr. ser. zool.* **57**, 1—361.

Cooke, H. B. S. (1963). Pleistocene mammal faunas of Africa, with particular reference to southern Africa. *In* "African Ecology and Human Evolution" (F. C. Howell and F. Bourlière, eds.), pp. 65—116. Aldine, Chicago, Illinois.

Deag, J. M. (1970). "The Apes of Barbary." 30-minute, 16 mm color-sound film. Produced in association with J. H. Crook and the University of Bristol, Audio-Visual Aids Unit.

Deag, J. M. (1973). Intergroup eccounters in the wild Barbary macaque *Macaca sylvanus*, L. *In* "Comparative Ecology and Behaviour of Primates" (R. P. Michael and J. H. Crook, eds.) pp. 315—375. Academic Press, London.

Deag, J. M. (1974). "A study of the Social Behaviour and Ecology of the Wild Barbary Macaque, *Macaca sylvanus* L." Ph. D. Thesis, University of Bristol.

Deag, J. M. (1977). Aggression and submission in monkey societies. *Anim. Behav.* **25**.

Deag, J. M. and Crook, J. H. (1971). Social behaviour and "agonistic buffering" in the wild Barbary macaque *Macaca sylvanus* L. *Folia Primatol.* **15**, 183—200.

Forbes, H. O. (1897). "A Handbook of the Primates." Edward Lloyd, London.

Frisch, J. E. (1959). Research on primate behaviour in Japan. *Amer. Anthrop.* **61**, 584—596.

Girod, C. (1961). Identification des cellules δ (thyrotropes) du lobe antérieur de l' hypophyse, chez le singe *Macacus sylvanus* L. *C. R. Soc. Biol.* **155**, 1043—1045.

Girod, C. (1962). Identification expérimentale des cellules ε, source de prolactine, dans l'antéhypophyse du singe *Macacus sylvanus* L. *C.R. Soc. Biol.* **156**, 845—846.

Girod, C., Domenech, A., and Slimane-Taleb, S. (1955). Action de l'ACTH et de la cortisone sur le leucogramme de la ratte albinos et du Magot d'Algérie. *C.R. Soc. Biol.* **149**, 1543—1547.

Heim de Balsac, H. (1936). Biogeographie des mammiferes et des oiseaux de l'Afrique du Nord. *Bull. Biol. Fr. Belg.* Suppl. 21, 1—447.

Hill, W. C. O. (1974). "Primates, Comparative Anatomy and Taxonomy", Vol. VII, Cynopithecinae. Edinburgh University Press, Edinburgh.

Hooton, E. A. (1942). "Man's Poor Relations." Doubleday, New York.

Joleaud, L. (1931a). Etudes de géographie zoologique sur la Berberie: Les primates: Le Magot. *Congr. Int. Geogr.* (Paris) **2**, 851–836.

Joleaud, L. (1931b). Le role des singes dans les traditions populaires Nord-Africanes. *J. Soc. Afric.* **1**, 117–150.

Jolly, C. J. (1965). "The Origins and Specializations of the Long Faced Cercopithecoidea." Ph. D. Thesis, University of London.

Kearton, C. (1924). "The Shifting Sands of Algeria." Arrowsmith, London.

Kenyon, E. R. (1938). "Gibraltar Under Moore, Spaniard and Briton." Methuen, London.

King. F. W. (1974). International trade and endangered species. *Int. Zoo. Yearb.* **14**, 2–13.

Kurten, B. (1968). "Pleistocene Mammals of Europe." Weidenfeld and Nicolson, London.

Lahiri, R. K. and Southwick, C. K. (1966). Parental care in *Macaca sylvana. Folia Primatol.* **4**, 257–264.

Leblond, C. P. and Chaulin-Servinière, J. (1942). Spontaneous beriberi of the monkey as compared with experimental avitaminosis. *Amer. J. Med. Sci.* **203**, 100–110.

Lepoutre, B. (1957–1961). Recherches sur les conditions édaphique de régénération des cèdraies marocaines. *Annls. Rech. for. Maroc.* **6**, 1–211.

Lepoutre, B. (1962–1963). Premier essai de synthèse sur le mécanisme de régénération du cèdre dans le Moyen Atlas marocaine. *Annls. Rech. for.Maroc.* **7**, 55–163.

Lindsay, S. and Chaikoff, I. L. (1966). Naturally occurring arteriosclerosis in non-human primates. *J. Atheroscler.Res.* **6**, 36–61.

Lydekker, R. (1894). "The Royal Natural History," Vol 1, pp. 117–122. Frederick Warne, London.

MacRoberts, M. H. (1970). The Social organization of Barbary apes (*Macaca sylvana*) on Gibraltar. *Amer. J. Phys. Anthropol.* **33** (N.S.), 83–100.

MacRoberts, M. H. and MacRoberts, B. R. (1966). The annual reproductive cycle of the Barbary ape (*Macaca sylvana*) in Gibraltar. *Amer. J. Phys. Anthropol.* **25**, (N.S.), 299–304.

Marion, J. (1953–1955). La régénération naturelle du cèdre dans les cèdraises du rebord septentrional du Moyen Atlas occidental calcaire. *Annls. Rech. for. Maroc.* **1**, 31–150.

Mellanby, K. (1968). The effects of some mammals and birds on regeneration of Oak. *J. Appl. Ecol.* **5**, 359–366.

Métro, A. (1958). "Atlas de Maroc—Notices Explicatives, Forets, "Comité National de geographie de Maroc, Rabat.

Morris, R. and Morris, D. (1967). "Men and Apes." Hutchison, London.

Napier, J. R. (1970). Paleoecology and Catarrhine evolution. *In* "Old World Monkeys" (J. R. Napier and P. H. Napier, eds.), pp. 53–95. Academic Press, New York.

Napier, J. R. and Napier, P. H. (1967). "A Handbook of Living Primates." Academic Press, London.

Panouse, J. B. (1957). Les mammiferes du Maroc. *Trav. Inst. Sci. Cherifien ser. zool.* **5**, 1–200.

Pujos, A. (1964). Les milleux de la cèdraie marocaine. *Annls. Rech. for. Maroc.* **8**, 1–283.

Southwick, C. H., Siddiqi, M. R., and Siddiqi, M. F. (1970). Primate populations and biomedical research. *Science*, **170**, 1051–1054.

Tappen, N. C. (1960). Problems of distribution and adaptation of the African monkeys. *Curr. Anthropol.* **1**, 91–120.

Taub, D. M. (1975). A report on the distribution of the Barbary macaque *Macaca sylvanus* in Morocco and Algeria. Unpublished report to the Fauna Preservation Society, The New York Zoological Society and the International Union for Conservation of Nature.

Temkin, O. and Straus, W. L. (1946). Galen's dissection of the liver and of the muscles moving the forearm. Translated from the "Anatomical Procedures." *Bull. Hist. Med.* **19**, 167–176.

Trouessart, E. L. (1900). La Faune des mammiféres de l'Algerie, du Maroc et de la Tunise. *Causeries Sci. soc. Zoo. Fr.* **1**, 353–410.

Whiten, A. and Rumsey, T. J. (1973). "Agonistic Buffering" in the Wild Barbary Macaque, *Macaca sylvana* L. *Primates*, **14**, 421–425.

Zeuner, F. E. (1952). Monkeys in Europe past and present. *Oryx*, **1**, 265–273.

9

The Lion-Tailed Monkey and Its South Indian Rain Forest Habitat

STEVEN GREEN and KAREN MINKOWSKI

I. Introduction

The rare lion-tailed monkey (*Macaca silenus*) lives only in the Western Ghat Mountains of South India, in the tropical evergreen rain forests known as *shola*. Although sometimes mislabeled with the Sinhalese name for langur, "Wanderoo," the lion-tailed monkey has never occurred in Sri Lanka except as introduced commercially by Arab horse dealers. Along with the tiger, snow leopard, rhinoceros, and Asiatic lion, it is one of India's most endangered species. Although never common, its numbers have been so reduced by accelerating habitat elimination and hunting that extinction is imminent. Only immediate and strenuous measures to preserve its disappearing habitat, and to protect the monkeys that remain, can save the lion-tailed macaque.

This handsome animal, with its full, almost white lionlike facial ruff, might be more aptly called the "lion monkey," which is the translation of the Sanskrit name *Singhalika* and its derivatives. The tuft on the tail is found consistently only in adult males, developing with sexual maturation. A subadult male whose tuft is just beginning to develop is shown in Fig. 1. Few adult females' tails are tufted. The ruff, on the other hand, characterizes all members of this species, first appearing on youngsters of 5 to 6 months' age. Infants look like other macaque species, a dark dingy gray-brown. Adult males are larger than females, but are still among the smaller macaques, averaging about 55 cm body length and 8 kg weight.

Macaca silenus is a uniquely important primate of profound scientific interest. It is not a local Indian race, like the Kashmir stag or Indian tiger, but a full, taxonomically distinctive species. Special reasons for preserving this unusual monkey relate to its singular role as the only exclusively arboreal forest-dwelling macaque.

The macaque monkeys, a genus of a dozen or so species (Thorington and Groves, 1970) living throughout most of South Asia and also in North Africa, are especially important to man. Their contributions to laboratory medical science are well known. Our knowledge of the evolution and biological regulation of animal societies has been dramatically enhanced by studies of wild and feral macaques. Investigations of the relationship between structure and regulation of social groups of primates and the habitat in which they live have increasing relevance to the fate of man as we attempt to delineate the factors modulating our own social behavior. Research on macaques has been the main basis for theorizing about the effects of different environments and habitats on the evolution and expression of social behaviors. [See, for example, Bertrand (1969) on *M. arctoides*; Carpenter (1942) on *M. assamensis*; Angst (1975) on *M. fascicularis*; Jay (1965) and Sugiyama (1965) for reviews of early work on *M. fuscata*; Lindburg (1971), Mukherjee and Mukherjee (1972), Southwick *et al.* (1961a,b), and Southwick and Siddiqi (1967) on *M. mulatta*; Bernstein (1967a,b),

Fig. 1. Subadult male *Macaca silenus* in a typical watchful stance. His temporal mounds have not attained the full size characterizing adult males, and his tail tuft is just beginning to develop.

Southwick and Cadigan (1972) on *M. nemestrina*; Rahaman and Parthasarathy (1967, 1969), Simonds (1965) and Sugiyama (1971) on *M. radiata*; Eisenberg *et al.* (1972) on *M. sinica*; and Burton (1972) and Deag and Crook (1971) on *M. sylvana.*]

Among the macaques only the lion-tailed monkey is an obligate rainforest dweller. Habitats for other macaques are typically open, lightly wooded ground, degraded secondary forest and forest edges, rocky coasts and cliffs, and commensal habitation with man. They are usually semiterrestrial. Some, such as *M. nemestrina* (Bernstein, 1967a,b) and *M. radiata* (Sugiyama, 1971), may spend over half their time in trees in a few habitats. We found that *M. silenus*, on the other hand, spends less than 1% of its time on the ground. It has been uniquely classified as the only truly arboreal macaque (Southwick *et al.*, 1970), thus offering the opportunity to examine the behavioral correlates of arboreal living and a forest canopy habitat and thereby measure and contrast the effects of hereditary and environmental influences on behavior.

Under the sponsorship of the Bombay Natural History Society and with the approval of the Government of India, we began investigating the ecology and behavior of the lion-tailed monkey in August 1973. With the permission of the Chief Conservator of Forests of Tamil Nadu, we studied lion-tailed monkeys in the least disturbed part of their range, the Ashambu Hills, the southernmost extent of the chain of Western Ghats. In addition to the focus on behavioral ecology and sociobiology, the research also encompassed habitat requirements, vegetational studies, factors leading to reduced numbers, and some facets of rain forest ecology. We observed the monkeys for 1767 hours from September 1973 to April 1975, usually following them from sunrise to sunset. We did not provision the monkeys or attempt to interfere with them in any way. Since April 1975, long-term observation has been continued by our local assistants, headed by Devadoss Michael, and by John F. Oates, who is engaged in a comparative study of the Nilgiri langur in the same area.

Following two identifiable troops of lion-tailed macaques over a continuous 19-month period led us to conclude that survival of this species depends on access to a very large and continuous expanse of undisturbed rain forest. In support of this finding, this report outlines the requirements of lion-tailed monkeys in terms of habitat structure and area necessary to ensure a viable breeding population. We report the results of our census and habitat survey undertaken in all areas where *M. silenus* has been reliably reported to dwell. We describe in detail the Ashambu Hills, the area we believe best suited for mounting an effective preservation campaign. Recommendations are presented for measures to be taken there against immediate threats to the region and to the monkeys. Guidelines are suggested for long-term multiple use conservation planning for this area. Appendix 1 highlights some of the arguments for conserving rain forests.

II.   Habitat

This overview of the lion-tailed monkey's habitat is principally based on detailed studies at $8°32'-38'N$ in the Ashambu Hills. Our survey of other localities in which the monkeys dwell indicates that their requirements are probably very similar in all areas south of ca. $11°30'N$, including the Nilgiris, Anamalais, and Cardamom Hills of Tamil Nadu (formerly Madras State) and Kerala, as well as the Ashambu Hills. Further north in Karnataka (formerly Mysore State), the rain forest structure differs floristically. The canopy there is generally abundant in *Dipterocarpus*, a genus entirely lacking in the more southerly forests. In these northern areas lion-tailed monkeys are much less dense, so it is likely that a larger area would be needed to preserve a breeding population than in the southern hills.

A. Structure, Area, and Use

1. *Shola* Vegetation

Although the Indian term *shola* properly applies to remnant patches of forest found in sheltered sites along watercourses among rolling grassy hills or downs, it is now generally used to connote any of the wet broad-leaved evergreen forest formations of the Western Ghats (Fig. 2). These occur in regions with a minimum annual rainfall as low as 175 cm, but usually above 250 cm. Their most luxurious growth occurs at the middle elevations of 500 to 1500 m altitude. Below 300–600 m the rain forest usually grades imperceptibly into the lower-lying moist deciduous type, but may infrequently occur as narrow belts of riverine forest to as low as 100 m. Above 1500 to 1800 m, they give way to more stunted montane forests (Blasco, 1971; Champion and Seth, 1968; Chandrasekharan, 1962; Kadambi, 1950).

All of the areas where lion-tailed monkeys have been reported contain the lofty dense evergreen forests characterized by a large number of tree species reaching 30–50 m or more in height and forming a dense canopy. The vast majority of trees have large simple leaves. Giant climbers, epiphytic ferns, mosses, and orchids are numerous at all levels from the canopy to the forest floor. The woody understories include saplings, smaller species of trees, shrubs, and often a tangle of cane or bamboolike reeds. Except in the ravines, the undergrowth is relatively free of herbaceous plants. In the southern *shola* forests, the plant communities favored by *Macaca silenus* are abundant in *Cullenia exarillata*, a very large, slightly buttressed tree with fluted bole, occurring at elevations from 600 to 1400 m in areas of over 3000 mm rainfall. The typical dominant canopy tree associations are *Cullenia-Palaquium*, *Cullenia-Calophyllum*, *Poeciloneuron-Cullenia*, and *Palaquium-Mangifera* [Champion and Seth, 1968 (Chapts. 6,8, and 9); Puri, 1960 (Chapt. 8)].

The Ghat forests are not uniform in structure. The highly variable terrain includes steep hillsides topped with rocky precipices reached by sharply ascending slopes. The major river valleys are usually very deep, and the monsoon torrents in the feeder streams have cut abrupt gorges. Between the major valleys are a few gently sloping interior hills with nearly flat stream banks. This landscape, together with edaphic conditions, factors of climate and exposure, especially to the two monsoons, and different biotic influences yield localized, distinct vegetation formations within each hill range.

Fig. 2. Streamside view of the wet tropical evergreen forest habitat of lion-tailed monkeys. This *Cullenia*-dominant *shola* in the southern Western Ghats is at 1200 m elevation in an area of 320 cm annual rainfall.

Although usually limiting their activities to the forested hillsides, the lion-tailed monkeys may invade the less extensive habitats within the *shola* regions on a seasonal basis, seeking, for example, the fruits of a gregarious palm (*Bentinckia coddapanna*) which grows on the narrow ledges of rocky precipices in the southernmost Ghats. They also make brief seasonal excursions into the lower lying semi-evergreen areas. These lower zones are subject to the heaviest hunting pressures and might be used more by the monkeys if adequately protected. One troop on the Sharavati River north bank (North Kanara District, Karnataka) reportedly is seen occasionally in the semievergreen and riverine *shola* forests at 200 m and even as low as 70–100 m.

Where the tall dense evergreen forests have been destroyed, the monkeys may still lead a precarious existence in some regions by moving between patches of remnant low-stature forests. Blasco (1971), Champion and Seth (1968), and Puri (1960) conclude that these low-stature scrubby forest patches and associated grasslands are degradation stages maintained by grazing and burning. Thus, they are of recent origin on an evolutionary time scale. There can be no doubt that *M. silenus* evolved in rain forest, and it is this habitat to which they are adapted.

The monkeys' varied diet can be obtained only from diverse flora and fauna of an undisturbed mature rain forest. Although their food consists mainly of fruits of the top and second story trees, the monkeys utilize every stratum of the forest, also eating flowers and fruits of climbers, small trees and shrubs, and leaves of reeds, grasses, and sedges (Table I).

Insects (adults, pupae, and larvae), lizards, tree frogs, and fungi are gleaned from foliage, exposed and snatched from inside dead wood, plucked from underneath bark, and picked from rotting log falls (Fig. 3).

2. Seasonal Ranging

Although the lion-tailed monkey's diet varies from month to month, *Cullenia exarillata* and *Artocarpus heterophyllus* (jack fruit) are important year-round foods (Table II). To obtain the fruits that form the bulk of their diet in a given month, the monkeys often range quite widely to utilize scattered aggregations of the tree species in fruit or to visit rare and isolated individual specimens. During March and April 1974, for example, one troop of monkeys moved along the ridges and higher slopes where fruiting *Syzygium* spp. abound. In May they occupied mostly the steep valleys where *Litsaea wightiana* occurs in local aggregations, emerging above the reed brakes along stream banks. In June the monkeys sought later ripening aggregations of other *Litsaea* species growing on the shallow wet soil at the base of precipitous cliffs. During July and August they covered yet another area, the more gentle slopes and wide valleys where fruiting *Artocarpus* trees were numerous, with rare forays onto the cliffs for palm fruit.

During a 1- or 2-month period, a lion-tailed monkey troop often foraged over an area only 100 to 200 hectares in size. [Sugiyama's (1968) map also shows ranges of this size for a period of less than eight weeks' study.] However, as our troops' diet shifted in composition throughout the year, the area they utilized also shifted and expanded to include new regions. In the apparently optimal, variable *Cullenia* rain forest of our study area, our main troop ranged over about 5 km² of continuous rain forest during a single year. Observations continued by

TABLE I

Classification of *Macaca silenus* Plant Foods by Forest Story[a]

Topmost canopy trees	*Cullenia exarillata, Calophyllum trapezifolium, Elaeocarpus munroii, Vepris bilocularis, Palaquium ellipticum, Ficus tsiela, Ficus retusa, Ficus talboti, Ormosia travancorica, Elaeocarpus tuberculatus, Litsaea oleoides, Cinnamomum sulphuratum, Hemicyclia elata, Symplocos sessilis, Aglaia bourdilloni, Canthium diococcum, Eugenia* spp. *Holigarna nigra*
Second story trees	*Artocarpus heterophyllus, Litsaea wightiana, Litsaea beddomei, Litsaea insignis, Diospyros peregrina, Diospyros nilagarica, Gordonia obtusa, Viburnum acuminatum, Actinodaphne tadulingami, Myristica beddomei, Gomphandra coriacea, Symphyllia mallotiformis, Canthium umbellatum, Elaeocarpus venustus, Rapanea daphnoides, Scolopia crenata, Drypetes oblongofolia, Eugenia* spp.
Third story	Small trees—*Jambosa mundangam, Macaranga roxburghii, Bentinckia coddapanna, Miliusa wightiana, Octotropis travancorica, Antidesma menasu,* Unidentified *Eugenia* sp. No. 1. Bamboo reed—*Ochlandra scriptoria*
Fourth story	Shrubs—*Chloranthus brachystachys, Psychotria congesta, Psychotria ostosulcata, Lasianthus cinereus, Saprosma corymbosum, Ardisia pauciflora, Calamus travancorica* Grasses—*Isachne gardneri, Oplismenus compositus* Sedge—*Scleria cochinensis*
Other	Large woody climbers—*Toddalia asiatica, Tetrastigma sulcatum, Randia rugulosa, Embelia adnata,* unidentified climber sp. No. 1, unidentified climber sp. No. 2. Twiners—*Aristolochia tagala, Dioscorea belophylla* Tree parasites—*Loranthus obtusatus, Loranthus elasticus, Viscum ramosissimum* Epiphytic orchid—*Luisia tenuifolia*

[a]Fungi are also commonly eaten. Other plants not eaten by the monkeys are also essential for their diet, since they host many animals the monkeys eat, for example, a species of caterpillar which lives primarily on the leaves of *Agrostistachys meeboldii.*

D. Michael and J. F. Oates after April 1975 indicate that the monkeys' range is expanding still further.

3. Water

Most of the world's tropical evergreen rain forests receive year-round precipitation. The *sholas* of the Western Ghats are unusual in that they regularly have one or two annual periods of drought (Blasco, 1971; Puri, 1960). During the

Fig. 3. Young nulliparous female lion-tailed monkey foraging for insects by picking at the bark of a dead *Calophyllum* tree. The bipedal posture is often assumed when hands are used for manipulation.

monsoons, the monkeys obtain water by licking leaf surfaces or by drinking from natural bowls in the forks of trees. Each troop's range includes perennial water-courses, and in the dry seasons the monkeys occasionally descend from the trees and drink from the banks of these permanent streams and rivers.

4. Competition

Two other arboreal mammals, the Nilgiri langur (*Presbytis johnii*) and the Malabar giant squirrel (*Ratufa indica*), are potential major food competitors with the lion-tailed monkey in the southern part of its range. The very abundant

TABLE II

Feeding Seasonality of One Troop of *Macaca silenus* for 12 Continuous Months, 1974–1975 (Plant Items Only)[a]

Species	March–April	May–June	July–August	September–October	November–December	January–February
Cullenia exarillata	xxxx	xxx	xxx	xxx	xxxx	xxxx
Artocarpus heterophyllus	xx	xxx	xxxx	xxxx	xx	xx
Elaeocarpus munroii	x	x	x	xxx	x	xx
Eugenia (10 similar spp.)	x	xx	xx	x	x	x
Vepris bilocularis	xx	xx	xxx	xxx	x	
Miliusa wightiana	xx	x		x	xxx	xxx
Toddalia asiatica	xx	x		x	xx	xx
Ochlandra scriptoria	xx	x	x	x		xx
Viburnum acuminatum	xx	xx	x	x	x	
Palaquium ellipticum	xx	xx	x	x	x	
Oplismenus compositus	x	xx	x	x	x	
Tetrastigma sulcatum	x	x		xxx	xxx	
Scleria cochinensis	xx	xx			x	x
Randia rugulosa	x	xx			x	x
Holigarna nigra	x	x	x	x		
Myristica beddomei	x	x	x			x
Jambosa mundangam	xxxx	x				x
Eugenia sp. No. 1	xxxx	x				x
Isachne gardneri	xx	xx				x
Gordonia obtusa	xx				x	xx
Embelia adnata (?)	x	xxx		x		
Ormosia travancorica	xx	x		x		
Symplocos sessilis	x	xx		x		
Octotropis travancorica	x				xx	xx
Psychotria congesta	x	xx			x	
Chloranthus brachystachys	x			x	x	
Litsaea wightiana	xxxx	xxxx				
Litsaea beddomei	xx	x				
Elaeocarpus tuberculatus	x	xx				
Canthium diococcum	x	x				
Symphyllia mallotiformis	x					x
Ficus retusa	xx					
Calophyllum trapezifolium	x					
Aglaia bourdilloni	x					
Drypetes oblongofolia	x					
Luisia tenuifolia	x					
Litsaea oleoides		xx	xxx			
Diospyros peregrina		xxx				x
Antidesma menasu		x		x	xx	
Actinodaphne tadulingami		x	xx			

Species	March–April	May–June	July–August	September–October	November–December	January–February
Unidentified climber No. 1		x		x		
Litsaea insignus		x				
Elaeocarpus venustus		x				
Loranthus elasticus			xx	xxxx		
Bentinckia coddapanna			x	x		x
Ardisia pauciflora			x		x	x
Cinnamomum sulphuratum			x			
Gomphandra coriacea			x			
Psychotria octosulcata				xx	xx	
Loranthus obtusatus				xxx	x	
Unidentified climber No. 2				xxxx		
Ficus tsiela					xx	
Hemicyclia elata				xx		
Rapanea daphnoides				x		
Diospyros nilagarica					xx	xx
Macaranga roxburghii					xxx	
Calamus travancorica					xx	
Lasianthus cinereus					x	
Saprosma corymbosum					x	
Scolopia crenata					x	
Viscum ramosissimum					x	
Dioscorea belophylla					x	
Aristolochia tagala					x	
Ficus talboti						x
Canthium umbellatum						x

[a]xxxx, Major food; xxx, commonly eaten; xx, seldom eaten; x rarely eaten.

fruits of *Cullenia* are a major item in the diet of the lion-tailed monkey year-round (Table II). Nilgiri langurs also eat these fruits, and giant squirrels often scavenge seeds from *Cullenia* fruits opened and abandoned by lion-tailed monkeys. *Cullenia* does not seem to be a limiting resource, and we have no evidence that competitive pressures from other arboreal mammals are limiting the lion-tailed monkey's population in the rain forests. Like other colobin monkeys, the Nilgiri langur is reported to eat mainly leaves (Horwich, 1972). Oates is currently examining the resource utilization of the Nilgiri langur in the rain forest.

In the northern Western Ghats, the common gray Hanuman langur (*P. entellus*) replaces the Nilgiri langur. Although studied extensively in dry and open habitats, its diet in rain forest is unknown.

Bonnet macaques (*M. radiata*) typically occur in the drier deciduous forests and scrub jungles at much lower elevations, but sometimes seasonally move up into the wet forests inhabited by lion-tailed monkeys. The few times we saw bonnets with *M. silenus*, no antagonism occurred, as recorded also by Sugiyama (1968) and Webb-Peploe (1947). However, interspecific resource competition with *M. radiata* may have prevented *M. silenus* from successfully occupying the lower elevation forests in which *M. radiata* are so well established.

Some seasonally abundant birds are potential competitors for fruits and insects. The large Jerdon's Imperial Pigeon (*Ducula badia*), for example, eats figs from the same *Ficus* trees as does the lion-tailed macaque, but we have not undertaken any systematic investigation of resource competition with the rain forest avifauna.

5. Predation

It would seem that terrestrial predators such as leopard, dhole (also called wild dog), or python would have negligible effects on a species as arboreal as the lion-tailed macaque, yet we observed noisy mobbing of the carnivores and very prolonged and close scanning of low vegetation and the ground before the monkeys ever descended. They show alarm responses to a few eagle species, especially *Ictinaëtus malayensis*, and we observed another eagle, *Spizaëtus nipalensis*, attempting to take an infant monkey, but it is unlikely that raptors are responsible for much mortality. The most destructive predator is man, hunting to kill or capture. Lion-tailed monkeys survive in very limited numbers, if at all, in areas where hunting has gone unchecked.

6. Disease

The most serious threat of disease can arise from the introduction of a virus, bacteria, or parasite to which the monkeys have little resistance. It has been proposed that captive lion-tailed monkeys from breeding groups in zoological parks be used to repopulate depleted areas. Since they may spread pathogenic organisms acquired in captivity to which wild monkeys would not have inherited defenses or acquired immunity, it is probably unwise to release such propagules into an area still containing a wild population. Similarly, pet or zoo bonnet macaques, langurs, or other primates should not be permitted entry to any forest inhabited by lion-tailed monkeys.

B. Viable Breeding Population

1. Population Size

A species' extinction can be averted only if there is at least one population of interbreeding individuals (a deme) numerically large enough to provide genetic diversity and geographically widespread enough to avoid succumbing to temporary and local adverse conditions and reductions in number.

Too small a deme lacks genetic diversity, and the inbred progeny are more prone to succumb to environmental fluctuations including shifts in climate or the appearance of new strains of disease-producing organisms. As Berry (1971) puts it, "So much variation will have disappeared when numbers decrease that the species may be so genetically rigid as to be effectively inviable." A deme may become too small either through reduction of numbers or by isolating the breeding units from one another. To preserve the lion-tailed monkey, a certain total number of individuals and troops of the species must survive, and these must be geographically located so that breeding interchange can take place. A minimum size is also necessary to allow the deme to survive chance fluctuations in reproductive success and the occasional environmental adversities which induce mortality or reduce reproduction.

To determine the minimum number of individuals required to form a viable deme able to survive population fluctuations and also sufficiently large to maintain genetic diversity, many things must be considered. The age structure of the population, mortality rates as a function of age, social breeding system, age to reproductive maturity, birth intervals, and number of young per litter must all be taken into account.

Macaca silenus bear their first young at about 5 years and probably have only one or two more young in their lifetime. Males begin to breed at about 8 years. (Ages were estimated on the basis of comparing dentition, body size, and morphology with captive *M. silenus* and wild *M. fuscata* of known ages.) They are not long-lived; our estimate for the age of the oldest female we have seen is 9 years. In an area where they are hunted, Sugiyama (1968) reported two young juveniles or infants per seven adult females in one troop, and three per ten in another. In our study area there were two surviving infants from three adult females in one troop and eight from eleven in another in a 1-year period. The interbirth interval in zoos is about 20 months (personal communication, L. Gledhill, Woodland Zoological Park, Seattle, Washington). Compositions of Sugiyama's troops and two of ours are as follows:

	Adult male	Subadult male	Mature female	Young
A	1	2	3	6
B	3	3	11	ca. 17
C	2	1	7	6
D	2	2	10	8

No study has set a definitive figure, but many experts quote a number on the order of 500 to 2000 as the minimum population of a deme which would not preclude survival of slowly maturing, short-lived, single-birth mammals such as these macaques.

2. Interbreeding, Local Extinctions, and Biogeography

Subadult and adult males left two of the *M. silenus* troops under observation, but we could not determine whether they joined another troop. In other macaque species males frequently leave their natal troop to join another, as demonstrated in rhesus (Lindburg, 1969) and Japanese monkeys (Koyama, 1970; Nishida, 1966). Encounters between lion-tailed monkey troops occur frequently, with males and females of different troops showing interest in each other, but adult males chase and herd adult females of their own troop away from adult males of the other troop. Intertroop mating might occur at such times. Genetic exchange between troops, either by migrating males or during troop meetings, can occur only as long as continuity of the forest habitat is maintained. If home ranges become discontinuous through reduction of troop density, or by cutting the forests, these mechanisms cannot operate.

The major southern hill ranges of the Western Ghats where lion-tailed monkeys now live each contain separated breeding populations. Monkeys do not cross the intervening gaps of kilometers of inhabited plains and low hills with degraded or cleared forests. Furthermore, within each hill range there may be separate demes if the population inhabits isolated areas of forest. In a few localized areas so disturbed that long-term survival is in any case unlikely, lion-tailed macaques have been reported to move between the very small patches of remnant forest (for example, in the Nilgiri plateau). Isolation of demes can arise when large gaps in the forest containing tea, coffee, wattle, eucalyptus, settlements, grassland, cultivation, or other non-*shola* habitats serve effectively to block the monkeys' movements. In over 1700 hours observing troops whose range borders on regions of tea, coffee, and eucalyptus, with which they have been familiar for over 30 years, we never saw them cross these areas. Even when these planted regions formed the shortest pathway from one *shola* to another toward which the monkeys eventually headed, they took the roundabout way through natural forests.

The population behavior of species distributed in disjoint regions of suitable habitat surrounded by inhospitable zones is the subject of island biogeography, a rapidly expanding area of ecology which received its impetus from MacArthur and Wilson (1963, 1967) and MacArthur (1972). Regions of *shola* forest isolated by other habitats are metaphorically islands, and the lion-tailed monkeys living in them are surrounded by distributional barriers analogous to the sea's role for terrestrial island dwellers. In proposing principles for designing wildlife reserves, the theories of island biogeography are being applied to such situations (e.g., Diamond, 1975; May, 1975; Sullivan and Shaffer, 1975). The guidelines which emerge reinforce the necessity for a large and continuous area, not only to allow genetic exchange, but also to permit long-term population survival in the face of localized adversity.

If through chance fluctuation in breeding success, disease, illegal hunting, or whatever reason, the troops in one area of a forest die out, then other monkeys from neighboring troops can eventually be expected to invade and repopulate the area, providing they have a pathway of suitable forested habitat. Only an area of forest sufficiently large to harbor monkeys far away from the cause of a population disturbance can protect them against extinction in that forest due to local calamity. Any small number of animals living in isolation from other members of its species will inevitably, at some point in time, die out and be eliminated from that area forever if there is no pathway for repopulation. Twenty separated areas of forest, each supporting one troop of monkeys, will at some point become 19 areas, then 18, and so on, as one misadventure or another occurs in each area, as it must in due course. Yet if a single area of forest twenty times as large supports twenty troops, then as a calamity occurs in one part or another of it, the adjoining troops provide a breeding nucleus for repopulation as long as the habitat remains undisturbed. Emphasis is, therefore, placed in this report on preserving an area which can support a large contiguous population of monkeys. The total number of *M. silenus* surviving today is not an important figure in itself, even if it were large. If it were distributed in numerous scattered populations, extinction could be confidently predicted. Survival requires at least one extensive area with a large interbreeding population.

3. Troop Size and Density: Area Required

We encountered troops of from six to thirty-four individuals, with twelve a typical size. Jerdon (1874) mentions troop sizes ranging from "12 to 20 or more," and this information has become hereditary, quoted by most authors succeeding him. Sugiyama (1968) counted two troops numbering sixteen and twenty-two. We shall take fifteen as our estimate of the average troop size.

Each troop of monkeys shares parts of its range with others if the area is well populated. There was frequent overlap with neighboring troops around the edges of the 5-km² range found in *Cullenia* forests, but the central core area of about 300 hectares was rarely entered by other troops. The overall density of troops in apparently optimal *Cullenia*-forest habitat is estimated at one per 400 hectares. The minimum area necessary to support a single deme of about 500 animals under the best conditions of mature climax *Cullenia*-forest is then 132 km² (495 animals at 15 per troop = 33 troops at 400 hectares each). If this area forms a continuous block, then it permits genetic exchange and is probably also large enough to prevent extinction due to local temporary adversity.

III.   Status of *Macaca silenus* Population and Its Habitat

Since recent information on the amount and distribution of suitable *shola* forests·and on the occurrence of *M. silenus* was not available, the Government of India, as part of its interest in conservation of this species, requested us to conduct a field survey and to recommend conservation measures.

A.   Methods of Survey

As a first step in our survey of the *M. silenus* habitat we examined available documents, including Government of India survey maps, private maps prepared for plantations, forest department working plans, and district gazeteers. Recent large-scale survey maps were not available to us for security reasons, so we were restricted to 50- to 60-year old Survey of India maps. These older maps impressed on us, however, the enormous areas of forest which have since disappeared. We also used recent Earth Resources Technology Satellite photographs filtered to show vegetation types (from NASA, United States), relief maps issued to airline pilots, and I.C.A.R. vegetation maps jointly produced by the Indian Council for Agricultural Research and the French Institute of Pondicherry.

In May 1974 and during April and May 1975 we visited all areas reported to contain lion-tailed monkeys. The actual boundaries of each forest were compared with those indicated by the state forest departments. In general, maps and figures in forest department working plans greatly overestimated the extent of the remaining Ghat evergreen forests. The sources of error were many: forests cleared then replanted with teak (*Tectona grandis*), *Bombax*, or exotics (principally *Acacia* and *Eucalyptus*), reduction in forest extent by fires since the earliest surveys, and wide-scale encroachment on forest lands for shifting cultivation were all frequently included with figures for evergreen rain forest. Vast areas recently filled and flooded for reservoirs were also sometimes still listed as forested. We determined the composition of the forest canopy by observation and by reference to forest department documents and to Aiyar (1932), Blasco (1971), Champion and Seth (1968), Kadambi (1950), Lawrence (1960), Rajasingh (1961), Ramaswami (1914), and Ranganathan (1938).

The locations of reliable sightings of lion-tailed monkeys in the last 50 years are shown in Table III. All published sightings known to us are included, as well as all reports by reliable observers which came to our attention. We have excluded hearsay and the ambiguous reports from areas south of 11°30′N referring to a "black monkey," a term often used interchangeably by local villagers for the black Nilgiri langur and lion-tailed macaques. Most casual observers cannot distinguish between *M. silenus* and *P. johnii*, both dark monkeys with light facial ruffs. When we questioned people who had supposedly seen *M.*

TABLE III

Areas Where *Macaca silenus* Have Been Reported in the Last 50 Years (from North to South)[a]

Name of general area and of specific locale	Approx. area center		Elevation[b] (m)	Status[c]
	Lat. N	Long. E		
Anshi Ghat[2]	14°56'	74°22'	300	−
Jog Falls—Sharavati River; Malemane[1]	14°16'	74°43'	500	+
Agumbe—Someshwar Ghat and steep western face into Kanara[2,13]	13°15'— 13°43'	74°55'— 75°15'	600	+
Kudremukh-Bhagavatti; GungaMola; Thunga-Bhadra[2]	13°14'	75°18'	900	+
Sakleshpur area of Hassan Ghat; Bisle[2,3]	12°42'	75°51'	900	?
Nilgiris—Wynaad; ghats near Gudalur[2,3]	11°21'	76°27'	1200	?
Nilgiris—Palghat; Silent Valley; Bhavani River Valley[3,10,11]	11°07'	76°26'	1000	+
Anamalai-Nelliampathi Hills; Top Slip; Karian *shola*; Chalakudi River Valley—Sholayar and Perambikulam; Punachi Range; Puthanur; Varagaliar; Manamboli; Chandramalai; Vellodai[3,4,8,9,11,13]	10°20'— 10°27'	76°35'— 77°05'	800— 1200	+
Cardamon Hills—Munnar High Range; Panniar[11]	9°59'	77°13'	1250	?
Cumban Valley—High Wavy Mountains[2,3,5]	9°38'	77°22'	1300	?
Periyar—Pamba; Pachakannan[2,3]	9°33'	77°06'	900	+
Periyar—S. shore; Mlaparra[3,4,11]	9°27'	77°20'	1050	+
Ramnad—Srivilliputtur—North Tirunelvelli; Puliyara[9]	9°01'	77°11'	550	+
Ashambu Hills—Kuttalam[2,7]	8°45'	77°10'	1200	−
Ashambu Hills—Papanasam; Kannikutty; Kattalamalai; Valaayar[3]	8°38'	77°17'	1000	+
Ashambu Hills; Singampatti, Kalakkadu, Manjolai, Mahendragiri, Kodayar, Sengalteri, Narakad[4,6,12]	8°34'	77°25'	900— 1300	+

[a] Key to superscript numbers: (1) H. R. Bhat, Indian Council for Medical Research's Virus Research, Centre Field Station at Sagar (personal communication); (2) Daniel and Kannan (1967); (3) Forest Department; (4) Green and Minkowski on survey trip; (5) Hutton (1949); (6) Karr (1973); (7) Krishnan (1971); (8) G. A. Kurup, Zoological Survey of India (personal communication); (9) J. F. Oates (in litt.); (10) Pruett (1973); (11) Sugiyama (1968); (12) Webb-Peploe (1947); R. Whitaker, Director of the Madras Snake Park (personal communication).

[b] Given by reference or determined by altimeter on survey.

[c] −, No indication of presence; +, present; ?, possibly present, but, if so, in a severely decimated state.

silenus in areas other than those listed here, they described the male Nilgiri langur's whooping call and bounding display. When presented with a variety of pictures of black primates, they occasionally pointed out black gibbons.

Sightings have occurred in six separate and distinct hill ranges south of 15°N in the chain of Western Ghats. The three most northerly are in Karnataka State, while the remaining three straddle the Kerala-Tamil Nadu border. Criss-crossing the forested areas of the Western Ghats, we covered over 4000 km by Jeep, boat, and on foot, visiting not only these areas of reported sightings, but adjacent forests as well.

B.   Results of Survey

The number of areas where lion-tailed macaques still occur continually declines. In 1974 Dr. G. A. Kurup of the Zoological Survey of India visited many locations where the monkeys were supposed to dwell. After interviewing local inhabitants he concluded (personal communication) that in most places where *M. silenus* had been seen frequently 10 or 20 years ago, none have been sighted in the last few years. Even areas where monkeys were reported only 5 years earlier yielded no sightings over the past 1 or 2 years. Our own survey and Kurup's investigation agree that they probably do not now occur between Goa and the Sharavati River, and so Krishnan (1971) is likely correct in stating the monkeys are extinct in the northern end of their former range. Some observations on each region are presented below.

1.   Anshi Ghat

One sighting was made 20 years ago a few kilometers south of Goa on the small forested hill forming the north bank of the Kalinadi River, not far from its outlet on the Arabian Sea. The suitable forest remaining today is very limited in extent and isolated from other forests. Some medium-stature forests transitional between evergreen and semi-evergreen remain near the hilltops at ca. 550 m elevation, but most of the area is under cultivation or is monospecific plantation forest. There are no local reports of a "black monkey."

2.   Jog Falls-Sharavati River

At least one troop of lion-tailed macaques ranges on the north bank of the Sharavati River below Jog Falls. The area is transversed by the Honavar-Belgaum road along which forest encroachment for slash and burn agriculture is extensive. Local people report a severe decline in numbers of monkeys since selective-felling operations began in 1958. The high forests lie between 200 and 500 m elevation and are transitional between semi-evergreen and evergreen. Since this is now probably the northernmost limit of *M. silenus*, and since the habitat differs markedly from *Cullenia*-dominant and *Dipterocarpus*-dominant forests further south, this area and its monkeys deserve protection. The area is not extensive enough to support a sizable population, however, since the forests bordering it have been largely felled.

3.   Shimoga, Chikmagalur, and Hassan Districts;
     South and North Kanara Districts

The upper reaches and eastward-stretching plateaus of the main range of the Western Ghats lie in Shimoga, Chikmagalur, and Hassan Districts, while the steep western face drops precipitously down to the coastal plains of South Kanara and North Kanara Districts. This range contains the largest remaining stand of evergreen rain forests in the Western Ghats, covering perhaps 800 km². At the northern end these forests lie on the western face, but toward the south they spread more and more easterly to cover most of the upper elevations. Three different locales have yielded more than one sighting of *Macaca silenus* in recent years; other reports come from scattered points in between.

a. AGUMBE—SOMESHWAR GHAT.   Lion-tailed monkeys have been sighted in the rain forests along the very steep western face of the Ghats. This precipitous slope rises from South Kanara and levels off above 600 m into gentle hills rolling eastward. Although the rain forests are extensive, only the very steep regions remain undisturbed. The forests in the flatter region are broken up by roads along which agricultural lands have been cleared for hundreds of meters on each side. Coffee and tea plantations occupy the more remote interior forests. The remaining evergreen forests are heavily worked for plywood. The regions too steep to be worked economically for timber are now partly incorporated into the Someshwar Sanctuary.

b. BHAGAVATTI VALLEY—KUDREMUKH.   Between 730 and 1100 m elevation, at the headwaters of the eastward-flowing Thunga and Bhadra Rivers, lie extensive rain forests interspersed with grasslands. These adjoin the more limited forests on the steep western face of the Ghats. Extraction of timber for railway sleepers is now disturbing the virgin forests, especially pockets of *Poeciloneuron*-dominant associations. Until recently lion-tailed monkeys were known to dwell here but are no longer sighted by local residents and forest department officers. Various schemes of selective felling have degraded most of the adjacent areas not too steep to be worked.

c. SAKLESHPUR AREA OF HASSAN GHAT.   The eastern region of the forested areas have been felled for coffee and tea. Although almost untouched as late as 1960, the western part of the evergreen rain forests is now being heavily exploited, especially for plywood. Along the Netravati River runs the Hassan—Mangalore highway which serves as a principal access road for timber extraction. A new railway project is also cutting through this area. Further south in the

Bisle—Kudamakal—Bhagumali area, where the slopes have proved too steep for timber extraction, the *Dipterocarpus* forests remain virgin. Although they may harbor lion-tailed monkeys, these forests border Coorg District from which hunting pressure is severe. We were not able to confirm the presence of the monkeys here, but according to local reports it is likely a few still remain.

Although extensive, these Karnataka forests nowhere comprise a single tract of undisturbed *shola* large enough to support a viable population of lion-tailed monkeys and remote enough from human interference to make their long-term preservation feasible. Roads crossing the main range of Ghats stretching north-ward from Coorg have made the forests vulnerable to timber exploitation and slash and burn agriculture. The large plywood and pulp industries of Karnataka continue to severely deplete the *sholas*. Nevertheless, every effort should be made to protect the monkeys living in the Sharavati River area since they occupy a habitat that is distinctly different floristically from elsewhere in their range (*Vateria indica*-dominant, with few *Dipterocarpus* and no *Cullenia*). A compact area of *Dipterocarpus* forest remains mostly intact at the Sakleshpur—Bisle southern part of this range touching Coorg, South Kanara, and Hassan Districts. For rain forest conservation, this region is excellent as a representative of the predominant type of floral structure in the northern part of the Western Ghats.

Lion-tailed monkeys have always been reported to be more dense south of 11°30′N. The *sholas* of this Cochin—Malabar—Travancore area lie along the hills forming the Kerala—Tamil Nadu border. These rain forests, mostly dominated by *Cullenia*-rich associations, were never as extensive as the *Dipterocarpus* forests. A greater fraction of the southern forests have been degraded or annihilated than the more northerly forests, as they occupy slopes more amenable to plan-tations and other agriculture. Nevertheless, because parts of them have re-mained more isolated, one hill range still contains *shola* well suited for preserving *Macaca silenus*.

4.   Nilgiri Hills—Wynaad

Lion-tailed monkeys are now confined mainly to the southwestern shoulder of these hills. The northwestern part of the hills (Wynaad) reportedly contained the monkeys a few years ago, but none have been sighted recently. There are a few current reports of one troop in the Ghat section near Gudalur in the central sector of the Nilgiris, but this area, as well as the Wynaad, has been virtually deforested. These hills were most extensively cleared by hill tribes prior to the mid-19th century, even before the introduction of coffee and tea. By 1960 the largest of the unaffected forest patches remaining on any of the plateau or gentle slope areas suitable for plantation was only 10 hectares in extent, with all such patches totaling about 40 km² (Bhadran and Achaya,

1960). These patches have since been further reduced by shifting agriculture, plantation expansion, and fires.

Only one substantial area of mostly undisturbed forest remains, a fairly steep and inaccessible region forming the southwestern corner of this hill range, lying in the Palghat and Nilambur forestry divisions (Kerala). The Attapadi forests lining the Bhavani River Valley are contiguous via a ridge with the Silent Valley forests of the Kundipuzha River catchment area and these, in turn, connect via another crest with the New Ambarampalam forests of the Kundah Hills and the Mannarghat forests. Together these form one compact block of about 90 km² of evergreen rain forests interspersed with large areas of grassland and surrounded by semi-evergreen, moist-deciduous, and degraded evergreen forests. Over 400 km² of recently nationalized private lands in this area have been overfelled and intensively hunted to the point where little wildlife remains.

Lion-tailed monkeys live in the Attapadi–Silent Valley regions containing *Cullenia*-dominant forests as well as other associations typical of the southern *shola* communities. The south end of these forests are fairly accessible and thus vulnerable to hunting pressure as well as selective extraction. Although most hunting appears to be for sambar or other game, some tribal peoples in the area hunt and eat lion-tailed monkeys. Grassland fires erode the forest margins, timber is being felled, and a dam under construction in Silent Valley will introduce further pressures. Damsite roads through forests in Kerala are quickly lined with squatters who clear for cultivation.

This area is important to protect since it harbors *M. silenus* and good habitat; however, several factors reduce the prospects of effective conservation. The area is not large enough to allow optimism about long-term preservation of its lion-tailed monkey population. Extraction of timber for railway sleepers, based on a long-term commitment, is underway. Regulating hunting in Kerala, where local people eat monkeys, is difficult and unpopular. Furthermore, land hunger proves to be especially pronounced in India's most densely inhabited state. Nevertheless, the Silent Valley–Attapadi are including the forests further north in New Ambarampalam are at present an important reserve for *M. silenus* and contain an area of *shola* which should be conserved. The entire region lies in Kerala and would make an ideal rain forest and wildlife sanctuary for that state if strict noninterference with forests and anti-hunting enforcement were effected.

5. The Nelliampathi–Anamalai–Palni–Cardamom Hills

This is the main range of the southern Western Ghats. Separated from the Nilgiris by the Palghat Gap to the north and from the Ashambu Hills in the south by the Ariankavu Pass–Shencottah Gap, this range includes sections known as Nelliampathi Hills, Anamalai Hills, Punachi Range, Palni Hills,

High Range, Cardamom Hills, Pandalam Hills, Varushanad Hills, and Srivilli-puttur Range. The eastern spurs of it are sometimes considered part of the Eastern Ghats. This range forms a "T" shape, the north end or crossbar being the Nelliampathi, Anamalai, and Palni Hills, from west to east, often all called the Anamalais. The upright part of the T is usually known as the Cardamom Hills, although local areas have their own names, such as Periyar, Cumban Valley, and High Wavy Mountains. The High Range, lying between the Cardamom Hills and the Anamalais, breaks the continuity of the forests since tea plantations now occupy the once forested main pass. The two parts will be treated sepa-rately.

    a. NELLIAMPATHI—ANAMALAI—PALNI.   The northern part of the hills contained about 400 km² of evergreen rain forests as recently as the mid-1800's. Around the turn of the century forests were cleared for tea, coffee, and cinchona. In the eastern spur, the Palni Hills, over 100 km² was cleared in the last 30 years alone, principally for horticulture and exotic tree plantations. What remains of the rain forests are a few strips of *shola* among moist protected valleys. Investigat-ing one such riverine forest where *M. silenus* had been reported (Vellodai Valley), we found only a very narrow belt of trees hugging a stream, unlikely to support many, if any, lion-tailed monkeys. There was no local knowledge of them.

    In central and western zones, the Anamalais proper, the I.C.A.R. map in-dicates this region to have contained the most extensive rain forests in South India before interference by man. The wetter regions of these forests, running from the High Range to the Hills' northern limits, were undoubtedly once well populated by lion-tailed monkeys. Many were captured for the pet and zoo trade, and hunting them for meat still continues. More than 200 km² of this forest have been cleared for tea, coffee, and cinchona. The forests which still remain tend to be compact, isolated patches. The largest of these, Karian *shola* of Tamil Nadu and adjacent forests of Kerala, contains less than 15 km² of ever-green rain forests with patchy distribution within semi-evergreen and moist-deciduous forests of another 40 km². These areas are being further cleared, especially for teak-planting schemes and exotic tree plantations, even in the newly constituted Anamalai Wildlife Sanctuary.

    In the northcentral area (Punachi Range—Valparai), *Macaca silenus* still inha-bit good *shola* of limited extent. One troop we observed near the Anamalai Sanctuary between Top Slip and Valparai was not at all shy, a fair indication that hunting there is limited. Further west across the Kerala border, in the Chalakudi River Valley, J. F. Oates saw a very shy *M. silenus* troop in a zone of active selection felling where the laborers report shooting the monkeys for meat (*in litt.*, report of a December 1975 survey visit by Oates to two low elevation forests). Deforestation, commercial trade, and local consumption have so reduced and isolated the lion-tailed monkey populations in this area of the hills that only heroic measures could assure their preservation.

b. CARDAMOM HILLS. The hills south of the High Range, including the High Wavy Mountains, the Cumban Valley, the Periyar Sanctuary and Catchment Area, the Rani forests, and Ramnad Srivilliputtur forests contain less than 75 km² of intact evergreen rain forests.

The forests of the northern part of Cardamom Hills have essentially all been underplanted with cardamom, and monkeys are shot there in the belief that they damage the crop. Nilgiri langurs, abundant here as recently as 10 years ago, have now been virtually eliminated by shooting.

The southern areas are intensively felled for timber and pulp. The rain forests within and adjacent to the Periyar Sanctuary are mostly isolated in moist protected valleys. Sightings of *Macaca silenus* have not been reported for more than 25 years in the High Wavy–Cumban Valley area northwest of Periyar Lake. The forests there are principally deciduous with only patchy evergreen regions and cannot form an important protection zone for any monkeys which may remain. On the west side of Periyar Lake, lion-tailed monkeys live along the boundary with the Rani forests where hunting is unregulated. A troop we observed on the south shore, also at the sanctuary border, showed all the reactions typical of a hunted troop—alarm calls followed by silent, rapid flight through the canopy. What little rain forest remains in the sanctuary is constantly eroded by poachers' and grazers' fires. New pressures for clearing these forests for eucalyptus are arising from the construction of a pulp factory. Further south, in a low-lying remnant forest not far from the Shencottah Gap, Oates (*in litt.*) saw a group of *M. silenus* in an area of plantation estates near the Tamil Nadu–Kerala border. Poaching was reported to be common here, and lion-tailed monkeys had been shot recently. In a nearby town, a captive juvenile lion-tailed monkey was seen tethered by the main road a few months earlier (J. Bland, personal communication).

Lion-tailed monkeys are not likely to survive in the Cardamom Hills–Shencottah Gap region for long unless they are strictly protected from hunters and the amount of rain forest still remaining does not decline further. There seems to be no portion of these hills suitable for a rain forest sanctuary and even the monkeys within and near the Periyar Sanctuary are not adequately protected.

After working in these hills in 1961–1963, Sugiyama (1968) concluded that the status of *M. silenus* was insecure. Our observations show that the situation has further deteriorated, but they are still found in the limited amount of remaining *shola*. Hunting, most pronounced on the Kerala side, continues to deplete the population. If hunting ceases and habitat despoilation halts, the monkeys possibly could reoccupy the rain forests running along the Kerala–Tamil Nadu border in these hills and maintain a precarious existence. The remaining habitat is so limited and patchy, however, that long-term survival of *M. silenus* in these hills is doubtful.

6. The Ashambu Hills

Of the 260 km² of forests standing here in the last century, about 160 km² remain today. These still form one continuous block, much of which toward the south is optimal *Macaca silenus* habitat. This is the only such large block of continuous *shola* remaining today. A number of breeding *M. silenus* troops inhabit the south central portion of the hills, particularly the Singampatti and Kalakkadu areas. If protected, these could provide a nucleus for repopulating the more northerly zones in which they have been hunted extensively. We believe that major conservation efforts must be concentrated in the Ashambu Hills, which contain both adequate habitat and a healthy, although limited, population of lion-tailed monkeys. This area is described in detail in Section IV.

Following our habitat survey we estimated the total lion-tailed monkey population. The estimates, given in Table IV, are derived from sampling limited areas at each location and extrapolating the results based on the extent and quality of habitat, severity of hunting pressures, and our experience attempting to regularly detect two well-habituated troops of monkeys over a 19-month period in familiar terrain. The figures also incorporate the recent reliable reports by trained field observers as given earlier in Table III.

TABLE IV

Estimates of 1975 *Macaca silenus* Population [a]

Ashambu Hills
Minimum number of troops: 6 (based on our sightings of troops which must be different)
Maximum number of troops: 14 (based on our sightings of troops which could be different)
Probable number of monkeys: 13 troops at 15 head each = 195

Silent Valley—Bhavani River Valley
Probable number of monkeys: 4 troops (range 2—8) at 15 head each = 60

All other locations combined
Minimum number of troops: 5 (based on recent sightings)
Maximum number of troops: 16 (based on recent and older sightings and assuming more than one troop in larger forest patches)
Probable number of monkeys: 10 troops at 15 head each = 150

Total number of wild *M. silenus* individuals in 1975
405 (range of estimates 195—570)

[a]The zoo population in 1975 was 275—300 (L. Gledhill, Woodland Zoological Park, who conducted a survey, personal communication).

In the time available for the survey, it was not possible to conduct strictly quantitative censuses based on sighting-record densities at fixed detection distances for various visibility conditions along measured random transects [see Southwick and Cadigan (1972) and Wilson and Wilson (1975a,b) for examples of forest primate censuses]. Such a census project for the entire population of this rare, shy, deep-forest species would take many months or even years. Nevertheless, we feel confident that our figures take account of the important variables and that the true population number is of the same order of magnitude as this estimate.

IV.   The Ashambu Hills

A.   Status of Ashambu Hills *Shola* Forests

The Ashambu Hills form the southern limits of the great irregular chain of Western Ghats. Southward from the Ariankavu Pass they straddle the Tamil Nadu—Kerala border from 8°56′N until just south of the highest point at the famous Agastyamalai peak (1869 m). Then at 8°34′N the hills make a bold sweep to the southeast for about 18 km to 8°29′N where once again they turn southward until terminating just 25 km north of Cape Comorin. From the low, lightly-wooded foothills flanking their base, a series of still densely forested undulating hills and wide valleys lead gradually upward to the rugged mountains whose line of peaks forms the majestic skyline visible from the plains below (Fig. 4). These forested slopes narrow into an isthmus of about 30 km² in the southeast sweep of the hills, forming a link which connects the 70 km² of good *shola* near the base of Agastyamalai with the 60 km² of similar forest at the south end of the hills. Other names applied to these hills and forests or portions of them include hills of South Travancore, Tirunelvelli Ghats (also spelled Tinnevelly), Kuttalam range (also spelled Courtallam), Sivasilam range, Ambasamudram range, Papanasam Hills, Singampatti Hills, Kalakkadu range (also spelled Kalakad), Sengalteri range, Nanguneri range, and Mahendragiri Hills.

Abundant precipitation, moderate elevation, and short dry spells have produced well-developed rain forest communities of tall evergreens in the *sholas* of these hills. From about 600 to 1000 m elevation, *Palaquium—Mangifera* associations, with a rich admixture of *Cullenia*, dominate. From 1000 to 1400 m, the dominant *Cullenia—Palaquium* association gradually yields to a *Cullenia—Calophyllum* one. Above 1400 m occurs a more stunted montane type of *shola* with the precipitous cliffs and crests often only scantily clad with palms and other shallow-soil vegetation.

The Ashambu Hills still contain a large area of mature climax forests, more than half of it essentially undisturbed, which can serve as catchment area

Fig. 4. *Shola*-clad slopes between two rocky ridges in the Ashambu Hills. The evergreen forest visible is less than one-fourth the range of a troop of lion-tailed morkeys in this area. The light-colored patches of low vegetation in the center are the bamboo reed *Ochlandra*, which grows about 2 to 4 m high. Taken from 1540 m elevation.

for important agricultural lands, reference forest for biological and environmental monitoring and land-use experimentation, and storehouse for unique plants and rare animals. It is the only preserve where *Macaca silenus* can certainly survive if protected. (See also Appendix 1.)

About two-thirds of the total forested area of the Ashambu Hills lies on the easterly Tamil Nadu side. In the northern area, some of the undisturbed evergreen forests also occupy the upper reaches of the western slopes, in Kerala, but the lower slopes in Kerala have been planted with rubber or so heavily worked as to alter their evergreen character and their continuity. The eastern and southern slopes of the Ashambu Hills, all in Tamil Nadu, are forested in one large continuous belt. This area was never occupied by tribes who clear-felled to any large extent, nor has it received the same attention from plantation and timber interests as the more northerly hills. Most of the undisturbed mature evergreen *sholas* are in the southern regions of the hills in Singampatti, Kalakkadu Reserve Forest, Papanasam Reserve Forest, and adjacent forests. Government Reserve Forests are public lands placed under jurisdiction of the state forest departments. A "Reserve Forest" signifies that use is regulated by the government rather than by private owners, not that it is designated for protection from human interference.

The upper elevation *sholas* in the Ashambu Hills receive about 3000 to 5000 mm rainfall annually. They are the sources of the innumerable streams coalescing into the river systems irrigating the surrounding agricultural plains. These rivers, especially the Tambraparni, irrigate some of India's most productive agricultural lands.

The hills are already well-known to bird watchers, orchid fanciers, and other naturalists. They have attractions for the casual visitor, as well, in their spectacular vistas. For example, from lookout points at the south, one can see at a glance the confluence at Cape Comorin of the Indian Ocean, the Arabian Sea, and the Bay of Bengal. One panoramic sweep takes in the coastal plains from the Cape up to the skyline crest forming the Kerala—Tamil Nadu border. This single view encompasses tea gardens in the foothills, which rise suddenly to densely forested slopes, and finally almost vertically to the rocky cliffs with the great bare spire of Agastyamali in the background.

The area contains numerous unique endemic plants, ranging from many species of colorful herbs like *Impatiens*, through gigantic trees such as *Gluta travancorica* and the only conifer indigenous to South India, *Podocarpus latifolia*. Conservation of this habitat, essential for the preservation of *Macaca silenus*, could also ensure survival of other endangered mammals, snakes, and the many *shola*-dependent birds. It is also an important wildlife refuge for populations of many animals now scarce elsewhere. Leopard, gaur, sloth bear, Nilgiri langur, and the Nilgiri yellow-throated marten occur in good numbers in the Ashambu Hills. Tiger and Nilgiri tahr were present until a few years ago, but have not been confirmed recently.

The Kanis, a hill tribe, were never numerous on the Tamil Nadu side, so the effects of their shifting agriculture on the eastern slope forests has been negligible. They were the sole inhabitants of the Ashambu Hills until coffee planters acquired *shola* in the mid-19th century. The best *sholas* of the Kuttalam area, in the north, were planted with coffee which was mostly later abandoned. This region now houses spice plantations and is seriously degraded. Another accessible region, Nanguneri, on the southeast slopes, was also once cultivated for coffee and other crops. The degraded forests there are slowly regenerating, but many are private lands and may soon be sold for tea plantations. The Papanasam region was worked briefly for timber in World War II, but not until the recent expansion of hydroelectric schemes and construction and extension of roads have serious incursions been made. This area has also been the site of silvicultural experimentation for many years. Although their character has undoubtedly been altered, these forests could regenerate to a climax stage if left undisturbed. The forests of the former Singampatti Zamindari, being remote and under independent jurisdiction, were never extensively worked. Even after much of them was leased and felled, the remaining areas were well protected since the access road was, until recently, private and guarded. The interest of management officials of the plantation in this area helped curb poaching the wildlife protection they encouraged has been largely responsible for the well preserved animal populations left in the Singampatti and neighboring forests.

Conservationists familiar with the Ashambu Hills have stressed that the wildlife and forests here have remained least disturbed of all in the Western Ghats chiefly because of their past inaccessability, rather than as a result of statutory protection.

However, the opening of the roads to Kodayar and Papanasam hydroelectric projects and the improvement of a road to Sengalteri now provide access to hitherto secluded areas, seriously threatening the integrity of this extensive area of good *shola* forest harboring the best population of lion-tailed monkeys.

B: Threats to Survival of *Macaca Silenus* and *Shola* in the Ashambu Hills

The area of suitable forest in the Ashambu Hills is near the minimum required for this species' long-term survival. The most serious long-range threat to the lion-tailed monkeys is reduction in extent and quality of rain forests. Hunting, of course, directly affects their numbers and cannot be tolerated by animals as scarce as *M. silenus.*

Several practices actively jeopardize the mature climax *sholas* in these hills. Wildlife and the rain forest ecosystem are also disrupted by activities which may not be overtly detrimental to animals and vegetation. A few activities menacing *M. silenus* and its habitat in the Ashambu Hills are enumerated below.

1. Clear-Felling

Although felling for plantation, eucalyptus, timber, and hydroelectric projects has not yet reached the same proportions as elsewhere in the Western Ghats, recent trends are highly disturbing. Most serious is the immediate danger to the lion-tailed monkeys' survival posed by expansion of plantation lands in the narrow central isthmus of rain forest connecting the major blocks to the north (Papanasam) and south (Kalakkadu). This crucial portion of forests, lying on both sides of the Manimuttar River in the Singampatti region, are under a long-term plantation lease from the state, which permits all but a small corner of them to be cleared for planting. The Singampatti forests provide continuity of the lion-tailed monkeys' habitat in the Ashambu Hills and allow interbreeding between the populations on either side. About half of the 30 km² isthmus has already been clear-felled.

Much of the Singampatti *sholas* which remain are cardamom plantation, in which the understories have been cleared, leaving only the top story as shade for the underplanted cardamom. The bulk of these areas and neighboring virgin *sholas* are presently slated to be clear-felled for timber revenue and replanted, first with eucalyptus, later with tea. These forests, along the banks of the Mánimuttar River, allow at least four troops of lion-tailed monkeys (probably over one-fourth their population in the Ashambu Hills) access to the perennial water supplies needed in the dry spells. The Manimuttar River Valley is also an area rich in the important *Artocarpus* food trees.

Contact between monkeys living in the Kalakkadu forests and those of Papanasam relies on the integrity of these Singampatti forests. These *sholas* support resident lion-tailed monkeys, bridging the ranges of troops on either side. Clear-felling of this connecting link of forest began in mid-1975. If it continues to be reduced in size, the area will be too small to support a resident troop, thus diminishing the potential for lion-tailed monkeys in the northern and southern parts of the Ashambu Hills to interbreed. The frequency of meetings between two troops, one living principally in the Singampatti link, the other in Kalakkadu, both of which frequented the areas being cut, decreased from 11 of the 60 days we were intensively observing the Kalakkadu troop just prior to felling, to zero in the 51 such days in the months afterward. Continued felling at Singampatti could divide the lion-tailed monkey population into two discrete demes, either by eliminating habitat for the centrally situated connecting troops or by replacing with tea and eucalyptus the forested access route for genetic exchange. As of early 1976, a temporary moratorium on felling has been arranged while the situation is under review by the plantation owners, the World Wildlife Fund-India, the Tamil Nadu and Central Governments, and the Bombay Natural History Society.

In addition to the effects of felling critically important links which preserve forest continuity, the major result of clearing operations is overall reduction of habitat with the resultant diminished carrying capacity of the forests. Removal of *shola* on even a limited scale can have devastating effects on wildlife. An area of only 30 hectares recently clear-felled in the Singampatti forests contained a concentration of fruiting *Litsaea wightiana* trees. These had provided an important food for two troops of lion-tailed monkeys for more than a 2 month period (Table II). Since the total number of troops is so low, the loss of any important dietary resource can have severe repercussions on the species' future in this area.

Other effects of clearing operations may be somewhat less obvious. Deforestation of one area can, for example, lead to irreversible changes in vegetation patterns elsewhere in the ecosystem if the disrupted ecological—meteorological precipitation cycles produce a decrease in rainfall (see Appendix 1). Another example of indirect ecosystem disruption occurred when seas of bamboo were cut or submerged in the last few years for the Kodayar dam project in the Ashambu Hills. The forest elephants of the region, for which the bamboo is a staple, then shifted their range, thus putting increased pressure on the remaining habitat. We saw elephants destroy saplings and small trees in regions where they were uncommon a few years ago (J. Bland, personal communication).

The total area man has deforested in the Ashambu Hills is not yet so extensive as to impair the region's potential for preserving *Macaca silenus*, but urgent steps must be taken to protect the remaining *sholas* from clear-felling.

2.   Selective Felling

Selective partial clearing of *shola* followed by underplanting the thinned forest with cardamom occurs throughout the Ashambu Hills. Cardamom plantations occupy private lands and land leased from the state Forest Department. Although these areas now help support the lion-tailed monkey troops living on their periphery, and provide important arboreal access routes between undisturbed forested areas, they can be considered as clear-felled from a long-term ecological viewpoint, since regeneration is prohibited (see Appendix 2).

Selective logging for timber extraction in broad-leaved evergreen forests destroys their character. All climbers are cut in a coupe before felling begins, resulting in increased light penetration to the forest floor. Shade-tolerant trees, including *Cullenia* and *Palaquium*, do not regenerate in as great a proportion as the climax composition (Aiyar, 1932; Kadambi, 1954). Gaps created by felling further influence natural regeneration by allowing partial desiccation of the soil as well as increased light. Although broad-leaved evergreen forest regeneration is a poorly understood subject, it is clear that unlike many deciduous or coniferous forests, the composition of *shola* changes under all felling and regeneration schemes so far attempted (Anon., 1960; Richards, 1966).

Rapidly growing deciduous trees are sometimes planted in the gaps produced by selective felling. Introducing deciduous trees for increased timber production alters transpiration characteristics of the forest. This may lead to decreased rainfall and further degradation of the forest (see Appendix 1). *Macaca silenus* and other wildlife adapted to *shola* conditions and dependent upon mature evergreen vegetation are adversely affected by regeneration, natural or artificial, which replaces the climax forest with one of a different character.

3.   Habitation and Access Roads

The construction of access routes to the upper dams at Papanasam and to the Kodayar hydroelectric project has generated pressures to use these roads for timber extraction, heretofore uneconomical, and to replace *shola* by teak in the lower areas and by eucalyptus up higher. Forestry operations have increased drastically in the Papanasam zone, and extraction has occurred along the road to Kodayar Dam. The newly improved road to Sengalteri may result in similar pressures on the interior part of Kalakkadu, the southern block of forest, until now one of the most inaccessible areas. The continuation of the Singampatti—Kodayar road going over the dam and through the Mahendragiri region would similarly open up the southernmost end of the hills. As well as being potential routes for timber extraction, all these roads allow access to poachers. Daniel (1971) also discusses the diminishing isolation of Kalakkadu.

In inhabited areas preservation measures are extremely difficult to enforce. Where settlements have sprung up at the damsites and elsewhere along these roads, the residents have clear-felled forest land for food cultivation, grazing, and firewood. Their activities, now expanding into the virgin forests around Kodayar, are the first step toward degradation of wet forests into dry scrub land. Collection of minor forest produce places futher pressures on the forests and wildlife by interfering with tree regeneration and animal food supplies. The wild gaur may suffer from competition with feral domestic cattle which are just becoming established in the upper *sholas*; gaur also risk the introduction of rinderpest or other livestock diseases, as occurred in Mudumalai Sanctuary in 1968 and Periyar Sanctuary in 1975. Pig, sambar, and other animals are commonly hunted and trapped for food. Allowing free access to the area by outsiders will aggravate these problems. Unless there is a major commitment to an active patrol force, it will become increasingly difficult to protect the forests and wildlife. The proposed resettlement in the Ashambu Hills of many Tamil refugees from Sri Lanka would place an insurmountable burden on efforts to protect this area.

4.   Hunting

Although *Macaca silenus* is nominally protected by the Indian Government's Wildlife Preservation Act of 1972, enforcement is inadequate at all levels. Probably most of the lion-tailed monkeys killed in the Ashambu Hills are victims of mistaken identity. Few people in these hills eat them, but they closely resemble Nilgiri langurs which are hunted for their flesh. Some indigenous systems of medicine have perpetuated the notion that the Nilgiri langur's flesh has medicinal value. In all the lower-lying and more accessible regions, especially in the northern Kuttalam—Sivasilam—Papanasam part of the Ashambu Hills, lion-tailed monkeys and other wildlife are heavily poached, primarily for meat, but also in part for the live animal trade. Both species of monkeys and their flesh can be bought in the markets near Shencottah and Sivasilam and at the camps and estates in the Papanasam forests.

Residents of the cardamom estate at Valayaar, in the north end of Papanasam forests, boasted to us of shooting leopard and elephant as well as monkeys and other animals. Both species of monkeys are recognized and hunted there. The northern area of the Ashambu Hills has been virtually depopulated of lion-tailed monkeys, even in the high interior forests, partly for sale to hill tribes in nearby Kerala who do eat them, and in part as casualties of the quest for Nilgiri langurs. Near the Kerala border in these forests, armed poachers are frequent. Even in the southern regions where hunting has always been less severe, lion-tailed monkeys were hunted and killed in the Singampatti forests during 1974 and 1975 by residents aware of their identity and protected status.

5.   Minor Forest Produce

Both legal and unauthorized collection of minor forest produce, especially honey, cane (*Calamus* sp.), various seeds and fruits, dead wood, and bamboo, disrupts the forests and affects the lion-tailed monkeys in many ways. Most noticeable is the number of tall trees illegally felled for honey located in high hollows. Harvesting scarce *Artocarpus* fruits deprives the lion-tailed monkeys of one of their most important year-round foods. The widespread felling of palm trees for their starchy heart reduces the supply of palm fruits, part of the monkeys' diet.

The lion-tailed monkeys and other animals eat their fruit of the cane plant, removal of which depletes their food supplies. Cane collectors mark and clear trails for extracting *Calamus* stalks by cutting numerous seedlings, poles, and shrubs as they wander through the forest. Often a dense covering layer of *Strobilanthus* and related shrubs occupies areas worked for cane and menaces normal forest regeneration. Near the temporary settlements erected by cane collectors, degradation of forests is severe because trees are felled for firewood and construction.

The fallen fruits of *Palaquium*, one of the most numerous and characteristic *shola* trees in these forests, are gathered and sold for processing the seed for oil. In only a very small region of the Kalakkadu forest near the Singampatti–Kodayar Road, we saw collectors, with and without permits, leave the forest day after day with headloads of gunny sacks containing *Palaquium* seeds and fruit. Many trees are slashed, even girdled, apparently to mark collection areas. The removal of such large numbers of seeds has serious consequences on forest regeneration (Aiyar, 1932).

Bamboo reeds (*Ochlandra* spp.) known locally as "eeta") are harvested commercially for processing in pulp mills. These reeds are collected in the accessible dense brakes along watercourses. The harvesters also cut saplings which grow among the reeds, thereby preventing regeneration of emergent trees. The natural boundaries of the reed belt are extended by cutting saplings and trees along its margins with open forest.

Cutting reeds along watercourses results in muddy eroded stream banks; the loss of soil further reduces the ability of the forest to regenerate. Eliminating reeds also deprives wildlife of needed resources. Depleting this cover decreases nest sites for birds, requires tigers to stalk in more open areas, and forces elephants to eat more leaves and shoots of the woody forest vegetation instead of the favored reeds.

Removing deadfalls for firewood robs the forest of the humus formed by rotting wood. It also destroys a niche for the beetle larvae and termites eaten by sloth bears as well as many birds, and eliminates the substrate for the fungi eaten by the lion-tailed macaques. The distillation of *arrack*, a locally produced spirit, is widespread in the forests near settlements; the fuel for this illegal operation is procured by cutting and drying poles or large trees, as well as collecting deadfalls. In areas near *arrack* stills or cane-cutters' settlements, the forests are open and virtually devoid of understory.

There are no tribal peoples in the area who traditionally harvest the minor forest products. All the procurement is done by outsiders, who establish temporary settlements near or inside the Government Reserve Forest, or by residents of the plantation, forestry, or damsite villages in the Ashambu Hills. Few areas of forest surrounding these habitations have remained unscarred.

V. Proposals for Preserving *Macaca silenus*

The lion-tailed monkey is on the verge of extinction. This is primarily due to extensive and widespread destruction of its habitat, and secondarily to hunting. Sufficient information is available about the type and extent of forest this monkey requires to allow planning for survival of a breeding population. The best remaining area suitable for preservation efforts has been located in the Ashambu Hills. These hills still contain enough mature *shola* climax forest, the monkeys' obligate habitat, to support a viable breeding population. The southern regions contain enough monkeys to provide a nucleus for repopulating the neighboring northerly forests of the Ashambu Hills where they have been more severely hunted.

A. Urgent and Immediate Steps

Six steps necessary to conserve lion-tailed monkeys and their Ashambu Hills habitat were proposed in a report to concerned government agencies and conservation organizations. They are not a substitute for a comprehensive conservation plan for the region, but the urgency of the situation indicated the necessity for immediate action on them while a long-term plan is being formulated and executed. Some of the steps are already covered by statutory provisions of the Wildlife Protection Act, but have not been adequately enforced. The first three points summarized below were indicated to be especially urgent, and we note with optimism that positive action has already been initiated on two.

1.   Stricter Regulation of Felling in Government-Leased Lands

Use of government lands under lease from the Forest Department should be regulated so as not to endanger the survival of *Macaca silenus*. In particular it is most important that the undisturbed forests of the Singampatti region, as well as those underplanted with cardamom, remain intact without any further reduction. Felling here is reducing the area available to the Singampatti troops which bridge the troops to the north with those to the south, and could destroy the continuity of habitat, thus eliminating the monkeys' forest pathway connecting the Kalakkadu, Singampatti, and Papanasam forests. In consultation with the World Wildlife Fund—India, Bombay Natural History Society, and State and Central Governments, in early 1976 the management of the plantation at Singampatti agreed to a temporary moratorium on felling. Proposals are under consideration by the plantation and the Tamil Nadu Forestry Department on how to resolve differences between conservation planning and economic interests.

2.   No Further Issue of Plantation or Felling Permits

Since the amount of undisturbed mature climax *shola* remaining in the Ashambu Hills is at a crucial lower limit for supporting a viable population of lion-tailed macaques, new permits or permit renewals for clear-felling or for cardamom leases anywhere in these hills should not be issued. Selective felling for research purposes could continue on small-scale experimental plots in areas where the forests have already been highly disturbed.

3.   Enforcement of Anti-Hunting Laws

Laws prohibiting hunting, possession, and sale of *Macaca silenus* (and *Presbytis johnii*, with which they are often confused) must be strengthened. These must be vigorously enforced, both in the forests and in the markets. In 1975 an animal dealer at the major pet and zoo market in Bombay whom we reported as stocking juvenile lion-tailed monkeys was raided by wildlife officers and is no longer in business.

4. Restricting Access to Ashambu Hills

Even before sanctuary status can be arranged for the area, protective provisions similar to those in national parks should be made for the Government reserve forests occupying most of the Ashambu Hills. Access to these forested areas should be restricted to those with legitimate occasion to be there. Residence should be permitted only to employees of concerns legally working in the forests, e.g., plantations, Electricity Board, and Forest Department. New settlements should not be permitted, nor should new access roads be opened.

Legal collectors of bamboo, cane, honey, *Palaquium* seeds, firewood, etc., should be restricted to carefully designated regions where these activities and their effects on the forest can be monitored by the Forest Department.

5. Stricter Regulation of Use of Privately Owned Lands

Private forested lands adjoining the government forests form an important part of the total remaining habitat. Their use, as well as that of leased lands, should be actively regulated to prevent deforestation or other abuse, or they should be acquired by the government. Current cardamom leaseholders must be supervised to see that maximum canopy cover is retained so that cardamom-planted forests have the best chance to regenerate at the expiration of the lease. New cardamom cultivation should not be permitted, or current leases renewed. It may be advisable for the state to purchase these lands as part of a protective buffer surrounding the climax forests.

6. Wildlife Guards

A force of patrolling forest and wildlife guards should be instituted to roam the area in order to prevent illegal poaching of animals, trees, and minor forest produce. They should frequent areas where lion-tailed monkeys live near settlements and are hunted. Along with enforcement of the wildlife, grazing, fire wood, and produce-collection laws, the guard force should also remove feral cattle from the forests and protect regenerating areas from fires. This force would supplement the guards manning the chaingate road check posts who are concerned principally with checking timber operations for royalties. The implementation of any conservation plan will be dependent on a well-trained and adequately equipped staff. Funding should include provisions for binoculars, walkie-talkies, and vehicles.

B. Long-Term Planning

While the above six points require urgent and immediate attention, a long-term comprehensive plan for preserving the lion-tailed monkeys and their *shola* habitat in the Ashambu Hills is also needed. The conservation plan requires a shift from the traditional forestry considerations of yearly revenue production and extraction rates to long-term implications of land-use policies and resource allocations. A balanced view would include watershed maintenance, potential for tourism and recreation, irreplaceability of endangered species, and necessity of reference forest (see Appendix 1).

The geography of the Ashambu Hills makes it feasible for the state of Tamil Nadu alone to be effective in establishing a wilderness reserve of *shola* forest on the eastern and southerly slopes of the hills. Ecosystems do not respect political boundaries, however, and although the *sholas* may be self-contained in Tamil Nadu, large mammals like gaur, elephant, and tahr make use of the passage over the crest of these hills into Kerala. Any rain forest conservation efforts would be enhanced if the central government could help coordinate plans with Kerala for protecting the western slopes. A plan for utilizing the unique resources which the Ashambu Hills offer might include detailed proposals covered in broad outline below.

A detailed set of vegetation and boundary maps should be prepared based upon a thorough botanical and topographical survey. A large core area of mature, undisturbed climax forest must be set aside under an inviolate set of rules protecting it from any and all interference. This area should probably include all the interior regions of Kalakkadu and Singampatti forests, the upper elevations of Papanasam forest, and all the undisturbed forests contiguous with them, a total of about 100 km^2. The partially disturbed 60 km^2 of forests surrounding this primary wilderness core should form a buffer zone with more peripheral areas which have been heavily worked, generally those most northerly and low-lying. These worked areas can serve as experimental zones, while the buffer region can be used for monitoring the course of natural regeneration. Regeneration of the buffer zone is essential to allow the lion-tailed macaque population to expand into it and reach an adequate size. Suggestions for the kinds of activities which might be appropriate to each region follow below.

1. Core Area

This region would be chiefly responsible for maintaining the ecological integrity of the Ashambu Hills. It should be the largest area. A program for regularly monitoring the environment might be coordinated with the Forest Department, the Bombay Natural History Society, and regional universities. Access to this region should be only for nonexperimental investigation or observation which does not interfere with any flora or fauna. As has been successfully attempted in some East African national parks and game reserves, a member of the Forest Department can be assigned to every investigator or research team desiring entry to the area. Cost of maintaining the staff would be borne by the scientists. This is an opportunity for foresters to gain valuable training in ecological, botanical, and zoological research, while ensuring there is no disturbance or collection.

2. Buffer Zone

The disturbed forests in this buffer zone should not undergo further degradation, and their course of regeneration should be monitored. Many of them lie along roads near dams and forest rest houses. They still harbor examples of beautiful *shola* vegetation, majestic scenery, and abundant wildlife. The potential of these areas for tourism and education should be explored, with the emphasis on "undeveloped" recreational tourism for the nature enthusiast, including guided walking tours through the forest. A touring party wishing to see lion-tailed monkeys could send word ahead. Guides could locate a troop in advance and lead the party to them on arrival, similar to the procedure for tourists wishing to see gorillas in dense African forests.

3. Outside Area

The peripheral regions can be used for investigating the long-range effects of various schemes of land-use management practice on the forests. Collection of forest produce and small-scale silvicultural experiments on selective felling regimes and regeneration practices should be confined to this area.

The first steps toward long-range planning for the conservation of *Macaca silenus* and its rain forest habitat were taken in early 1976 when part of the southern end of the Ashambu Hills was established by the Tamil Nadu government as the Kalakkadu Forest Sanctuary for the lion-tailed monkey. We hope that this is the beginning of an effective program to conserve all the remaining habitat in these hills and that such efforts will continue to be promoted by the Forest Department and approved by the government.

Efforts should also be made to protect the lion-tailed monkeys and the forest they inhabit in the Silent Valley—Bhavani River Valley region, a refuge of secondary, but significant, value. If two sanctuaries were to be established, one here and one in the Ashambu Hills, the lion-tailed monkeys' chances for survival will be enhanced and the grim conservation picture brightened.

VI. Summary

The lion-tailed monkey (*Macaca silenus*), one of the world's rarest and most beautiful animals, inhabits only the evergreen rain forests and similar formations, all known as *shola*, which clad the upper elevations of the Western Ghats in South India. As an obligate dweller of such forests, this species faces imminent extinction due principally to habitat destruction. Its demise is being hastened by hunting. The monkeys' varied and seasonal diet leads them to range quite widely and live at low densities. To maintain at least one viable breeding population with sufficient genetic diversity and able to survive temporary fluctuation in number requires a large continuous block of suitable habitat. At least 130 km² of typical southern rain forest rich in *Cullenia* trees is needed; a larger area of the *Dipterocarpus*-dominant forests found north of 11°30′N would be required. Large blocks of *Cullenia*-rich forests once occurred south of 11°30′N but have been destroyed over most of the Ghats.

In addition to wildlife preservation, the *shola* forests are valuable for many other reasons, including their role in ensuring adequate water supplies to the plains. These forests are also important for erosion prevention, for their tourism potential, and as urgently needed control areas for reference and experimentation which are necessary for wise land-management planning.

Nearly 160 km² of forests lie in the Ashambu Hills, the southernmost range of the Western Ghats, along the eastern and southern slopes in Tamil Nadu, and across the crest on the western slopes in Kerala. This is the area most suitable for a conservation effort to save the lion-tailed monkey. Only the Ashambu Hills contain both a good population of these animals and also a large area of mature, virgin, climax vegetation, the bulk of which is the unique, yet vanishing, *Cullenia*-rich rain forest of the southern ghats. In addition to lion-tailed monkeys, other wildlife also maintain adequate breeding populations in the least accessible regions toward the south of the Ashambu Hills. Given effective protection, and if the continuity of the rain forests remains unbroken, the monkeys could repopulate the entire hill range to a level sufficient to prevent their extinction. The total area remaining is critically low, however, and division of the forest into separate blocks and all clear-felling operations must cease immediately. Effective measures must be taken against hunting and habitat destruction. The degraded forests in the vicinity should be allowed to regenerate. The area is ideal for establishing a core zone of primary wilderness and reference forest with outlying regions available for tourism and for small-scale experimental land-use, forestry, and silvicultural practices.

It is urged that immediate steps be taken to set aside in their entirety the rain forests of the Ashambu Hills for the practical needs of future generations and as the only way to save the lion-tailed monkey from extinction.

VII. Appendix 1: Preserving India's *Sholas*

Rain forest conservation in the Western Ghats has implications far beyond requirements for perserving the lion-tailed monkey. The economic health and vitality of the regions for which the *sholas* serve as watershed and catchment areas necessitate the conservation of large areas of climax rain forest as a matter of the utmost practical concern. The full value and significance of the *sholas* cannot be measured solely by production of forest revenues.

Pressures to garner annual revenues from the timber and lands of India's forests have led many foresters to call for new ways to utilize evergreen tree timber and to make extraction in remote areas economical. To produce lease revenue, underplanting government forests with cardamom has been advised, even though inhibition of *shola* regeneration in cardamom plantation is well known (Anon., 1960, pp. 162—3; see also Appendix 2). It has also been urged that evergreen rain forests be made more productive of timber by introducing deciduous elements, but without a full consideration of the possible effects such a plan could have on water supplies. Reforestation projects have concentrated on establishing rapidly growing, uniform plantations of exotic acacias, eucalyptus, and conifers, although these trees do not serve the same beneficial functions in conserving rainfall as do the evergreen rain forests they replace (Blasco, 1971).

In addition to water conservation arguments advanced by biologists, ecologists, and conservationists, some of the more important reasons for conserving evergreen rain forests, including several ideas propounded by Struhsaker (1972), are summarized here.

A.   Rainfall

Moisture-laden clouds from the Bay of Bengal (October through December) and the Arabian Sea (May through July) pass over the evergreen forests of the Western Ghats after first crossing the lowlands. Above the mountains they encounter a zone of higher humidity and cooler temperature than over the surrounding plains and low hills, so more rainfall is released over the forests then elsewhere. Transpiration from the enormous leaf surface area presented in a multistoried rain forest cools the atmosphere and maintains its moisture. Thus the *shola* vegetation, which requires heavy rainfall, also actually helps produce it by participating in the meteorological—ecological cycles of precipitation—percolation—transpiration—evaporation.

Deforestation can drastically alter rainfall patterns. Localized effects of forests on precipitation patterns are documented by Puri (1960, Chapter 21) who describes the mechanisms and presents data illustrating the inability of agricultural crops to preserve rainfall. Further amplification is found in Warren (1941) and McCann (1959, pp. 31—34).

Although the effectiveness of forests on modulating localized weather patterns is clear, their potential effect on climate over large areas is poorly understood, and some meteorologists dispute the existence of any relationship. [See Richards (1966, p. 405) who notes the controversy and supplies references.] Randhawa [1945; also summarized in Puri (1960)], however, showed that many areas of northern India which are now deserts were flourishing tropical evergreen rain forests as recently as 2000 years ago. He argues that the deterioration in rainfall and eventual drying out were due entirely to deforestation by man. The processes which eventually lead to rainfall reduction, erosion and floods, progressive desiccation, and finally barrenness are generally so slow that our attention is attracted only when they are too late to alter. Fosberg (1973) has also noted that, after destruction, large areas of tropical forests cannot reestablish themselves and are usually replaced by savanna. Between 1951 and 1973, India lost over 34,000 km² of forests, primarily to agriculture, and secondarily to river valley projects, settlements, and industrial plants, roads, and miscellaneous development schemes (Chakravarty, 1974).

Forests need not be completely denuded of vegetation to disrupt the meteorological—ecological cycle. Legris and Blasco (1969, pp. 57—59) present evidence that reduction in rainfall in the Palni Hills (of the Western Ghats) has accompanied replacement of *shola* by agricultural crops. Other non-*shola* vegetation, whether exotic evergreens like wattle or eucalyptus, or even indigenous economically valuable non-evergreens like teak (*Tectona grandis*) or sal (*Shorea robusta*) can likewise severely affect water supplies. Non-evergreens are not capable of high transpiration rates with the first rains since they lack their full complement of foliage at these times. Thus they cannot maintain atmospheric humidity to the same degree as evergreens, and rainfall may be reduced.

Grassland or single-layer plantations such as tea, eucalyptus, or food crops present a much smaller leaf surface area than does a natural forest. Moreover, leaves of these plants transpire much less than those of typical *shola* vegetation. [See Richards (1966, p. 185) and also Walter (1971, pp. 109, 119, 120) who gives these transpiration rates for leaves in mg/gm/min: tea (*Thea* or *Camellia sinensis*): average 1.5; *Eucalyptus citriodora*: 3.4; cardamom (*Elettaria cardamomum*): 1.6; pineapple (*Ananas sativa*): 0.035. Rain forest trees average 8.9, with a range for seven species of 2.6 to 22.0.] Temperatures are higher in these areas, and atmospheric moisture content is lower, hence humidity greeting the monsoon clouds is less and rainfall may be reduced. The modulating effects of forest vegetation on temperature as compared with cultivated or deforested lands is documented in Puri (1960, Chapt. 21). He also notes that during the monsoons the humidity over grasslands which neighbor *sholas* is about one-half that above the forests themselves.

Shola forests are climatic climax formations dependent upon high rainfall as well as helping to produce it. Over a period of time the rate of decreased rainfall may accelerate as a consequence of the initial reduction in precipitation as the composition of neighbouring undisturbed forests, with vegetation dependent on high rainfall, begins to deteriorate. [See Richards (1966) and Champion and Seth (1968).]

Large parts of Kerala, Tamil Nadu, and Karnataka are dependent for their agricultural water supplies largely on the monsoon precipitation falling in the Western Ghats. Destruction of Ghat *sholas* or alterations in their character inevitably endangers water supplies to the regions for which these forests serve as catchment areas. Krishnamoorthy (1960) predicted, for example, that hydroelectric projects flooding many forested valleys in Kerala would threaten rainfall. Fourteen years later, Balaram (1974) noted that the dam-regulated irrigation projects, based on predeforestation rainfall predictions, are now faltering due to reduction of precipitation in the Ghat catchment areas.

B.   Floods and Erosion

Two other effects of deforestation are well known and need be mentioned only briefly: erosion and silting. Unprotected hilly regions are especially vulnerable to erosion of their surface soil which accompanies rapid runoff from concentrated rainfall. The *sholas* of the Western Ghats entrap soil in their extensive root systems and absorb water in their porous, humus-rich soil and in plant tissues. Thus the area is buffered against monsoon flooding since the runoff into drainage streams is mediated by being spread out over a longer time. In combination with an ancient system of flood-control anicuts and dams, the *sholas* of South India have prevented the floods which wreak havoc in severely deforested North India. If vegetation is removed, whether for extraction or agriculture, or density of the root system is reduced, by replanting eucalyptus or underplanting with cardamom, for example, then the *shola* humus layer is degraded, porosity of the soil reduced, and the entrapping effect of a myriad of roots is removed. Puri (1960) gives these figures for the increase in rate of annual soil loss in comparison with normal forest: forest with poor ground cover, e.g., cardamom, 20 times greater; grassland, 130 times greater; agricultural crops, 3250 times greater. He also notes (p. 259) the difficulty of reforestation once the erosion process is established due to changed soil conditions.

C.   Biological Storehouse

The luxurious tropical evergreen and semi-evergreen rain forests are the earth's richest and most diverse biological communities. Their flora have served as reservoirs of genetic diversity and foci of evolutionary activity from which much of the rest of the world's flora derive. Each region's rain forests are unique.

Those on the Indian subcontinent are today restricted to the Indo-Malayan types of the most northeasterly regions and the unique *sholas* along the Western Ghats. In the middle elevations of the Western Ghats alone grow at least 442 endemic species of dicotyledons, plus uncounted numbers of grasses, orchids, and other monocots [Blasco, 1971 (Appendix I)].

These *sholas* serve as a vast natural storehouse of many plant and animal species found nowhere else. Members of this endemic flora form unduplicated raw material for research into the medicinal properties of plant compounds; they can supply seed banks for unique tree species which tomorrow's foresters may find need to propagate. New uses are frequently discovered for many organisms in medicine, pest control, and the breeding of economic plants and animals. Two examples from Indian flora of the Western Ghats are the climax evergreen rain forest trees, *Hydnocarpus* spp. whose seeds contain an oil used in treating leprosy, and the herbaceous shrub, *Rauwolfia serpentina*, from which is produced the famous drug for combating hypertension. Preserving representative blocks of *shola* ecosystems in a completely undisturbed state is the only way to ensure survival of irreplaceable plant and animal resources. [See also Richards (1966), p. 406.]

D.   Wildlife Refuge and Tourism

The reeds and grassy downs and the swamps and mountainous crags within the *shola* habitat supply the forage and lairs for much of India's remaining wildlife. The Nilgiri tahr, for example, now very rare, lives exclusively in the grasslands which border the *sholas* as part of the Ghat rain forest ecosystem. Persecuted animals like elephant, leopard, tiger, sambar, python, and king cobra were once much more widely distributed, but now occur almost exclusively in the inaccessible parts of the hill forests. For those sharing an appreciation of living things, it is clear that *sholas* should be conserved for their uniqueness, inherent beauty, and as the only way to save many endangered species in their natural habitat.

Carefully planned tours centered around the unrivaled scenic beauty, diversity of plant communities, and wildlife abundance of the Indian *sholas* remain to be developed. As Struhsaker (1972) noted in discussing the tourism potential of African rain forests, a portion of visitors are keenly interested in rain forest plants and animals. Among tourists who now visit the wildlife parks in India, many would also be interested in entering the completely different *shola* habitat. Amateur naturalists, bird watchers, orchid fanciers, and visitors looking for something novel would appreciate the opportunity for guided walks through the rain forests. In many African countries, specialty tour operators cater to this clientele; this is an area where India's tourism potential has not yet been realized.

E. Monitoring, Research, and Control Area: Land Utilization Practices and Forest Management

Increasing demands for timber and pulp, and for land for agricultural, industrial, and hydroelectrical schemes, have so far been met without fully considering the possible ecological consequences. It is imperative that at least one large block of minimally disturbed mature evergreen climax rain forest be preserved for gauging the effects of various land utilization practices. Only an environmental monitoring station in undisturbed forest could, for example, provide the control data necessary to determine whether rainfall and temperature trends elsewhere are part of regional changes in weather patterns or if they are localized effects which can be ascribed to local land utilization practices. Struhsaker (1972) lists other examples of the uses of reference forest, including ecological effects of agricultural chemicals and soil degradation in cultivated areas.

The reference area must be large enough to be stable and should be surrounded by protected forests to help secure the integrity and stability of the regional ecosystem. In the peripheral zone, small plots can be managed under a variety of different experimental regimes to determine the best practices to adopt on larger scale in other forests. Richards (1966) examines the factors to be considered in managing secondary succession and emphasizes our current ignorance of these processes.

VIII. Appendix 2: Cardamom Cultivation in *Sholas*

Elettaria cardamomum is a 2-m tall leafy plant cultivated for its seed pods. A perennial herbaceous monocot of the ginger family, it grows wild in the wet evergreen *sholas* of the Western Ghats and is also raised in plantations. For its cultivation, all herbs, shrubs, small trees, saplings, and climbers are removed from the forests and the trees forming the topmost canopy are thinned, usually leaving about 85% of them standing for shade. Although forest regeneration is prohibited by the tending operation in cardamom fields, which includes weeding, the fields appear to be forested areas to the casual glance and are popularly considered to be the same as the virgin climax forests from which they derive. They are classified in forest statistics with undisturbed forests even though they are in a state of degradation in which their ability to function in soil and water conservation is diminished.

The leaf area of cardamom presents only a small fraction of the leaf area of the vegetation destroyed to promote its cultivation. Furthermore, the cardamom leaf's transpiration rate is low [1.6 mg/gm/min versus an average for rain forest trees of 8.9 mg/gm/min (see Appendix 1)]. The cooling and humidifying effects of *shola* are therefore greatly reduced when transformed into cardamom fields.

Clearing the dense underwood vegetation also destroys the intricate network of roots supplied by these smaller plants, rendering the soil in cardamom fields more vulnerable to erosion. As the removed vegetation no longer contributes to leaf litter and soil humus, the moisture retention capabilities of the soil is reduced. Rainfall storage in plant tissues is also decreased. The soil surface tends to get baked in the dry spells, and the runoff during rainfall leads to rivulets joining to form the scarred erosion gulleys characteristic of cardamom-planted hills.

Forest degradation is exacerbated by the increased incidence of windfall losses of the remaining canopy trees. Among 50 *Cullenia* trees we arbitrarily selected for phenological investigation, one was lost to windfall over a 1-year period, yielding a loss rate of 2%. Shallow-rooted species like *Calophyllum* are even more prone to windfall. In older fields, the loss rate accelerates since the gap produced by one windfall encourages others by admitting the force of the winds into the forest. The gaps are not closed since tree regeneration is prohibited by removal of all new seedlings. For shading the cardamom, these gaps are filled by introducing or encouraging small, second-growth trees like *Erythrina indica, Clerodendron* sp., *Macaranga* sp., or *Mallotus* sp. When a field is finally abandoned, the covering layer contains few large trees, having been thinned by wind and felling. The ensuing regeneration is of different composition than normally regenerating *shola* due to the changed soil characteristics, the influence of excess light penetration, and to the paucity of seeds from the few remaining large trees. An entire succession sequence must occur before the climax formation is once again attained, a process which can easily take 100 years.

Areas have been leased for cardamom cultivation in the mistaken belief that the fields are equivalent to undisturbed forested land yielding a minor produce and that they still serve the ecological function of forests. Yet the role of the forests in conserving water and soil is disrupted by cardamom cultivation, and the forests are left in a degraded state in the early stages of secondary succession when a field is abandoned after lease expiration.

Acknowledgments

This publication is an abridged version of a report of our surveys and research originally prepared at the request of the Government of India for an advisory memorandum on the status and future of *Macaca silenus*. It is based on a behavioral and ecological study funded by United States Public Health Service Research Grant MH 24269, with partial funding from National Science Foundation Grant GB 33102X.

We thank the Government of India's Ministry of Education and Ministry of Agriculture for approval of the project and the Chief Conservator of Forests of Tamil Nadu, K. A. Bhoja-Shetty, for permission to conduct the intensive study. S. Haidar, Director of the Prime Minister's Secretariat, and M. K. Ranjitsinh, Director of Wildlife Preservation of the Government of India, are especially thanked for their aid and encouragement. Sponsorship by the Bombay Natural History Society and support from its officers and staff, including Salim Ali, President, Z. Futehally and A. N. D. Nanavati, Honorary Secretaries, J. C. Daniel, Curator, R. Grubh, Assistant Curator, S. Hussein, Ornithology Assistant, R. Hedge, Accountant, and J. S. Serrao, Librarian, are gratefully acknowledged. Advice and support from Z. Futehally, D. Variava, and R. Whitaker, officials of World Wildlife Fund—India, are most appreciated.

The management and staff of the Bombay Burmah Trading Corporation rendered valuable assistance and generous hospitality throughout the field study. We especially thank: P. Vissanji, J. B. Gibbons, N. H. Sethna, J. and E. Bland, M. C. and P. Muthanna, P. D. and R. Jothikumar, V. K. and S. Krishnamurthy, A. S. and R. Gill, J. and H. Kariappa, and P. V. Kuruvilla for their courteous help and encouragement and Messrs. Cardoz, Xavier, and A. Michael for their kind and competent help with logistics. N. Sankaran of Natesan Agencies generously offered logistic support for the research, and A. K. Mahadevan most ably executed it. For assistance at the study site, we thank our most able aide-de-camp and chief assistant, D. Michael, and also Devadoss, Aruloppan, Shanmugham, Rajasekaran, Devadass, Ruth, and D. Esther. J. M. Johnson, Divisional Forest Officer, Tamil Nadu, and G. P. Kuruvilla, graduate student through the Bombay Natural History Society, were associated with the project and were most helpful throughout. J. Rappaport provided excellent liaison with the Field Research Center in Millbrook, New York. The Center's Director, P. Marler, was unfailingly helpful throughout the study.

P. Marler, J. F. Oates, and T. Struhsaker visited the study site and offered valuable advice on the research and conservation efforts. Many of their suggestions have been incorporated into this report. We also thank them for their comments on this manuscript. Oates has most generously devoted time to monitoring the lion-tailed monkeys and providing information on their conservation while conducting his research in the Ashambu Hills.

Forest and wildlife officials of Kerala, Karnataka, and Tamil Nadu are thanked for permitting the survey and census work in their respective states, and for aiding it with kindness and hospitality. G. A. Kurup of the Zoological Society of India kindly related to us the results of his 1974 survey of the lion-tailed monkey. H. S. Sehgal of the Forest Research Institute and K. N. Subramanian of the Botanical Survey of India arranged for us to use their herbarium collection. H. R. Bhat of the Virus Research Centre, Sagar, provided generous hospitality and led us to the areas furthest north inhabited by *Macaca silenus*. S. M. Mohnot, Jodhpur University, was a most gracious host during our visit and is spearheading the conservation efforts by Indian academic primatologists.

Last we thank our forest friends Lief, Sam, Stump, Moshe, Little Jack Macaque, and all the others who taught us so much during the many pleasurable hours we observed them.

References

Aiyar, T. V. V. (1932). The *sholas* of the Palghat Division —A study in the ecology and silviculture of the tropical rain forests of Western Ghats. Parts I and II. *Indian Forester*: pp. 414–432; 473–486.

Angst, W. (1975). Basic data and concepts on the social organization of *Macaca fascicularis*. *In* "Primate Behavior: Developments in Field and Laboratory Research" (L. A. Rosenblum, ed.), Vol. 4, pp. 325–388. Academic Press, New York.

Anonymous, (ed.). (1960). "Proceedings of the All-India Tropical Moist Evergreen Forest Study Tour and Symposium." Forest Research Institute, Dehra Dun.

Balaram, C. (1974). Problems of 'kole' cultivation. *Indian Express Suppl., Kerala Montage, Cochin, Dec. 25, 1974*, p. I.

Bernstein, I. S. (1967a). Intertaxa interactions in a Malayan primate community. *Folia Primatol* **7**, 198–207.

Bernstein, I. S. (1967b). A field study of the pigtail monkey (*Macaca nemestrina*). *Primates* **8**, 217–228.

Berry, R. J. (1971). Conservation aspects of the genetical constitution of populations. *In* "The Scientific Management of Animal and Plant Communities for Conservation" (E. Duffey and A. S. Watt, eds.) pp. 177–206. Blackwell, Oxford.

Bertrand, M. (1969). The behavioral repertoire of the stumptail macaque. *Bibl. Primatol.* **11**, Karger, Basel.

Bhadran, C. A. R. ånd Achaya, T. (1960). The tropical evergreen forests of Madras State. *Proc. All-India Trop. Moist Evergreen Forest Study Tour Symp.*, pp. 128–151. Forest Research Institute, Dehra Dun.

Blasco, F. (1971). "Montagnes de Sud de l'Inde. Forêts, Savanes, Écologie." Tome X, Fascicule 1. Travaux de la Section Scientifique et Technique. Institut Français de Pondichéry, Pondicherry, India.

Burton, F. D. (1972). The integration of biology and behavior in the socialization of *Macaca sylvana* of Gibraltar. *In* "Primate Socialization" (F. E. Poirier, ed) pp. 29–62. Random House, New York.

Carpenter, C. R. (1942). Societies of monkeys and apes. *Biol. Symp.* **8**, 177–204. [Reprinted *In* "Primate Social Behavior" (C. H. Southwick, ed.). D. Van Nostrand, Princeton, New Jersey, 1963.]

Chakravarty, F. (1974). Indian forests—A story of neglect. *Sunday Stand. Mag. Kerala*, Dec. 29, 1974.

Champion, H. G. and Seth, S. K. (1968). "Forest Types of India." Govt. of India, Delhi.

Chandrasekharan, C. (1962). Forest types of Kerala State. *Indian Forest. No. 9*, pp. 660–674; *No. 10*, pp. 731–747; *No. 11*, pp. 837–847.

Daniel, J. C. (1971). The Nilgiri tahr, *Hemitragus hylocrius* Ogilby, in the High Range, Kerala and the southern hills of the Western Ghats. *J. Bombay Natur. Hist. Soc.* **67**, 535–542.

Daniel, J. C. and Kannan, P. (1967). The Status of the Nilgiri Langur [*Presbytis johnii* (Fishcher)] and Lion-Tailed Macaque [*Macaca silenus* (Linnaeus)] in South India— A Report. Mimeographed report of the Bombay Natural History Society.

Deag, J. M. and Crook, J. H. (1971). Social behaviour and 'agonistic buffering' in the wild Barbary macaque *Macaca sylvana* L. *Folia Primatol.* **15**, 13−200.

Diamond, J. M. (1975). The island dilemma: lessons of modern biogeographic studies for the design of natural reserves. *Biol. Conserv.* **7**, 129−146.

Eisenberg, J. F., Muckenhirn, N. A., and Rudran, R., (1972). The relation between ecology and social structure in primates.*Science* **176**, 863−874.

Fosberg, F. R. (1973). Temperate zone influence on tropical forest land use: a plea for sanity. *In* "Tropical Forest Ecosystems in Africa and South America" (B. J. Meggers *et al.*, eds.) pp. 345−350. Smithson. Inst. Press, Washington, D. C.

Horwich, R. H. (1972). Home range and food habits of the Nilǵiri langur *Presbytis johnii. J. Bombay Natur. Hist. Soc.* **69**, 255−267.

Hutton, A. F. (1949). Notes on the snakes and mammals of the High Wavy Mountains, Madura District, South India. Part II-Mammals. *J. Bombay Natur. Hist. Soc.* **48**, 681−694.

Jay, P. (1965). Field studies. *In* "Behavior of Nonhuman Primates" (A. M. Schrier, H. F. Harlow, and F. Stollnitz, eds.), Vol. II, pp. 525−591. Academic Press, New York.

Jerdon, T. C. (1874). "The Mammals of India." John Weldon, London.

Kadambi, K. (1950). Evergreen, montane forests of the Western Ghats of Hassan District, Mysore State. *Indian Forest. Jan.*, pp. 18−30; Feb., pp. 69−82; Mar., pp. 121−132.

Kadambi, K. (1954). *Cullenia excelsa*, Wight. (*C. zeylanica*, Gardner, *Durio zeylanicus*, Gardner). *Indian Forest.*, pp. 442−445.

Karr, J. R. (1973). Ecological and behavioral notes on the lion-tailed macaque (*Macaca silenus*) in South India. *J. Bombay Natur. Hist. Soc.* **70**, 191−193.

Koyama, N. (1970). Changes in dominance rank and division of a wild Japanese monkey troop in Arashyama. *Primates* **11**, 335−390.

Krishnamoorthy, K. (1960). The evergreen forests of Kerala. *Proc. All-India Trop. Moist Evergreen Forest Study Tour Symp.* pp. 123−125. Forest Research Institute, Dehra Dun.

Krishnan, M. (1971). An ecological survey of the larger mammals of peninsular India. *J. Bombay Natur. Hist. Soc.* **68**, 503−555.

Lawrence, C. A. (1960). The vegetation of Kanyakumari District (Cape Comorin). *J. Bombay Natur Hist. Soc.* **57**, 184−195.

Legris, P. and Blasco, F. (1969). "Variabilité des Facteurs du Climat: Cas des Montagnes du Sud de l'Inde et de Ceylan." Tome VIII, Fascicule 1. Travaux de la Section Scientifique et Technique. Institut Français de Pondichéry, Pondicherry, India.

Lindburg, D. G. (1969). Rhesus monkeys: mating season mobility of adult males. *Science* **166**, 1176−1178.

Lindburg, D. G. (1971). The rhesus monkey in north India: an ecological and behavioral study. *In* "Primate Behavior: Developments in Field and Laboratory Research" (L. A. Rosenblum, ed.), Vol. 2, pp. 1−106. Academic Press, New York.

MacArthur, R. H. (1972). "Geographical Ecology." Harper and Row, New York.

MacArthur, R. H. and Wilson E. O. (1963). An equilibrium theory of insular zoogeography. *Evolution* **17**, 373−387.

MacArthur, R. H. and Wilson E. O. (1967). "The Theory of Island Biogeography." Princeton University Press, Princeton, New Jersey.

McCann, C. (1959). "100 Beautiful Trees of India." D. B. Taraporevala Sons, Bombay.

May, R. M. (1975). Island biogeography and the design of wildlife preserves. *Nature (London)* **254**, 177–178.

Mukherjee, R. P. and Mukherjee, G. D. (1972). Group composition and population density of rhesus monkey (*Macaca mulatta* [Zimmermann]) in Northern India. *Primates* **13**, 65–70.

Nishida, T. (1966). A sociological study of solitary male monkeys. *Primates* **7**, 141–204.

Pruett, C. (1973). A trip to Silent Valley—March 1972. *J. Bombay Natur. Hist. Soc.* **70**, 544–548.

Puri, G. S. (1960). "Indian Forest Ecology," Vols. I–II. Oxford Book and Stationery, New Delhi.

Rahaman, H. and Parthasarathy, M. D. (1967). A population survey of the bonnet macaque (*Macaca radiata* Geoffroy) in Bangalore, South India. *J. Bombay Natur. Hist. Soc.* **64**, 251–255.

Rahaman, H. and Parthasarathy, M. D. (1969). Studies on the social behavior of bonnet monkeys. *Primates* **10**, 149–162.

Rajasingh, G. J. (1961). A contribution to the knowledge of tropical wet evergreen forests—the *sholas* of Papanasam Hills in Madras State. *Indian Forest.* 5 Feb., pp. 77–86.

Ramaswami, M. S. (1914). A botanical tour in the Tinnevelly Hills. *Recent Botan. Surv. India* **6**, 105–171. *Records Bot. Surv. India* **6**, 105–171.

Randhawa, M. S. (1945). Progressive desiccation of Northern India in historical times. *J. Bombay Natur. Hist. Soc.* **45**, 558–565.

Ranganathan, C. R. (1938). Studies in the ecology of the *shola* grassland vegetation of the Nilgiri plateau. *Indian Forest.* **64**, 523–541.

Richards, P. W. (1966). "The Tropical Rain Forest." Cambridge Univ. Press, Cambridge.

Simonds, P. E. (1965). The bonnet macaque in South India. *In* "Primate Behavior: Field Studies of Monkeys and Apes" (I. DeVore, ed.) p. 175–196. Holt, New York.

Southwick, C. H. and Cadigan, F. C. Jr. (1972). Population studies of Malaysian primates. *Primates* **13**, 1–18.

Southwick, C. H. and Siddiqi, M. R. (1967). The role of social tradition in maintenance of dominance in a wild rhesus group. *Primates* **8**, 341–353.

Southwick, C. H., Beg, M. A., and Siddiqi, M. R. (1961a). A population survey of rhesus monkeys in villages, towns, and temples of Northern India. *Ecology* **42**, 538–547.

Southwick, C. H., Beg, M. A., Siddiqi, M. C. (1961b). A population survey of rhesus monkeys in Northern India. II. Transportation routes and forest areas. *Ecology* **42**, 698–710.

Southwick, C. H., Siddiqi, M. R., and Siddiqi M. F. (1970). Primate populations and biomedical research. *Science* **170**, 1051–1054.

Struhsaker, T. T. (1972). Rain-forest conservation in Africa. *Primates* **13**, 103–109.

Sugiyama, Y. (1965). Short history of the ecological and sociological studies on non-human primates in Japan. *Primates* **6**, 457–460.

Sugiyama, Y. (1968). The ecology of the lion-tailed macaque [*Macaca silenus* (Linnaeus)]—a pilot study. *J. Bombay Natur. Soc.* **65**, 283–292.

Sugiyama, Y. (1971). Characteristics of the social life of bonnet macaques (*Macaca radiata*). *Primates* **12**, 247–266.

Sullivan, A. L. and Shaffer, M. L. (1975). Biogeography of the megazoo. *Science* **189**, 13—
17.

Thorington, R. W., Jr., and Groves, C. P. (1970). An annotated classification of the Cercopithecoidea. *In* "Old World Monkeys: Evolution, Systematics, and Behavior" J. R. Napier and P. H. Napier, eds.) pp. 629–647. New York, Academic Press.

Walter, H. (1971). "Ecology of Tropical and Subtropical Vegetation." Oliver and Boyd, Edinburgh.

Warren, W. D. M. (1941). The influence of forests on climate. Sal regeneration *de novo*. *Indian Forest.*, June, pp. 292–301.

Webb-Peploe, C. G. (1947). Field notes on the mammals of South Tinnevelly, South India. *J. Bombay Natur. Hist. Soc.* **46**, 629–644.

Wilson, W. L. and Wilson, C. C. (1975a). Methods for censusing forest-dwelling primates. *In* "Contemporary Primatology" (S. Kondo, M. Kawai, and A. Ehara, eds.), Basel, Karger.

Wilson, W. L. and Wilson, C. C. (1975b). The influence of selective logging on primates and some other animals in East Kalimantan. *Folia Primatol.* **23**, 245–274.

10

Population Dynamics of Rhesus Monkeys in Northern India

CHARLES H. SOUTHWICK and M. FAROOQ SIDDIQI

I. Introduction

The rhesus monkey (*Macaca mulatta*) has been the most important and widely used nonhuman primate in biomedical work throughout the world. It played the key role in the development and production of polio vaccine, it provided an important basis for analyzing the Rh factor in human blood, and it has been used in a wide range of medical research programs from studies on arteriosclerosis to zoonotic diseases. It has, in fact, been the standard laboratory research monkey for more than 60 years.

339

Although the rhesus has long been famous for its extensive distribution and great abundance throughout Asia, particularly India, the question has naturally arisen as to whether its populations could survive the combined onslaughts of exploitation for biomedical programs, competition from expanding human populations, and changing habitat conditions. The rhesus populations of India have sustained a trapping removal of more than a million animals over the last 20 years, with annual exportation rates varying from over 100,000 monkeys per year in the late 1950's, to approximately 50,000 per year in the mid-1960's, to 20,000 in 1975. These populations have also experienced trapping and harassment as agricultural pests, and losses of habitat through deforestation and intensified land use.

This research was begun in 1959 to obtain basic data on the abundance and habitat distribution of the rhesus monkey in northern India, to study its ecology and behavior, to address the question of how its populations were responding to these decimating pressures, and to consider methods of conservation.

The first 10 years of this research did indeed find evidence that the rhesus populations of India were declining. Field surveys throughout the Gangetic basin and Himalayan foothills indicated that the number of rhesus groups were declining, group sizes were getting smaller, and the remaining populations showed a serious shortage of juveniles (Southwick et al., 1961a,b 1965; Southwick and Siddiqi, 1966, 1968; Southwick et al., 1970). Juveniles, of course, were the individuals most intensively trapped for commercial export.

These studies also found a very significant shift in the attitudes of many villagers of India, from traditional views of reverence and protection for rhesus monkeys, to new attitudes which considered them as agricultural pests and nuisance problems. Most villagers were anxious to get rid of monkeys—relatively few still tolerated them in the vicinity of their villages and croplands. These changing attitudes seemed to be the inevitable result of greater agricultural pressures and the need for more food. We felt that this social and cultural change represented the most basic threat to the long-term conservation and survival of the rhesus monkey in India.

Population studies were planned with two different approaches: (1) extensive population surveys which would sample a broad area of northern India via roadside, canal bank, village, and forest surveys at 5-year intervals, and (2) a more intensive population study of a limited area in Aligarh district of western Uttar Pradesh. The latter study would census the same groups every year for a period of 15 to 20 years. This paper presents data on this more intensive study.

A program of systematic field censusing began in October of 1959 with a population of seventeen groups of rhesus monkeys in typical rural habitats of the western Gangetic basin. By 1961, these original seventeen groups had increased to twenty-two, and we began a more systematic program of censusing each group three times a year.

This paper presents data on the changes in this population from 1961 to 1975, and on its basic demographic attributes of natality (birth rates) and mortality or disappearance rates.

II. Study Area

All of the rhesus groups of this study lived in an area of approximately 800 km², within 25 km of the city of Aligarh in western Uttar Pradesh. This is in the western Gangetic basin, 130 km southeast of New Delhi at lat. 28°N, long. 78°E.

The land is flat and intensively cultivated with crops of wheat, bajra, ragi, maize, barley, millet, rice, gram or pulse, sugarcane, and mustard. The study area is well watered, bisected by a main branch of the Gangetic canal with many small feeder canals. Groves of mango and papaya are common. Native trees are scattered across the landscape, especially along roadsides, canal banks, and in and around villages and towns. Some of the most important trees are neem (*Melia azadirachta*), jamun (*Eugenia jambolina*), sheesham (*Dalbergia sissoo*), pipal (*Ficus religiosa*), banyan (*F. bengalensis*), imle or tamarind (*Tamarindus indica*), acacia (*Acacia catechu*), and a variety of palms and shrubs.

Human population density is moderately high and typical of the Gangetic plains. The city of Aligarh had a population of approximately 150,000 people when this study began in 1959, and now is about 300,000 people. Many villages in the range of 500 to 1500 people dot the countryside. Human population density is about 300 people per km². Livestock density is also high, with abundant populations of cattle, water buffalo, goats, and some swine.

The climate is subtropical monsoon, with the rainy season from June to September. Occasional winter rains may fall, but normally there is very little rain from September throughout the winter, spring, and early summer. Annual rainfall averages about 80 cm (31.5 inches). January temperatures average 7°−21°C (44°−70° F) and July temperatures average 27°−36°C (81°−96° F). April and May are the hottest and driest months of the year, and daytime temperatures over 38°C (100°F) are typical.

The rhesus groups for this study represented most of the monkeys within Aligarh district except those living in Aligarh city itself and other sizeable towns nearby. The sample did not include, for example, the Aligarh temple monkeys which were the focus of earlier behavioral studies (Southwick *et al.*, 1965). Most of the groups lived in typical rural habitats, along roadsides, canal banks, in or near villages, in mango groves, or in small forest patches. Table I lists the name and location of 22 groups which were censused three times annually beginning in 1961. The groups varied in size from 6 to 36 monkeys, with an average of 17 monkeys per group, typical of rhesus in agricultural areas of northern India.

TABLE I

Group Locations and General Habitat of Original Population Sample

Group no.	Group name or location	General habitat	Original group size—October 1961	Fate of group
1	University Farm	Rural agricultural	10	Extinct 1966
2	Govern. Press Bldg.	Urban park and compound	8	Now 15 monkeys[a]
3	Cemetery group	Rural agricultural	6	Now 13 monkeys
4	Chhatari group-A	Rural agricultural	16	Now 6 monkeys
5	Chhatari group-B	Rural agricultural	33	Now 106 monkeys
6	Sumera Fall Jungle-A	Canal bank forest patch	14	Extinct 1966
7	Sumera Fall Jungle-B	Canal bank forest patch	19	Now 5 monkeys
8	Dauthara village	Rural agricultural village	8	Extinct 1965
9	Qasimpur Canal-A	Canal bank agricultural	24	Extinct 1974
10	Qasimpur Canal-B	Canal bank agricultural	21	Extinct 1965
11	Bajgarhi Bridge	Canal bank and roadside	18	All trapped 1970
12	Barautha Village	Rural agricultural village	36	All trapped 1974
13	Harduaganj	Lakeside, edge of small town	34	Extinct 1963
14	Barauli Bridge-A	Canal bank and roadside	16	Extinct 1975
15	Barauli Bridge-B	Canal bank and roadside	26	Extinct 1968
16	Nanau Bridge-A	Canal bank and roadside	19	Now 7 monkeys
17	Nanau Bridge-B	Canal bank and roadside	11	Now 22 monkeys
18	Sindhauli Village	Rural agricultural village	9	Extinct 1966
19	Agra road	Rural agricultural roadside	10	Extinct 1967
20	Delhi road school	Rural school yard and grove of trees	12	Now 4 monkeys
21	Satha Mango grove	Rural mango grove	13	Extinct 1966
22	Jawan Village	Rural village and mango grove	14	Now 7 monkeys
		Total population	377	185

[a] As of October, 1975.

Fig. 1. Chhatari rhesus group B in school yard compound.

The groups were remarkably stable in their locations over the years. We could return to the same area year after year, often the same specific trees, and find the same group of monkeys, except, of course, for those groups which were trapped to the point of extinction.

Chhatari-do-Raha

Chhatari-do-Raha was one location within the study area which deserves special mention because the rhesus groups there were partially protected by local people. At most of the other locations, local people did not make any special effort to protect the monkeys from trappers. The rhesus monkeys of Chhatari-do-Raha live in two groups inhabiting a rural school yard and grove of trees along the road from Aligarh to Anupshahr. Trees in the school yard and roadside are primarily neem, jamun, sheesham, pipal, banyan, and tamarind. Figure 1 shows the school yard which forms the core area of the home range of the larger group, Chhatari-B, and Fig. 2 shows the roadside trees which form the core area for the smaller group, Chhatari-A. This also represents typical agricultural land around Aligarh.

The school yard borders a crossroad with a bus stop where passenger buses between Aligarh and Anupshahr frequently stop. A small tea shop at the crossroad caters to travelers (Fig. 3). Passengers, as well as pedestrians, feed the

Fig. 2. Chhatari rhesus group A along roadside south of school yard.

Fig. 3. Tea shop and group B in corner of school yard.

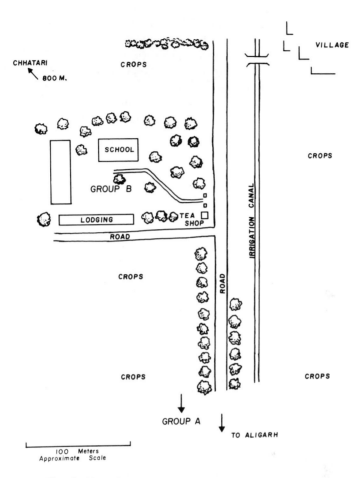

Fig. 4. Map of main habitat features of Chhatari groups.

monkeys with peanuts, gram nuts, wheat, rice, or miscellaneous bits of food. Paralleling the Aligarh–Anupshahr road on the eastern side is an irrigation canal, approximately 10 m wide. A small village of several hundred people is about 200 m northeast of the school yard, and one of over 800 people is about 800 m to the northwest. A map of the area is shown in Fig. 4.

The school yard, roads, and villages are all surrounded by agricultural fields of wheat, sugar cane, millet, bajra, and pulse. The monkeys often venture into these fields, but are usually repelled by "rakhwallas" guarding the crops. The rakhwallas are often children or elderly people who attempt to keep livestock, monkeys, dogs, and birds from feeding on the crops.

The most significant feature of Chhatari, which distinguishes it from most other locations in Aligarh district, and in fact from thousands of similar locations throughout the Gangetic plain, is the protection given to the monkeys by local people. Trapping of the monkeys is strictly controlled, so that only limited numbers are removed, mainly when the monkeys move too far from the school yard and crossroad. The protection is afforded primarily by two resident chowkidars (guards and overseers) of the school yard compound, who prohibit all trapping of monkeys in the immediate area of the school yard. When some monkeys venture too far, or these chowkidars are absent or otherwise occupied, a few monkeys may be trapped. The security and survival of the monkeys depend upon these two men who maintain the traditional attitudes of Hinduism toward monkeys, and who enjoy sufficient status in the community to have their views prevail. Neither depletion nor wholesale removal of the groups has been allowed, and the main group at Chhatari has increased considerably over the past 15 years. Just recently, in 1975, the resident chowkidars have not been able to prevent trapping of group A, which usually resides about 400 m away from the school yard. Over half this group was trapped in 1975, and henceforth it cannot be considered a protected group.

III.   Methods

All groups were first censused once in the fall of 1959 and again in the spring of 1960, bur regular systematic counts more than once a year were begun in July of 1961. Since then, all groups have been counted three times per year: in March and April, just before the annual birth season of the monkeys; in June and July, just after the birth season; and in October and November, just after the monsoon and before the winter. Censuses at these times provide the best data on the maximum count of the year (July), the minimum count of the year (March), the number of young born, and the number of deaths or disappearances during key seasons, especially the monsoon and winter.

Censuses are done by careful visual inspection of each home range area, and complete counts are made of all individuals in each of four classes: (1) adult males, (2) adult females, (3) juveniles, including subadults, and (4) infants. The criteria used in defining these age classes have been described in previous publications (Southwick et al., 1965; Southwick and Siddiqi, 1968). With good luck, an accurate group count can be made in 1 day or less for each group, but if larger groups are sometimes scattered, it is occasionally necessary to visit an area several consecutive days to obtain a completely accurate count.

We did not trap or mark any of the animals in this study, due primarily to the public relations problems this would have created. In each group, particularly the smaller ones, it was possible to recognize many individuals on the basis of natural appearances. Due to the very stable and consistent location of each group over the years, we had no problems recognizing the same groups. If a group could not be found on any given visit, repeated visits were made, and a thorough search of the area surrounding the home range was undertaken. Usually the group could be found, but some groups disappeared without a trace, and in subsequent discussions with local people we usually found out that the entire group had been trapped, or that the last individual had been found dead, in cases of groups that dwindled slowly over a period of years.

Estimates of annual mortality and loss rates are obtained from the population counts in the following ways. Infant monsoon mortality is simply the loss of infants from the July count to the October-November count divided by the number of infants present in July. Annual infant mortality is the loss of infants from July of one year to March of the next year, divided by those present in July of the first year. This may be represented by,

$$I_l = (I_{j1} - I_{m2})/I_{j1}$$

where I_l equals infant loss per year, I_{j1} equals the number of newborn infants in July of one year following the birth season, and I_{m2} equals the number of infants present the following March just before the next birth season. This is actually a 9-month mortality, but since weaning begins by 8 or 9 months of age, and the next infants are born shortly thereafter, this realistically represents the infants' "year."

Juvenile loss per year, J_l, is based on the following calculation:

$$J_l = [(0.67 J_{m1} + I_{m1}) - J_{m2}]/(0.67 J_{m1} + I_{m1})$$

where J_{m1} equals the juveniles present in March of one year, I_{m1} equals the infants present in March of the same year, and J_{m2} equals the juveniles actually present in March of the subsequent year. The use of the factor 0.67 is based on the fact that, in any given year, about two-thirds of the juveniles will remain as juveniles, whereas one-third of them will pass into adulthood (the juvenile period is approximately 3 years, from ages 1−4).

Adult loss per year, A_l, is calculated on the following,

$$A_l = [(A_{m1} + 0.33 J_{m1}) - A_{m2}]/(A_{m1} + 0.33 J_{m1})$$

where A_{m1} is the number of adults actually present in March of any one year, J_{m1} is the number of juveniles present in the same March, and A_{m2} is the number of adults present in the subsequent year. The factor of 0.33 is again based on the premise that one-third of the juveniles in any given year will become adults; that is, they will pass into adulthood shortly following the March census. Although the above methods involve these generalizations and simplifications, each number in the formulas, other than the factors of 0.33 and 0.67, are based on exact field data.

Annual turnover is estimated by multiplying the annual mortality of each age class by its percentage composition in the population and summing the totals. That is, annual turnover, T_a, is calculated on the following,

$$T_a = \Sigma\, I_l I_p + J_l J_p + A_l A_p$$

where I_l equals infant loss, I_p equals infant percentage composition in the population, J_l equals juvenile loss, J_p equals juvenile percentage composition, and A_l equals adult loss and A_p equals adult percentage composition. In other words, the annual turnover is based on the sum of age-specific mortality rates. If the annual turnover is 20%, and the population is stable with no immigration, this means that 20% of the individuals in a population each year would be expected to die and be replaced each year.

IV.   Results and Discussion

In the initial years of this study, the rhesus monkey population of Aligarh district increased from 337 monkeys in 17 groups in 1959 to 403 monkeys in 22 groups by July of 1962. Subsequently, however, the population showed an irregular decline to a low of only 163 monkeys in 10 groups by March of 1970. Since

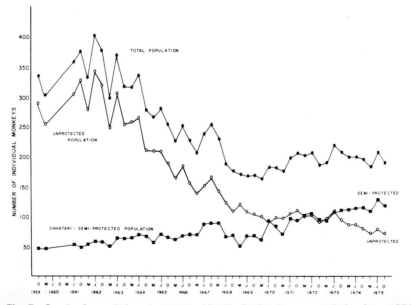

Fig. 5. Graph of population changes in Aligarh district rhesus population from 1959 to 1975.

1970, the population has recovered slightly, and it now numbers 185 monkeys in 9 groups. These population changes are shown in Fig. 5.

The best overall representation of what has happened to the population over the past 15 years, however, is its irregular decline. The total population has declined 50% from 337 monkeys in October of 1961 to only 185 in October of 1975 (Table II). The number of groups declined 59% from 22 to 9.

Within this general pattern of decline, however, two cohorts of the population have shown opposite trends. Most of the groups which have not had any protection from local people have declined 77.7% from 328 monkeys to only 73, and these groups have decreased in number from 20 to 7 (Table II). Another cohort of the population, consisting of two groups at Chhatari-da-Raho, has received some protection from local people, and has increased from 49 monkeys to 112, an increase of 128.5% (Table II).

With these distinctly different trends prevailing, all subsequent data on population attributes are divided into two sets: (1) the Aligarh district unprotected population, referring to the twenty original groups which have not had local protection, and have declined nearly 78%, and (2) the Chhatari-do-Raha groups, which have had some local protection, and have increased more than 128%.

The complete loss of groups has usually been a gradual process, with a group slowly decreasing over a period of years because too many animals were trapped or killed each year. For example, the University farm group went from eight monkeys in 1961 to seven by 1963, to three by 1964, two by 1965, and the last one died in 1966. In 1962, villagers near the group reported that trappers regularly captured young monkeys each year. In 1964, five monkeys were shot by a Sikh farmer because they fed upon his crops. The history of this group is typical

TABLE II

Changes in the Aligarh District Rhesus Population, 1961–1975

Cohort of population	October 1961	October 1975	Percentage change
I. Total Aligarh district population			
Number of individual rhesus	377	185	−50.9%
Number of groups	22	9	−59.1%
II. Unprotected population			
Number of individual rhesus	328	73	−77.7%
Number of groups	20	7	−65.0%
III. Semiprotected population (Chhatari-do-Raha)			
Number of individual rhesus	49	112	+128.5%
Number of groups	2	2	0

of many others in the unprotected population and it has been previously detailed in a separate publication (Southwick and Siddiqi, 1967).

Occasionally, an unprotected group has been trapped and removed all at once. For example, the group at Bajgarhi bridge was entirely trapped and removed in 1970, and the group at Barautha village met a similar fate in 1974. Thus, by either gradual or abrupt means, 13 groups out of 22 have become extinct over the past 15 years (Table I).

A. Population Size and Structure

The average population each year, obtained by averaging three different census counts, declined from 316 monkeys in the unprotected population in 1961 to 76 in 1975. The adult sex ratio, as indicated by the number of adult females per adult male varied from 1.9 in 1961 to 1.0 in 1975 (Table III). This indicated that females were declining more rapidly than males in unprotected groups.

TABLE III

Aligarh District Unprotected Population: Population Size, Sex Ratio, and Percentages of Infants and Juveniles

Year	Average total population	Adult females per adult males	Percentage infants in population	Percentage juveniles in population	Percentage immature
1959	288	—	—	—	—
1961	316	1.9	34.2	7.3	41.5
1962	314	1.7	29.3	10.9	40.2
1963	270	1.6	28.0	14.9	42.9
1964	245	1.7	27.9	16.2	44.1
1965	203	1.6	25.9	15.8	41.7
1966	168	1.5	24.8	21.0	45.8
1967	152	1.6	24.6	19.7	44.3
1968	125	1.5	25.3	19.5	44.8
1969	111	1.5	25.8	15.9	41.7
1970	97	1.3	21.6	20.6	42.2
1971	105	1.4	21.3	21.0	42.3
1972	99	1.5	24.2	15.5	39.7
1973	102	1.4	24.2	19.6	43.8
1974	86	1.2	21.3	25.2	46.5
1975	76	1.0	20.6	27.2	47.8
15-year average (1961–1975)		1.5 ± 0.06	25.3 ± 0.9	18.0 ± 1.3	43.3 ± 0.60

There were significantly more adult females per adult male in the semi-protected population at Chhatari.

The percentage of infants in the unprotected population varied from 20 to 34%, averaging 25.3% over the 15-year period. The percentage of juveniles varied from 7 to 27%, and averaged 18.0%. The percentage of the total population represented by immature animals averaged 43.3% (Table III).

All of these population parameters, except percentage infants, differed significantly in the semi-protected population at Chhatari. The average population increased from 52 in 1961 to 119 in 1975, showing an increasing trend in all but 3 years. The adult sex ratio showed an average of 2.7 females per male, almost twice as many females per male as in the unprotected population (Table IV).

The percentage of infants in Chhatari averaged 27.7, not significantly greater than in the unprotected population, but the percentage of juveniles averaged 26.2, which was considerably greater. The percentage of the population at Chhatari represented by immatures was 53.9%, significantly greater than the

TABLE IV

Semiprotected Population (Chhatari-do-Raha Groups): Population Size, Sex Ratio, and Percentages of Infants and Juveniles

Year	Average total population	Adult females per adult males	Percentage infants in population	Percentage juveniles in population	Percentage immature
1959	49	2.3	32.6	0	32.6
1961	52	2.7	34.0	12.6	46.6
1962	57	2.3	33.3	8.8	42.1
1963	59	2.9	29.9	16.4	46.3
1964	67	2.5	26.4	28.8	55.2
1965	64	2.5	25.9	31.1	57.0
1966	67	1.9	24.0	30.5	54.5
1967	81	1.8	22.3	36.8	59.1
1968	75	1.8	22.3	36.6	58.9
1969	62	2.1	27.3	25.7	53.0
1970	80	3.8	30.5	28.9	59.4
1971	92	3.2	26.7	28.9	55.6
1972	101	3.3	31.1	25.8	56.9
1973	105	3.6	28.7	29.9	58.6
1974	115	3.3	24.3	40.5	64.8
1975	119	3.2	24.1	37.5	61.6
16-year average (1959–1975)		2.7 ± 0.16	27.7 ± 0.97	26.2 ± 2.81	53.9 ± 2.05

43.3% in the unprotected population. We have previously proposed that 50% immatures is approximately the break-even point for macaque population; that is, if the population shows fewer than 50% immatures, the population is not likely to maintain itself; if it shows over 50% immatures it is more likely to be stable or increasing (Southwick and Siddiqi, 1977).

B. Natality

The natality or live birth rate of the unprotected population varied from 53% per year to 89.6%, and averaged 76.4% over the 15-year period (Table V). This is quite high considering the difficult circumstances under which these animals were living and the degree of harassment to which they were subjected.

The semiprotected population showed an even more remarkable natality, with annual birth rates ranging from 72.7 to 100%, averaging 90.7% over a 15-year period (Table VI). This is exceptionally high compared to birth rates of other

TABLE V

Aligarh District Unprotected Population: Annual Natality and Infant Mortality

Year	Annual natality[a]		Infant mortality	
	Ratio	Percentage	Monsoon[b]	Annual[c]
1961	104/116	89.6	0	28.6
1962	96/121	79.3	0	31.0
1963	88/101	87.1	20.4	20.4
1964	74/89	83.1	24.3	24.3
1965	54/73	74.0	14.8	25.7
1966	44/54	81.5	9.1	30.3
1967	36/48	75.0	0	13.9
1968	31/41	75.6	12.9	12.9
1969	27/36	75.0	0	6.9
1970	16/30	53.3	0	15.0
1971	24/36	66.7	0	3.8
1972	24/36	66.7	4.2	4.2
1973	27/35	77.1	18.5	32.1
1974	19/23	82.6	10.5	10.5
1975	16/20	80.0	12.5	—
Average 1961—1975 $\bar{x} \pm$ S.E.		76.4 ± 2.3	8.5 ± 2.2	18.5 ± 2.7

[a] Ratio and percentage of adult females producing one infant per year; live birth rate.
[b] Loss of infants from July to October.
[c] Loss of infants from end of one birth season to beginning of the next birth season.

TABLE VI

Semiprotected Population (Chhatari-do-Raha Rhesus Groups): Natality and Infant Mortality

Year	Annual natality[a]		Infant mortality	
	Ratio	Percentage	Monsoon[b] (%)	Total estimated annual[c] (%)
1961	16/20	80.0	0	0
1962	21/23	91.3	19	33.3
1963	21/24	87.5	14.3	14.3
1964	16/22	72.7	6.2	6.2
1965	18/20	90.0	0	16.7
1966	17/19	89.5	5.9	5.9
1967	20/21	95.2	0	0
1968	16/20	80.0	0	43.7
1969	21/22	95.4	0	0
1970	27/27	100.0	7.4	44.4
1971	34/35	97.1	26.5	26.5
1972	34/35	97.1	8.8	11.8
1973	32/34	94.1	0	9.4
1974	27/29	93.1	0	3.7
1975	31/32	96.9	6.5	—
15-year averages (1961–1975)		90.7 ± 2.0	6.3 ± 2.1	15.4 ± 4.2

[a] Ratio and percentage of adult females producing one infant per year.
[b] Loss of infants from July to October.
[c] Loss of infants from end of one birth season to the beginning of the next birth season.

populations. The rhesus population of Cayo Santiago averaged 78% natality in the early 1960's (Koford, 1965), and the rhesus of La Parguera averaged 73% natality from 1964 through 1972 (Drickamer, 1974). Two populations of the toque monkey (*Macaca sinica*) in Sri Lanka averaged 50.8% and 68.8% natality over a two to four year period from 1968 to 1972 (Dittus, 1975). Most confined colonies of macaques in the United States have averaged less than 60% despite high quality food and veterinary care (Nat. Acad. Sci., 1975). Even though the Chhatari population lived in a competitive agricultural environment, and have been forced by human pressures into a relatively small and limited home range, they are so remarkably well adapted to this environment that their reproduction approaches the theoretical maximum. The partial protection they receive enables them to achieve a significantly better birth rate than that of unprotected groups.

C.   Infant Mortality and Juvenile Loss Rates

Infant mortality during the monsoon period, when the infants were normally 1—5 months of age, varied from 0 to 24.3% in the unprotected population, and averaged 8.5% over the 15-year period. The estimated annual mortality varied from 3.8 to 32.1% and averaged 18.5% for the unprotected population (Table V).

In the semiprotected population of Chhatari, infant mortalities were slightly less. Monsoon mortality averaged 6.3% and annual mortality 15.4% (Table VI).

Most of this loss probably reflects true mortality, though we cannot eliminate the possibility of illegal trapping. Normally monkeys under 4 pounds in body weight and under 1 year of age should not be trapped, but we have seen some trapping of infants as early as 7 to 8 months of age. However, this has been relatively rare.

The great variation of infant loss between years is probably attributable to one of three factors: (1) variable disease mortality, (2) variable mortality associated with weather or food supplies, or (3) variable losses incurred from illegal trapping of infants. Additional research is needed to distinguish among these possibilities.

Infant mortality in both populations (15.4—18.5%) was higher than that of the rhesus colony on Cayo Santiago (8—9%), but in the same range as that at La Parguera (17—19%) (Drickamer, 1974). Drickamer (1974) showed that infant mortality varied according to the parity of the mother (primiparous births suffer higher mortality than do subsequent births), the social rank of the mother (infants of low-ranking mothers experience higher mortality than infants of high-ranking mothers), and month of birth [infants born in the peak of the birth season (May—June) experience less mortality than those born early or late].

Juvenile loss was more variable than infant mortality, ranging from 31.4% to a high of 76.6% in any one year in the unprotected population. It averaged 57.6% in the unprotected population (Table VII). In the semiprotected population of Chhatari, juvenile loss ranged from 0 to 75% and averaged 32.0% (Table VIII).

Although we could not separate true mortality from trapping removal, we believe that trapping probably accounted for the majority of juvenile loss. Once young rhesus reach the age of 1 year, they are probably immune to most of the common respiratory and enteric diseases in the area, and their chances of mortality from disease or starvation are probably slight as long as they remain in the group in the same home range. Furthermore, the chowkidars have not reported seeing any dead juveniles, nor have we ever found juvenile skeletal material in the area. Similarly, Lindburg (1971) found no evidence of juvenile mortality in his study of forest groups at Dehra Dun. In temple sites in Aligarh and Kathmandu, on the other hand, where trapping is prohibited, we have frequently found skeletal remains of juvenile rhesus. Thus, we conclude on circum-

TABLE VII

Aligarh District Unprotected Population: Annual Loss Rates of Juveniles and Adults

Year	Percentage loss juveniles	Percentage loss adult males	Percentage loss adult females	Percentage loss total adults
1962	70.0	11.2	24.0	19.5
1963	59.4	14.5	4.4	8.4
1964	76.6	19.4	19.8	19.7
1965	65.9	29.4	27.2	28.3
1966	75.7	22.3	9.5	14.4
1967	31.4	18.4	21.7	20.6
1968	68.0	5.5	18.6	13.6
1969	57.1	35.3	26.1	30.0
1970	41.0	0	10.8	6.4
1971	50.5	13.2	2.2	7.1
1972	58.0	20.6	7.0	12.5
1973	51.4	6.1	24.7	17.5
1974	44.2	26.3	37.1	32.4
Average 1962–1974	57.6 ± 3.8	17.1 ± 2.8	17.9 ± 2.9	17.7 ± 2.4

TABLE VIII

Semiprotected Population (Chhatari-do-Raha Groups): Annual Loss Rates of Juveniles and Adults

Year	Percentage loss juveniles	Percentage loss adult males	Percentage loss adult females	Percentage loss total adults
1962	75.0	32.2	0	8.8
1963	29.4	0	11.3	9.1
1964	53.3	30.0	16.7	20.6
1965	24.0	14.0	10.3	12.5
1966	32.1	0	9.5	2.9
1967	0	18.4	13.2	15.4
1968	58.1	38.2	17.9	26.6
1969	38.1	57.1	0	6.7
1970	10.0	2.5	0	0
1971	9.1	13.6	4.6	6.7
1972	57.1	43.7	5.5	17.3
1973	11.4	20.0	9.3	12.0
1974	18.2	45.1	25.7	31.6
Average (1962–1974)	32.0 ± 6.4	24.2 ± 5.1	9.5 ± 2.1	13.1 ± 2.5

stantial and tentative evidence that most of the loss of juveniles in forest and agricultural habitats is due to dispersal or trapping, whereas that in temple and urban habitats may be disease related.

Juvenile loss in the unprotected population was moderately high every year, never falling below 30%, whereas it varied more in the semiprotected population, being less than 12% in 4 years (Tables VII and VIII). In 4 other years at Chhatari, more than 50% of the juveniles disappeared (1962, 1964, 1968, and 1972). Three of these years in which the greatest juvenile loss occurred were years in which a large increase in the population occurred during that year or in the preceding year (1964, 1967, 1972), suggesting that the villagers were using trapping as a way of regulating the population; that is, as a management effort. It is very apparent at Chhatari when too many juveniles are present—they cause considerable nuisance in the school yard, in surrounding fields, and at the tea shop and bus stop. In October of 1975 it was obvious that the Chhatari-B population was somewhat too large—over 100 monkeys in a relatively small area cause enough local consternation that they invite some judicious removal. Unfortunately, we could not obtain any reliable information from the chowkidars or villagers on the number of juveniles trapped from year to year, possibly because they feared this would result in more trapping or some form of governmental taxation.

If we assume that most of the juvenile loss in a well-managed rhesus group is due to dispersal or trapping, and that true juvenile mortality in a healthy population can be kept to a very low percentage, these data from Chhatari suggest that about 30% of a juvenile population could be harvested annually, while still maintaining breeding stock and maximizing reproduction.

D. Adult Loss

A striking difference appeared in the patterns of adult loss in the two population cohorts. Total adult loss rates per year were only slightly higher for the unprotected (17.7%) than for the semiprotected population (13.1%), but the mortality pattern between the two sexes was significantly different. In the unprotected population, both sexes had a similar annual loss rate—17.1% for males and 17.9% for females. In the semiprotected population of Chhatari, however the adult male loss rate was more than twice as great as that for females (24.2% for males; 9.5% for females).

We believe the Chhatari pattern is more typical of rhesus groups, in which a higher annual loss rate of adult males is due to the greater dispersal mobility of males and their tendency to move between groups (Drickamer and Vessey, 1973; Sade, 1975). Drickamer (1974) observed, "The group-changing process, the higher levels of aggression, and the rigors of a solitary existence all lead to a higher mortality rate among males over the age of four years." In one cohort of the La Parguera rhesus population, males from 4 to 8 years of age showed an annual mortality of 22.0%, whereas a similar cohort of adult females showed no mortality in the same age span (Drickamer, 1974).

One of the most surprising aspects of the data on adult loss was the fact that the annual loss of males in the semiprotected population was actually higher (24.2%) than in the unprotected population (17.1%). This suggests that the monkeys themselves were expelling adult males to a greater extent in the semiprotected groups, or at least dispersal and/or mortality of males was higher in these groups than in the smaller, unprotected groups. It seems logical that the adult males in the smaller groups (where there is often only one or two males) did not disperse as much as in the Chhatari groups, where there may be from 8 to 11 adult males in Chhatari-B. The adult males in the smaller groups are in a more secure position socially, and they are also relatively free from trapping because they often tend to be aggressive and imposing individuals. They are, in fact, somewhat intimidating to local people when they have an absolutely uncontested dominance position. Trappers have told us that they prefer not to try to trap big males. For this or other reasons we have the anomalous situation that the loss of adult males is higher in semiprotected groups than unprotected groups.

With adult females, the opposite and more expected situation prevails—adult females have a higher loss rate in unprotected (17.9%) than in semiprotected groups (9.5%). The females definitely benefit from partial protection as do the juveniles.

E. Annual Turnover

Annual turnover, calculated by combining the age class loss rates per year with the known age structure of the population, was estimated to be 25.3% for the unprotected population and 18.7% for the semiprotected population of Chhatari. This means that 18–25% of the populations would be expected to die in any given year. This is population turnover, and not behavioral or social turnover in the sense of group-changing behavior.

These turnover rates can be compared to an estimated annual population turnover of 28.4% for the baboons of Amboseli, calculated from the mortality data of Altmann and Altmann (1970), and an annual turnover of 29.9% for the lions of the Serengeti, calculated from the data of Schaller (1972).

F. Annual Population Increase

The rhesus population of Chhatari increased from 49 monkeys in October of 1959 to 112 in October of 1975, an average annual increase of 5.6%. Macaque populations that are completely protected and provisioned with food have shown considerably higher rates of population increase. From 1960 to 1964, the rhesus population of Cayo Santiago Island increased at an annual rate of 16% (Koford, 1966). The rhesus population of La Parguera, first established in 1962, increased at the annual rate of 13.4% during the first 10 years (Drickamer, 1974). The Japanese macaques of Mt. Takasaki increased from 220 monkeys in 1953 to 1400 in 1974, an average annual increase of 10.2% over a 20-year period (Itani, 1975).

New World monkey populations can also show high rates of increase under favorable conditions. The howler monkey population of Barro Colorado Island increased from 239 monkeys in 1951 to 814 in 1959, an average annual increase of 16.4% (calculated from data in Carpenter, 1962).

These data demonstrate the capacity of monkey populations to increase substantially if provided with suitable habitat, food, and protection. We feel the Chhatari population would have increased at an average rate of 10 to 16% per annum if no monkeys had been trapped. If such increase had occurred, however, excessive numbers of monkeys would have created a serious agricultural problem for the villagers. The best course for both the monkey population and the protection of agricultural crops has been to trap limited numbers of monkeys at planned intervals.

G. Comparison of Semiprotected with Unprotected Rhesus Populations

The partial protection afforded the rhesus population at Chhatari affected most of the population parameters considered in this study. It affected population trends and size most noticeably, permitting the Chhatari population to increase while the unprotected population decreased. It also permitted higher rates of natality (90.7 compared to 76.4%), lower mortality and loss rates for all age classes, especially juveniles, and lower annual turnover rates (Table IX). Thus, the semiprotected status of the Chhatari monkeys permitted better reproduction and better survivorship of all age groups, with the exception of adult males. Apparently the social system of the monkeys accelerated the dispersal and loss rates of adult males in the semiprotected state in comparison to the totally unprotected state.

V. Comments on Conservation

These data clearly demonstrate the positive values of partial protection for rhesus monkey populations in an agricultural environment. With partial protection, a rhesus population can increase, show outstandingly high natality, possess favourable sex and age ratios, and show relatively low mortality and loss rates.

TABLE IX

Demographic Comparisons between the Chhatari Population and Unprotected
Groups in Aligarh District

Population trait	Chhatari groups (semiprotected)	Unprotected groups
Natality [percentage birth rate per year among adult females (15-year averages)]	90.7 ± 2.0	76.4 ± 2.3
Percentage immature animals in population (15-year average)	53.9 ± 2.05	43.3 ± 0.6
Infant mortality/year (15-year average)	15.4 ± 4.2	18.5 ± 2.7
Juvenile loss/year (13-year average)	32.0 ± 6.4	57.6 ± 3.8
Adult loss/year (13-year average)	13.1 ± 2.5	17.7 ± 2.4
Annual tunover (%)	18.7	25.3

Without protection, rhesus populations in agricultural India do not seem able to maintain themselves. Their natality is good, but not as high as in a semiprotected population, their sex and age structure is not favorable, and they show higher mortality and loss rates.

We believe that without protection the rhesus population of northern India will continue to decline. Furthermore, we believe this will occur whether or not rhesus monkeys are trapped for commercial export and biomedical use throughout the world. Even without such utilization, the same agricultural and human pressures will continue to operate against the rhesus monkey, and it will be eliminated from intensely cropped landscape as it already has been in most of West Bengal where human population densities are very high.

The potential solutions we can see to this problem are the following:

1. In forest lands of northern India, certain preserves and sanctuaries must be established where the trapping of monkeys is strictly controlled. Under proper management and enforcement, this would guarantee the survival of forest-dwelling rhesus monkeys, and would still permit a regulated harvest to occur.

2. In agricultural areas of the Gangetic plains, similar controlled trapping programs must also be established, but, in addition, it is necessary to convince local people that they should protect and manage monkeys in a way similar to the Chhatari situation. This will probably require some subsidized compensation for crop damage or for feeding monkeys. At the very least, local people

should participate in the financial benefits gained from the sale of the juvenile monkeys removed from the population each year. In other words, the villager should be shown that the monkeys can be an economic asset, of greater value than the agricultural damage which they commit. This was proposed several years ago by Bermant and Chandrasekhar (1971). The monkeys at Chhatari do relatively little damage, when kept within reasonable numbers, because they receive most of their food from artificial feeding by local people.

The monkeys of Chhatari demonstrate the need for regulated trapping. Without any trapping removal, this population would be totally out of hand. Although 32% of the juveniles disappear each year, probably most of them through trapping, this is necessary to keep the population in balance. We suggest the approximate figure of 30% harvest rate for juveniles as a reasonable starting point in a managed trapping program. Actual experience may increase or decrease this percentage depending upon the response of the population.

We believe the data from Chhatari demonstrate that local primate management by villagers in India can be an effective conservation method. Unfortunately, every good habitat setting in India does not have local people with the same animal conservation values as the chowkidars at Chhatari. Unless more people of India can be converted to similar values, we doubt if the rhesus can survive much beyond this century in the open agricultural lands of the Gangetic plains. Except for monkeys in forests, parks, cities, and temples, most of the rhesus monkeys of rural India seem to be destined in the downward direction of the unprotected population of Aligarh district.

VI. Summary

A natural population of rhesus monkeys in the agricultural land of the western Gangetic Basin in northern India has been censused since 1959. Two cohorts within this population have shown opposite population trends: (1) an unprotected cohort declined from 328 monkeys in 20 groups to only 73 monkeys in 7 groups; (2) a semiprotected cohort increased from 49 monkeys in two groups to 112 in two groups.

The semiprotected groups at Chhatari showed a more favorable sex and age structure (more females per male, and over 50% immatures in the population), higher natality (90.7 compared to 76.4%), lower infant mortality (15.4 compared to 18.5%), lower juvenile loss (32.0 compared to 57.6%), lower adult loss (13.1 compared to 17.7%), and lower population turnover per year (18.7 compared to 25.3%). The partial protection afforded the rhesus groups at Chhatari by local people thus had a favorable impact on population conservation and trends, birth rates, mortality, and turnover rates.

Unless rhesus monkeys receive some protection from local people we feel they will be eliminated from most agricultural areas of India within 25 years. The main pressure will come from the increasing need for more food in India, and the changing social customs of the villagers of India who no longer tolerate crop depredation by monkeys. The population at Chhatari demonstrates, however, that local conservation and management efforts by villagers can be very effective. We feel it will be necessary to establish more sanctuaries and parks for wildlife in the forests of northern India, and to convert more villagers to a practice of protecting and managing monkeys in agricultural environments through educational and economic incentives.

Acknowledgments

This work was begun when the senior author was a Fulbright postdoctoral Fellow at Aligarh Muslim University. Its early establishment was due to the cooperation and assistance of Dr. M. Babar Mirza, then Chairman of the Department of Zoology, Dr. M. Rafiq Siddiqi, and Mr. Mirza Azher Beg. We have enjoyed the support and encouragement of many colleagues in India, including Drs. S. M. Alam and J. A. Khan of Aligarh University, Drs. K. K. Tiwari and R. P. Mukherjee of the Zoological Survey of India, Dr. R. K. Lahiri, Chief Wildlife Officer of West Bengal, Drs. Donald Lindburg, Melvin Neville, and Phyllis Jay Dolhinow when they were affiliated with the California Primate Center, Dr. Mireille Bertrand, and Dr. and Mrs. John Oppenheimer. Dr. B. C. Pal and Dr. M. Y. Farooqui contributed to the field work. Since 1960, the field work has been supported by U.S.P.H.S. Grants No. RO 7 AI−10048, MH−18440, and RR−00910 to Johns Hopkins University.

References

Altmann, S. A. and Altmann, J. (1970). "Baboon Ecology," 220 pp. University of Chicago Press, Chicago, Illinois.

Bermant, G. and Chandrasekhar S., (1971). Rescue plan for Indian monkeys. *Science* **171**, 628−29.

Carpenter, C. R. (1962). Field studies of a primate population. *In* "Roots of Behavior" (E. L. Bliss, ed.), Chapt. 21, pp. 286−294. Harper, New York.

Dittus, W. P. J. (1975). Population dynamics of the toque monkey, *Macaca Sinica*. *In* "Socioecology and Psychology of Primates" (R. H. Tuttle, ed.) pp. 125−151. Mouton, The Hague and Paris.

Drickamer, L. C. (1974). A ten year summary of reproductive data for free-ranging *Macaca mulatta. Folia. Primatol.* **21**, 61−80.

Drickamer, L. C. and Vessey, S. H. (1973). Group-changing behavior of free-ranging rhesus monkeys. *Primates* **14**(4), 359−368.

Itani, J. (1975). Twenty years with Mt. Takasaki monkeys. *In* "Primate Utilization and Conservation" (G. Bermant and D. G. Lindburg, eds.), Chapt. 10, pp. 101−126. Wiley, New York.

Koford, C. B. (1966). Population changes in rhesus monkeys: Cayo Santiago 1960—64. *Tulane Stud. Zool.* **13**, 1—7.

Lindburg, D. G. (1971). The rhesus monkey in north India: an ecological and behavioral study. *In* "Primate Behavior: Developments in Field and Laboratory Research" (L. A. Rosenblum, ed.), pp. 1—106. Academic Press, New York.

National Academy of Sciences-National Research Council. (1975). "Nonhuman Primates: Usage and Availability for Biomedical Programs," 122 pp. Washington, D.C.

Neville, M. K. (1968). Ecology and activity of Himalayan foothill rhesus monkeys. *Ecology* **49**, 110—123.

Sade, D. S. (1975). Population dynamics in relation to social structure on Cayo Santiago. *Yearb. Phys. Anthropol.* in press.

Schaller, G. B. (1972). "The Serengeti Lion: A Study of Predatory-Prey Relations," 480 pp. University of Chicago Press, Chicago, Illinois.

Southwick, C. H. and Siddiqi, M. R. (1966). Population changes of rhesus monkeys (*Macaca mulatta*) in India, 1960—1965. *Primates* **7**(3), 303—314.

Southwick, C. H. and Siddiqi, M. R. (1967). The role of social tradition in the maintenance of dominance in a wild rhesus group. *Primates* **8**, 341—353.

Southwick, C. H. and M. F. Siddiqi. (1977). Demographic characteristics of semi-protected rhesus groups in India. *Yearb. Phys. Anthropol.* in press.

Southwick, C. H. and Siddiqi, M. R. (1968). Population trends of rhesus monkeys in villages and towns of northern India, 1959—65. *J. Anim. Ecol.* **37**, 199—204.

Southwick, C. H., Beg, M. A., and Siddiqi, M. R. (1961a). A population survey of rhesus monkeys in villages, towns and temples of northern India. *Ecology* **42**, 538—547.

Southwick, C. H. Beg, M. A., and Siddiqi, M. R. (1961b). A population survey of rhesus monkeys in northern India: II. Transportation routes and forest areas. *Ecology* **42**, 698—710.

Southwick, C. H., Beg, M. A., and Siddiqi, M. R. (1965). Rhesus monkeys in north India. *In* "Primate Behavior: Field Studies of Monkeys and Apes (Irven DeVore, ed.), pp. 111—159. Holt, Rinehart and Winston, New York.

Southwick, C. H., Siddiqi, M. R., and Siddiqi, M. F. (1970). Primate populations and biomedical research. *Science* **170**, 1051—1054.

11

The Gelada Baboon:
Status and Conservation

R. I. M. DUNBAR

I. Introduction

The taxonomic affinities of *Theropithecus gelada* Ruppell 1835 (Fig. 1) were the subject of much dispute during the century following its "discovery." Some authorities regarded it as an aberrant member of the genus *Papio*, while others asserted that it was a member of the Asian genus *Macaca* and some even considered that its affinities lay properly with the genus *Cercopithecus* (see review in Hill, 1970). Most recent authorities, however, consider it to be most closely related to the *Papio* baboons, while recognizing its unique generic status (Hill, 1970; Jolly, 1972). The ancestral theropithecines appear to have differentiated from the basal *Papio* stock at some point in the Pliocene, some

363

Fig. 1. An adult male *Theropithecus gelada.*

4 million or so years ago, when they invaded a grassland biotope along the margins of lakes and rivers. During that time span from the late Pliocene through the Pleistocene for which fossil evidence is available, the genus seems to have ranged from as far north as Ternifine on the Mediterranean coast of Africa to Swartkrans and Makapan in South Africa (Jolly, 1972). Jolly distinguishes only three good species within the genus as a whole: the early *T. darti* of southern Africa, the later *T. oswaldi* of eastern Africa, and the recent *T. gelada* of Ethiopia. (However, recent findings indicate that a fourth species, *T. brumpti*, occupied a limited range in southern Ethiopia during the early part of the

Pleistocene; G. Eck, personal communication.) All of these species are characterized by an extreme skeletal adaptation to a highly terrestrial way of life and a dental specialization for a diet based on leaves, roots, and seeds of grasses.

Of the known species of the genus *Theropithecus*, only one (*T. gelada*) exists today. All of the other species had become extinct by later Pleistocene times, perhaps as recently as 50,000 years ago in the case of *T. oswaldi* (Jolly, 1972). Tappen (1960) observed that the gelada today occupy a retreat habitat on the mountainous plateau of Ethiopia, and he attributed this to the fact that they were forced to take refuge on the inaccessible highlands as a result of competition from the more successful *Papio* baboons. However, *Papio* baboons and gelada differ radically in diet and competition occurs between them only to a very limited degree (Dunbar and Dunbar, 1974). Furthermore, Jolly (1972) noted that the gelada show many specializations for life at high altitudes and it seems probable that the species colonized its present habitat at a very early stage in its evolutionary history. Jolly (1972) remarked on the fact that the disappearance of the larger theropithecines from the African scene some 50,000 or so years ago occurred at a time when dramatic changes were taking place on Africa's savanna grasslands. During this period, many giant plains species also became extinct (Cooke, 1963; Martin, 1966). Jolly attributed this to the disruption of the natural habitat by early man, partly as a result of the newly discovered control over fire and partly due to the increased efficiency of his hunting methods. The large, slow moving species which became extinct at this time would have been easier prey for emergent man than the smaller, faster species, which have survived to this day (Martin, 1966). The remains of large numbers of theropithecines have been found at some early hominid camp sites (e.g., Olorgasaillie in Kenya). Thus, it seems that the human penchant for hastening the extinction of our fellow animals may have an ancient origin.

As the last surviving member of a formerly wide-spread genus, the gelada baboons are of particular interest from the ecological point of view. It is the only primate which is a true graminivore, capable of competing successfully with the ungulates on the open grassland. Furthermore, Jolly (1970) has argued that the anatomical and dental differences between the gelada and the *Papio* baboons closely parallel those between man's ancestors and the apes. He was led to suggest that the gelada might provide an heuristically more fruitful model for investigating the evolution of human behavior and social organization than either the apes or the *Papio* baboons which anthropologists have hitherto favored. Irrespective of the validity of this comparison, the gelada possess a complex social system which is of considerable interest to social ethologists.

II. Distribution and Natural History

A. Distribution

Figure 2 summarizes the present distribution of *T. gelada*. The locations where gelada are known to occur lie exclusively within what is often known as the Ethiopian Amhara plateau, an area of volcanic uplift lying north and west of Addis Ababa. These highlands, which slope roughly northeast to southwest from an altitude of 4500 m down to around 1500 m, are dissected by extensive gorge systems, the deepest of which (e.g., the Blue Nile) may be as much as 1500 m.

The present distribution of the gelada is somewhat anomalous, since no

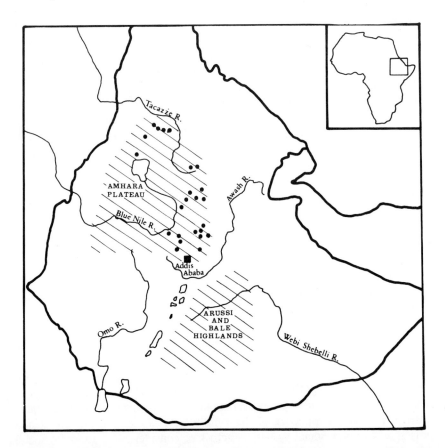

Fig. 2. Locations where *Theropithecus gelada* are known to occur in relation to the main mountain massifs in Ethiopia.

populations are found in the Arussi and Bale highlands which lie east of the Rift Valley to the south of the Amhara plateau (see Fig. 2). Both of these areas are similar to the latter in altitude, climate, vegetation, and topography and would seem to be ideal habitats for the gelada. Since fossil records indicate that theropithecines abounded both in the Omo Valley near the Kenya border and in the Afar to the east of the Amhara plateau during the Pliocene and Pleistocene (G. Eck, C. Jolly, personal communications), the absence of gelada from the highland areas to the south is surprising. We do not know for certain whether gelada were ever found in the Bale and Arussi mountains, but the suitability of the habitat suggests that they could have occurred there at some time in the past. Indeed, Stark and Frick (1958) quote native reports that they did once occur there, although such reports must be treated with caution. If gelada did occur in this area in the past, then the entire population has, for reason or reasons unknown, disappeared.

B. Ecology

The Amhara highlands may be classed as montane grasslands, although much of their present-day vegetational profile reflects extensive destruction of the habitat for agricultural purposes over the past few centuries. Relict patches of *Acacia* woodland and *Juniperus*, *Podocarpus*, and *Olea* forest testify to the fact that many of the better-drained areas once supported a luxuriant tree cover. Open grasslands must still have been common, however, both on higher ground above the tree line and on well-watered plains at lower altitudes where the drainage is often poor. Tree cover, of course, would always have been poor on the steeper gorge sides, although extensive patches of forest must have occurred on the wider shelves where they may still sometimes be seen today. The gorge sides were thus probably predominantly grass covered with bush cover being common wherever the soil was of sufficient depth to support it.

The gelada are found almost exclusively on these slopes. Our studies have yet to reveal an area where the gelada range further than 2 km inland from the gorge rim (Crook, 1966; Dunbar and Dunbar, 1975). Although the gelada commonly forage on grasslands on the plateau top, they depend on ready access to the steep gorge sides both for sleeping sites and for refuge from predators. This has probably always been the case, since the gelada are singularly ill-adapted to an arboreal way of life (Jolly, 1972). Even where they occur in or near forested areas, they only rarely climb into trees. Indeed, their clumsiness when climbing into and out of trees is striking when compared with the more arboreal *Papio* baboons.

The gelada's dependence on gorge sides probably accounts for their survival as a species while their congeners in eastern and southern Africa have followed the road to extinction. These gorges have not only provided a refuge from the incursions of man and protection from the more conventional predators, but also a habitat to which the species is particularly well adapted.

Some 90–95% of the species' diet consists of the leaves, roots, and seeds of grasses; the remaining 5–10% is made up of the leaves, fruits, and flowers of a variety of herbs, shrubs, and bushes and the very occasional insect (Dunbar, 1977). Figure 3 shows part of a gelada herd feeding on grass roots during the dry season. Their success in exploiting a habitat where grasses are the most abundant form of vegetation is indicated by the fact that they usually account for a large proportion of the herbivore biomass throughout the highlands. A typical section of gorgeside, for instance, is capable of supporting three times the numbers of gelada as it is of the more omnivorous *Papio anubis* baboons (Dunbar and Dunbar, 1974).

This dietary specialization is reflected in a number of facets of gelada ecology. First, the gelada are able to maintain exceptionally high densities for an open-country primate, 70–80 animals per km² being typical. Even in optimum habitats, *Papio* baboons may reach maximum densities of only 30 to 35 animals per km², while the desert baboon, *Papio hamadryas*, typically averages only 1.8 animals per km² (Kummer, 1968). Second, since the gelada are almost totally

Fig. 3. Part of a gelada herd feeding in grassland at Sankaber in the Simien Mountains.

dependent on grasses as a source of food, the number of animals in a given area is directly dependent on the quantity of grass available. Our data indicate that 1 hectare of grass cover available in the dry season is able to support approximately 2.5 gelada. Third, since grass tends to be rather evenly distributed (particularly in comparison with fruiting bushes and trees), the animals need to travel only relatively short distances each day in order to find sufficient food. Isolated one-male groups may travel on average as little as 600 m a day, but even large herds of 300 to 400 animals very rarely travel more than 2 km. More detailed discussions of these aspects can be found in Crook (1966), Crook and Aldrich-Blake (1968), and Dunbar and Dunbar (1974, 1975).

C.   Social Organization

It is not necessary in the present context to discuss in any detail the social organization of the gelada: a detailed account can be found in Dunbar and Dunbar (1975). A brief summary of the main features, however, may be given in order to clarify some aspects of the following discussion.

The gelada population is divided up into bands of some 30–250 animals. Each band consists of a number of one-male reproductive units and 1–3 all-male bachelor groups. Each band occupies a home range which overlaps to varying extents with the ranges of neighboring bands. Combined herds of up to 600 animals occasionally form in these areas of home range overlap. The band is held together by a centripetal tendency among the constituent units. The reproductive units of the band probably share a common recent ancestor, and familiarity and kinship bonds give the band some degree of cohesion. Nonetheless, individual units may sometimes leave the herd and move alone.

Within the reproductive units themselves, the females form the core. Indeed, to a large extent it is the strength of the bonds between the females that keeps the unit together. The male, although a conspicuous member of the unit, holds remarkably little real power. Even if he is removed, the unit remains together as a cohesive social entity under the leadership of one of the females. The bonds between individuals are expressed through a variety of friendly behaviors such as grooming and contact calling. In fact, the gelada have an unusually large and complex behavioral repertoire. A large proportion of these behaviors are related to friendly interactions and provide the fundamental means whereby the social group is held together.

The young males leave their natal groups as subadults to form all-male bachelor groups. Later, as young adults, they return to the reproductive units in search of young females with whom to form their own harems, although not always within the parent band. The young males acquire females in one of two

ways: either they attack and defeat a unit leader and take over all his females,
or they join a unit as a follower to build up bonds with one or two of the units
younger females. In the latter case, the unit, for a time, becomes a two male
team until fission ultimately gives rise to two independent units.

III.   Status

Unlike the *Papio cynocephalus* superspecies baboons, the gelada are not
hunted or trapped to any significant extent for commercial purposes, nor have
they been in the past. This can largely be attributed to the apparent preference
of medical and other researchers for the more plentiful and easily obtained
Papio and *Macaca* species. An unknown, but probably small, number are shot
by farmers, ostensibly for crop raiding, and some immatures are caught and
sold as pets in the larger towns. These factors probably have an insignificant
effect on the population size.

There are, nonetheless, a number of causes for concern. First, large numbers
of adult males are shot in the southern part of the species' geographical range.
Second, there is extensive destruction of the natural habitat for agricultural
purposes as a result of the ever-increasing demand for land as the human
population increases. Before embarking on a discussion of these two points,
however, some attempt to estimate the numbers of gelada in existence seems
desirable.

A.   Population Size

Any attempt to estimate the total number of gelada baboons in Ethiopia is
hampered by the topographical structure of the habitat. A simple estimate
based on the overall area of the highlands and the typical densities found would
be totally misleading, since the animals occur only in the vicinity of gorges of
sufficient depth. However, two estimates of the population size are available
and at least provide minimum and maximum figures.

During the course of a recent aerial survey of livestock and land use on the
Amhara plateau, Watson *et al.* (1973) censused baboons. On the basis of a strip-
census technique, they estimated the existence of a total of 440,000 baboons on
the highlands. The authors observe that this figure is almost certainly an underestimate since they experienced great difficulty in observing baboon groups
unless they were moving. A second problem is that they did not distinguish
between *T. gelada* and *Papio anubis*, both of which occur on the plateau. However, the numbers of anubis baboons on the plateau are probably small relative
to the numbers of gelada, and the fact that the authors remark that the majority
of baboons they observed were on or near gorges suggests that the quoted

figure can probably be taken as a reasonable lower limit on the numbers of gelada.

A second estimate can be derived using the Watson *et al.* (1973) data on land-type areas in conjunction with the density of gelada observed by us during the course of our field studies. Watson *et al.* (1973) estimate that there is a total of 44,200 km² of gorges, escarpments, and broken mountains on the Amhara plateau. Our field studies in a number of areas indicate that population densities vary between 15 and 70 animals per km² (see Table II).* However, overall densities throughout the Amhara plateau are probably no greater than 20 animals per km². First, much of the plateau is under intensive cultivation where population densities tend to be low. Second, the land types identified by Watson *et al.* (1973) include large sections of habitat not normally utilized by gelada (S. Sandford, personal communication). This is the case not only in respect of the larger gorges which have relatively wider floors than those on which our density estimates were based, but also those sections of plateau top which are too far from a cliff line to be accessible to gelada. A density of 20 animals per km² would thus represent a maximum figure; the true overall density could well be considerably lower. These two figures, however, would indicate a maximum population size of 884,000 animals.

These two figures, then, provide an upper and a lower limit on the numbers of gelada baboons that exist today. Bearing in mind the reservations regarding the accuracy of both estimates, we can probably say that the total population size is in the region of 600,000 animals.

Thus, the numerical size of the gelada population gives no immediate cause for concern over the survival of the species. However, as already mentioned above, two aspects of the general situation are disturbing, and it is to a discussion of these that we shall now turn.

B.   Shooting of Males

Every 8 years the local Galla tribesmen in the southern part of the plateau (Shoa, Gojjam, and Wollo provinces) shoot large numbers of males for their capes. These are used to make "lion-mane" headdresses and capes which are worn on ceremonial occasions by the men. The origin of this practice appears to lie in the ubiquitous custom of the pastoral peoples of eastern Africa that a young man must demonstrate his manhood by killing a lion. The Galla in this area are the settled, agricultural descendants of the nomadic hordes that invaded the Ethiopian plateau from the south in a series of waves during the

---

*These densities are based on the total area sampled in each case and include sections of habitat not utilized by gelada. The figure of 70 to 80 animals per km² given in Section II,B. is based on the ranging areas only and reflects the carrying capacity of the home ranges.

sixteenth to eighteenth centuries (Greenfield, 1965). In their newly occupied lands, it seems, lion, if not already in short supply, were soon exterminated. It was perhaps natural that, as the lion headdress came to be more a form of ceremonial dress than a proof of male prowess, the Galla should turn to the more easily obtained (and less dangerous) gelada for a source of "lion" manes.

As a result of this, large numbers of adult males are removed from the breeding population. In fact, the life expectancy of males in these southern areas may be little more than 10–12 years, compared with the 20 or so which animals probably achieve in the less disturbed areas to the north. Consequently, the proportion of adult males in the southern population is commonly very low (Crook, 1966). This can be seen from Table I which gives the compositions of three populations: the Sankaber area of the Simien Mountains in the north where shooting does not occur, and the Bole Valley and the upper Kassam gorge in the extreme south where it does. The following age–sex classes are distinguished: mature males (ca. 8 + years), young adult males (6–ca. 8 years), subadult males ($3\frac{1}{2}$–6 years), adult females ($3\frac{1}{2}$+ years), and immatures (less than $3\frac{1}{2}$ years old). It can readily be seen that the proportion of adult males in the population is substantially lower in the two southern areas where shooting occurs than in the Simien. In fact, only one adult male (and a young adult at that) was found in the Kassam gorge population.

The immediate effect of this is that reproductive units are substantially larger in the south than in the north. The mean size of the 68 reproductive units in our Simien study area was 10.0 (range 3–23), whereas at Bole 1 year after an age change the mean size was 17.1 (range 8–28). Likewise, Crook and Aldrich-Blake (1968) found a mean group size of 15.1 (range 7–30, $N=31$) at Debra Libanos in the south 2 years after the previous age change. In Simien,

TABLE I

Comparison of the Compositions of Populations of Gelada Baboons in Areas Where Shooting Does and Does not Occur

| | Percentage animals in population | | | | | |
	Mature males	Young adult males	Subadult males	Adult females	Immatures	Total animals
Sankaber[a]	9.1	2.9	10.9	34.5	42.5	339
Bole[b]	2.2	5.5	7.2	32.6	52.5	181
Kassam[b]	—	0.7	5.1	(94.2)		136

[a] No shooting.
[b] Regular shooting.

86.5% of the units were led by mature males, compared with only 40% at Bole. Indeed, one of the Bole units was led by a subadult male, and Crook and Aldrich-Blake (1968) also report a case of a unit led by a subadult male in the Debra Libanos area. Hence, in these areas, a small number of males are making a disproportionate contribution to the species' gene pool.

The long-term effects of this is that selective factors other than natural processes are determining the breeding capabilities of the males. In fact, in areas where shooting is common, some young subadults may be contributing heavily to the reproductive output of the population. We can only speculate about the effects that this may have on the species as a whole, both in terms of its fitness and its morphological appearance. Presumably, for instance, the generally disastrous effects of inbreeding will be more serious. This phenomenon might also in part explain the apparent existence of two subspecies, namely, *T. gelada gelada* in the north and *T. gelada obscurus* in the east and south. The distinction between the two subspecies is considered, at least by some more recent authorities (e.g., Hill, 1970; Jolly, 1972), to be of questionable validity. The distinction was in part based on differences in pelage color in mature males with *T. g. obscurus* being darker than *T. g. gelada*. However, among males there are marked changes in pelage coloration over age, with younger males having darker capes than older males (see Dunbar and Dunbar, 1975). Hence, the fact that males from the south tend to be considerably younger than those in the north may in part account for the supposed subspecies differences.

In addition to the purely morphological effects, the periodic removal of males has a marked effect on the animals' social behavior. Such shooting occurred in the Bole Valley in 1971. During our study there in 1972, we found the population to consist of a small number of very large units (mean size of seventeen animals) with a few small all-male bachelor groups. Attacks on reproductive units by all-male groups were extremely common, perhaps ten times more common than they were in the Simien. Two years later the mean group size had dropped to 12, the number of reproductive units had almost doubled, and attacks on reproductive units were less frequent. One particular male, for instance, maintained a harem of eight females from May 1971 until July 1972. During the next 2 years he lost all but three of his females to a number of bachelor males. It seems that following the shooting that occurs when the Galla have an age change, the few remaining males in the population divide up the leaderless harems among themselves, perhaps by combining two or more units. During the succeeding years, the juvenile and young subadult males mature: consequently, there is intense competition for females. Since large units are unstable, the harem-owning males lose their more peripheral females to these males as time passes. Ultimately, the imbalance in the composition of the population is probably reduced until the population begins to resemble those in undisturbed areas in structure and organization. By this time, however, the next age change year has come round, and the population is once again disrupted. It is consequently unlikely that any of the gelada populations in these provinces can be considered as typical of the species from the behavioral point of view.

C.   Habitat Destruction

The second factor which is having a serious effect on the gelada is the destruction of the habitat over the past few centuries. This destruction has been of two main types.

First, almost all the natural forests on the Amhara plateau have been destroyed within recent times (Brown, 1972). Mostly this has involved the cutting down of forest trees by the local people for construction material and fuel and to create more ploughland. Because of the subsequent lack of timber, eucalyptus has been planted in many areas to replace the natural forest since this fast-growing tree provides a more rapid return on investment. The use of this species has a number of unfortunate consequences which result in further deterioration of the habitat (Gebregziabher, 1974). Since the species most commonly used (*Eucalyptus globulus*) is shallow rooted, it absorbs much of the ground water, thereby preventing the growth of a herb layer. This is exacerbated by a tendency for the trees to be planted as close together as possible and by the fact that the trees continue to grow during the dry season at a time when the habitat can least afford to surrender excess quantities of ground water. Consequently, there is no ground level vegetation to absorb the rainfall and runoff is heavy, taking with it large quantities of top soil. Indeed, the long, pointed eucalyptus leaves actually increase top soil loss by turning even fine rain into large drops which fall unchecked onto the bare ground at high velocities (Gebregziabher, 1974).

The second form of habitat destruction involves intensive cultivation of grass- and scrubland. The highlands are densely populated and pressure on the land is extremely high; indeed, Watson *et al.* (1973) estimated that all the usable land on the highlands is already under cultivation. As a result of this, the local peasants are ploughing increasingly steeper slopes every year (Fig. 4). In fact, in certain areas, some of the slopes now being ploughed are too steep to permit the use of oxen and ploughing has to be done by hand (Fig. 5). Thus, with the ground level vegetation removed, soil erosion, both from heavy rain during the wet season and from wind during the dry season, is considerable. It is estimated that the rate of top soil loss in some parts of the Simien is of the order of 1 cm per year: this means that bedrock will be reached in less than 50 years (Gilbert, 1974; Klotzli, 1975). In many of these areas the possibility of reclamation is very small since ploughing invariably continues until the ground is so eroded that crops will simply no longer grow. By then the natural vegeta-

Fig. 4. Ploughing on steep slopes in the Simien Mountains.

tion also has difficulty in growing and the possibility of regeneration becomes less likely as time passes. In addition, these problems are exacerbated by overgrazing, mainly by herds of cattle and sheep (Klotzli, 1975).

Of course, the top soil loss has been going on for millennia: it was precisely this soil, washed down by the Blue Nile and Tacazze rivers, that permitted the ancient Egyptians to inhabit and cultivate the otherwise inhospitable Nile delta. What is more critical today is the *excess* quantities of soil which are now being lost due to the removal of the vegetation which would normally retain much of it.

As the natural grasslands are destroyed, the gelada are forced to make increasing use of those parts of the habitat which are too steep for cultivation. These areas tend to support a poorer grass cover, and, indeed, may to a large extent, consist of bare rock faces. Since these areas are able to support much lower densities of animals than the richer areas on the plateau top, the overall numbers of gelada must decline. In part this effect may be mitigated by the baboons' ability to coexist with man to some extent. In many agricultural areas, gelada are able to glean a passable existence in the dry season by picking over recently abandoned threshing floors (Crook, 1966). However, the overall quantity of food resources available in these habitats is naturally markedly

Fig. 5. Hand ploughing on slopes too steep for the use of oxen.

reduced, and the number of animals such areas can support must be substantially lower than in undisturbed areas. Being dependent on the availability of grasses, the gelada are less adept at coexisting with man than the more omnivorous *Papio* baboons. The effect of this can readily be seen from Table II which gives data on population density for a number of localities. Of these areas,

TABLE II

Comparison of Population Densities of Gelada Baboons in Agricultural and Nonagricultural Areas

Locality	Habitat conditions	Number of animals	Area sampled (km²)	Density per km²
Sankaber	Undisturbed	762	11.0	69.3[b]
Geech (East)	Undisturbed	430	11.9	36.1[b]
Bole	Disturbed	243	8.4	28.9
Geech (West)	Heavily cultivated	112	4.9	22.9[b]
Debra Libanos[a]	Heavily cultivated	ca. 300	19.0	15.8

[a] Based on Crook and Aldrich-Blake (1968).

[b] These values differ from those given in Dunbar and Dunbar (1975) due to the fact that the ground scale given for the aerial photographs was incorrect. The correct scale has been used in the present case.

Sankaber and the eastern section of the Geech plateau in the Simien Mountains were relatively undisturbed. Although the Bole Valley itself did not include any cultivation, the plateau top bordering the gorge was extensively cultivated. On the other hand, both the western section of the Geech plateau and the Debra Libanos area were heavily settled with considerable cultivation and widespread habitat destruction and erosion. It can be seen that the relatively undisturbed areas were able to support between two and three times as many gelada as the more heavily disturbed areas. The particularly high density found at Sankaber reflects the fact that the area was somewhat unusual in that it consisted of a narrow ridge less than $\frac{1}{2}$ km wide. The gelada were consequently able to move easily from one side to the other without being far from a cliff line. At Geech, on the other hand, the plateau was some 2 km in width, and a large section of the interior was never entered by the gelada. Thus, in disturbed areas, densities are typically in the region of 15–20 animals per km^2, while in undisturbed areas they exceed 30 animals per km^2.

This effect can be directly attributed to the destruction of the natural grasslands in areas under intensive agricultural use. It may also be exacerbated by a tendency for farmers to trap and shoot crop raiders whenever the opportunity arises. (Farmers are currently permitted to shoot crop raiders providing they report the numbers killed to the local authorities.) Crop raiding by gelada, however, is probably less significant than that by the more destructive *Papio* baboons, who are also known to kill livestock in some areas (Aldrich-Blake *et al.*, 1971). Unlike the latter, gelada show relatively little interest in cultivated fields until harvest time approaches when the crops are in an advanced stage of seeding. In the Bole Valley area, for instance, we have observed *Papio anubis* baboons feeding voraciously in fields of half-grown crops while gelada fed on grass around the edges of the fields, apparently oblivious to the presence of the crops only a few meters away. However, once harvesting and threshing are complete, the gelada are often able to dig over the harvested fields and abandoned threshing floors unmolested (Crook, 1966; personal observation).

Nonetheless, gelada suffer considerable harassment from farmers in areas where cultivation is so widespread that little undisturbed land surface is available to them. In such areas, the dry season may impose conditions of severe food shortage on the gelada and they may have little alternative but to forage in and around cultivated fields. Indeed, in some of the more densely populated areas in the south of their range, crop raiding by gelada has reportedly reached such proportions in the past few years that the Ethiopian Government Wildlife Conservation Organization has received a large number of complaints from farmers. As a result, the Organization has been obliged to shoot a number of animals in these areas and to permit farmers to shoot crop raiders. To some extent, these increases in crop raiding may be due to the drought which has badly affected this part of the countryside, in particular, during the past few years. The height of the dry season coincides with the period around harvest time. Consequently, in those areas where habitat destruction has been extensive, the unusually dry conditions experienced during drought and semidrought years may lead to a dramatic increase in crop raiding by gelada.

Elsewhere, where substantial sections of gorge side remain uncultivated, the animals will commonly respond to harassment from farmers (usually in the form of stone throwing) by retreating to the more precipitous and inaccessible parts of the gorge. In contrast, *Papio* baboons will often continue to raid fields despite the efforts of the farmers to prevent them from doing so. The tendency for gelada to retreat down the gorgeside can readily be illustrated in the case of the Bole Valley. During most of the year, the gelada in this area range on the steep upper sections of the gorge and on the plateau top (Dunbar and Dunbar, 1974). However, during that period of the year when the crops are growing, the animals tend to make more use of the lower 300 m of gorgeside. This can be seen from Table III which gives the frequencies with which individual units were observed on the upper and lower 300 m of the gorge during visits to the area in June, September, and October 1974. June corresponds to the period of ploughing on the plateau top prior to seed planting in July once the rains have started; at this time of year, the baboons are not harassed by the farmers. During September, the last month of the rains, most crops are in an advanced stage of growth with seeding due to occur by October and harvesting by December; throughout this period, the baboons are harassed continuously. It can be seen from the Table that of the 17 sightings of units in June, only two (11.8%) were on the lower slopes, compared with 18 of the 28 sightings (64.3%) in September, and 11 of 26 sightings (42.3%) in October. The tendency for the gelada to spend a great deal of time low down during September can in part be attributed to the heavy mists which cover the upper cliff

TABLE III

Frequencies with which Gelada Units Were Observed on the Upper and Lower 300 meters of the Bole Valley Before and During the Crop-Growing Season

Month	Number of sightings		Percentage on lower slopes	Agricultural conditions	Climatic conditions
	Upper 300 m	Lower 300 m			
June	15	2	11.8	Ploughing	Dry season
September	10	18	64.3	Crops growing	Wet season, heavy mist on upper slopes
October	15	11	42.3	Crops growing	End of wet season; no mist

faces during the rains. However, since the gelada continued to remain low during October when mists did not occur, it is clear that the animals are responding in large part to the harassment meted out to them by the farmers on the plateau top.

A second effect of increasing cultivation is that *natural* populations of gelada will gradually become more difficult to find, let alone study. This is perhaps less serious in terms of the survival of the species per se than in terms of the survival of the species under natural conditions where valid studies can be made of its ecology and behavior. From the scientific point of view, the latter is only marginally less serious than the former. While the anatomist and the physiologist are only rarely concerned with either the origins of their animals or the conditions under which they live, the ecologist and ethologist cannot expect to make useful contributions to their sciences unless studies of natural, undisturbed populations can be undertaken. Studies of captive and semicaptive groups have repeatedly underlined the dangers of such studies in the absence of adequate comparative data from free-ranging populations occupying habitats which are relatively free from human interference.

IV. Conservation

Wildlife conservation is a relatively recent phenomenon in Ethiopia. A separate government department (the Wildlife Conservation Organization) was established as recently as 1965 to deal with these problems. Up to the time of writing, only two National Parks had officially been created, although a number of other areas had been earmarked as suitable conservation areas and five of these are due to be gazetted as National Parks in due course.

It is fortunate that of the two established parks, one, the Simien Mountains National Park, contains a substantial population of gelada living in some of the least disturbed habitat on the Ethiopian plateau. However, this is likely to remain the only conservation area for gelada baboons. The species has also been placed under total protection, and may not be hunted or trapped except under special license for educational or scientific purposes, although, as already noted, the shooting of crop raiders is permitted. Although the future of the species would thus seem to be reasonably assured, there are a number of causes for concern

First, since wildlife conservation is a recent innovation in Ethiopia, there has been considerable antagonism toward it from the local people who view the establishment of National Parks as an encroachment on their traditional land rights. This has particularly been the case on the Amhara plateau where good agricultural land is now at a premium and, due to the use of outmoded and inefficient farming techniques, is becoming increasingly difficult to find. The laws regarding land use and land rights are complex, thus exacerbating the inefficiency of the agricultural methods which the highly conservative and traditional peasantry practice (Gebregziabher, 1972). There remains the considerable risk, therefore, that, as pressure on the land increases, public opinion might force the government to remove the gelada from the protected list, perhaps permanently. The equally unique and almost certainly less plentiful hamadryas baboon (*Papio hamadryas*), for instance, is classed as vermin and may be killed or trapped without license. The reason for this apparently lies in its reputation as a crop raider. However, since this baboon inhabits semi-arid steppe country, it is questionable as to just how much crop raiding it actually does except in those areas on the periphery of its range which border on the more heavily populated and agricultural Amhara and Arussi highlands.

There is, consequently, the disturbing prospect of the gelada being confined to a single small enclave due to the encroachments of settlement and agriculture. This is perhaps inevitable, and in realistic terms in the long run the establishment of efficiently run National Parks must be viewed as the chief means whereby the protection of particular species and their habitats can be ensured. While there will, no doubt, always remain populations of gelada outside the Simien Park, these will inevitably be either heavily disturbed or in areas which are completely inaccessible. From the present point of view, there is considerable interest therefore in maintaining the Simien Mountains National Park as a viable conservation area.

This point leads into the second problem. Even though the local people have different names for each of the three baboon species in Ethiopia (the gelada, the hamadryas, and anubis baboons), they do not always differentiate clearly between them. Baboons, in general, are regarded by most people as destructive pests which must pay the price of all crop raiders. In point of fact, the gelada are protected in the Simien Park more by virtue of the fact that the Park exists to protect the last remaining population of the Walia ibex (*Capra ibex*) than by virtue of the fact that they are a species which itself requires protection. The likelihood that the Park may be abandoned if the Walia ibex become extinct is an ever present possibility which must be borne in mind.

Both of these problems might be mitigated by intensive educational effort. However, such effort is confronted by the third major problem, namely, the purely practical consideration of the topography of the Amhara plateau. The extensive gorge systems, some of which may be as much as 1.5 km deep, make the terrain exceptionally rugged and hamper communications to a considerable extent. The Blue Nile and other large gorges, for instance, have acted as natural barriers to communication between the various sections of the plateau throughout the history of Ethiopia and, indeed, they have played a considerable part in the country's political history (Greenfield, 1965). These problems have considerably hampered road building activities and even today there are only a few roads accessible to motor vehicles in an area of over 100,000 km². Away from these roads, access to the interior is possible only by mule and is sometimes even limited to travel on foot. Until recently, for instance, access into the Simien Mountains National Park itself has been possible only by mule caravan, although a dry weather, four-wheel drive track has now been built. These considerations clearly make it difficult for wildlife protection measures to be enforced since the adequate patrolling of any part of the country is difficult. It likewise makes it extremely difficult to provide practical aid of any kind, and, hence, educational programs, whether connected with improving agricultural techniques or disseminating conservation concepts, become arduous undertakings.

Finally, there is the question of financial resources. Like many African countries, Ethiopia's economy is primarily agricultural and the Gross National Product is consequently low. With pressing demands for more extensive educational and health facilities and the general improvement of the human condition, the country can ill-afford to devote large sums of money to wildlife conservation. Indeed, a sizable proportion of the money spent on conservation in Ethiopia is derived from international sources.

V. Conclusions

Although the status of the gelada provides little cause for concern at the present moment, we should not dismiss the species as one which requires no further attention from the conservationist. The distribution of the gelada is exceptionally limited, and it would be a great pity if the species were to go the way of its congeners simply due to an excess of optimism and a concern for other species which are in greater danger of extinction. The gelada are highly specialized and are probably unable to withstand the kind of encroachment from humans with which many of the baboons and macaques have been able to cope.

Three immediate recommendations come to mind. First, the Ethiopian Wildlife Conservation Organization requires material support to help with the management of the Simien Park. Second, a considerable improvement in the situation could be made by a greater effort in public education in conservation. This applies not merely to the conservation of particular species of animals, but to the conservation of the habitat. In part, at least, this must involve the introduction of improved agricultural methods and forest management techniques (see Gebregziabher, 1974). Finally, the protection of the species would be greatly improved if a second area were reserved for its protection. In point of fact, this need not involve the creation of a new Park or Game Reserve, since it might in any case be difficult to find a suitable area within the species' present geographical range. An easier solution, and one which the Wildlife Conservation Organization in Ethiopia has already adopted in the case of some endangered species, might be to translocate a breeding nucleus of between 100 and 200 animals to a suitable area which is already a National Park. The obvious candidate for this is the Bale Mountains Proposed National Park in the highlands to the southeast which is due to be gazetted as a full park in due course. Although this area is currently outside the known distribution limits of the gelada, the habitat is very similar topographically, climatically, and vegetationally to that on the Amhara plateau. This area has the additional advantage that the human population density is considerably lower than it is in the gelada's present geographical range. Furthermore, the people in this area are primarily pastoralists rather than agriculturalists, and, as we have noted, it is the habitat destruction created by intensive cultivation that is most damaging to the gelada.

Acknowledgments

I am grateful to the National University, Addis Ababa, the Ethiopian Government Wildlife Conservation Organization, and the Bole Valley Society for their cooperation and support. The fieldwork was made possible by grants from the Science Research Council, London, and the Wenner-Gren Foundation for Anthropological Research. I am particularly indebted to my wife, Patsy, for her assistance in the field and in the preparation of the manuscript, and to Dr. J. H. Crook for his support and encouragement at all times.

References

Aldrich-Blake, F. P. G., Bunn, T. K., Dunbar, R. I. M., and Headley, P. M. (1971). Observations on baboons, *Papio anubis*, in an arid region in Ethiopia. *Folia Primatol.* **15**, 1−35.

Brown, L. H. (1972). "Conservation for Survival: Ethiopia's Choice." Haile Selassie I University Press, Addis Ababa.

Cooke, H. B. S. (1963). Pleistocene mammalian faunas of Africa with particular reference to southern Africa. *In* "African Ecology and Human Evolution" (F. C. Howell and F. Bourlière, eds.), pp. 65−116. Viking Fund Publ., New York.

Crook, J. H. (1966). Gelada baboon herd structure and movement: a comparative report. *Symp. Zool. Soc. London* **18**, 237−258.

Crook, J. H. and Aldrich-Blake, F. P. G. (1968). Ecological and behavioural contrasts between sympatric ground-dwelling primates in Ethiopia. *Folia Primatol.* **8**, 192−227.

Dunbar, R. I. M. (1977). Feeding ecology of gelada baboons: a preliminary report.

In "Primate Ecology: Studies of Feeding and Ranging Behaviour in Lemurs, Monkeys and Apes" (T. H. Clutton-Brock, ed.). Academic Press, London.

Dunbar, R. I. M. and Dunbar, E. P. (1974). Ecological relations and niche separation between sympatric terrestrial primates in Ethiopia. *Folia Primatol.* **21**, 36—60.

Dunbar, R. I. M. and Dunbar, E. P. (1975). "Social Dynamics of Gelada Baboons." (*Contrib. Primatol.* Vol. 6). Karger, Basel.

Gebregziabher, T. B. (1972). Soil conservation in Ethiopia. I. Conservation in its ecological perspective—agricultural highlands. *Walia* **4**, 32—38.

Gebregziabher, T. B. (1974). Soil conservation in Ethiopia. II. Some problems of afforestation and forest management. *Walia* **5**, 12—14.

Gilbert, M. G. (1974). Plants in danger. *Walia* **5**, 6—8.

Greenfield, R. (1965). "Ethiopia: A New Political History." Pall Mall Press, London.

Hill, W. C. O. (1970). "The Primates," Vol. VIII, Cynopithecinae. Edinburgh University Press, Edinburgh.

Jolly, C. J. (1970). The seed-eaters: a new model of hominid differentiation based on a baboon analogy. *Man* **5**, 5—26.

Jolly, C. J. (1972). The classification and natural history of *Theropithecus* (*Simopithecus*) (Andrews, 1916), baboons of the African Plio-Pleistocene. *Bull. Brit. Mus. Natur. Hist. Geol.* **22**, 1—123.

Klotzli, F. (1975). Simien—a recent review of its problems. *Walia* **6**, 18—19.

Kummer, H. (1968). "Social Organization of Hamadryas Baboons." Karger, Basel.

Martin, P. S. (1966). Africa and the Pleistocene overkill. *Nature (London)* **212**, 339—342.

Stark, D. and Frick, H. (1958). Beobachtungen an äthiopischen Primaten. *Zool. Jahrb.* **86**, 41—70.

Tappen, N. C. (1960). Problems of distribution and adaptation of the African monkeys. *Curr. Anthropol.* **1**, 91—120.

Watson, R. M., Tippett, C. I., Tippett, M. J., and Marrian, S. J. (1973). "Aerial Livestock, Land-Use and Land Potential Surveys for the Central Highlands of Ethiopia." Mimeo Rept. Ethiopian Government Livestock and Meat Board, Addis Ababa.

12

The Baboon

S. S. KALTER

I. Introduction

Any consideration of the baboon and its immediate future in the wild and in captivity necessitates a review of its present status. Of the vast number of monkeys and apes known to exist throughout the world, the baboon has come to be regarded almost as a distinct entity, separated from apes, on one hand, and monkeys, on the other. The expression, "the baboon" as an animal distinct from "a monkey," is frequently encountered. This somewhat unique status suggests that the baboon possesses qualities and characteristics worthy of more than cursory examination. It may be difficult to imagine that an animal, which today exists as "vermin" or at best a nuisance in its natural habitat, at one time had a role in society akin to that of a deity.

Interestingly, while the rhesus monkey (*Macaca mulatta*) is, by far, the most extensively employed nonhuman primate in biomedical research today, there is probably as much information available on the baboon as there is on any other nonhuman primate used as an experimental animal. In spite of this, the baboon still is not widely used in the laboratory; approximately 4000 per year compared to the 50,000 to 75,000 *Macaca* species are used annually. In 1966, Dr. K. R. L. Hall, University of Bristol, England, stated, "Because of their natural adaptability, it is not surprising that the common baboon has been found by many medical and scientific workers to be better suited for their purposes than other species." In effect, the present indifference to the baboon may be indirectly responsible for its continued presence in relatively large numbers in the wild. However, current restrictions on the use of other simians, especially the macaque, may increase use of the baboon and cause a decimation of its numbers.

II. General Considerations

Baboons have persisted in a prominent role for thousands of years as evidenced by early Egyptian hieroglyphics and art, Greek literature, and continual reference throughout civilized writings until this day. Various tales, many fictitious, have been related concerning human relationships with the baboon. Even today, mention is made of the use of baboons in positions of domestication. How much is truth or fiction is difficult to determine. Those who have studied this animal will attest to its intelligence and potential for accomplishing a variety of ascribed tasks.

One of the more common beliefs is that baboons will attack humans without provocation. Little documentation for this is available, though much may be found regarding "threats" to human by baboons. Wounded or endangered animals present a somewhat different situation. The author was charged by a badly wounded male baboon which made it necessary to kill the animal at a distance of only a few feet. However, over years of field study most experiences with baboon aggression failed to materialize into anything more menacing than threat displays and distant charges. It would appear that much of what is stated in the literature regarding interrelationships between baboons and humans is anecdotal and requires validation.

"Baboon" is a loosely employed term that includes a variety of "terrestrial" simian species. Some of these species, while belonging to the same superfamily as baboons (*Cercopithecoidea*), are considered by most authorities as separate genera. We will restrict this discussion to the genus *Papio* ("true baboons"), including such species as *anubis* (olive), *cynocephalus* (yellow), *papio* (guinea), *ursinus* (chacma), and *hamadryas* (sacred) baboons. Brief mention will be made of more distant relatives of the true baboons, such as the *Mandrillus* (drills and

mandrills) and *Theropithecus* (gelada) baboons. All three genera have characteristics in common, although each has major distinguishing differences: *Mandrillus* have rather short tails; the mandrill (*M. sphinx*) face is multicolored; the drill
(*M. leucophaeus*) has a jet-black mask; the gelada baboon (*T. gelada*) does not
show any pronounced genital changes resulting from the menstrual cycle, but
both male and female geladas have brightly colored bare skin patches on the
chest and neck. It is this skin area that changes color during menstruation in the
female. True baboons (*Papio* sp.) are characterized by size, sexual dimorphism,
and relatively long faces. All females show a pronounced external swelling of
the sexual skin as a characteristic of the menstrual cycle. The *Papio* coat is dense
and males have a well-developed mane. The tail is moderately long and held in
a characteristic U-curve.

A. Taxonomy

Most taxonomic sources agree to classifying baboons into three genera
as given in the following tabulation.

Superfamily
  Cercopithecoidea
    Family
      Cercopithecidae
        Subfamily
          Cercopithecinae
            Superspecies (group)
              Papio cynocephalus
              Genus
                Papio
                Species
                  P. anubis
                  P. cynocephalus
                  P. papio
                  P. ursinus
            Superspecies (group)
            P. hamadryas
              Genus
              Papio
                Species
                  P. hamadryas
              Genus
              Mandrillus
                Species
                  M. sphinx
                  M. leucophaeus
              Genus
                Theropithecus
                Species
                  T. gelada

These designations are all provisional, and since extensive overlapping occurs along with hybridization (especially among the savanna *Papio* sp.), distinct speciation is at times difficult and perhaps empirical. Thus, it is probable that intermediate forms occur and the four species in the superspecies *P. cynocephalus* may be racial variants.

Consideration must be given, however, to the existence, in certain remote geographical areas where reproductive isolation occurs, of distinct "species" in this superspecies. For example, the "chacma" baboon of South Africa appears to be quite distinct from the others. For simplicity, many individuals prefer to use the superspecies or all-inclusive group designation for *P. cynocephalus* thus avoiding the necessity of involvement with varieties or races.

B. Habitat

The various "baboon" species are widely distributed throughout Africa south of the Tropic of Cancer (south of the Sahara desert). Although habitat differences are marked, the baboon has adapted admirably to a considerable range of geographical variations and conditions.

In his extensive studies on the baboon, Dr. Hall indicates that the distribution and habitat of baboons is as follows: *P. cynocephalus* or the yellow baboon is found mainly in woodland and woodland savanna of Mozambique, Tanzania, central Rhodesia, Malawi, forests of Ratanga, eastern Angola, and overlaps with the chacma (*P. ursinus*) of southern Africa. The yellow baboon resides in trees at night and is considered to be the most arboreal of the baboon species.

Papio ursinus or the chacma baboon is found principally in southern Africa including South Africa, Rhodesia, Bechuanaland, West Angola, and southern Zambia. Southern Africa encompasses a wide range of climatic conditions, including snow in certain parts of the mountainous areas with temperatures at times below freezing, temperate regions of forests and shrubs, vast arid areas, and grassy highlands. The chacma baboon will reside in sea cliffs, but can also sleep in trees.

Both *P. anubis* and *P. papio* have similar habitat preferences. Less information, however, is available for the guinea baboon than for the olive baboon. Because of the wide area covered by *P. anubis*, their range is extremely variable and encompasses tropical rain forests, dense forests, tree clusters, and savanna grasslands. In parts of West Africa, some live in cliffs while others may find holes to sleep in. Their territorial distribution includes subdesert and rocky, arid areas. *Papio anubis* has a wider vegetation scope than *P. papio*. Both species will sleep in trees.

Papio hamadryas is found principally in Ethiopia and Somalia. For the most part, the area is arid, consisting of subdesert or wooded steppes with a variable amount of trees. These baboons prefer to spend their nights in steep cliffs.

Both *Mandrillus* species are found in the central equatorial forests of West Africa. A certain amount of overlapping occurs between the two species, with *M. sphinx* preferring the mountains of south Cameroon and *M. leucophaeus* extending into the forest-savanna mosaic where it may come into contact with *P. anubis*. Although both species will climb into trees at night, they ordinarily are terrestrial. Gelada baboons are most frequently found in the high mountains and areas where trees are scarce, but may also be found in the same general area as hamadryas baboons. These baboons usually sleep in caves.

Food and water availability will influence selection of baboon habitation. A major portion of the baboon's diet is made up of vegetation, but it can be extremely variable, consistent with the animal's ability to adapt to local conditions. Arthropods and small animals are included to a lesser degree in the baboon's daily food intake. It is this predominantly vegetarian existence, as well as the ability to live in proximity to man (where it raids planted crops as a source of food), that has made the baboon the nuisance it is considered to be.

C.   Ecological Considerations

Many factors contribute and will continue to influence the baboon's future in its present native habitat. Although it is distributed over most of Africa, the baboon's natural balance is becoming more and more disturbed by man's inroads into those territories originally inhabited by animals. Thus, as man's agricultural and industrial needs expand, the different baboon species are forced to change their natural living patterns with a marked alteration in ecological considerations. Unfortunately, data are not available on all baboon species, e.g., baboons of western Africa have been incompletely studied. Even those species that have been studied in some depth do not provide complete information on regional variations. Nonetheless, the basic ecological needs of baboons are the same regardless of their territorial distribution.

D.   Human Interactions

Dr. G. Bourne's (1974) "Primate Odyssey" provides numerous references to baboons and their historical relationship to humans. It would appear that the baboon is unique among simians in that it was deified by the ancient Egyptians. Carvings and hieroglyphics of the baboon still exist in Egypt, depicting these animals performing various tasks, particularly in relation to the sun god. Curiously, use of the baboon by humans over the years apparently was not diminished as reference is continually made to these animals in paintings, songs, and religious ceremonies from a number of different countries. Baboons have served extensively as pets, but as with other simians, this should not be encouraged.

The ability of the baboon to be a successful scavenger has aided in its developing a capability to associate closely with man. This proximity to humans has resulted in a certain amount of familiarity associated with a corresponding loss of fear. Further, contact with man has permitted development of anthroponoses and an alteration in the disease and infection status of the baboon. Extensive studies concerning the health of the baboon have made it one of the monkey species most understood in this regard. Such studies have encompassed not only laboratory aspects of the baboon's health, but fiield studies have provided data on natural infections of the baboon in the wild. In addition, histopathologic studies of the baboon, both in the wild and in captivity, provide baseline information unique to simians. Recently, comparative immunologic studies have developed a profile on the baboon demonstrating its similarity to man.

The results of these studies indicate that there are no data to support the concept that human contact has resulted in the devastation of baboons, either in Africa or in captivity. Infection does occur, as evidenced by isolation of various microorganisms, as well as the presence of antibody to a wide variety of these agents. Seroconversions following importation may be observed in colonies of animals as a result of contact with their human handlers. It is of interest to note that occurrence of major disease outbreaks, as seen among other species of monkeys and apes following colonization, has rarely if ever been reported. Similarly, shedding of various organisms to any extent has not been reported. Thus, B virus seen in *Macaca* sp., hepatitis virus as carried by newly imported chimpanzees to man, and the occurrence of natural poxvirus (monkeypox, Yaba,; Yaba-like) seen in several different simian species, are generally not observed in the baboon. Measles infection, which may result in a fatal illness in marmosets and in other simians also is not seen in baboons. Yellow fever, reported as devastating certain simian populations, does not appear to be a major factor in baboon fatalities. Limited serology studies have failed to show widespread evidence of antibody among these animals, indicating lack of contact with the virus. This has been interpreted by some investigators as suggesting the possibility of increased susceptibility with high mortality rates since antibody to yellow fever may be found in baboons in geographical areas other than those studied. Three facts mitigate increased susceptibility: (1) large numbers of dead baboons have not been found in areas where yellow fever exists, (2) antibodies in baboons tend to follow the same general patterns seen in human populations in endemic areas, and (3) infectivity studies with large numbers of different infectious agents support the thesis that these animals are more resistant than many of the other similarly studied simians. Whether or not this resistance is a reflection of the baboon's immunologic competence, which may provide some form of protection not seen in the other simian species, is unknown at this time. Both humoral and cell-mediated immunologic capabilities of the baboon closely approach those of man.

If interactions with humans do account for any losses among baboons, it would appear that they would be the result of direct action rather than anything induced, such as human diseases. In many areas, baboons are shot and poisoned. According to most calculations, these activities account for less than 5000 animals per year and no evidence exists to demonstrate to any reduction in baboon numbers where this activity is carried out. In many areas of Africa, the baboon is by far the most conspicuous of the simian species associated with man. As a consequence, it is this genus that receives the brunt of man's ire. It is noteworthy that in South Africa, where the chacma baboon has been killed in large numbers (with the assistance of the government), baboons are now recognized as a valuable asset to biomedical research. The government is now proposing legislation to safeguard these animals for medical research purposes.

E.   Nonhuman Interactions

In considering the future and conservation of baboons, it may be asked if the baboon is subject to predation by animals other than man? Although various mammalian predators have been suggested as dangerous to the baboon, most data do not support this theory. In the author's experience, no evidence has ever been obtained, either directly or indirectly, to suggest that the many recognized predators account for any but an occasional baboon. Even these are usually very young or "lost" animals or those that are debilitated by injury or age. Actually, predators, such as lions and leopards, are frequently run off by attacking baboon males when they approach a troop. Although there are reports (or reviews) that the leopard is responsible for killing numerous baboons, this does not seem to be substantiated in observations reported by B. L. Mitchell (Mitchell et al., 1963) in the Kafue National Park (northern Rhodesia).

F.   Trapping and Holding of Baboons

Current evidence suggests that baboons are still relatively numerous, but there does appear to be some decrease in number in newly populated and cultivated areas. Whether this reflects actual decreases in numbers or the migration of baboon troops to less congested areas is not clear. At this time, use of the baboon for research purposes does not constitute a major threat. Decreased availability of other species such as the rhesus monkey, as well as the developing recognition that the baboon offers many desirable features, will undoubtedly change this present limited usage. However, as a result of practices employed by various commercial trappers for the capture and holding of baboons, large numbers of animals die needlessly. Because of these undesirable procedures, man becomes the major predator of the baboon.

1.   Trapping

Commercial trappers use different methods for capturing baboons, but the baited trap is the most frequently employed. Since baboons travel in troops, it is deemed most expedient to attempt to trap an entire troop by utilizing as many baited traps as possible at one time. This procedure is most effective but it is time-consuming and requires large numbers of traps for capturing the volume of animals required for a commercial operation. This type of trap rarely harms the baboons and, except for the unhygienic conditions and psychological impact, does not cause any recognized major loss among them. Because of their generally unsanitary conditions from one trapping to the next, these traps are instrumental in passing microorganisms from one animal to another. It is at this time that the individual baboon first comes into contact with alien microflora and fauna. Death and debilitation also may result from the prolonged periods of time the captured animals may spend in their traps in the field. The native trappers only give cursory consideration to such necessities as food, water, and cleanliness. It is also of interest to note that the baboon soon recognizes the threat of the trap. If all the animals in a troop are not captured within two to three trap settings, then it is best to remove the traps from the area. The animals not captured will make efforts to destroy the traps or simply not respond to the bait. Since the baboon's eating habits are such that the males have the first choice of food, consideration also must be given to the fact that the firtst trap setting will generally capture the dominant males unless sufficient numbers of traps are used to include both the males and females. Other methods, such as killing adults in order to obtain young animals or use of nets and snares, are not practiced on baboons. Individual natives may resort to snares for purposes of food or elimination of a nuisance.

Large losses among baboon populations still occur at the holding facilities. Here the general level of cleanliness is so poor that even the baboon, with its high resistance to infection, succumbs in great numbers. Although precise numbers are not available, it is highly probable that at least one-half of any group of captured baboons will die either at the holding facility or as a result of being held at one of these facilities. Deaths are usually the result of diarrhea, pneumonia, or tuberculosis. In addition to the unsanitary conditions, malnutrition and dehydration add to the difficulties of the baboon at this stage. Wastage of these animals can, therefore, be considerably reduced by improving conditions in the trapping and holding operation. A number of suggestions may be made that could provide guidelines for improving present operational procedures: (1) trapping and holding should be done by licensed dealers only, (2) veterinary or other qualified personnel trained in public health should be associated with these operations, (3) routine cleaning (and sterilization) of traps and equipment should be requisite, (4) traps and cages should comply with regulations established for holding the animals, (5) different species should be maintained separately, (6) reference serum samples should be collected and maintained immediately following capture, (7) bacteriology (stools) and tuberculosis testing should be performed on each animal, and (8) permanent records should be developed on each animal.

2. Transport

Shipping conditions have been improved considerably, but a number of corrective measures should be instituted which would immeasurably enhance the chances of the animal's survival and the maintenance of a healthy state. To minimize the time in transit, appropriate and careful planning of all shipments is important. These plans should take into consideration mandatory, as well as emergency, stopovers, and necessary arrangements for food and water should be made. Two extremely important adverse elements generally enter into transportation of baboons and other animals. The first is contact with alien species en route, as well as at animal hostels in airports or other transportation centers. While frequently well meaning, these hostels are not equipped to properly handle different animal species arriving at a particular airport from different parts of the world. As a consequence, an interchange of microorganisms from all types of exotic and foreign animal and avian species ensues. Second, temperature changes may be considerable and the chill factor extensive, especially at airports in northern areas. Animals originating from tropical climates are frequently disembarked at European or American airports in the dead of winter within a period of less than 24 hours after shipping. The losses from such actions may be diminished by proper planning.

3. Holding in Importing Countries

In spite of the marked improvements made in most importers' holding facilities, there is much that can be done to further correct current inadequacies. A fundamental need is the development of appropriate training programs for veterinarians in the care and management of nonhuman primates. The present pool of knowledgeable veterinarians is extremely limited and usually committed. There is a continuing need for adequately trained supervisory personnel to maintain proper holding facilities. These limitations result in breakdowns of the procedures required to minimize endangering the health of animals. The primary losses at this stage result from infectious diseases (tuberculosis, diarrhea, and pneumonia) rather than dehydration and malnutrition.

Similar problems are met at the various research centers utilizing these animals. Few animal facilities are physically organized for proper health maintenance of the various simian species. Correction of these centers would require costly changes, frequently considered unnecessary or given a low priority. The animal, regardless of the species, consistently reflects colony management (or mismanagement). Probably because of their overall resistance and weeding out of undesirable animals at the exporters, and in spite of this encounter with extraneous organisms, baboons do not demonstrate a high number of fatalities. However, seroconversions and illnesses do occur and are easily recognized. Facility limitations are a major problem among many other simian species, as noted by the increased deaths resulting from the horizontal transmission of organisms within a colony.

The question is frequently asked concerning advisability of vaccination against infectious agents for prevention of losses and transmission of disease among colony animals. In the baboon, this measure does not appear to be necessary. However, baboons arriving from endemic yellow fever areas in Africa may have the propensity for carrying this virus and, therefore, are vaccinated or quarantined to minimize this danger. A word of caution, however, should be noted applicable to all situations. Yellow fever vaccine is a live virus vaccine and extremely labile. Therefore, its employment is attendant upon appropriate handling and the prevention of inactivation. An extensive study of yellow fever-vaccinated baboons has already shown that, in effect, many of these animals were given inactivated and, therefore, ineffective vaccine.

Other vaccines appear to be unnecessary, although the use of smallpox vaccine to prevent monkeypox may be given some consideration under appropriate circumstances. Experimentally, smallpox virus in newborn baboons does produce fatalities although infections with this virus have not been seen in nature. Antigenically, monkeypox is only distantly related to Yaba and Tanapox viruses and vaccination with vaccinia virus or infection with monkeypox virus will not protect against Yaba or Tanapox viruses. It would appear that the baboon is only infrequently infected with these viruses in nature, although baboons are found in areas where these viruses have been recovered.

Most enteroviruses, such as the coxsackie-, polio-, and echoviruses, produce nonsymptomatic infections usually detected by the development of seroconversions to one or another of these viruses. Poliovirus, known to cause a paralytic disease as well as fatalities among chimpanzees and man, has not been observed to produce similar types of infection in the baboon. Vaccination of baboons against poliovirus, therefore, is not considered necessary. On the other hand, paralytic disease as a result of coxsackievirus has been noted in the human and baboon population in Kenya.

Bacterial vaccines, while available, have not been used to any extent in the baboon or other simian. As in man, the efficacy of these vaccines may be questioned, but could be used where proved to be efficacious. Use of BCG for prevention of tuberculosis has been suggested, but not employed. As with the use of anti-tuberculosis chemotherapy, the possibilities of masking active disease in the colony negates its general usage. Similarly, mycotic and parasitic infections exist in baboons, especially in those living in the wild. Mycotic infection is usually ignored, since fungi and yeasts do not seem to be responsible for any extensive disease process. Parasites of all types are found in the baboon, but most are eliminated by appropriate use of drugs during quarantine. Under clean colony conditions, reinfection rarely occurs and parasites become self-limiting. However, it is not unusual for laboratory personnel to encounter various blood parasites or filariae in cultivated kidney cells.

4. Quarantine

Effective quarantine measures ensure a better laboratory-grade animal. Perhaps most important is the stabilizing effect quarantine has on an animal. Following all that has transpired since its capture, maintenance in one area for a relatively long period of time undoubtedly permits the animal to regain some composure. Quarantine also offers the veterinary staff an opportunity to carefully inspect each animal, to collect appropriate specimens for baseline studies, to make comparisons with previously collected materials, to do a variety of necessary tests (such as blood studies and T.B. testing), and to properly evaluate the animal. Losses of animals, particularly baboons, at this time are minimal. This is not necessarily true for other simians; for example, extensive losses of marmosets occur during quarantine.

A virus shedding study completed in baboons may be of value to all individuals interested in the use of nonhuman primates. In order to ascertain how long animals shed viruses following the stresses of capture and transport, multiple samples were collected on baboons immediately upon capture in Africa and then for 6—12 months thereafter. Baboons "in the bush" were noted to shed viruses infrequently, i.e., 5% of the captured animals were found to be positive (principally adeno / or enteroviruses). Upon arrival in our laboratory some 24—48 hours after capture, approximately 60% of the animals were found to be excreting virus, mainly in the stool. The original levels of 5% virus shedding were not regained again until approximately 6 months later (Kalter and Heberling, 1968). This observation not only points out the inherent dangers of transport to the health of the animals, but emphasizes the need for similar studies on other species of simians.

An effective program of testing and sampling of animals during quarantine has been developed for the baboon. It is adaptable to all simian species, but was primarily developed as a result of experiences with the baboon and chimpanzee. It is recognized that quarantine periods of months may be necessary for the complete restoration of an animal to original field levels of virus shedding. However, practical aspects must be considered and the animals made to "earn their keep." In most instances, continued surveillance will act as a control mechanism and should be pursued. Thus, we have developed a 30- to

40-day quarantine program (shown in the following tabulation) that follows this schedule:

Days 1–3 after arrival:
1. Tuberculin test
2. Blood drawn for complete blood count as well as for reference serum
3. Throat and rectal samples collected for bacteriology, virology, and parasitology studies
4. Anthelminthics administered
5. Institution of a permanent record file on each animal including observation of daily heat cycle

Days 14–17 of quarantine:
1. Tuberculin test
2. Stool samples collected for parasitologic examination
3. Anthelminthics administered
4. Tattooing performed

Days 30–35 of quarantine:
1. Tuberculin test
2. Blood drawn for complete blood count as well as for reference serum
3. Throat and rectal samples collected for bacteriology, virology, and parasitology

Day 35 + :
Animals released from quarantine if there are three consecutive negative tuberculin tests and no other disease problems

---

Continued surveillance and monitoring of these animals in the colony is helpful for long-range determinations.

G. Breeding

After a suitable quarantine period, desirable animals may be placed into breeding groups. Experiences at Southwest Foundation for Research and Education (SFRE), spanning some 15 years of successful breeding of baboons, indicate that this species breeds exceptionally well in captivity. Both timed pregnancies (individual cage mating) and group matings may be achieved with minimum effort. Similar success in breeding large numbers of baboons also has been achieved at the primate center in Sukhumi, in the Soviet Union. Smaller breeding groups have been successfully established in many major research centers.

In the SFRE colony, approximately 25 animals, i.e., twenty to twenty-five females and one or two males, compose a good family group. Some care in selecting animals to make up this grouping is necessary. Housing of these colonies may vary depending upon local conditions. Details for the baboon and chimpanzee breeding facilities at the SFRE have been well documented. Larger "families" of baboons are maintained at the Sukhumi facility, evidently with similar success.

The newly imported female may develop a period of transient infertility.

However, at least 50% of the females will conceive during their first year of captivity. Once the breeding group has been established (generally within a year), each female will produce at least one live infant per year. With careful husbandry, selection of animals, and removal of newborns shortly after birth, two pregnancies per year may be routinely accomplished, since the gestation period for baboons is approximately 185 days. Optimal mating time may be predicted on the basis of sex-skin cycles. Ovulation is estimated retrospectively as the third day preceding deturgescence of the sex skin. Maintenance of careful records, including the cycling of each female, is important for developing an efficient breeding colony. Undesirable baboons should be culled and replaced with more productive animals. Dr. A. G. Hendrickx (1971) has published details on sex skin cycles, menstrual cycles, and individual cage mating of baboons. Briefly, all baboons have an estrous cycle of 30 to 35 days, menses vary from 3 to 7 days, gestation is approximately 185 days, and births occur all year with some peaking in the early fall.

Embryo transfer, recently perfected by Dr. D. C. Kraemer *et al.* (1976), will undoubtedly enhance the productivity of animals possessing special characteristics which make them valuable as models for human disease. This procedure has been highly successful in cattle breeding and permits the transfer of preferred embryos to genetically less desirable animals for development to term.

Experience with genetically selected colonies of baboons is somewhat limited. Available information does not appear to suggest any contraindications. A limiting factor is that sexual maturity for a female takes approximately 5 years. This time period would preclude any desired short-term studies, but could be of value in planning long-range experiments. However, when all factors are considered, breeding of baboons by genetic selection of animals is worthy of consideration.

No difficulties have been encountered in separating the newborn baboon from its mother and raising it either conventionally or under germ-free conditions. For practical routine purposes, the delivery may be natural. The pregnant baboon is removed from the group and individually caged 3–4 week prior to parturition and allowed to give birth naturally. Removal of the pregnant animal immediately prior to parturition is not desirable since the mother may either kill the newborn immediately after birth or allow it to die of neglect. Early removal of the pregnant female permits the animal to acclimate to its new environment and eliminates this problem. Permitting delivery to occur in the colony is generally satisfactory, but may, at times, result in fighting among the animals in that group. Vaginal parturition usually occurs at night and the mother eats the placenta promptly after its expulsion. If prompt removal of the infant is necessary, then "continuous" supervision is recommended. After giving birth, the mother is permitted about 30 minutes to remove the afterbirth and clean the infant; the mother is then sedated and the infant removed.

Cesarean parturition may be performed if the infant is to be raised germ free or if experimental needs require such a procedure.

Postnatal care of the infant is identical to that given a human baby. The infant is dried with a sterile towel, the umbilical cord is then cut 1 inch from the abdominal wall, tied with sterile surgical silk, and disinfected. If necessary, mucus in the respiratory passage is aspirated. The baby, following natural birth, is given an antiseptic bath (Phisohex) only if soiled with fecal material. Prior to placing the infant in a standard hospital incubator, it is given a complete physical examination, measured, and a permanent record made. If the animal is to be raised germ free, the procedure is the same except that the animal is maintained in a germ-free isolator. The cesarean section is performed in a sterile plastic sack, placed over the sterilized abdominal wall of the mother and is attached to the germfree unit. In all instances, humidity and temperature are controlled.

Feeding of infants may vary among institutions, but our experiences have shown that the commercial formulas available for human infants are very satisfactory for baboons. We have found that two parts Similac liquid (Ross Laboratories) to one part water, containing 26.7 calories per 30 ml, was somewhat analogous to baboon breast milk. Some adjustments may be needed on an individual animal basis, particularly if diarrhea or failure to gain weight is noted. The baboon infant adapts rapidly to bottle feeding and generally is "trained" within a week to self-feed. This is in contrast to the chimpanzee infant, which may take several weeks or months to self-feed. Strained foods are given shortly after birth and solid food may be introduced in the form of fresh fruit within a month. When the infant is freely taking water and solid food, formula feeding is discontinued.

III. Problems

The baboon offers little in the way of problems. Its general adaptability to most conditions in the wild, as well as in captivity, has resulted in this species remarkably successful existence. Current practices for reducing baboon populations, at least for the present, do not appear to be causing any major inroads in existing number. However, extensive and well-organized extermination practices associated with expansion of human populations may change this situation. Further, continued need for simians in the laboratory, decreased availability of other species, and an increased recognition of the availability of the baboon for investigations could change the status of the baboon.

Because of their scarcity, the gelada, mandrill, and drill baboons have been little used and should be excluded from general discussions concerning baboons as laboratory animals.

Initial costs for baboons, though somewhat higher than macaques or most of the New World monkeys, are not excessive. Most charges are dependent upon a weight basis and animals may be procured at prices competitive with the rhesus monkey. Maintenance costs for a baboon in captivity are similar to those for most larger monkeys. Caging does not need to be special and those cages that are suitable for adult macaques are satisfactory for baboons.

The size of baboons, which some investigators consider to be a detraction or perhaps a problem, varies among the different species. Mandrills are considerably larger than most *Papio* sp., although the chacma baboon may approximate them. Male chacma baboons may be in excess of 35 kg, with females about one-half that size. The other species will average 22 and 12 kg for male and female animals, respectively. Inasmuch as most handling of baboons is done with sedated animals, their size should not be a major consideration. Further, a great deal of usage involves infants and young adults, which are considerably smaller than full-grown animals. Although most investigators prefer to use a sedated baboon, it is always exciting to see the female animal handlers at the Sukhumi primate facility, capture and hold an unsedated adult baboon (male or female) for an investigator.

When developing breeding colonies, the social structure and hierarchy of the animals brought together to form the colony must be considered. In initiating a colony, and until the animals in that colony become established, they must be carefully observed. Unless food is provided in excess, timid animals may be deprived of access to food containers. Males will fight with other males for dominance, resulting in lacerations, contusions, and occasional fractures. Females, too, will develop a hierarchy by fighting. Generally, deaths are infrequent if proper observation is provided to remove troublemakers. Sufficient space for escape is frequently all that is necessary to minimize major problems. Extraction of canine teeth from obstreperous males will greatly reduce major damage.

Despite the vast amount of available biological information, investigators are generally reluctant to use baboons. The rhesus monkey, for no reason other than historical precedent and in spite of numerous shortcomings, is by far the most extensively used laboratory nonhuman primate.

IV. Uses

The baboon has achieved notable success in a variety of biological areas where it has served as a model for studies of human illnesses. Its similarity and parallelisms to man in many biological parameters make it an outstanding candidate for continuing, and probably expanded, employment. Unquestionably, there are other nonhuman primates that may be singled out as more advantageous in the pursuit of a particular or specific investigation. There are many investigators who wish to use the great apes for research. Limited populations of apes, however, preclude such use. Therefore, it is deemed practical to use the more populous simians, such as baboons, for research purposes.

It is its wide usefulness in a variety of investigations that places the baboon in a coveted position. Further, a number of field investigations have indicated that there does not appear to be, as yet, a limitation on numbers. In fact, several surveys of baboon populations have demonstrated general increases in population sizes. As stated above, increasing population by humans or cultivation for farming in certain geographical sections will result in baboon population shifts more than a decrease in the number of animals.

Contrasting the desirable characteristics of the baboon (availability and similarity to man) with general limitations being placed on other nonhuman primate species may make one examine where and how baboons (*Papio* sp.) have been or may be utilized in the laboratory. Employment of simians in the laboratory must be viewed critically. Two major categories exist in which monkeys and apes are used. The first is in studies in which the nonhuman primate is considered essential or irreplaceable for the successful conduct of the experiments. In the second category are those investigations in which monkeys and apes, although not essential, are considered highly or uniquely valuable.

A.   Essential Category

1.   Testing and Evaluating of Vaccines

Vaccines are tested for their effectiveness and safety in a variety of animal models. At present, insufficient data are available regarding the baboon's usefulness as a model for vaccine testing. However, there does not appear to be any evidence that the baboon would not be satisfactory for such purposes since seroconversion occurs in the baboon in response to poliovirus vaccine. The rhesus (or another macaque) and the African green monkey are generally used for poliovirus (neurovirulence) testing. Interestingly, of the many monkeys and apes susceptible to poliovirus, only the chimpanzee is known to develop infection similar to that of man. Other vaccines such as yellow fever, measles, mumps, and rubella (where the agent may be neurotropic) can be reliably tested in monkeys.

Presently, limited susceptiblity to hepatitis virus is recognized in the chimpanzee and marmoset. However, other species have also been suggested to be susceptible to this virus and it is expected that a hepatitis vaccine may soon be available. Although a satisfactory model for safety testing of such a vaccine is not available, it will unquestionably be tested in some type of monkey or ape.

As indicated, rubella vaccines are generally safety tested for neurovirulence, but a test for production of congenital malformation is not included.

Known variation in susceptiblity among simian species to infectious agents emphasizes the difficulties in developing vaccine safety testing procedures. Model systems, and eventually vaccines, need to be developed which are specific for each infectious agent. Recognition of quiescent disease in test animals also is essential to the satisfactory performance of the test itself.

2. Evaluating New Pharmaceuticals

Numerous drugs and other pharmaceuticals require various types of tests in order to ascertain potency, toxicity, and other physiological activities prior to public release. Unquestionably, there is a definite need for the use of monkeys as a means for predicting potential toxic reactions to new drugs. It is retrospectively recognized by the scientific community that Thalidomide would never have been used had it been tested in nonhuman primates prior to human employment.

Suggested use of cell cultures as a substitute for the intact animal obviously falls short. Cell cultures offer much in the way of a test system for various purposes but this limited system cannot substitute for an intact organism when a host response is needed.

Resistance and response of an organism to various biological or pharmacologic materials is considerably diverse among different animal populations. There are data, however, that clearly demonstrate that the nonhuman primate is preferable to the dog for predicting the metabolic fate of drugs in man. Again, Thalidomide had been tested prior to release in a variety of animals (nonprimates) without any evidence of producing malformations. Current concepts in drug testing in animals suggest that the baboon closely approximates man phylogenetically. There are obviously certain limitations to the concept. In a recent editorial, D. F. J. Ingelfinger (1975) stated, "Considerable interest has developed in using nonhuman primates, especially baboons, in the hope that their parahuman structure and behavior might betoken a drug metabolism and sensitivity closely similar to that of man." In all fairness to Dr. Ingelfinger, he also very carefully points out limitations in the use of baboons (or other animals for that matter) and suggests (at least for the cited studies) that, eventually, tests will need to be made on human, not baboon, subjects.

3. Preparation of Primary Cell Cultures

Cells in culture are a highly efficient mechanism for investigations of a multitude of cellular problems. Cell cultures are recommended for all studies that preclude use of the intact animal but still need a living environment. One application for living cells, generally "monkey" kidney cells, is that of virus diagnosis. Living cells are required for the isolation of viruses from patient material and primary monkey kidney cells manifest the greatest susceptibility of currently available cell systems. It must be emphasized, however, that there is no single, all-purpose, or universally susceptible cell. Many different primary cell systems, including baboon kidney cells, have been examined in the vain hope of finding the one all-purpose diagnostic system. Current procedure in most virus diagnostic laboratories is to use a primary monkey kidney cell line and human fetal lung line. These cells are supplemented by additional systems (including intact animals) if it is warranted.

The question is frequently asked, especially when working with exotic species, if it would not be most appropriate to utilize cells from the host under study. The idea is that the virus will more than likely proliferate in cells from the same host species that was infected. In essence, this would require use of gorilla cells for gorilla viruses, chimpanzee cells for chimpanzee viruses, etc. There are arguments pro and con for this concept. Of course, this concept would not be valid when considering xenotropic, endogenous viruses which will not proliferate in cells from their host of origin.

Other considerations need clarification in use of primary cells, regardless of the host: (1) the presence of unwanted viruses, either in the occult form or as contaminants, and (2) the presence of vertically transmitted endogenous viruses or genomic material. In the first instance, there have been some 50−60 viruses recovered from simian tissue grown in culture. These are usually counterparts of human viruses but are antigenically distinct. Most are not highly infectious as disease production is generally rare. However, several (most notably, *H. simiae* or B virus) are highly lethal to man and other animals. B virus is found only in macaques in nature, principally in the *M. mulatta* (rhesus monkey). There are other viruses that may be recovered from monkey tissue that also have a capacity to infect humans. Another herpesvirus (*H. saimiri*), found naturally in tissues of squirrel monkeys (*S. sciureus*) produces a fatal lymphoproliferative disease in New World monkeys (marmosets, owl monkeys, etc.). Thus far, there is little evidence to indicate any pathogenicity in humans.

The potential of viruses or their nucleic acid in tissue cells is just beginning to be understood, but the significance is still obscure. Recent demonstration in this laboratory of a vertically transmitted (xenotropic, endogenous) type C virus (oncorna- or retravirus) in primates, including humans, has revitalized the quest for a viral relationship to cancer and the potential of such viruses in cell cultures to be used perhaps for vaccine production. We have recently isolated this virus from baboon placenta as well as embryonic and fetal tissue. It is known that type C viruses from other animal species produce tumors. Preliminary studies with the baboon type C virus suggest that it will produce tumors, under appropriate conditions, in a variety of animal hosts including monkeys and apes, as well as baboons.

Vaccine development may be included in this section. The same general concerns expressed above obviously are applicable to the development of any product that will be administered to humans. Regardless of the animal source of cells, the health of the animal must be continuously monitored, sick animals must be diagnosed, and dead animals necropsied to determine the cause of death.

Nonetheless, in spite of the limitations, cells in culture are one of the more important contributions resulting from the use of nonhuman primates in the laboratory. It is highly conceivable that developments in the not too distant future will permit the use of cells maintained in series rather than primary cells for vaccine preparation as well as for the isolation of viruses. Suggestive data on the use of such cells as WI—28 and others emphasize this possibility.

B. Nonessential Category

1. Viral and Chemical Carcinogenesis

It is premature to speculate on the final role of nonhuman primates, least of all the baboon, in the study of viral and chemical carcinogenesis. This area of research is currently too fluid to predict which of the many pathways investigators utilizing monkeys and apes will take in the unceasing search for the cause and eventual cure of cancer.

Less than 10 years have passed since Dr. L. Melendez and his associates (1969) at the New England Regional Primate Center demonstrated that a primate herpesvirus (*H. saimiri*), recovered from its natural host, a normal squirrel monkey (*Saimiri sciureus*), produced malignant lymphomas when inoculated into other New World monkeys. Thus, it was established that primate viruses were able to induce malignant disease in other primates, a fact heretofore only demonstrable in lower animal forms or birds. Approximately 3 years ago, it was shown by the author and his collaborators (Kalter *et al.*, 1973) at SFRE that type C particles, the etiologic agent in neoplasms of various animals and avian species, were also present in primates including man. Though first recognized in baboon placenta, placentas from other primates were observed to harbor this agent. This virus has now been isolated from baboon tissue and studies are underway in a number of laboratories attempting to ascertain whether or not the baboon type C virus, as well as similar agents, will produce tumors. This finding, so clearly demonstrated in baboon tissues, is responsible for the reactivation of interest in the possible role of RNA (ribonucleic acid) viruses in the etiology of cancer. Our laboratory (Heberling *et al.*, 1976) also has been responsible for demonstrating the presence of unrelated (to the baboon virus), morphologically similar viruses in tissues of a New World monkey (squirrel monkey).

Nonhuman primates, including baboons, have been used in many types of cancer research. Naturally occurring tumors, benign and malignant, may be observed in all species of monkey and ape. These tumors are similar to those found in man and other animals and, when malignant, induce a fatal illness in the involved animal. The baboon is recognized to have its share of natural tumors.

Investigators throughout the world have used the baboon in programs designed to produce tumors by chemicals, radiation, and viruses. The success of many of these experiments has clearly shown that the baboon may serve as a model for certain studies. As in other biologic reactions, simians vary in their susceptibilities to one or another carcinogen. In attempting to unravel the mechanisms involved in these species differences, immunologic studies have shown that immunologic competence varies among the studied species. These studies do not suggest that immunologic parameters alone are responsible for susceptibility to tumor development; they do, however, vividly demonstrate that the immunologically compromised host appears to be more highly susceptible. These findings are parallel to those noted in humans and augment the suggestion that part of the host response to cancer-inducing agents is immunologically associated. This appears to explain partially the baboon's resistance to oncogenic agents (as well as other microorganisms) and the high susceptibility of the marmosets to some of these same agents.

A number of the early studies on viral oncogenesis were attempted in the baboon. An avian tumor-producing virus (morphologically similar to the baboon type C virus mentioned above), Rous sarcoma virus, known to cause tumors in birds and rodents, was shown to induce malignancies in the baboon and other simian species. These studies were highly significant in that they demonstrated, for the first time, the ability of tumor viruses not only to cross the so-called species barrier, but to do this between primate species.

More recently, research workers in the Soviet Union (Dr. B. A. Lapin, 1973) claim to be able to produce a "leukemia-like" disease in hamadryas baboons by inoculation with human patient material. While these animals do develop an illness, it is not clear what these investigators are inoculating into the baboons and whether or not the inoculated material is responsible for the clinical picture that develops. The relationship of this syndrome to the possible activation of the endogenous virus, similar to that reported from our laboratory, is currently under investigation.

Chemical carcinogenesis has not been studied as extensively in the baboon as in other species. A number of baboon prostate glands were inoculated with dimethylbenzanthracene (DMBA). Some 3 years later, oncorna-like virus particles were observed (by electron microscopy) in prostatic tissue of some of these animals. No evidence of neoplasia was noted. The baboon had also been used (with negative results) by Drs. J. E. Hamner III of the National Institutes of Health and O. M. Reed of SFRE (Hamner and Reed, 1972) to determine the oncogenic capacity of betel quid ingredients by chronic exposure of the buccal mucosa to these materials.

2. Psychology and Behavior

Human psychological and behavioral patterns are often based upon data obtained from nonhuman primate studies. The baboon has contributed its share in developing knowledge of how humans react to their environment. Extensive studies, both in the field and under captive conditions, provide data on *Papio* sp. behavioral patterns, but adequate sampling has been limited principally to Kenya and South Africa.

Because of the widespread distribution of the baboon, vast differences in ecologic conditions exist and these would be expected to influence the behavioral responses of baboons. Such effects have been noted. Forest and savanna baboons have been compared in their reactions to environmental pressures. Forest baboons are under less pressure and, as a result, have less social stimuli than the savanna baboon. Comparisons of captive and wild baboons show many patterns of similarities, although a few obvious differences may be noted.

The extremely dominant role of the baboon male in the social structure, however, does not make this animal the most suitable for generalized theories on primate behavior. On the other hand, the baboon is principally terrestrial living and most successful in this regard. As such, it is one of the animals of choice for students of evolutionary problems confronting early man.

It would appear from an examination of the literature that psychological and behavioral studies on the baboon have lagged behind studies in other species, e.g., the chimpanzee and rhesus monkey. This may be true in terms of total numbers of experiments, but the findings obtained with baboons, as well as their preferential phylogenetic position, have yielded data of equal and perhaps, at times, greater significance when compared with other species. It should be noted that in the past the baboon has only infrequently been included in comparative studies, even though we have noted that the baboon is an extremely rapid learner when compared with other primate species.

3. Reproductive and Developmental Biology and Fertility

The baboon is prominently used in reproductive biology research. Its natural adaptability and the establishment of highly successful breeding programs with baboons have led to its early use for such studies. Data are available on the baboon in studies on gametogenesis, origin of germ cells, implantation, placentation, angiogenesis of the amnion, embryogenesis, organogenesis, hormonal patterns and influence, neurologic influence, pathology, abnormalities, contraception and contraceptive devices, embryo transfer, artificial insemination, immunologic factors, genetics, and comparative reproduction.

There is a vast amount of literature on the above subjects and details will not be pursued herein. As repeatedly emphasized, generalizations about nonhuman primate responses are misleading and must be avoided. Therefore, all of these subjects must be regarded in terms of specific questions. In many instances, some generalizations between baboons and macaques may be made, but there are major differences; for example, the baboon placenta is like the human (monodiscoid), whereas the rhesus is not (bidiscoid). The mature ovum of the baboon is structurally very similar to that seen in humans. The diploid ($2n$) number of baboon chromosomes is 42. Baboon embryo development parallels that of the human in all respect except for the tail region. Gestation takes approximately 185 days and the animals breed all year.

4. Cardiovascular Disease

The baboon has been widely used in cardiac research and is one of the few primates studied under field conditions with this parameter in mind. It has been suggested that cardiovascular disease is uncommon in nonhuman primates, although arterial disease has been observed in baboons in the wild. This opinion may reflect lack of sufficient data on this point and/or failure of the animals to survive in nature to the age where disease may be prevalent.

Electrocardiographic patterns have been published and various abnormalities in comparison with human standards described. Because of its size and ease in maintenance, this animal is ideal for studies on hypertension, experimental valvular lesions, renal disease, pulmonary disease, and other coronary artery diseases. Use of the baboon by many investigators in experimental atherosclerosis has been extensive and rewarding. Cholesterol transport and lipid metabolism in the baboon appears to be similar to that in man, although much in the way of experimental study is still required. Atherosclerotic lesions, remarkably similar to those occurring in man, are seen in baboons in the wild and can be experimentally induced in this animal.

The blood vessels of the baboon are structurally similar to those of man. As in man, it has also been demonstrated that differences in structure and function between different parts of the same blood vessel may be noted in the baboon. This difference may account for development of disease in one area in contrast to another.

Hypertension is not naturally seen in simians and usually must be produced by experimental manipulation. Dr. B. A. Lapin and co-workers (1966), however, in the Soviet Union have described hypertension in male baboons resulting from stress due to social and sexual deprivation.

5. Immunology, Including Transplantation

The phylogenetic similarity of nonhuman primates to humans make these animals ideal for understanding the various immunologic parameters of primates in general. Numerous studies have been done on the baboon defining its immunologic capability. Drs. J. Moor-Jankowski and A. S. Wiener (1969) have reported extensively on the composition of baboon blood. In general, baboon blood is similar to that of humans and it may be typed using human ABO typing sera. No type "O" baboon blood has been described. Structural and bio-chemical differences may be noted between human and baboon erythrocytes. Serum from one species will agglutinate erythrocytes from another, although these heteroagglutinins vary in strength and from species to species. Immuno-globulins are present, both major (IgG, IgA, and IgM) and minor (IgD and IgE), as in humans, and are responsible for humoral antibody reactivity. Anti-body surveys are extremely helpful in indicating past antigenic experiences of the host animal. The antibody profile of the baboon is well documented in the literature.

Immunoglobulins and other serum proteins have been used for demonstra-ting comparative evolutionary and phlyogenetic relationships not only in regard to man, but also in an attempt to characterize simian populations. A classifica-tion scheme (immuno-taxonomy), based on immunologic analysis rather than anatomic similarities, has been developed demonstrating a certain relatedness among primates. It is not within the scope of this report to develop all the pathways in which immunologic studies may be used to provide information. In addition to the overall taxonomic and genetic information that immunology may provide, immunologic data offer an opportunity to pursue such studies as comparative molecular biology, protein chemistry, and breeding. Further, immunology offers such pragmatic applications as host response including trans-plantation and response to other foreign materials as well as the detection of immunologic variations existing among species.

In all such studies, the baboon has been utilized extensively and has enjoyed a rather prominent role. Interesting, and of some consequence, is the position of the baboon among the Cercopithecoidea. Thus far, immunologic studies would suggest that the genus *Papio* would make an excellent model for compara-tive studies and be useful in many experiments which are more difficult to perform in primates of the family Hominidae.

Perhaps one of the more active areas in which the baboon has been used has been in organ transplantation. Here, principally because of its size and because of the limited availability of the perhaps more coveted apes (chimpanzees), the baboon has contributed not only in research, but as a donor for certain heterotransplants as well. As a donor to humans, the baboon may not be as desirable as the chimpanzee. For research purposes, however, the baboon appears to be the animal of choice among investigators who have made com-parative studies utilizing a variety of animal species. Detailed data are available on isohemagglutinin and heterohemagglutinin titers following transplantations from baboon to man. Increases in these antibodies may be observed during rejection.

More recently the baboon has been used for studies on cell-mediated immunity. Here, too, the baboon shows its "relatedness" to the human in its responsiveness. Thus, in many comprehensive experiments designed to determine why the baboon is more resistant than other nonhuman primates (for example, New World monkey species) to various infectious agents, including a number of viral oncogenic agents, it was found that the baboon relates well to humans in its ability to respond to antigenic stimuli.

The baboon has been recommended for studies evaluating antilymphocyte serum and antilymphocyte globulin prior to use in humans. Autoimmune disease needs to be studied in simian models. Current information indicates that many of the nonhuman primates would be ideal for such purposes. It would appear that the baboon offers much in the way of an experimental model for all phases of immunologic studies. Preference for the baboon is based on many features—availability, size, previous information, limited availability of apes, minimal need for postoperative care, easy access for specimen selecting, rapid learning, and ease of training, etc.

6. Neurophysiology

Neurophysiologists have not made as much use of nonhuman primates for understanding human function as we would expect. It is rather evident that gradations exist among animal species, especially in cortical development and in similarities of the central nervous system in approximating man. This ascension toward human development is quite apparent when comparing the various cortical complexities of simians, from prosimians through New World, Old World, lesser apes, great apes, and man. It would appear, however, that increased usage of simians is imminent for neurophysiologic research. Certain disorders cannot be simulated in most lower animal models, particularly in the study of perceptual deficits which require a trained model generally not available other than in monkeys and apes.

Baboons have been used to ascertain effects of trauma and physical stress, especially those caused by high velocity accidents. One would expect that the nonhuman primate would offer an excellent opportunity to examine brain and spinal cord damage. Such studies would, by their very nature, exclude the use of humans. More important than physical stress is the impact of other forms of stress on the central nervous system. These could be studied in nonhuman primates. The baboon has been used in the Soviet Union as the animal of choice for experimental destruction of neurons and resulting motor disturbances. It has been suggested by Soviet investigators that most human neurologic disturbances could be readily reproduced in baboons and other simians.

From studies on neurophysiological mechanisms involved in human and baboon reflexes, it would appear that the human and baboon have many similarities permitting the baboon to serve as a model for further experimentation. Accordingly, investigators in the Soviet Union also suggest that the baboon may develop myocardial infarction and other types of cardiac involvement as a result of neurosis due to "disturbance of the diurnal stereotype" (disruption of the day/night cycle). Baboons have also been studied as a model for naturally occurring epilepsy.

An atlas of the baboon brain is available and a number of tissue reactions to implanted materials have been recorded for this animal. Also, baboons have been used in order to ascertain the chronic effects of various drugs on neurologic responses. However, much more study is necessary in the mapping of interconnections of basal ganglia and specific pathways in the primate brain. It is difficult to comprehend use of any other animal model system to collect these data and correlate them with human activities. There is need for detailed studies of the monkey and ape central nervous systems. Recently a study on the normal EEG of baboons has been undertaken; only limited similar information is available on other simians. Studies on EEG patterns have been done in baboons in the laboratory and in their natural environment. The results suggest that environmental influences affect the pattern of sleep within the limits of other factors, such as genetics and learning. Convulsions, while infrequently observed in baboons, are noted, but the cause(s) is as yet undetermined. The baboon is under study in attempts to understand demyelinating diseases, especially multiple sclerosis. Apparently the baboon is resistant to the virus responsible for causing the demyelinating disease called "Kuru."

7. Viral, Parasitic, and Other Infectious Diseases

In the search for the etiology of disease in human populations, monkeys and apes have been most used in attempts to either isolate the causative agent or reproduce the clinical disease. Two major pathways of investigation have resulted from such studies: (1) the investigation of natural disease and/or reservoirs of infection, and (2) laboratory studies in experimental hosts. Because of the primate's response to the antigenic insult resulting from the above two investigative approaches, another parameter has developed with parallel importance, namely the immunologic response. Although the advent of antibiotics has fortunately reduced the severity of bacterial disease, it is the usage of these antibiotics that has prevented further understanding of the pathogenic mechanism of most infectious agents. The realization that there is still a need for development and testing of vaccines, evaluation of drug efficacy and toxicity, studies on development of resistance to antibiotics, host response, etc., has resulted in an upsurge of interest in use of animal, particularly simian, models. Of some consequence is the realization that simians in general have a microbial flora and fauna of their own. A number of these agents are capable of infecting and/or producing disease in humans and other animals. In turn, monkeys and apes are susceptible to a number of human (and other animal) infectious agents, responding with infections that range from inapparent to highly fulminating fatal disease. We will attempt to highlight these and show the involvement of the baboon in these infectious processes.

The baboon is one of the few simian species that has been studied in its native habitat (Kalter and Heberling, 1971; Kalter, 1973a,b). A number of different geographic and ecologic areas have been included in the various field trips by SFRE personnel in Africa, principally Kenya. During these situations, animals were trapped (see above) and sampled (blood, throat, and stool) immediately upon capture. These materials were then subjected to a variety of laboratory procedures designed to provide information on the presence of naturally occurring bacteria, fungi, parasites, and viruses. Local citizens, various wild animal species as well as domestic animals, and soil samples provided comparative information. The results of these field studies have been reported in detail.

These studies were continued on many baboons that were transported back to the United States (San Antonio, Texas). The results of these experiments have shown that the baboon, like other simians, has a characteristic microbial profile. In many instances, the agents are decidedly simian in origin, although no evidence is available as yet which would demonstrate a species-specific flora and fauna. Many organisms recovered from the baboon are identical to those found in humans, while others may be considered as antigenically distinct counterparts. Therefore, depending upon numerous ecologic and other geographic factors and individual animal habits, one finds, at one time or another, organisms similar to those of the human population living in the same general geographic area. Alterations in flora and fauna develop as a result of captivity, frequently with the loss of an organism encountered in nature. Inasmuch as reinfection is held at a minimum during captivity under proper husbandry and management, a healthy colony may be developed. Such a colony has many advantages. One major disadvantage is that newborns of such animals are often highly susceptible to organisms that are normally relatively innocuous.

Of more than academic interest is the failure to find B virus (*Herpesvirus simiae*) in baboons. This lack of one of the most virulent (in humans) of simian viruses is a source of comfort to those employing the baboon for research purposes. Other herpesviruses are present in baboons, but to date appear to be of limited pathogenicity for other animals, including humans. In addition, the baboon appears to be free of any oncogenic herpesviruses similar to those found in such New World monkeys as the squirrel monkey (*Saimiri sciureus*) and the spider monkey (*Ateles fusciceps*).

As mentioned above, perhaps one of the most exciting findings with regard to nonhuman primates, and specifically the baboon, has been the observation and isolation of a virus with all the biologic characteristics of recognized RNA (ribonucleic acid) oncornaviruses (type C) from the baboon placenta. These studies, done in conjuction with Drs. R. L. Heberling, T. Barker, and J. Eichberg, have provided an opportunity to explore virus—tumor relationships and ascertain whether or not viruses are involved in human tumor development. If, as predicted, a virus is responsible for certain forms of cancer, its presence and ability to produce a cancer must be demonstrated. This finding of a baboon type C virus permits such studies. Experiments are underway at SFRE and results suggest that these type C viruses are able to produce tumors under appropriate circumstances.

The presence of type C viruses in normal baboon placentas suggested examination of other baboon tissues. Generally, these are negative as determined by electron microscopy. However, type C viruses can be isolated, or their presence "chemically" determined, in baboon tissues. It was important, therefore, when the following two additional observations were made. In collaboration with Dr. R. E. Kuntz and his co-workers (Kalter *et al.*, 1974), the same type C virus was observed in bladder tumors induced in baboons by a parasite, *Schistosoma haematobium*. At this time, the relationship of virus to parasite to baboon is still obscure. However, in a collaborative study with Dr. S. Shain (Kalter *et al.*, 1975), who has been attempting to produce prostatic cancer in baboons by various chemical carcinogens, we have seen oncornaviruses in "normal" prostatic tissue. Again the relationship of virus to tissue is not understood. In these later studies, there was no evidence of neoplasia although the animals had been receiving chemicals known to be carcinogenic.

The baboon has been extensively studied for the presence of ecto- and endoparasites, both in its natural habitat and following captivity. Dr. Kuntz and his associates (Kuntz, 1966) have prepared a listing of the various parasites found in baboons. It is evident from these studies that the parasitic fauna varies considerably in baboons from different habitats and geographic areas. Space does not permit detailed description of the various parasites and the type of infections produced. Ectoparasites are uncommon in the baboon, although some arthropods may be found.

An interesting experiment performed in baboons and of value to humans has been performed by Drs. E. J. Goldsmith and B. H. Kean (1969). These investigators were able to produce schistosomiasis in baboons by the subcutaneous inoculation of cercariae. They were then able to remove virtually all the adult worms from the portal blood and the animal was freed of infection by use of extracorporeal hemofiltration. This procedure was so successful in baboons that it has been applied with equal success to humans.

The mycology of the baboon has been well described, principally through the efforts of Dr. Y. Al-Doory (1967). Again, many factors influence the presence of these organisms and their differentiation from air and soil contaminants may be a problem and must be carefully evaluated. The results do suggest that infection with this group of organisms may exist, but probably they do not contribute extensively to the general health status of the animal.

Bacterial infections of the baboon may present several severe problems to both colony animals and attending humans. The bacterial flora of the baboon has been well established through the efforts of Dr. M. Pinkerton (Pinkerton *et al.*, 1967). Many of these agents have been demonstrated to be capable of producing infection and/or disease with varying chemical responses noted. The intestinal organisms (*Salmonella, Shigella*) can and do produce varying degrees of infection in the baboon and may be passed on to man with resulting disease. Tuberculosis may be devastating to baboons, as well as other primates, and is always a threat. The source of this agent is usually the human but could be from other animal and avian species. However, tuberculosis is rare in baboons in the wild. Other bacteria may be present in the baboon and the ability to produce infection and/or disease, as in humans, is dependent upon many host and environmental factors.

8. Pharmacology (Other than Drug Testing)

Use of baboons in pharmacologic research has been alluded to above. It is apparent from these studies, as well as other experiments, that species differences exist and are to be expected. In planning such studies, therefore, it is incumbent upon investigators to recognize these differences and make the appropriate selection based upon preferred responsiveness. The physiologic response of the baboon to drugs is significantly different from that of man but the baboon still more closely approximates man than do lower animals. In addition to drug testing (toxicity), the nonhuman primate may be used to advantage in such pharmacologic programs as: (1) drug metabolism, (2) antibiotic and chemotherapeutic (including antineoplastic) effects, (3) endocrinologic (including hormonal) effects, (4) behavioral and psychological effects, (5) teratology, (6) central nervous system effects, (7) drug dependence and liability, and others. Use of monkeys for toxicologic testing is not a requirement at this time; however, it is encouraged.

The baboon has been successfully used in providing information on the effects of drugs in reproductive biology. Numerous studies by Dr. A. G. Hendrickx and D. C. Kraemer (1970) and by Dr. J. W. Goldzieher *et al.* (1974) attest to the usefulness of the baboon for such investigations. Also, as previously mentioned, the occurrence in baboons of a syndrome resembling human epilepsy has permitted use of this species as a model for studies on the effects of therapeutic agents. Use of baboons for the evaluation of drugs useful in treating atherosclerosis has been performed by several investigative groups.

It would appear that the baboon has been long neglected as a primate model for pharmacologic studies. Dr. I. Geller (personal communication) of this institution who has used simians other than baboons for many years, recently indicated that this has been a serious omission. Recent studies in his laboratory with baboons clearly demonstrate that the baboon's rapid learning and perceptual abilities are ideal for many behavioral studies. Using self-administration techniques, Dr. J. V. Brady (communicated by Dr. I. Geller) of The Johns Hopkins University has found the baboon to be a good experimental model for measuring liabilities of new and unknown compounds. Dr. Geller (personal communication), in collaboration with Dr. B. Beer and Dr. Clody of the Squibb Institute, has noted that major tranquilizers will produce extrapyramidal symptoms in the baboon as quickly as in humans. It was also noted that tranquilizers ineffective in humans also were ineffective in baboons.

9.   Congenital Malformations

Baboons have not been used for studies on causes of congenital anomalies in spite of the vast amount of data suggesting their usefulness. The excellent breeding record of the baboon in captivity as well as its ability to withstand the rigors of laboratory procedures make the baboon an ideal model for birth defect studies. We may focus our attention on two major factors contributing heavily to recognized congenital malformations: (1) chemicals (drugs), and (2) infections. In both these situations, prescreening of drugs and testing of infectious agents in suitable animal hosts would do much to provide a reliable indication of potential effects in humans.

Several years ago, Dr. A. G. Hendrickx (1970) clearly demonstrated the teratogenic effects of Thalidomide in baboons. In collaborative studies with Dr. Hendrickx (unpublished data), we were able to demonstrate a clear parallelism between the baboon and human with regard to the teratogenic effects of rubella virus. Subsequent studies in association with Dr. R. Ackermann of the University of Cologne Neurology Clinic, have shown that the pregnant baboon responds much the same as humans following infection with the virus causing lymphocytic choriomeningitis (unpublihed data).

It would appear from the above, as well as numerous studies on baboons infected with viruses, that the baboon should be given greater consideration for use in contemplated studies on the causation of congenital malformations. The baboon offers many advantages over the macaque in this context.

10.   Dentistry

The baboon has succedaneous and heterodont dentition and no differences exist between the male and female deciduous dentition. Dr. O. M. Reed (1965) of San Antonio has reported on the dentition and eruption patterns of baboon teeth and stated, "The baboon dentition is well developed, relatively free of malformations or abnormalities, free from dental decay in the natural state, but capable of developing dental disease under modified experimental conditions." As part of his studies with the baboon, Dr. Reed has developed an ability for age determination by studying baboon dental eruptions. In studies at Brown University, Dr. M. Povar (1967) and collaborators developed a procedure for substituting plastic teeth for natural teeth. This methodology has been helpful in evolving procedures for use in humans with other dental materials.

Information on dentition of baboons as seen in nature and in captivity may be noted in the literature. The prevalence of cavities in these animals has been reported. Dr. W. W. James (1960) of Birmingham University, England, has provided a source book, "The Jaws and Teeth of Primates," in which the teeth of baboons are described. According to Dr. B. M. Levy (1970) of the University of Texas Dental Science Institute, dental research using rodents is questionable and substituting nonhuman primates "will greatly accelerate the accumulation of relevant data that can be applied to human dental problems."

11.   Ophthalmology and Other Sensory Research

For ophthalmologic studies, as with others, a suitable model must share the desired characteristics of human. According to Dr. F. A. Young (1973), Washington State University, Pullman, Washington, "The nonhuman primate meets these criteria (for visual characteristics) better than any other animal." Dr. A. G. Hendrickx (1971) in his book "Embryology of the Baboon" provides developmental information on the eye based on his studies of this species. The baboon has not been used extensively in research done in this field. It is to be noted, however, that research in this area utilizing primates in general has not been extensive. Corneal and vitreous transplant studies have been done using homo- and heterotransplants involving man and baboon. The baboon has been used with much success by surgeons in South Africa as a possible mechanism for storing human corneas. Because the structure of the baboon cornea is very comparable to that of man, the baboon has been used in microscopy of the anterior chamber.

12.   Experimental Surgery

Baboons are one of the more popular experimental animals used in surgical research transplantation experiments in primates. The University of Stellenbosch Primate Colony in South Africa has been one of the forerunners in the use of baboons in surgical research. In a recent review of "The Cape Chacma Baboon in Surgical Research," Drs. J. H. Groenewald and J. J. W. Van Zyl (1973) describe various surgical research projects involving the animal: mitral valve replacement, cardiac transplantation, corneal transplantation, lung transplantation, auxiliary *ex vivo* extracorporeal liver perfusion and hepatic assist, immunology, and immunosuppressive therapy after tissue transplantation, antilymphocyte serum, drug evaluation, liver transplantation, organ preservation, cardiodynamic studies, and normal anatomic and physiological studies and values. These authors discuss the many advantages that the baboon offers over those of the dog and state, "Any surgeon who has had the opportunity of working with both types of animal will have no doubt in his mind as to the numerous advantages that the baboon offers."

Dr. C. R. Hitchcock and his group in Minneapolis, as a consequence of studies initiated in 1959, developed an interest in the use of the baboon as a model for many kinds of surgical research. In a review of experimental surgery in primates and standard laboratory animals, Hitchcock (1969) provides comparative surgical data and concludes that "the baboon has only begun to be utilized to the degree that is justified at this time."

Numerous other surgical groups have demonstrated the value of the baboon for a multitude of surgical procedures. Baboons have been employed as one of the surgical experimental animals of choice at the Laboratory for Experimental Medicine and Surgery in Primates (LEMSIP) in New York.

13. Environmental Programs

In this rapidly moving civilization of ours, with its ever changing mores and insults to our physiology, the present-day environment undoubtedly exerts a marked influence on our very existence. The expressed opinion that our environment is a direct source of carcinogens has stimulated development of protocols to examine this potential. Inasmuch as the environment has the potential for exposing man to various degrees of hazards, its detailed exploration becomes essential. Use of nonhuman primates, which has been successful in studies of space environment, needs to be expanded in the study of man's environment. The Food and Drug Administration is concerned with the overall problem of environmental contamination and its potential danger to man and has initiated programs utilizing primates as animal models.

Baboons are presently under study for use in evaluating many of the materials currently found in the environment. The results of such studies will do much to provide a capability for detecting effects of noxious materials, including their potential for oncogenesis. Dr. H. McGill (personal communication) of The University of Texas Health Science Center at San Antonio and of SFRE is currently using baboons trained to smoke cigarettes to determine effects of smoking on health. A wide range of other exploratory experiments utilizing the baboon for monitoring environmental hazards are currently being developed.

V. Conclusion

In our attempts to provide an insight into the current status of the baboon not only as an experimental animal, but as another primate actively enjoying its participation in our world, we have per force drawn upon our own personal experience with this animal. As a consequence, we have omitted much of the experience of others in the attempt to develop an overview of the problem—present and future. Space and time were the major influences preventing any in depth discussion. No attempt was made, therefore, to review the literature. Much has been written about the baboon and several symposia (Vagtborg, 1965, 1967), dedicated principally to "The Baboon in Medical Research," have provided details concerning specific experiments involving the species. Similarly, "The Baboon—An Annotated Bibliography," (Shilling, 1964) provides a comprehensive listing and abstracting of references pertaining to the baboon. Included in that bibliography are several hundred references provided by Dr. B. A. Lapin of the Soviet Union and translated from Russian into English.

Present limitations on availability of nonhuman primates exphazize the need to reevaluate our long-term goals for the use of nonhuman primates. Depletion of certain species is evident. Current practices for providing the simian species for medical research are completely unsatisfactory and undoubtedly will become worse. The breeding of several species for general distribution rather than "in-house" research has been initiated, but, for several years to come, these colonies will be unable to supply desired numbers. The baboon, which is recognized to be a resistant animal living and breeding well in captivity, is suggested as a host for experimentation requiring a primate model in which humans cannot be used.

Historical use of the rhesus monkey has unfortunately placed this particular animal in a preferred position with regard to other simian species. In certain instances, this may be desirable. However, as a consequence, an unwarranted dependency has been placed on this one species. The accumulated scientific evidence emphasizes the difference in responsiveness among species, but nonetheless urges the use of nonhuman primates wherever indicated as necessary. Restrictions on availability of rhesus monkeys would support a reevaluation of the usefulness of this species, especially in view of the demonstrated effectiveness of other simians, such as the baboon.

Finally, the use of primates for "necessary" research is recommened. It behooves us to give careful consideration to the need for a simian model for man and to the selection of species in the laboratory. Even where sufficient numbers of animals exist, their future will be jeopardized if appropriate plans are not made to breed them in captivity or develop other suitable programs for supplying sufficient numbers before depletion results in extinction.

References

Al-Doory, Y. (1967). *In* "The Baboon in Medical Research" (H. Vagtborg, ed.), Vol. II, pp. 731–739. University of Texas Press, Austin, Texas.
Bourne, G. H; (1974). "Primate Odyessy." Putnam, New York.
Goldsmith, E. I. and Kean, B. H. (1969). *Ann. N. Y. Acad. Sci.* **162**, 453–458.
Goldzieher, J. W., Joshi, S., and Kraemer, D. C. (1974). *Acta Endocrinol. (Copenhagen)* **75**, 90–118.
Groenewald, J. H. and Van Zyl, J. J. W. (1973). *In* "Nonhuman Primates in Medical Research" (G. H. Bourne, ed.), pp. 270–279. Academic Press, New York.

Hall, K. R. L. (1966). *In* "Some Recent Developments in Comparative Medicine" (R. N. T-W-Fiennes, ed.), pp. 49–73. Academic Press, New York.

Hamner, J. E., Jr., and Reed, O. M. (1972). *J. Med. Primatol.* **1**, 75–85.

Heberling, R. L., Barker, S. T., Helmke, R. J., Smith, G. C., and Kalter, S. S. (1976). *Abstr. Annu. Meet. Amer. Soc. Microbiol.* p. 216 (S69).

Hendrickx, A. G. (1970). *Proc. 3rd Intern. Congr. Primatol., Zurich, 1970* **2**, 230–237.

Hendrickx, A. G. (1971). "Embryology of the Baboon." The University of Chicago Press, Chicago, Illinois.

Hendrickx, A. G. and Kraemer, D. C. (1970). *In* "Reproduction and Breeding Techniques for Laboratory Animals" (E. S. E. Hafez, ed.), pp. 316–335. Lea & Febiger, Philadelphia, Pennsylvania.

Hitchcock, C. R. (1969). *Ann. N. Y. Acad. Sci.* **162**, 393–403.

Ingelfinger, D. F. J. (1975). *New Engl. J. Med.* **293**, 500–501.

James, W. W. (1960). "The Jaws and Teeth of Primates." Pitman Med. Pub. Co., Ltd., London.

Kalter, S. S. (1973a). *In* "Primates in Biomedical Research" (G. H. Bourne, ed.), pp. 67–165. Academic Press, New York.

Kalter, S. S. (1973b). *Primates Med.* **8**, 171 pp.

Kalter, S. S. and Heberling, R. L. (1968). *Nat. Cancer Inst. Monogr.* **29**, 149–160.

Kalter, S. S. and Heberling, R. L. (1971). *Bacteriol. Rev.* **35**, 310–364.

Kalter, S. S. Helmke, R. J., Panigel, M., Heberling, R. L., Felsburg, P. J., and Axelrod, L. R. (1973). *Science* **179**, 1332–1333.

Kalter, S. S. Kuntz, R. E., Heberling, R. L., Helmke, R. J., and Smith, G. C. (1974). *Nature (London)* **251**, 440.

Kalter, S. S., Shain, S. A., Smith, G. C., McCullough, B., Heberling, R. L., and Dalton, A. J. (1975). *J. Nat. Cancer Inst.* **55**, 1237–1241.

Kraemer, D. C., Moore, G. T., and Kramen, M. A. (1976). *Science* **192**, 1246–1247.

Kuntz, Robert E. (1966). *Primates* **7**, 27–32.

Lapin, B. A. (1973). *In* "Unifying Concepts of Leukemia" (R. M. Dutcher and L. Chieco-Bianchi, eds.), pp. 263–268. S. Karger, Basel.

Lapin, B. A., Yakovleva, L. A., and Cherkovich, G. M. (1966). *In* "Some Recent Developments in Comparative Medicine" (R. N. T-W-Fiennes, ed.), pp. 195–212. Academic Press, London.

Levy, B. M., Dreizen, S., Hampton, J. K., Taylor, A. C., and Hampton, S. H., (1970). *In* "Medical Primatology" (E. I. Goldsmith and J. Moor-Jankowski, eds.), pp. 859–869. Karger, New York.

Melendez, L. V., Hunt, R. D., Garcia, F. G., and Fraser, C. E. O. (1969). *Lab. Anim. Care* **19**, 378–386.

Mitchell, B. L., Shenton, J. B., and Uys, J. C. M. (1963). *Symp. Afr. Mammal.*, Salisbury S. R.

Moor-Jankowski, J. and Wiener, A. S. (1969). *In* "Primates in Medicine" (W.I.B. Beveridge, ed.), Vol. 3, pp. 64–77. Karger, Basel.

Pinkerton, M., Boncyk, L. H., and Cline, J. A. (1967). *In* "The Baboon in Medical Research" (H. Vagtborh, ed.), Vol. II, pp. 717–730. University of Texas Press, Austin, Texas.

Povar, M. L. (1967). *In* "The Baboon in Medical Research" (H. Vagtborg, ed.), Vol. II, pp. 871—873. University of Texas Press, Austin, Texas.

Reed, O. M. (1965). *In* "The Baboon in Medical Research" (H. Vagtborg, ed.), Vol. II, pp. 167—180. University of Texas Press, Austin, Texas.

Shilling, C. W. (1964). "The Baboon—An Annotated Bibliography." George Washington University, Washington, D. C.

Vagtborg, H. (ed.). (1965). "The Baboon in Medical Research," Vol. I. University of Texas Press, Austin, Texas.

Vagtborg, H. (ed.) (1967). "The Baboon in Medical Research," Vol. II. University of Texas Press, Austin. Texas.

Young, F. A. (1973). *In* "Nonhuman Primates and Medical Research" (G. H. Bourne, ed.), pp. 354—379. Academic Press, New York.

13

The Guereza and Man

How Man Has Affected the Distribution and Abundance of *Colobus guereza* and Other Black Colobus Monkeys

J. F. OATES

I. Introduction

At present, it is generally recognized that there are four species of African
black colobus monkeys, members of the subgenus *Colobus* (Rahm, 1970; Thorington and Groves, 1970; Dandelot, 1971). Three of these species (*angolensis, guereza,*
and *polykomos*) have coats with different patterns of black and white hair, while
the fourth (*satanas*) is totally black. Populations of all four species are threatened
in varying degrees by man. Although I will comment on each species, the
emphasis of this chapter will be on East African populations of the black-and-
white colobus monkey *Colobus guereza* (formerly known as *C. abyssinicus*), since
these are the animals with which I am most familiar.* "Guereza" is the Amharic
name for this colobus in Ethiopia (R. I. M. Dunbar, in litt.) and was incorporated
into the Latin name by Rüppell (1835).

The guereza is a large, handsome monkey (Fig. 1). Adult males range in
weight from about 9.0 to 14.5 kg, and adult females from about 6.5 kg to 10.0
kg.† A cape or mantle of long white hairs extends from the monkey's shoulders,
down its flanks, and across the lower back. The distal part of the tail bears long
white hairs, while shorter white hairs form a ring around the dark facial skin.
The length and fullness of the cape, the proportion of white to black on the
tail, and the total length of the tail vary between and, to a more limited extent,
within local populations (Fig. 2). The guereza's hindlimbs are relatively long,
its nose slightly overhangs its upper lip and, like other *Colobus* species, its thumb
is greatly reduced (Napier and Napier, 1967). As in other members of the Old

---

*The term "East African" is used here to describe the territories of Kenya, Tanzania, and
Uganda.
†Data from specimen labels in the American Museum of Natural History, the British Museum
(Natural History), the Powell-Cotton Museum, and the U.S. National Museum of Natural History;
from Dandelot and Prévost, 1972 and from personal observation.

Fig. 1. *Colobus guereza occidentalis:* adult female with 10-week-old infant, Kanyawara, Kibale Forest Reserve; February 1972.

World subfamily Colobinae (the "leaf-eating monkeys"), the forestomach of the guereza is modified as a fermentation chamber in which microorganisms help to digest the monkey's cellulose-rich food (Bauchop and Martucci, 1968).

The guereza has been exposed to a wide range of threats from man. Together with the other black colobus species, it has long been hunted for its attractive pelt, which has been sought after as decoration both locally and internationally. It has also been killed for its meat and because of genuine and suspected crop raiding, and many of its habitats have been altered or destroyed. After describing the general distribution of the species and outlining some relevant features of its ecology, this chapter will be concerned largely with an examination of the effects of habitat disturbance on the guereza and with a review of the trade in its skins. Detailed conservation proposals are not formulated, but some recommendations on forest preservation and the control of the skin trade are made.

My own involvement with this species began with a field study lasting from October 1970 to March 1972. The study was concentrated at Kanyawara (0°34'N, 30°21'E; 1500 m elevation) in the Kibale Forest Reserve, West Uganda (see Section IV, B). Here I investigated guereza ecology and social organization to provide information for a comparison with the red colobus monkey (*C. badius*), which was being studied by T. T. Struhsaker (see Struhsaker and Oates, 1975). During this period I also examined populations of *C. guereza* and *C. angolensis* in many other localities in Uganda, Kenya, and Tanzania. Detailed

Fig. 2. Adult guereza photographed in dry forest at ca. 2000 m near the summit of Mt. Warges, North Kenya, in July 1974. The bushy white tail tuft on the Warges animals (arrow), taking up about two-thirds of the total tail length, is much larger than that in West Ugandan animals (see Fig. 1). This 1974 observation seems to be the first record of the continuing existence of the isolated Warges population since the collection of the first (three) specimens there in 1911 and 1912 by Edmund Heller and the game warden A. Blayney Percival (the specimens are now in the U.S. National Museum and the British Museum). Heller (1913) named the population as a subspecies, *percivali*, but its distinction from other upland Kenya animals, usually called *kikuyuensis*, seems questionable.

results of this investigation are presented elsewhere (Oates, 1974). A return visit was made to Kanyawara in October–November 1973, followed by limited survey work in Kenya (including an investigation into the skin trade) and another visit to Kanyawara in July–September 1974.

II.   Where the Guereza Lives

A.   Geographical Distribution

The guereza has the most wide-ranging geographical distribution of all the black colobus species. This distribution is shown in Fig. 3, in terms of occupation of a grid formed by 1° lines of latitude and longitude. *Colobus guereza* has been recorded in 106 squares of this grid, an area of approximately 1,280,000 km². The animals do not, of course, inhabit more than a small fraction of the area of the great majority of the squares. Populations in a few squares, plotted from old records, may be extinct.

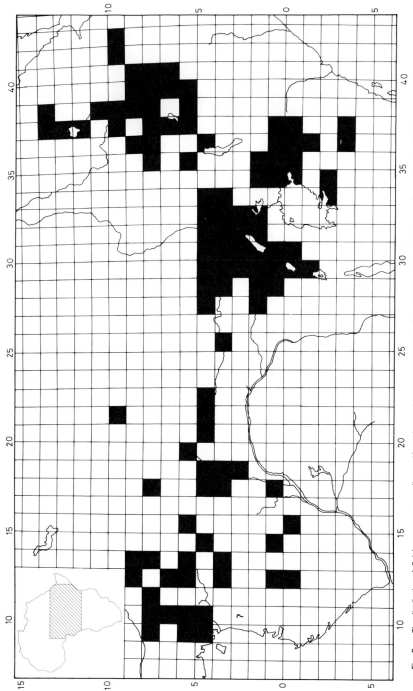

Fig. 3. The distribution of *Colobus guereza*, according to 1° quadrants of latitude and longitude. Records (indicated by shaded squares) came from the following sources: personal observation; locality data with specimens in the American Museum of Natural History, the British Museum (Natural History), the Muséum National d'Histoire Naturelle (Paris), the National Museum of Natural History (Smithsonian Institution), the Odontological Museum of the Royal College of Surgeons (London), and the Powell-Cotton Museum (Birchington); data published by J. A. Allen (1925), Bahuchet (1972), Brown and Urban (1970), Butler (1966), Dandelot and Prévost (1972), Dunbar and Dunbar (1974), Kingdon (1971), Rahm and Christiaensen (1960), Rosevear (1953), Sanderson (1940), Schwarz (1929), and Swynnerton and Hayman (1951). Additional information from Ethiopia was supplied by Robin Dunbar (personal communication), and from the C.A.R. by Serge Bahuchet (personal communication). Queried is the record of Basilio (1962) for Rio Muni.

423

Guerezas have been recorded from the following territories (sources of information are given in the legend of Fig. 3): Cameroon, Central African Republic, Congo Republic, Ethiopia, Gabon, Kenya, Nigeria, Rwanda, Sudan, Tanzania, Uganda, and Zaïre. Basilio (1962) reports that the species is known in the area of Mikomeseng, Rio Muni (Equatorial Guinea), but I have located no specimens from that territory in collections. Geographical extremities of distribution are given in Table I.

Of the other black colobus, *C. polykomos* occurs in West Africa from Guinée to West Nigeria, *C. angolensis* occurs in Zaïre, Tanzania, and the southeastern corners of Kenya and Uganda, and *C. satanas* occurs in Macias Nguema Biyoga (formerly Fernando Poo), Cameroon, Gabon, and Rio Muni. In West Equatorial Africa, *C. guereza* apparently overlaps with *C. satanas*, and in the eastern Congo Basin with *C. angolensis* (Tappen, 1960; Rahm, 1970; museum specimens). However, despite large-scale geographical overlap and sightings of two species within one mile (1.6 km) of each other (Tappen, 1960), there is no recorded observation of either *satanas* or *angolensis* actually sharing a habitat with *C. guereza* (attribution of museum specimens to the same locality cannot be taken as firm evidence of habitat sharing).

B.   Environmental Distribution

The guereza is basically arboreal and is restricted to habitats with trees. However, it occupies a very wide range of such habitats, and although individuals spend the majority of their time above ground, they will frequently feed and travel on the ground in areas where trees are not densely packed.

TABLE I

The Extremities of *Colobus guereza* Geographical Distribution

Direction of extremity	Location	Source of record
Most northerly	North of Adai Arkai, Ethiopia (13°50′N; 38°00′E)	R.I.M. Dunbar, in litt.
Most southerly	West of L. Jipe, Tanzania (3°35′S; 37°45′E)	Neumann, 1896[a]
Most easterly	Harrar, Ethiopia (town = 42°10′E; 9°18′N)[b]	BMNH Spec. No. 51.713, collected by Sir F. Colyer, 1951
Most westerly	Fotabong, Cameroon (9°55′E; 5°35′N)	Sanderson, 1940

[a]No recent record known from this locality.
[b]"Harrar" might refer to the town or province of that name. R.I.M. Dunbar (in litt.) informs me that there are, or were, quite large forests within 50–100 km of the town.

The species has been recorded over a wide altitude range, from many low-lying areas in West Equatorial Africa (D. McKey, in litt., reports visiting a coastal swamp forest in Cameroon where a hunter claimed to have shot guerezas) up to the highest forested levels in the mountains of East Africa and Ethiopia. Dunbar and Dunbar (1974) observed guerezas at 3300 m in *Hagenia* forest on Mt. Gaysay, Ethiopia, and Guest and Leedal (1954) note finding frozen guereza remains at 4680 m in a cave in a rock wilderness on Kibo, Kilimanjaro, Tanzania (they suggest that the animal got there by itself, but I think it could perhaps have been carried to the cave by a leopard; the condition of the remains is not reported).

Table II records the localities in which I have observed guerezas in East Africa, indicating the broad vegetation type in each case (details of vegetation types are given in the Appendix). The majority of habitats in which guerezas were observed were colonizing, riparian or upland forest, and single groups were quite often seen in small patches of colonizing or riparian growth. A survey of locality data for the whole of Africa (those used in Fig. 3) also reveals a prevalence of records from river and lake sides and from upland areas.

III.   How the Guereza Lives

A.   Group Size and Structure

Guerezas live in relatively small, highly cohesive social groups. In East Africa as a whole I found a range in group size of from three to fifteen, with a central tendency to a group size of nine in large forest blocks. In riparian forest and small forest patches, groups tended to be smaller, with a central tendency to seven. Typically, groups had a single fully adult "dominant" male (though often a few other large males were also present) together with three to four adult females and their young offspring (for further information see Schenkel and Schenkel-Hulliger, 1967; Marler, 1969; T. J. Kingston, 1971; Leskes and Acheson, 1971; Dunbar and Dunbar, 1974; Oates, 1974).

B.   Population Dynamics

During 1971, at least 13 births occurred to 21–23 adult females in six guereza groups in my study area at Kanyawara, Kibale Forest. This is a rate of about 0.6 per female per year, or one birth to each female every 20 months. At a secondary study site at Chobe, Kabalega National Park, West Uganda, six adult females in two groups produced three infants during 1 year (0.5 per female). In both areas a few other births might have occurred, with the infants dying before I sighted them. Dunbar and Dunbar (1974) have presented very similar

TABLE II

Localities in which *Colobus guereza* Was Observed in East Africa in 1970—1974

Locality	Map reference	Vegetation type in which Guerezas observed[a]	Altitude (m)
Kanyawara, Kibale Forest Reserve, West Uganda	0°34'N; 30°21'E (Forest Station)	Moist lowland high forest and colonizing moist lowland forest	1500
Kigami, Kibale Forest Reserve, West Uganda	0°38'N; 30°23'E (Tea estate)	Colonizing moist lowland forest	1460
Dura River Bridge area, Kibale Forest Reserve, West Uganda	0°28'N; 30°23'E (Dura Bridge)	Moist lowland high forest	1220
Itwara Forest Reserve, West Uganda	0°49'N; 30°25'E (Kijura, west edge forest)	Colonizing moist lowland forest	1300—1500
Rwentuha River, Toro District, West Uganda	0°04$\frac{1}{2}$'N; 30°23$\frac{1}{2}$'E	Woodland	1190
Near Katego, Toro District, West Uganda	0°03'S; 30°24'E	Permanent swamp forest	1405
Wasa River, Toro Game Reserve, West Uganda	0°56'N; 30°22'E (Ntungama)	Riparian forest	700
Budongo Forest Reserve, West Uganda	1°41'—1°54'N; 31°27'—31°40'E	Moist lowland high forest and colonizing moist lowland forest	855—1190
Bugoma Forest Reserve, West Uganda	1°15'N; 31°02'E (Karwata Fort)	Colonizing moist lowland forest	1100
Kalinzu Forest Reserve, West Uganda	0°23'S; 30°06'E (Lubare Ridge)	Colonizing moist lowland forest	1495
Kasyoha-Kitomi Forest Reserve, West Uganda	0°7$\frac{1}{2}$'S; 30°18'E (Kitaka Mine)	Colonizing moist lowland forest	1220
Kaizi River headwaters, Maramagambo Forest, Rwenzori (Queen Elizabeth) National Park, West Uganda	0°31'S; 29°56'E	Moist lowland high forest, colonizing moist lowland forest and permanent swamp forest	1220

Locality	Map reference	Vegetation type in which Guerezas observed[a]	Altitude (m)
Lake Nyamusingire, Rwenzori National Park, West Uganda	0°17′S; 30°03′E (Ranger post)	Woodland	1005
Ishasha River, Rwenzori N.P., West Uganda	0°37′S; 29°40′E (Ishasha Camp)	Riparian forest	975
Chobe, Kabalega (Murchison Falls) N.P., West Uganda	2°15′N; 32°09′E (Chobe Lodge)	Relic riparian forest and wooded grass-land	945
Rabongo Forest, Kabalega N.P., West Uganda	2°05′N; 31°52½′E (Ranger post)	Relic patches of moist lowland high forest	975
Fajao, Kabalega N.P., West Uganda	2°17′N; 31°41′E	Riparian woodland	670
Bwindi Forest Reserve, South-west Uganda	1°03′S; 29°46½′E (Ruhiza Forest Station)	Colonizing moist upland forest	2590
Kapkwata, Mt. Elgon Forest Reserve, East Uganda	1°21′N; 34°38′E (Forest Station)	Moist upland forest and dry upland forest	2165–2470
Kadam Forest Reserve, East Uganda (inc. in Debasien Animal Sanctuary)	1°50′N; 34°43′E (Geol. Survey camp)	Dry upland forest and riparian woodland	1525–1830
Limuru, Central Kenya	1°05′S; 36°41′E	Colonizing moist lowland forest	2210
Naro Moru track, Mt. Kenya N.P., Kenya	0°10′S; 37°10′E	Moist upland forest	ca. 2450– ca. 2750
Mt. Warges, North Kenya	0°57′N; 37°23′E	Dry upland forest	ca. 2000
Arusha N.P., North Tanzania	3°05′–3°15′S; 36°45′–36°55′E	Moist upland forest, colonizing moist upland forest and dry upland forest	1370–2135

[a]Vegetation types are described in the Appendix, Section X.

data from Bole, Ethiopia, where twelve females were estimated to have had an average of 0.5 births/female/year.

At Kanyawara, adult females formed 29–35% of group members in March 1972. A similar proportion of females was found by Dunbar and Dunbar: of 119 animals in eighteen groups counted in the Bole Valley, 30–40 (25–34%) were adult females. Marler (1969) recorded an even higher proportion of adult females in fourteen accurately counted groups from the Budongo Forest, West Uganda; of 117 individual guerezas, 51 (44%) were females. A population with 30% of adult females and 0.5 births/female/year will have a potential annual increment of 15%.

Although guerezas have a potentially high rate of population increase, infant mortality is also relatively high. In January 1971 there were six infants (animals under 1 year of age) in the six Kanyawara groups noted above. Between then and March 1972, sixteen to seventeen more infants were born but nine disappeared, including six of those born since January 1971. Two of the latter six were lost before coat color change was complete (guerezas are born with a pure white coat; the change to the adult pattern takes place between 12 and 20 weeks after birth—Oates, 1974; Horwich and Manski, 1975). Dunbar and Dunbar report that of eight infants born at Bole, three died before coat color change was complete.

Guerezas at Kanyawara reacted with alarm to the close approach of raptorial birds such as the crowned hawk-eagle, *Stephanoaetus coronatus*. White infants are known to have fallen to the ground during "transfers" between older monkeys, and falls are probably common among older infants as they start to move independently of their mothers. Haddow *et al.* (1947) record that only 4% of juvenile guerezas in the Semliki Forest, Uganda, had immunity to the yellow fever virus, compared with 100% of adults, and they suggest that infants become susceptible near the end of their first year when they lose maternal immunity. Disease, falls, and predation are, therefore, probably major mortality factors among young animals, and the additional possibility of killing by adult males cannot be discounted. Such mortality has been described for the Asian colobines *Presbytis entellus* (Sugiyama, 1965; Mohnot, 1971; Hrdy, 1974) and *P. senex* (Rudran, 1973). There is circumstantial evidence for similar behavior among guerezas at Kanyawara.

Apart from infants, four or five other animals were thought to have disappeared from the six Kanyawara groups (original total size 58) between January 1971 and March 1972. This may have been due to death or emigration. Over the same period, six males joined one group. However, they may not have been immigrants to the population. They could well have been in the area, unnoticed, before they joined the bisexual group. In fact, these Kanyawara data do not relate to a well-circumscribed population. The six groups ranged into areas almost certainly used by other groups, while one study area group, which was not successfully counted until September 1971, has been excluded from the analysis.

Dunbar and Dunbar (1974) report only two disappearances among 47 animals older than precolor change infants in five groups at Bole between May 1971 and September 1972, assuming no replacement of disappearing adults or subadults by similar immigrants during a 1 year break in observation.

C.   Reactions to Man

Guerezas were found to be shy of man in every locality they were observed; groups reacted warily even in National Parks, where people were often seen and where little or no felling or hunting had presumably occurred for some years.

On being disturbed by the approach of an observer on foot, members of an unhabituated guereza group would generally move rapidly away, especially if disturbed from a low level in their habitat. An adult male sometimes lingered and used intermonkey threat gestures at the observer. Occasionally, males would produce roaring vocalizations on disturbance; these vocalizations were also produced in response to the close approach of a raptor or upon hearing the alarm calls of other animals.

A common response of animals after the initial dash was to enter a tall, usually well-foliated, tree and to remain quietly in the crown for a long period; hiding rather than prolonged flight occurred. Once disturbed, animals not habituated to the observer were therefore very difficult to locate in thick forest, despite their size and striking coat pattern. With luck, bits of fur of a few animals might be spotted, and the white tail tips were often a telltale sign. In open, riparian vegetation, relocation was easier, although here a group might descend to the ground and move rapidly and quietly away, concealed by low growth, only appearing above ground far from the point of disturbance. This was an effective way of eluding an observer and also, presumably, a hunter.

Even after long periods of observation, guereza groups remained more wary than other monkey species under study in the Kibale Forest. Although such wariness might be the result of past hunting, it is possible that the animals are instinctively "shy." I have noticed that some zoo guerezas will give threats and move to another part of their cage when closely watched—not a reaction of many zoo monkeys.

IV.   Human Interference with Guereza Habitats in East Africa

A.   Introduction

Human interference with plant communities has had effects on the populations of many, perhaps most, animal species. This interference ranges from the removal of a few plants to the wholesale destruction of a community and its replacement by another. This interference may be indirect, changes being brought about by human activities in another place or through the introduction, protection, or destruction of particular animal species.

To examine the effects on a guereza population of the controlled alteration by man of its forest habitat, I made a series of census walks at three sites in the Kibale Forest at different stages of timber exploitation. For comparative purposes, similar walks were made in a variety of habitats in other parts of Uganda. Many of these habitats had been affected in some way by man. However, I found no evidence of men hunting guerezas in any of these places, though hunting might have been occurring in several at a low level (and must have occurred in many at some time in the past). I believe the variations that were revealed in guereza abundance were due largely to factors other than hunting.

B.   The Kibale Forest

1.   Description and Census Routes

The Kibale Forest Reserve covers 560 km² of Toro District, West Uganda, and was first gazetted as a Crown Forest in 1932. Only 60% of the reserve is dominated by trees, the remainder comprising grass-covered hilltops and nonwoody swamp vegetation. The trees form a moist high forest (the average annual rainfall at Fort Portal, close to the northern edge of the forest, was 1485 mm in the years 1904—1971). The forest lies on the boundary of the lowland and upland vegetation zones, rising from an altitude of 1160 m in the south to an altitude of 1620 m in the north. Further information about the reserve is given in Uganda Forest Department Working Plans (Osmaston, 1959; B. Kingston, 1967) and by Wing and Buss (1970).

The Working Plans for the reserve propose a controlled exploitation of the forest for timber, and the planting of the unforested hills with exotic conifers. Both these processes are now well-advanced in northern sections of the reserve. For exploitation, the reserve is divided into a number of compartments, of between about 150 and 600 hectares. Exploitation of forested compartments is planned on an 80-year cycle (see also Struhsaker, 1972, 1974). Large, commercially valuable trees ("desirables") are cut for timber. In the process, many other large trees are cut or knocked down, and left to rot. Some years (1—7) after this selective felling, the compartment is "refined," old and defective desirables, together with undesirable "weed" species, are poisoned with hormonal arboricides (2,4-D and 2,4,5-T). The plan is that a forest dominated by even-aged valuable trees will develop to a point where felling is practicable in 80 years after the original felling.

Although forest is not totally removed by this process, its structure and composition are almost completely altered in a very short time. A high, multi-layered structure containing a great variety of plant forms of different ages is replaced by a dense, low tangle of second-growth species, out of which rise a few medium and large trees.

To assess the relative densities of the guereza and other monkey species in parts of the forest at different stages of this exploitation, repeated walks were made (at intervals of at least 6 days) along the following three routes:

1. 4000 m of trail through my main guereza study area at Kanyawara. Most of this walk was through unfelled forest in compartment 30, but the route also encompassed areas where felling had occurred in 1969: a small section of compartment 30 and the edge of compartment 14. These felled areas were unrefined. The typical vegetation of middle hill slopes in compartment 30 was a forest of many tree species, both evergreen and deciduous, with no one species obviously dominant (Fig. 4). There was a discontinuous "B" stratum (Richards, 1952) at 15 to 25 m through which the crowns of large isolated emergent trees ("A" stratum) reached to over 30 m. Toward the hilltops, the variety and height of trees decreased and toward the valley bottoms the shrub layer thickened until the forest gave way to low-stature swamp vegetation (a fuller description is given in Oates, 1974). Exploitation in compartment 14 had most affected the middle and lower hill slopes, where felling and ancillary track-clearing operations had produced extensive areas where no trees reached above 6 m. Below this height sprawled a dense tangle of shrubs and climbers. There were scattered groves of medium-sized trees reaching to 15 m and the fast-growing colonizing tree *Trema orientalis* was common. A few "undesirable" large trees remained, growing to 30 m. The ground was littered with decaying logs and stumps.

2. 3850 m of dirt track running roughly northeast from Kanyawara through compartments 14 and 15. The vegetation in compartment 15, which was felled in 1968–1969 and was also unrefined, closely resembled that in compartment 14. Tall *Pennisetum* grass grew on both sides of this route for about 400 m and on just one side for an additional 275 m.

3. 2000 m of dirt track running from Kigami Tea Estate (0°38′N, 30°23′E; 1460 m elevation) into compartment 7, which had been felled between 1957 and 1960 and refined in 1964. In 1971, this compartment carried a dense tangle of growth reaching somewhat higher than in compartments 14 and 15 (Fig. 5). A few large trees remained standing, but trees were generally less common than in compartments 14/15. A few scattered trees reached to around 15 m but very few patches of forest remained. This was the site of studies in 1965 by C. B. Koford, who performed census work along the same route and whose results are compared below with mine.

2. Census Methods and Results

Full details of the census technique employed are given by Oates (1974) and Struhsaker (1975). The census was a slow walk (with frequent pauses) during

Fig. 4. Moist high forest, Kanyawara, Kibale Forest Reserve; compartment 14 (selectively felled in 1969) to the left of the track and compartment 30 (unfelled) to the right. Photographed in August 1974.

which the habitat was carefully scanned for evidence of primates. When primates were sighted, or heard at close range, the observer stopped, and then usually remained in approximately the same place for 10 minutes. In this period the identity of all visible individuals was noted.

The sightings have been assessed in terms of *clusters*. A *cluster* is any number of primates of any number of species, within 50 m of one another at some point during the observation period. Any individuals separated by a horizontal dis-

Fig. 5. Regenerating vegetation at Kigami, Kibale Forest Reserve; originally high forest, felled in 1957—1960 and refined in 1964. Photographed in December 1971.

tance of at least 50 m (or by an obvious physical barrier) throughout the period belonged to a separate cluster. In the poor visibility conditions, accurate group counts were impossible in the short space of time given to each cluster. However, in the case of *C. guereza*, the number of separate clusters in which the species was recorded probably corresponds closely to the number of social groups that were observed. At least at Kanyawara, individual guerezas spend most of their time within 50 m of members of their own social group and at least 50 m away from members of other groups.

The results of these walks are presented in summary in Table III. *Colobus guereza* was the most commonly encountered species in terms of numbers of records in separate clusters, and it was encountered at a similar frequency in each area. Considering the differences in length of the walks, *Cercopithecus ascanius* (the redtail) and *Colobus badius* (the red colobus) were much less frequently encountered in the completely felled areas, compared with the main study area. Differences were less great for *Cercopithecus mitis* (the blue monkey). Three other species were infrequently encountered.

However, felling and refining operations had produced considerable differences in visibility between the three areas. These differences are indicated by a comparison of the mean distance at which clusters were first sighted.

TABLE III

Comparison of Results of Primate Census Walks Made in Three Areas of the Kibale Forest Reserve, West Uganda, in 1971–1972

Census area	Length of walk (m)	Number of walks	Total search time (minutes)	Total primate contact time (minutes)	Mean estimated distance at which clusters first sighted (m) [a]	Total number clusters encountered	No. clusters encountered per walk: mean and range (in parentheses)	Number of distinct clusters in which a particular species was recorded: Mean and range (in parentheses) per walk							Mean no. C. guereza sightings/km	Mean no. C. guereza sightings per minute of search time
								Colobus guereza	Colobus badius	Cercopithecus ascanius	Cercopithecus lhoesti	Cercopithecus mitis	Cercocebus albigena	Pan troglodytes		
Kanyawara main study area (compartments 30/14)[b]	4000	15	2779	1331	44	112	7.5 (4–12)	3.3 (2–7)	2.9 (1–5)	2.9 (0–5)	0.1 (0–1)	1.7 (1–4)	0.3 (0–1)	0 (0)	0.83	0.018
Kanyawara felled area (compartments 14/15)[b]	3313[c]	7	737	459	76	44	6.3 (2–12)	3.7 (1–8)	1.9 (0–4)	1.3 (0–3)	0 (0)	1.0 (0–3)	0.1 (0–1)	0 (0)	1.12	0.035
Kigami (compartment 7)[d]	2000	7	310	253	56	26	3.7 (2–7)	1.6 (1–2)	0.4 (0–1)	0.9 (0–4)	0 (0)	1.0 (0–2)	0 (0)	0.1 (0–1)	0.80	0.035

[a] Horizontal distance in straight line from observer.
[b] Walks always commenced at about 0740 hours.
[c] Total walk length of 3850 m corrected to account for areas of grassland uninhabited by monkeys.
[d] Six walks commenced in late afternoon (when C. B. Koford conducted most of his census work).

Compared with the main study area, long-distance visibility was better in felled forest and sightings were obtained at distances further from the path.

3. Estimation of Guereza Density

In the main Kanyawara study area, the mean number of guereza sightings per census was 3.3. I believe that walks revealed most groups within a mean distance of 50 m on either side of the census route. On that basis, the census may be taken to have surveyed 36 hectares of forest, if allowance is made for overlap zones. An average of 3.3 groups in that area represents a mean density of 0.092 groups per hectare. The average number of individuals in Kanyawara study area groups in March 1972 was 10.7, giving a mean population density of 1.0 individuals/hectare. At times, guereza density rose well above this mean figure. On one census walk (February 29, 1972), all seven groups known to use the study area (most also ranged outside it) were sighted, representing a density of 2.1 individuals/hectare within the census strip.

The Kanyawara compartments 14/15 and Kigami compartment 7 census routes probably surveyed a larger area of forest on either side of the path than did the route through the main study area. These census strips are estimated to have been 150–200 m wide rather than 100 m. Therefore, although 1.12 guereza sightings per kilometer were made in compartments 14/15, the density of guereza groups was probably lower than in the study area (0.83 sightings/km). If the census strip was 150 m wide, 1.12 sightings/km would be equivalent to 0.075 groups/hectare. Similarly, the 0.80 sightings/km at Kigami would represent 0.053 groups/hectare. Since 150 m is a minimum estimate of census-strip width, these figures are maximum estimates of density.

Also to be taken into account in comparing census data is the possible variation in group size between areas. In compartments 14/15, eight groups believed to have been counted with fair reliability contained an average of 7.9 animals, while five groups counted with similar reliability at Kigami contained an average of 8.6. Applied to the above estimates of maximum numbers of groups per hectare, such figures would give a density of 0.6 individuals/hectare in compartments 14/15 and 0.5 individuals/hectare at Kigami. It must be remembered that a large number of estimations have been made to derive these values, while the census data from felled forest are based on a small number of walks.

4. Comparisons with Other Studies

The results from my three census routes may be compared with those obtained, using a similar technique, by Struhsaker (1975) at Kanyawara in 1970–1972. Struhsaker made 44 census walks along a 4020-m route entirely within unfelled forest in compartment 30 at Kanyawara, and eleven walks along a 6100-m route in felled forest. The latter passed through compartments 13 (felled in 1968 and refined in 1968—1969 and again in 1970—1971), 12 (felled in 1967—1968 and refined in 1973, after Struhsaker's walks were conducted), and 17 (felled in 1966 and not yet refined by the end of 1974). My results may also be compared with those obtained by C. B. Koford along the same 2000-m route at Kigami in 1965 (Koford, 1972).

In compartment 30, Struhsaker found a somewhat higher density of *C. badius* than is indicated in the results from the areas I surveyed, and he found a much lower *C. guereza* density (only 0.2 groups were encountered per kilometer). His results from felled compartments 12, 13, and 17 are similar to those which I obtained in compartments 14 and 15. In the felled area, Struhsaker encountered an average of 4.9 guereza groups per walk, or 0.8 per kilometer. Estimating that his census surveyed 100 m on either side of the path, Struhsaker derived a density of 0.04 groups/hectare from his felled area data. With a group size of 8, this would be equivalent to 0.3 individuals/hectare.

Koford's data were collected in a different way from mine. He attempted to count all monkeys within 100 m of the track on both outward and return journeys, which were made either on foot or in a vehicle. However, he did not endeavor to recount animals on the return journey which he believed were sighted on the outward section. Koford did not attempt to describe "clusters." He reports a mean of 24.6 individual guerezas per census from twenty-three round trips along the 200-m route. He counted an average of 5.3 members in 89 guereza "groups" sighted at Kigami (including single animals). On that basis he averaged 4.6 groups per round trip. However, Koford is unlikely to have seen all group members during his census periods.

Even after making allowance for the extra groups his method would be likely to count, Koford's data suggest a greater density of guerezas at Kigami in 1965, the year after compartment 7 was refined, than in 1971—1972, when I averaged 1.6 sightings· of guerezas as members of distinct clusters (probably equivalent to separate groups) on one-way walks. I never definitely saw more than three separate guereza groups during a census, while on just the outward or return part of his survey Koford regularly saw five or six groups and once as many as seven (C. B. Koford, in litt).

5. Discussion: The Effect of Forest Exploitation on a
   Guereza Population

On first examination, the census results suggest that commercial exploitation of the Kibale Forest reduces overall primate densities, but that the guereza itself is much less affected than other species. This opinion has been stated by Struhsaker (1972), who suggests that felling may in fact benefit the guereza. The results from Struhsaker's census work within unmodified areas of compartment 30 certainly indicate strikingly fewer guerezas than in my immediately adjacent study area, which contained the edge of felled compartment 14 (together with felled and unfelled sections of compartment 30). However, I believe that some of this difference in density was due to factors other than felling, for guerezas within my study area did not concentrate their time in the felled forest and, in 1970 (the year after felling took place), groups were occupying stable ranges, which they apparently knew well. Work by Struhsaker in another unfelled area at Dura Bridge, to the south of Kanyawara, confirms that the population concentrations of different primate species are far from homogeneous throughout the Kibale Forest (Struhsaker, 1975).

In the absence of human interference, the primary influence on guereza population levels through the Kibale Forest may be the abundance of major food trees, although other factors, such as special nutritional requirements, may also play a significant role (Oates, 1977). The guereza's food supply, which is derived mostly from forest-edge and middle-story trees, is probably less affected by large-tree felling than that of species such as red colobus, which obtain much of their food from those large trees. Selective felling may indeed increase the density of guereza food trees over a short period of time (though it may remove other requirements, such as sleeping sites). However, the complete removal of large trees over wide areas, the end result of felling plus refining, is almost certainly deleterious even to the guereza.

The high guereza densities which Koford found at Kigami in 1965 could reflect an increase in guereza density (at the expense of other species, which he encountered much less frequently), in the years following selective felling in 1957—1960, and before the results of refining in 1964 had been felt. However, my data from the same place suggest that guerezas have declined in numbers between 1965 and 1971—1972, a period during which most large trees disappeared.

Apart from the possible long-term effects of the alteration of forest structure and composition on guereza population density, the exploitation of the Kibale Forest has other, more immediate, results. A number of monkeys, including guerezas, are killed during the tree-felling operations, and the disruptions which must be made to the ranging patterns of guereza groups (which occupy small ranges that include defended areas) may be expected to cause great social disturbances and interference with reproduction.

C. Other Areas of East Africa

1. Census Results

Table IV presents the results of some census walks made in Ugandan localities outside the Kibale Forest, using the technique described in IV,B,2. Great variations are apparent in the number of guereza sightings. The highest number of sightings (greater than 0.04 per minute of search) were obtained from one

TABLE IV

Guereza Sightings on Primate Census Walks in Some Ugandan Localities Outside Kibale Forest Reserve

Locality	Date	Vegetation type[a]	Approximate length of walk (m)	Search time (minutes)	No. of distinct clusters in which C. guereza sighted	No. of C. guereza sightings per minute of search time
Kaizi River head-waters,	January 20, 1971	LHF,CLF	Undet.[b]	258	4	0.016
	January 20, 1971	LHF,CLF	3000	84	1	0.012
Maramagambo Forest, Rwenzori National Park	January 21, 1971	LHF,CLF,SF	Undet.	135	5	0.037
	January 21, 1971	LHF	Undet.	115	0	0
	January 21, 1971	LHF,CLF	Undet.	173	8	0.046
Lake Nyamusingire, Rwenzori N.P.	January 24, 1971	W	Undet.	66	3	0.045
	July 25, 1971	W	Undet.	90	4	0.044
Ishasha River, Rwenzori N.P.	July 22, 1971	RF	Undet.	126	1	0.008
	July 23, 1971	RF	4000	123	3	0.024
Chobe, Kabalega N.P.	June 17, 1971	WG,RF(r)	6450	118	3	0.025
	October 21, 1971	WG,RF(r)	5500	76	3	0.039
Rabongo Forest, Kabalega N.P.	February 17, 1971	LHF(r)	Undet.	165 (Forest) 111 (Grass)	5	0.030 (Forest)
Budongo Forest Reserve	May 24, 1971	CLF	3450	66	3	0.045
	May 26, 1971	CLF	2700	70	3	0.043
	May 27, 1971	LHF	4750	137	1	0.007
	May 28, 1971	CLF	5900	141	0	0

Itwara Forest Reserve	LHF,CLF	March 23, 1971	5500	241	0	0
	CLF	September 24, 1971	3350	122	0	0
Bwindi Forest Reserve	UF(bamboo)	October 10, 1971	4100	115	0	0
	CUF	October 10, 1971	3300	87	0	0
	UF,CUF	October 11, 1971	Undet.	278	0	0
Kadam Forest Reserve	DUF,RW	January 19, 1972	1700	155	5	0.032

[a] Key to abbreviations (the following vegetation types are described in detail in the Appendix, Section X.) LHF, moist lowland high forest; CLF, colonizing moist lowland forest; UF, moist upland forest; CUF, colonizing moist upland forest; DUF, dry upland forest; RF, riparian forest; SF, permanent swamp forest; W, woodland; RW, riparian woodland; WG, wooded grassland; (r), relic of vegetation type.

[b] Undet., undetermined.

walk along the eastern edge of the Maramagambo Forest, from two walks through the woodland beside Lake Nyamusingire, and from two walks in the Budongo Forest (one through a felled area, the other near the forest edge). Walks into the main body of the Maramagambo, Itwara, and Bwindi Forests produced no sightings but, on the other hand, neither did one walk through a felled section of the Budongo Forest nor two along a road following a ridge through the Bwindi Forest, where guerezas certainly occur (personal observation).

It is difficult to draw reliable conclusions from the results, because of the likely biases caused by variations in visibility between habitats. A guereza density of at least one individual per hectare, however, is possible in those areas where census walks yielded more than 0.04 sightings/minute. The highest number of guereza sightings on any census at Kanyawara (where mean density was estimated to be 1.0/hectare) was seven, representing 0.037 sightings/minute. An extrapolation from Kanyawara would not be justified for an area where group sizes were very small, but groups in Maramagambo and Budongo were similar in size to those in Kibale. The census results suggest that high guereza densities are more typical of forest edge (colonizing forest), riparian forest, and woodland than of mature high forest. This was also the impression gained from other, unsystematic, observations throughout East Africa and supports the idea that limited felling of mature forest might actually lead to increased guereza densities.

However, guerezas are not absent from mature forest. The 4750 m walk through mature *Cynometra* forest in the northern part of Budongo produced one sighting of guerezas (0.007 sightings/minute of search time), while a different group was spotted as I retraced my steps. This was in an unfelled compartment at least 4.5 km from the margin of the forest. In mature high forest, compared with more open and lower-stature vegetation, it is probably more difficult for a ground-based observer to detect arboreal animals. This could have contributed to the low number of sightings.

2. Comparisons with Other Studies

Data available from other studies seem to confirm the evidence that guereza densities are highest in colonizing and riparian forest situations. T. J. Kingston (1971 and in litt.) has estimated a minimum density of 40 guereza groups/km^2 in the Kinale (or Kinari) area of the Kikuyu Escarpment Forest, Kenya, based on a *maximum* number of thirteen groups seen within 100 yards (91 m) of one side of a 3.6-km cut line in a set of four walks. The Kinale vegetation may be classified as colonizing moist upland forest; it is at about 2600 m and is described by Kingston as "relatively disturbed."

Kingston's data suggest a mean group size of just over five animals (excluding solitaries) at Kinale and in similar forests. However, this is based on short-term counts; it is likely that actual group sizes were somewhat greater. An average group size of five would give a density of about two individuals/hectare. If groups were actually larger, two/hectare is an underestimate of guereza density in this area. On the other hand, it is unlikely that groups restricted their movements to within 90 m of Kingston's path, so that the maximum density recorded at any one time in the census strip probably does not correspond to long-term average density in that area.

An even higher guereza density has been found in a remnant patch of the same Kikuyu Escarpment Forest near Limuru. In this valley-bottom growth, dominated by colonizing tree species such as *Polyscias kikuyuensis*, six groups counted by Schenkel and Schenkel-Hulliger (1967) contained 41—48 individual guerezas. These occupied an area of about 10 hectares (measured from their sketch map), or a minimum of 4.1/hectare.

For Ethiopia, Dunbar and Dunbar (1974) report a guereza density of 1.4/hectare in gallery forest in the Bole Valley and in lakeside forest at Lake Shalla. In part of the Bole gallery forest they estimate a density of 3.7/hectare, based on seven groups occupying barely overlapping ranges totaling 14 hectares in area. Lower densities were estimated for other habitats such as "bush" in the Bole Valley, dry upland *Juniperus* forest, and upland *Hagenia* forest.

In support of the view that guerezas are not abundant in mature forest, L. S. B. Leakey (quoted in Harrisson, 1971) has stated:

> . . . all evidence known to me in Kenya and Tanzania is that *Colobus* occupy the fringe areas around the big forests. and seldom penetrate more than a mile from the edge inwards. . . .

However, on the evidence presented above from Budongo it seems that guerezas do occur well inside mature forest, albeit at low densities. Gautier and Gautier-Hion (1969) report the species (as *C. polykomos abyssinicus*) from a homogeneous forest of large trees at Belinga in Gabon, and Haddow *et al.* (1947) report it (as *C. polykomos uellensis*) as the most common of monkeys in the Semliki Forest, Uganda, particularly in the "denser areas of lowland rain-forest." It must be remembered, of course, that there is probably no high forest in Africa that has been completely undisturbed by man, and even where human activity has been slight, watercourses and tree-falls produce edge effects. Kingdon (1971) indicates that much of the central and western part of the Semliki Forest was once cultivated, while the low-lying and poorly drained eastern section is subject to flooding during the rains.

3. The Effects of Agriculture and Forestry

Low-intensity crop growing may alter high forest in such a way that guereza densities are increased, if the damage results in the appearance of areas of colonizing forest. However, unless human interference occurs at a very low level, or is spread patchily over a large area, its result is often the gradual erosion of forest and its replacement by lower-stature, grass-dominated plant communities, commonly sustained by regular fires. In such situations, guerezas (and other forest animals) may survive in forest patches in places (such as mountains and valleys) that are difficult to cultivate or raise stock. Although a network of forest fragments, with small gaps between, seems to provide suitable conditions for guerezas, small and isolated patches are unlikely to favor the long-term survival of their inhabitants.

Although several East African Forest Reserves are managed in the same way as Kibale, with a seminatural form of regeneration encouraged, many areas of forest that must once have supported guereza populations have been completely removed and replaced by man's crops. One example in Uganda is the tea estates immediately outside the boundaries of Forest Reserves (including Kibale). Another is formed by the overcrowded hills of Kigezi District, where the once forested land is now covered by a patchwork of small fields and where trees of any kind are rare. There are, in addition, Forest Reserves like that on Mt. Elgon, where, after initial felling, forest is totally cleared by the "taungya" system (Synnott, 1968): local people are allowed to cut and burn the remaining vegetation and then to grow crops; among the crops, exotic conifers are planted; when mature, these conifers will form a forest uninhabitable by guerezas. A forest of exotic conifers and *Eucalyptus* has now almost totally replaced the original vegetation of the Mafuga Forest Reserve in Kigezi District. Over vast areas of the Kenya Highlands, there are also plantations of exotic trees.

4. Man and Game

Plant communities are obviously affected not only by direct human actions but also by climatic changes and by the activities of animals which inhabit the communities. Some animal species, in addition to man, can produce profound changes in a plant community.

The hippopotamus and elephant have been held responsible for the large-scale alteration of plant communities in East Africa. Laws *et al.* (1970) attribute an increase of small trees and bushes beside the Nile in Murchison Falls (now Kabalega) National Park in Uganda to erosion and hippopotamus overgrazing, and they relate the destruction of much of the woodland in the same park to high elephant densities. The remaining high forest in the park, probably once much more extensive, is also being threatened by elephant damage at its margins, so reducing its resistance to fire (see also Buechner and Dawkins, 1961). The heavy concentration of elephants in the Kabalega Park is the result of the removal of traditional human hunting inside the park and the growth of human settlement outside it. In the park, riparian forest along the Nile and high forest remnants near Rabongo Hill are inhabited by guerezas. These guerezas are threatened by the elephant-induced destruction of their habitat.

Many East African National Parks contain guereza populations, but the theoretical preservation of the species within such a park does not guarantee its survival in the face of changes in the ecosystem. Most of the parks containing guerezas are not, however, so immediately threatened by drastic change as is Kabalega.

Human impact on wild animal populations may affect guereza population density in less obvious ways. For instance, the tree-species composition of East African forests may be modified by elephant activity, and the present structure of many forests probably reflects the status of elephant populations over a long period of time (see Laws *et al.*, 1970). East African elephant populations have long been influenced by man's hunting (Spinage, 1973), while forest tree-species composition will influence the density and dispersion of primate populations.

V.   Hunting and Trapping of Guerezas in East Africa and Ethiopia

A.   Skins

1.   Local usage

For many centuries, the striking coat of the guereza must have made the animal a target of hunting by African peoples seeking ornamental dress. Johnston (1886) describes Chagga soldiers of the Kilimanjaro district wearing striking war costumes which included guereza skin caps, mantles, and tails. Johnston also notes similar dress worn by Masai warriors (Fig. 6). Blixen (1937) has recorded "bold cavalier-like cockspurs" of colobus skin displayed at Kikuyu dances. Haddow (1952) reports colobus fur headdresses worn at accession ceremonies in Bunyoro (and associated Ugandan Kingdoms) and colobus fur headdresses and costumes worn at Watutsi ritual dances in Rwanda. Ullrich (1961) refers to Masai, Wamera, and Warusha wearing the skins and mentions colobus skin shield-coverings in Ethiopia. In the Kenya National Museum there is a pair of Samburu anklets made of guereza skin decorated with beads, and I have seen a man wearing a guereza skin hat at Namalala near Mt. Debasien, Karamoja. Koford (1972) has seen guereza tail fly whisks near Mt. Elgon.

African people capture monkeys in a number of ways, but shooting with a gun has probably been the most widely employed technique in eastern Africa in recent times. In Uganda's Semliki Forest, bows and arrows are employed (Struhsaker, in litt.).

2.   Past Overseas Demand

The heaviest hunting of guerezas for their skins has, however, probably resulted not from local use but from overseas demand. According to Sanderson

Fig. 6.   A Masai warrior wearing a guereza-skin cape (from Johnston, 1886, by permission of Routledge and Kegan Paul Ltd.).

(1957), guereza skins were greatly prized in Central Asia at least as long ago as the time of Marco Polo's journeys (1271–1295) and were used as shoulder capes in decorative mantles. Marsden and Wright's (1908) edition of Polo's "Travels" contains no reference to such skins, although there is a description of the extravagant use of "lion" (probably tiger and leopard), ermine, and sable skins in the covering and lining of tents in the court of Kublai, the Grand Khan.

The overseas demand for colobus skins increased greatly in the mid-nineteenth century. Wilcox (1951) records that long, black, silky monkey fur was first displayed at the Great Exhibition in London in 1851:

> A furrier who liked its lustrous quality made use of the pelts in costume, especially for muffs and short capes. . . . Often the whole pelt of both black and white hair was fashioned into feminine capes. This brings to mind a picture of the coachman's dress of the late nineteenth and early twentieth centuries, seen especially in large cities. With his glossy stovepipe hat and his smartly cut paletot, he wore a cape of monkeys furs, or sometimes one of bearskin, and over his arm he carried a carriage blanket of the same fur.

Bachrach (1946) says that the fur was used mainly in narrow strips to trim cloth coats, and G. Dyke (in litt.) tells me that it was also used to trim dresses and nightdresses, as well as being used for rugs and wall decorations.

Black colobus were the only monkeys whose skins were used in appreciable quantities by the fur trade. The majority of skins in this trade were West African *C. polykomos*, of which those from the area that is now Ghana and the Ivory Coast (*C. p. vellerosus*) were most in demand, because of the length and glossiness of their hair (Poland, 1892; Brass, 1925; Bachrach, 1946). The other species were used less commonly, but Poland noted that the skins of *C. guereza* were "extremely rare and much esteemed as a fur." The pelts of all species were generally dyed to darken the skin and enhance the blackness of the hair.

Monkey fur was never a staple of the fur trade and demand fluctuated with fashion (G. Dyke, in litt. 1975). According to statistical data in Brass (1925), 1,750,000 skins of "Afrikanische Affen" (these would have been almost entirely *Colobus*) were auctioned between 1871 and 1891 in London, which was the great *entrepôt* of the world fur trade until at least the 1930's. Annual auction figures rose from 19,814 skins in 1871 to 223,599 in 1889. Most of these were, as I have noted, *C. polykomos* and not, as suggested by Sanderson (1957), *C. guereza*. Sanderson has stated that the trade in colobus skins declined after 1892, but on the evidence of Brass's statistics he seems to be wrong in suggesting that, with 175,000 skins reaching the European market, 1892 was the peak year. Germany was apparently the ultimate destination of the majority of skins, but they were also used in the United States, Canada, and Italy (Poland, 1892).

Poland reports that in the early years of the fashion, *C. polykomos* skins fetched 1 pound each. By 1892 some were sold for as little as 1 shilling (1/-), although *C. guereza* skins could still fetch 10/- to 15/-.

Shortly after the first world war there was some renewed fashionable interest in monkey fur; Brass reports a trade of 30,000 to 40,000 skins per year with prices ranging from 20/- to 30/- in 1923. By 1930, however, demand had dwindled to nothing (G. Dyke, in litt.).

The impact on black colobus populations of such large-scale hunting is not known with any accuracy, but it must have had effects in many parts of Africa and some populations may have been extinguished. In 1900, Schillings (quoted in Ullrich, 1961) had difficulty obtaining colobus from Kale "oasis" for the Berlin Natural History Museum because the animals had been "almost exterminated" by the Askaris of Moshi Station. G. M. Allen *et al.* (1936) stated that guerezas had been exterminated in the Sipi Forests by Bugishu hunters (Sipi is on the northwest flank of Mt. Elgon, on the edge of the present Bugisu District, Uganda). Kingdon (1971) ascribes the absence of guerezas from the forests of Mengo and Busoga Districts in Uganda to the 19th century demand for skins.

The species was apparently found there in recent times as it is "a well-known class totem and the skins were formally worn at feudal courts." Kingdon notes a very old skin, reputed to originate from the Mabira Forest, Mengo, in the possession of a witch doctor in Kyaggwe.

3.   The Skin Trade Today: Sales to Tourists

At the time of my investigation (1974), large numbers of guereza skins were still to be found on sale in Kenya and Ethiopia, with foreign tourists the major targets of the sellers. This trade has been described by Mittermeier (1973). Most of the skins are sold in the form of circular rugs to be used on the floor or as wall hangings. The rugs are a sewn composite of the back skin (including the white mantle) of several animals (see illustrations in Mittermeier, 1973 and in Dunbar and Dunbar, 1975). Most rugs on sale in Kenya contain skin from five animals, but Mittermeier reports one rug of 49 skins. R. I. M. Dunbar (in litt. 1975) informs me that rugs of 25 to 45 skins are common in Ethiopia and that 75-skin rugs are not unknown.

Mittermeier surveyed 70 "tourist shops" in Nairobi and Mombasa, Kenya, and Dar-es-Salaam and Arusha, Tanzania, in August and September 1972. He found no colobus skins in the Tanzanian shops, but reports that 24 of 60 Kenyan shops had skins in stock. He counted skins from 5002 monkeys in the goods on display in these shops (1530 in eleven shops in Mombasa and 3472 in thirteen in Nairobi). One shop told him that it had 15,000 single skins and 500 five-skin rugs in stock in addition to 626 skins on display. Without independent confirmation, such "in stock" claims have to be viewed with caution and cannot fully justify Mittermeier's estimate that at least 27,500 guerezas had been killed to provide the stocks in Kenya at that time.

As a check on the situation in 1974, I surveyed many shops in Nairobi during 4 days in September. I located guereza skins on display in sixteen shops. A few single skins were available, and there were many rugs of five skins and a few of more (the largest number of skins I counted in one rug was 26). One shop also had a number of coats made of guereza skin on display, each of which I estimated to contain remains from fifteen animals. In most shops I was able to make accurate counts of the number of skins on public display, in others I made conservative estimates. The total number of skins assessed to be on display in Nairobi was 982. Half of these were inside a single shop: *Jewels and Antiques*, Kimathi Street. The actual total on display was almost certainly within the range 950—1100, and I am confident that I did not miss any shop in central Nairobi displaying a large number of skins.

I was quoted prices of 100 to 200 Kenya shillings for single skins (about U.S. $ 15—30). Five-skin rugs were quoted at 200/- to 500/-, with most at about 350/- (compared with 160—373/- found by Mittermeier). The prices of larger rugs varied from 800/- for a fourteen-skin rug to 1500/- for an eighteen-skin rug

(Mittermeier found that rugs of 15 to 49 skins varied from 650—2000/-). Bargaining could possibly have reduced these prices.

Although "in stock" statements are likely to be unreliable, no Nairobi shop in 1974 indicated that it had anything like 17,500 skins in stock. At the shop which had the most skins on display, a salesman suggested that there were about 500 other *rugs* in stock. Mary Sue Waser enquired at the same shop 2 months later and was told by a salesman that he thought one thousand *skins* were in stock in addition to those on display (however, the same man gave greatly exaggerated estimates of the number of skins contained in rugs and coats on display) (M. S. Waser, in litt.).

Beryl Kendall carried out a survey of Mombasa shops for me in late January 1975: 201 guereza skins were displayed for sale in seven shops, made up into five- and twenty-skin rugs; prices seemed lower than in Nairobi, with single skins quoted at 70—150/- and five-skin rugs at 180—250/- (B. Kendall, in litt.). In Malindi, to the north of Mombasa, I noticed several colobus skins for sale in souvenir shops in July 1974. These were not counted.

The total number of skins on display in Nairobi and Mombasa in 1974—1975 is therefore considerably smaller than that recorded in 1972 by Mittermeier, with prices, at least in Nairobi, slightly higher.

The number of tourists visiting Kenya has dropped since 1972, probably due to general world monetary inflation and, in particular, to the rising cost of travel. The Nairobi *Standard* reported on September 6, 1974 that there were only 397,700 visits to Kenya in 1973, compared with 444,330 in 1972 (a 10% drop). However, the first quarter of 1974 showed a 5% increase over the same period in 1973. A decrease in the level of tourism presumably affects the demand for goods such as skins, but other factors also influence the level of retail stocks. Several traders told me that colobus skins have been more difficult to obtain in the last few years. Two traders related this to difficulties in getting Ethiopian imports since the species became protected in that country. One trader was obviously prepared to negotiate with me on the importation of skins from Uganda when he learned I had been studying guerezas there.

One of the most useful indicators of the impact on guereza populations of these tourist sales would be turnover figures, as skins in stock could have been held for months, or years. Accurate turnover figures are, however, very difficult to obtain. Although I asked traders for information, many were understandably reluctant to provide it. Some refused to give any information and most claimed not to keep records. Some claimed to sell only one rug every 3—4 months, but one claimed sales of seven to eight rugs daily at some times of year. One trader claimed to have had twelve five-skin rugs in stock on August 28, and he had only three left on September 3. Another said that in the previous month stocks had been reduced from 200 to two rugs. A very crude estimate, based on a judgement of what I was told by traders, is that the annual Nairobi turnover is 20,000 or fewer skins. It is worth noting that during my survey I never saw anyone examining or purchasing colobus skins and never saw anyone with a skin outside a shop. In Mombasa, figures quoted by traders suggest a monthly turnover of some 700 skins at the height of the December—January tourist season (B. Kendall, in litt.).

In Addis Ababa, which, with Nairobi and Mombasa, has been one of the chief sites of colobus skin sales to tourists, one dealer told Dunbar and Dunbar (1975) that he had 4000 rugs in stock, which the Dunbars estimate would have contained 100,000 skins. The Dunbars guess at 200,000 skins in Ethiopian shops, probably accumulated over 5 years. Despite the efforts of the government Wildlife Conservation Organisation, the skin trade in Ethiopia is apparently very large and lucrative.

Much smaller sales occur to tourists elsewhere in East Africa, including Uganda. People (including pygmies) on the road through the Semliki Forest, Uganda, will attempt to sell single skins to visitors, but it is unlikely that more than a dozen skins are sold annually in this way—even when many tourists are visiting the country. At present, the number of tourists is small. M. S. Waser (in litt.) reports finding two guereza skins for sale in shops in Kampala, Uganda in late 1974.

4. Present Export Trade

Although monkey furs are not presently as popular in the western world as they were in the 19th century, they are still exported from East Africa for clothing and decoration. A Mombasa dealer told Mittermeier (1973) that he sold about 1000 five-skin rugs each year to firms in the United States (especially in Baltimore, Maryland) and Germany, and in 1972—1973 there seems to have been renewed fashionable interest in the skins for clothing. There was a large trade on the London fur market for about 18 months, with big sales to the Italian market and some to France and Germany (W. Ridge, personal communication). Up to 10,000 skins per month were fetching up to 4—5 pounds each (U.S. $ 9—12), when demand was at a peak. Most of the skins apparently came, as in the past, from West Africa (and belonged to *C. polykomos*), but I have seen documentary evidence that thousands of East African (and/or Ethiopian) guereza skins were also traded. A coat of guereza skins made in the Netherlands was pictured in the London *Evening Standard* in April 1973 (Fig. 7); this was quoted to cost around 600 pounds (about $1400). Although the skins were a popular gimmick for a time, the trade soon collapsed (at least in London), leaving unsaleable warehouse stocks.

While I was in Kenya in 1974, evidence of a continuing export trade was published in the Nairobi *Standard* on July 10. Two brothers had been charged with unlawfully possessing 26,200 colobus skins and attempting to export them to Switzerland. The magistrate dismissed the charges on the evidence of a game warden from Moyale (a town lying on Kenya's border with Ethiopia)

Fig. 7. Guereza-skin coat illustrated in the London *Evening Standard* on April 19th 1973 (London Express News and Feature Services).

that the accused had told him that the skins were from Ethiopia and that, after checking, he was satisfied and had issued the necessary permit. This is despite the fact that the species is protected under the laws of both countries (see Section VI).

I do not know whether these skins eventually reached Switzerland and, if so, how and where they were marketed. Some may have entered the East African tourist market. However, the size of this single cargo, carried in one lorry, indicates the very recent existence of a large trade in guereza skins.

R. I. M. Dunbar informs me (in litt. 1974) that, according to the Ethiopian Government Wildlife Conservation Organisation, official exports of skins and rugs are very small. Most go to Japan and none to Kenya. However, most Nairobi dealers claim that their goods originated in Ethiopia, and the above court case suggests this may be true. [The Nairobi dealer who told Mittermeier (1973) that he had 15,000 skins in stock, claimed that his skins came from the Wajiri District of North Kenya. Since this district is arid, with hardly any potential guereza habitat, it is more likely that the skins came from Ethiopia: Wajir lies on the Ethiopian border and carries the road from Moyale to Garissa, from which there are roads to Nairobi and Mombasa.]

There may be a continuing demand for guereza skins in the very long-established markets of the Middle East. C. B. Koford says that even in the 1960's, according to W. T. Roth of the Baltimore Zoo, thousands of guereza pelts were probably being exported to Arab countries through Somalia (Koford, 1972).

B. Meat

Monkey meat has long been a normal part of the diet of forest-dwelling hunting people in Africa. Throughout much of West and Central Africa, hunters still take a heavy toll of monkeys, including black colobus species, for food. The extra value of the pelt has no doubt made these species especially susceptible to meat hunters, and the two different uses of the animals are not easily separated in a consideration of hunting pressures.

Many pastoral and agricultural peoples in East Africa (such as the Batoro who live around the Kibale Forest) traditionally do not eat monkey meat. This has saved some guereza populations from meat hunting. However, there are monkey eaters in East Africa; Haddow (1952) believes that the majority of smaller Bantu and Nilotic tribes in Uganda have in recent times used monkeys as food. The Bakonjo of the Ruwenzori Mountains eat monkeys, and colobus are very rare on the Ruwenzori slopes (personal observation). Some Bakonjo hunted in the Kibale Forest until 1966. Haddow reported in 1952 that in Bwamba County, West Uganda, monkeys were still eaten by a large section of the African populace and that black-and-white colobus was one of the favored food species. Seventy years ago, Major Powell-Cotton (1904) reported that the Dorobo of the Mau Forest (on the eastern edge of the Kenya Rift Valley) lived "for the most part" on guereza flesh and he complained that, despite this, he had to pay 1 pound for every guereza he shot beyond the two permitted on his 50-pound shooting license.*

C. Crop Damage

In many areas, guerezas are killed in the belief, or with the excuse, that they have raided human crops. Since guerezas will travel and feed on the ground, such crop raiding probably occurs occasionally at forest edges. For instance,

---

*Shooting of hundreds of specimens for museum collections has been another form of human hunting pressure on guerezas, though one of less significance than hunting for trade or meat.

L. S. B. Leakey (quoted in Harrisson, 1971) has reported guerezas eating the leaves, but not the tubers, of sweet potatoes. However, I have never witnessed crop raiding myself.

In East Africa, guerezas have been shot by forestry officials in the belief that they damage planted conifers. L. S. B. Leakey (op. cit.) reported more than 100 guerezas shot in this way in one small forest in Kenya. I have heard of small numbers shot for the same reason on Mt. Elgon in Uganda. In fact, although guerezas may enter conifer plantations, there is no evidence that they eat conifers. Leakey confirmed this by an examination of stomach contents. *Cercopithecus mitis* is probably the monkey responsible for most conifer damage; large numbers of this species are shot in Kenyan and Ugandan conifer plantations (for Kenya, see Omar and de Vos, 1970).

Killing guerezas for allegedly damaging crops is obviously a potential source of skins for commerce.

D. Medical Research and Zoological Collections

The use of colobus monkeys in laboratory research seems to be negligible. ILAR statistics for the United States in 1969 (tabulated in Harrisson, 1971) show 25 *Colobus*, of unspecified origin, used in research.

Many colobine monkeys do not do well in captivity, but the black colobus species survive better than most and often breed. Scanning through volumes of the *Int. Zoo Yearb.* reveals that *C. guereza* was bred in at least twenty zoos (mostly in Europe and the United States) in the years 1968 to 1973. Black-and-white colobus births reported from eight other zoos probably refer to *C. guereza*. The species is probably represented in many more collections. Although captive breeding may provide some of the zoo stock, I suspect that a large percentage of zoo animals have been caught in the wild. The few zoo guerezas I have seen (as live specimens and as photographs) are eastern rather than western African in origin. I have not been able to investigate this subject, however, in any detail.

VI. Legal Protection of Guerezas In East Africa and Ethiopia

A. Kenya

By law, only one guereza may be hunted per person per year under license, except in cases of crop damage or for research purposes. To possess skins, a permit is required from the Game Department and such a permit is issued if the skin: (a) is a legally obtained hunting trophy, or (b) can be shown to have come from outside Kenya, or (c) comes from an animal found dead or damaging crops (Licensing Officer, personal communication). The Game Department auctions skins confiscated from unlicensed holders. Trapping for "recognized institutions" may be carried out under license. Licenses are also required for the export of skins and live animals. As L. S. B. Leakey (in Harrisson, 1971) has pointed out:

> It is impossible to trace the skins of animals that have been caught illegally because a limited number come on to the market legally each year so that nearly all dealers in hides have a number of legal skins to which they can refer.

The most serious loophole in the present arrangements is that it is apparently lawful to possess Ethiopian skins, even if they have been illegally obtained.

The Kenya Game Department informs me that two sale permits for colobus monkey skins were issued in 1973, covering 33,500 Ethiopian skins (Licensing Officer, in litt. 1975). One of these permits was for 26,000 skins (I think it could perhaps refer to the consignment referred to in Section VA 4 above). The Game Department has been unable to provide me with statistics on the number of permits of legal possession and export issued for monkey skins, as I am told that this would involve asking dealers to surrender their records. However, in the offices of a London fur trader I have seen a large number of Kenya Game Department skin-possession permits issued to dealers between January and March 1974 and covering 6600 skins. Most of these permits were issued to large Nairobi retailers.

In addition, the Game Department informs me that 1671 live "monkeys" were exported from Kenya in 1973 (Licensing officer, in litt. 1974). I am told that the records do not state the monkey species. Earlier information supplied to T. T. Struhsaker (personal communication) is that 1506 "colobus monkeys" were exported in 1970. However, enquiries with one of the importers listed (who have actually never imported colobus) suggest that there has been confusion in these records with baboons or vervet monkeys.

In Kenya, guerezas occur in the following well-protected areas: the Aberdare, Mt. Elgon, Mt. Kenya, and Lake Nakuru National Parks, and the Masai Mara Game Reserve.

B.   Tanzania

Black-and-white colobus are well-protected in Tanzania, where *C. angolensis* is more widespread than *C. guereza*. They can only be hunted under Presidential license; no licenses have been issued since 1972 for crop raiding, trapping, or the export of live animals; skins cannot legally be marketed (Director of Game, in litt. 1974). However, poaching still occurs by local people who use the skins in traditional ceremonies.

Colobus guereza occurs only in northern Tanzania, where populations are found in the Arusha and Serengeti National Parks and in the Kilimanjaro Game Reserve.

C. Uganda

Black-and-white colobus are protected by law in Uganda and no cropping or control hunting is permitted. However, trapping for export "or other uses" is possible with the Minister's Special Permit. Ten monkeys a year, or fewer, are captured under these conditions (Chief Game Warden, in litt. 1974).

As I have noted, the monkeys are hunted for food in the Semliki and Ruwenzori areas (where both *C. guereza* and *C. angolensis* occur) and here legal protection seems ineffective. The same may apply in other areas, such as Kigezi. The Game Department admits that poaching occurs to obtain skins for use in traditional dances, and one Nairobi dealer told me that he obtained a few of his skins from Uganda.

Guerezas occur in the Rwenzori (formerly Queen Elizabeth) and Kabalega (formerly Murchison Falls) National Parks, and in the Toro Game Reserve, Debasien Animal Sanctuary, and Kigezi Gorilla Sanctuary.

D. Ethiopia

R. I. M. Dunbar (in litt. 1974; see also Dunbar and Dunbar, 1975) informs me that the guereza is classed as a protected animal in Ethiopia, though special trapping licenses can be bought by dealers in the skin trade. Dealers pay a capitation fee and skins and rugs receive an official stamp. In 1974 there were only three licensed dealers in guereza skins in Ethiopia. Unstamped skins can be confiscated: in recent years the Wildlife Conservation Organisation has made many raids and confiscated large numbers of skins (7000 were seized in 1973, according to a report in *Oryx* **12**, p. 304, February, 1974). These confiscated skins are later auctioned. However, many unlicensed skins were still on sale in 1974. Clearly, a situation in which many skins are legally traded allows large-scale abuses, as in Kenya. However, the issue of skin licenses may be halted.*

Exports of skins from Ethiopia must have official permission, and according to the Wildlife Organisation none go legally to Kenya (see Section V,A,4).

Guerezas occur in the Simien Mountains and Awash National Parks, though only in very small numbers in the latter (R. I. M. Dunbar, in litt. 1975). They also occur in at least four proposed National Parks. Most of the skins traded do not

---

*Guerezas may now be legally hunted in Ethiopia only for scientific or educational purposes and the skin trade is said to be "approximately 80% under government control" (L. Berhanu, 1975).

come from near these areas, and there are no plans to establish reserves in the "finest colobus habitat in Keffa and Illubabor" (Dunbar and Dunbar, 1975). A large-scale slaughter of animals by unlicensed hunters in the latter area in recent years is reported by F. Duckworth, Warden in the Wildlife Conservation Organisation (*Oryx* **10**, pp. 306–307, February 1974).

VII. Guereza Status in West and Central Africa and the Sudan

A. West Africa: Nigeria, Cameroon, and Gabon

Colobus guereza occurs only in a small area in the east of Nigeria: the upper Benue Valley and some of its drainage areas. The species' status in this area is unknown (S. K. Sikes, Benue Plateau State Wildlife Unit, in litt. 1974). It is not found in the only major sanctuary in the vicinity, the Yankari Game Reserve (Sikes, 1964). Hunting of monkeys (and of other game, both large and small) for their meat occurs at a very high level in most of the forested areas of Nigeria.

M. Kavanagh (in litt. 1975) tells me that all colobus are completely protected by law in Cameroon. Guerezas probably occur in much of the riparian forest in the Guinea savanna up to 9°N, an area relatively devoid of people, and they are found in the National Parks of Faro, Benue, and Bomba Njida. Kavanagh reports that a poacher who shot guerezas in North Cameroon recently was heavily fined. The rapid expansion of the road network and cotton industry in North Cameroon might pose a threat to the guereza's habitat. D. McKey (in litt. 1975) reports evidence of guerezas from the moist forest zone near Douala and Yabassi in Cameroon: he has seen skins hung out by the roadside at Bonendale, near Douala.

T. T. Struhsaker (in litt.) informs me that in Cameroon, monkeys are most commonly hunted with muzzle-loading shotguns ("Dane" guns). Such guns are also common in Nigeria (Happold, 1972; personal observation). Wire snares are also used in Cameroon, but not usually for colobus.

C. M. Hladik (personal communication) has informed me that the guereza has a widespread distribution in Gabon, though its density is not known and hunting of monkeys is common. The guereza is not well protected in any area.

B. Central Africa: Rwanda, Zaïre, Central African Republic, and Congo Republic

Guerezas once occurred in northern Rwanda, but little forest now remains in that area. Alan Goodall (personal communication) reports that there are large sales of black-and-white colobus skins, mostly as rugs, to tourists in

Rwanda, but the skins apparently come mainly from eastern Zaïre. It is not certain whether the skins are of *C. guereza* or of *C. angolensis*. Sales are from shops in Kigali and Gisenyi, and from the street in both these towns and in Goma.

Verschuren (1975) reports that hunting (especially by "gangs"), rather than habitat destruction, is the main threat to wildlife in Zaïre outside the National Parks. Hunting is most severe along roads and rivers and is largely uncontrolled. There is no specific information on the guereza. Verschuren says that the situation in Zaïre National Parks is generally very good. The guereza probably occurs in both the Parc Garamba and the Parc Virunga (my extrapolation from Verschuren's map) and also in the "untouched" Ituri Forest.

S. Bahuchet (in litt., 1975) informs me that the guereza is widespread in the western and southern parts of the Central African Republic (C. A. R.) where the wooded savannas are broken by gallery forest and forest islets. These islets often contain plantations of coffee beneath the trees. The guereza also occurs in the semideciduous high forest in the southwestern corner of the territory. Bahuchet believes that the species is not rare, although it is not often seen.

The majority of the human population of the C. A. R. still hunt, but using traditional weapons (bows and crossbows) since guns, cartridges, and hunting permits are expensive and the people poor. The trade in hunted meat and skins is controlled by license. However, wildlife protection laws are difficult to apply. The guereza apparently suffers less from hunting than *Cercopithecus* and *Cercocebus* species. Bahuchet only rarely saw dried skins of guerezas in villages and never saw the animals freshly killed. BaBinga Pygmies in the southwest of the C. A. R. do hunt monkeys, but this may be a recent phenomenon due to the demand for meat from Bantu peoples. In some areas it seems that the BaBinga traditionally do not hunt monkeys (Bahuchet, 1972).

Deforestation in the south of the C.A.R., though presently a slow process, is a big problem, and one from which the guereza suffers. There are no forest areas where the fauna is protected, but guerezas do occur in savanna reserves in the north, including the very rich St. Floris National Park (according to the World Wildlife Fund's "Conservation Programme 1975–1976" this park is threatened by invasions of nomads and by poaching).

For the Congo Republic, G. Vattier-Bernard (in litt. 1975) tells me that *C. guereza* is abundant in the north and east of the country (in the basins of the Alima, Sangha, and Likouala Rivers). Although it is absolutely protected under a 1962 law, it is hunted for its flesh and coat.

C. Sudan

Colobus guereza occurs in the south of the Sudan. Butler (1966) notes the species to be common in the forests of the Imatong and Dodinga Mts. and "across to Lui" on the west bank of the Nile. He also states that guerezas are "in danger of extermination since the natives, particularly the Latulea of the east bank, are very fond of their meat and use their skin for ceremonial dress."

VIII.   Other Black Colobus

A.   *Colobus angolensis*

This species might be called the "long-whiskered colobus"; it has notably long, white cheek whiskers. Despite its Latin name it probably does not occur in Angola; the major populations live in Zaïre to the south of Congo River and in Kivu. The species also occurs in Tanzania, and there are pockets of distribution on the southern coast of Kenya and in the Sango Bay forests and Ruwenzori Mountains of Uganda. There are past records from southern Rwanda and from Burundi (see Rahm and Christiaensen, 1960).

The status of *C. angolensis* in Zaïre is unknown, but it apparently occurs in many sparsely populated areas, such as the Ituri and Maniema forests, where hunting by man, although it must take place, is probably at a relatively low level (see Verschuren, 1975).

In Tanzania, the species is probably well protected (see Section VI, B) although it is found mostly on isolated mountain ranges and does not occur in any National Park or Game Reserve. Many years ago, Allen and Loveridge (1933) reported the use of ingenious bag nets by the Wahehe of the Ukinga Mts., Southwest Tanzania, to catch *C. angolensis*.

In Kenya *C. angolensis* occurs in the Shimba Hills National Reserve and in the small forests of Jadini and Shimoni on the coast. Most of Jadini (which is not a government reserved forest) is apparently divided up into private lots, in many of which extensive felling and clearance is under way for hotel developments, while a wide new road cuts a swathe through the forest. *Colobus angolensis* was still present in Jadini in 1974 (personal observation), but its future outlook must be poor. Prospects seem better in the forest at Shimoni to the south (W.L.N. Tickell, in litt.).

In Uganda, the numbers of *C. angolensis* on the Ruwenzori Mountains seem to be very small, perhaps as a result of Bakonjo hunting. It is even possible that the species is extinct on the Ruwenzori; I have come across no recent record. Numbers seem to be fairly large at Sango Bay, where there are several isolated patches of seasonal swamp forest, of which the most extensive (Malabigambo) is contiguous with the larger Minziro Forest across the border in Tanzania. There is still one large tract of forest in southern Rwanda, but little forest may remain in Burundi (Alan Goodall, personal communication).

B. *Colobus polykomos*

The so-called "ursine colobus" may have suffered most at the hands of man. It has been the major target of the international trade in monkey fur (see Section V, A, 2), and it inhabits a part of Africa where human population density is high (so that forests have been extensively degraded by cultivation) and where "bush meat" is of major economic importance. West African forests have also been heavily exploited for timber and when felled they are often replaced by cultivation. The ursine colobus may have been affected less than some other West African forest monkeys by forest clearance for, like the guereza, this species seems to thrive in riparian forest in the savanna and in patches of dry forest, well outside the major moist forest blocks (Booth, 1956, 1960).

According to Booth, this species has been more heavily persecuted than any other primate in Ghana because of demand for its meat and fur. It is rare where people are common and has been exterminated over large areas of the country. However, one area in Ghana where it is common has recently been gazetted as a National Park. This is the forests of the Bia Tributaries, where the ursine colobus was formerly hunted, despite its theoretical complete legal protection (Jeffrey, 1970, 1975). Jeffrey reports that wildlife laws in Ghana have been strengthened and their enforcement improved.

In Sierra Leone, Lowes (1970) described the erosion of forest by farming and timber exploitation, and the relentless hunting of all monkey species (except in certain Muslim areas). At that time, colobus were on a proposed list of prohibited animals for hunting and Lowes added in a footnote that several forest reserves had been created "where no hunting will be allowed." Recently, however, Wilkinson (1974) reported that "the government seems intent on felling all the remaining primary forest" and that the black-and-white colobus is under increasing hunting pressure from the tourist trade for its skin.

In Liberia, Robinson (1971) relates that monkeys are much more heavily persecuted than in Sierra Leone and "are seldom seen except in remote areas." He notes a trade in dried monkeys from eastern Sierra Leone to Liberian mining camps and Monrovia markets.

Struhsaker (1972) reports that only one large tract of relatively undisturbed forest remains in Ivory Coast; this is the Tai Reserve, in which *C. polykomos* is common.

Colobus polykomos has been recorded from the southwestern corner of Nigeria. There is a specimen in the British Museum (No. 1938.11.22.2) collected in 1937 from the Okpara River in Abeokuta Province (the label states that the species was fairly common in fringing forest on the river), and Hopkins (1970) notes a 1955 record of the species from the Olokemeji Forest Reserve, west of Ibadan. However, D. C. D. Happold (in litt. 1974) reports that colobus are now very rare or nonexistent in western Nigeria.

This species has also been recorded in the territories of Dahomey, Togo, and Guinée, but I have no data on its status in those areas. It probably does not occur in Gambia as stated by Rahm (1970). Schwarz (1929) says that the British Museum type of "*C. leucomeros*" from "River Gambia" was almost certainly not from that locality, and I have been able to find no other records.

C. *Colobus satanas*

Of the four "black colobus," the only all-black species has the most limited distribution. In Cameroon it seems to be restricted to the evergreen forests of coastal areas (D. McKey, in litt. 1975), although there are museum specimens collected some years ago which are labeled with localities far inland (for instance, in the Powell-Cotton Museum, Birchington, are specimens collected in 1932 from Batouri District, Cameroun, at 4°15′N, 14°15′E). *Colobus satanas* is one of the preferred meats and seems to be extensively hunted (McKey notes seven animals shot from a single group in one afternoon). McKey reports that the species is very rare wherever there are people, although it is common in the interior of the Douala-Edéa Reserve where there has been little hunting. Unlike the guereza, *C. satanas* is probably very vulnerable to any form of forest felling.

Colobus satanas also occurs on the island of Fernando Poo (the type locality, now called Macias Nguema Biyoga) and in Rio Muni. Sabater Pí (1973) records sightings made in 1966–1968 from these territories, once Spanish, which now comprise the Republic of Equatorial Guinea. There has been concern about the survival of the Fernando Poo monkey fauna, including *C. satanas*, since Equatorial Guinea became independent (*Oryx* **10**, pp. 146–147, December 1969). There was certainly intensive hunting of monkeys in the south of the island in 1964, when I failed to see a single live wild monkey in the forests near Moka (3°22′N, 8°40′E). Struhsaker (1972) notes that "virtually all commercially valuable timber has been removed" in Rio Muni, but Sabater Pí (1973) reports that *C. satanas* occurs in the "dense forests" of Matama and Okorobiko where the vegetation is "practically intact" due to difficulties of exploitation.

Schwarz (1929) lists specimens of *C. satanas* in the Berlin Museum from Corisco Bay and Lambaréné, Gabon. There is also a flat skin in the Powell-Cotton Museum, obtained in 1927, from Odzala in what is now the Congo Republic (0°45′N, 14°45′E, close to the Gabon border); this skin could have been brought from elsewhere. However, Rahm (1970) states that the species probably occurs in some areas of the Congo Republic. I have no other knowledge of the distribution or status of *C. satanas* in these territories.*

*I have been informed by Dr. G. Vattier-Bernard (in litt., 1976) that the species does occur in the Congo, where it is absolutely protected by law.

IX. Conclusions

It would be impossible, with present knowledge, to produce an accurate estimate of the number of guerezas in the wild today. I would be reluctant to estimate the size of the guereza population even in the area I am most familiar with, the Kibale Forest of Uganda. In that forest, guereza density is certainly very variable from place to place. However, I do not have sufficient data on the extent of variation to allow extrapolation to a total population size, based on the theoretically habitable area of the forest. Although there is some evidence that guerezas are most abundant where there are many colonizing tree species, it is far from certain that all areas of colonizing forest in Kibale support a high density of guerezas and that all areas of mature forest support a few. For Uganda as a whole, not only is there variation of population density from place to place, but available census data are even more limited in extent than those from Kibale, while no estimation of habitable area can readily be made—it would not be reasonable to base an estimate on the area of high forest present, or on the area of Forest Reserves.

I am particularly reluctant to hazard large-scale estimates of population because such estimates, liable to great error, tend to become widely used. They may be used only by those wishing to protect a species, but they could also become the basis of exploitation schemes. Inaccurate figures could, therefore, be very dangerous.

Brown and Urban (1970) produced a population estimate of 500,000 guerezas for the whole of Ethiopia. They based this estimate on the sighting of just four groups, totaling seventeen animals, in 0.25 square miles of forest near Badagur, and extrapolated from this using an estimate of the total forested area of Ethiopia. Already, these figures have become part of "knowledge" and have been quoted in three publications dealing with guereza conservation: by Leakey (in Harrisson, 1971), by Mittermeier (1973), and by Dunbar and Dunbar (1975). Leakey has suggested that the estimate is too high, while Dunbar and Dunbar think it may be too low. Dunbar and Dunbar judge that Brown and Urban may have underestimated average group size by 50%, so that the population of the southwest Ethiopian forests may be double Brown and Urban's estimate of 200,000. My own view is that Brown and Urban's data give no reliable basis for a population estimate, and that at present we really do not know how many guerezas there are, in any country.

Even without accurate data on the size of guereza populations, it is clear that the species has undergone a decline which has been more marked in some areas than in others. Before the influence of western industrial civilization spread through Africa, the guereza probably maintained its numbers well in the face of human pressures. It seems to be an ecologically flexible species, able to exploit a wide variety of wooded habitats, and prepared to cross open ground to occupy new areas (see Dunbar and Dunbar, 1974). It has a relatively high potential population growth rate, and it is a difficult animal for a hunter to approach with a clear view. Although hunting for skins and meat must have taken a steady toll in many places, low levels of cultivation in forest areas may have actually increased the amount of suitable habitat available for the species, leading to an increase in numbers. However, there can be little doubt that in the last 100 years the guereza population, together with populations of most of the larger African mammals, has fallen steeply.

The first of the serious pressures on the species came from the international fur trade in the latter half of the nineteenth century. The spread of firearms must have accentuated the amount of destruction that took place and some local populations of guerezas may have been wiped out. Since the beginning of this century, the fur trade has probably had a smaller impact on guereza numbers. Of equal or greater significance has probably been the rapid growth of the human population in tropical Africa and the expansion of western practices of agriculture and forestry. These factors have led to the widespread destruction and encroachment of existing plant communities, often with the complete removal of suitable guereza habitat; East African examples have been given above and Dunbar and Dunbar (1975) describe extensive forest clearance in Ethiopia, where many guereza habitats have been totally destroyed.

The guereza is not in imminent danger of extinction throughout its range. There are several rarer and more seriously threatened primate species in Africa, including at least one colobus: the Zanzibar red colobus, *Colobus kirkii*. Forested land is disappearing at a rapid rate in the tropics of Africa, as in other parts of the world, and if present trends continue it may not be long before the guereza is found only in National Parks and some other reserves.

Even those guerezas inside protected areas are not entirely safe. Certainly, forest reserves are not a guaranteed refuge for the animals, especially where these are subject to clear felling and planting with exotic tree species. Even when some form of natural regeneration is encouraged after selective felling, allowing the possibility of an actual increase in guereza numbers during some phases, the long-term future of the species can still not be assured. It may be that at the end of an 80-year exploitation cycle in the Kibale Forest, the density of guerezas in any one area will differ little from that which existed at the start of the cycle. On the other hand, the production of even-aged stands of commercially valuable trees, the goal of the present operations, could well create unfavorable conditions for the species. So far, there is no evidence as to the long-term effects of felling and refining in Kibale: a strong argument for the maintenance of untouched Nature Reserves within the forest, against which changes may be measured.

In National Parks, an emphasis on the preservation of big game may lead to habitat changes that are highly deleterious as far as the guereza is concerned. Even if the guereza should be well protected in a number of Parks, its disappearance from areas outside the Parks will produce a population fragmented into a number of small units with no genetic contact. In the case of the guereza, however, this may not be too dissimilar to the recent "natural" situation, in which many populations have been restricted by climatic and vegetational factors to isolated mountain areas.

One cause of the guereza's present decline could certainly be remedied. Some suitable guereza habitat outside National Parks and Game and Forest Reserves is likely to remain in existence for some time, particularly on mountains and along rivers in regions of sparse human population. If commercial guereza hunting in these areas could be stopped, the outlook for the species would be much more promising. Efforts must obviously be made to enforce hunting laws, but it may be difficult to stop all hunting of guerezas for local uses in remote areas. However, much of the incentive for guereza hunting, particularly in the areas where the animal is most threatened (such as Ethiopia), comes from the demands of the tourist trade and, more sporadically, the international fur market. If strong demands exist from these money-rich sources, almost any regulation that African governments introduce is likely to be circumvented in one way or another. It would be much better if demand could be stopped. This could be accomplished by banning or limiting the importation of all black colobus skins into those countries where they are in demand (largely the industrial West and Japan). Such a ban would apply to returning tourists as much as to fur traders and would surely cause no great economic problems for any country. It could perhaps be effected by adding black colobus species to Appendix II of the Convention on International Trade in Endangered Species of Wild Fauna and Flora, which covers species which may become threatened with extinction if trade is not regulated to avoid overexploitation, and species with which they may be confused (King, 1974). The ban should be accompanied by publicity from conservation organizations, aimed at discouraging any desire to possess monkey skins.

Demand may not be very high at present. Colobus coats have again gone out of fashion and the number of tourists visiting Ethiopia, in particular, has probably fallen considerably. Fashion is notoriously fickle, however, and long-term patterns of tourism difficult to predict. It would be a pity if a ban on the import of monkey skins had to wait for the next occasion when demand for black colobus reaches 150,000 skins in a year, which may well be the level of sales attained during 1972 and 1973.

The West African species, *Colobus polykomos* and *Colobus satanas*, are almost certainly rarer and more seriously threatened than the guereza. A ban on the trade in black colobus skins should aid *C. polykomos*, but the continued existence of both species will ultimately depend on whether an adequate set of National Parks or similar reserves is established in which they, and other primates, are protected from hunting and from the destruction of their habitats.

X.   Appendix: Vegetation Types

The terminology used in this chapter to describe vegetation is based, with modifications, on the system proposed at the 1956 Phytogeography Meeting at Yangambi (and described in Langdale-Brown *et al.*, 1964). Some of the modifications bring the classification more into line with that used by Lind and Morrison (1974). Obviously, a broad classification conceals many important local variations, but I believe it gives a useful general view of guereza distribution according to environment. The vegetational classification is as follows:

(i) *Moist lowland high forest*: this is the Low and Medium Altitude Moist Forest of the Yangambi classification. It is a closed tropical forest with several strata, including an upper canopy at 20 to 35 m and, above that, a layer of "emergents" 40—60 m high. Woody vines are characteristically present and grasses are absent, or represented only by broad-leaved species. This is the vegetation type often called Tropical Rain Forest (Richards, 1952). (In East Africa I mostly observed guerezas in medium altitude forest, at 900 to 1500 m; however, the species occurs in more low-lying forests in western and central Africa, and in the Semliki Forest of West Uganda, which lies between 670 and 760 m.)

(ii) *Colonizing moist lowland forest*: I use this term to describe the previous vegetation type in the process of regeneration after disturbance, or at its margins where it is reaching out into other types of vegetation.

(iii) *Moist upland forest (lower zone)*: similar to high forest at low and medium altitudes, but differing in the habit and smaller height of the trees, and often including a greater abundance of epiphytic bryophytes; in East Africa mostly between 1500 and 2500 m.

(iv) *Colonizing moist upland forest*: the previous vegetation type in the process of colonization after disturbance.

(v) *Dry upland forest*: a closed stratified forest growing in relatively dry conditions. Trees of the upper stratum are reduced in size and deciduous or sclerophyllous; there is a discontinuous grass stratum; in East Africa mostly between 1500 and 3000 m.

(vi) *Riparian forest*: moist high forest confined to river banks and dependent on the river for water. This is sometimes known as gallery forest.

(vii) *Permanent swamp forest*: forest growing on permanently water-logged ground.

(viii) *Woodland*: an open community of small- or medium-sized trees, often reaching a height of about 18 m, without many layers, and with the tree crowns more or less touching; a sparse grass layer is present. (I have not restricted this category to deciduous trees as does the Yangambi classification.)

(ix) *Riparian woodland*: woodland confined to river banks and dependent on the river for water.

(x) *Wooded grassland*: a formation containing grass at least 80 cm high in a continuous layer dominating the lower stratum; usually burnt annually. Trees and shrubs are scattered or grouped, but have a cover of less than 20%. (I did not observe guerezas residing permanently in such vegetation, but at Chobe they did exploit wooded grassland for food, while spending most of their time in relic riparian forest.)

Acknowledgments

For allowing field studies in Uganda, my thanks are due to the Office of the President, the National Research Council, the Chief Conservator of Forests and the Scientific Advisory Committee of Uganda National Parks. The 1970–1972 studies were supported by the U.S. National Science Foundation under Grant GB 15147.

I am most grateful to Dr. Robin Dunbar for providing me with much firsthand information about the guereza and the skin trade in Ethiopia and for very helpful criticisms of the manuscript. I am also grateful to Dr. Serge Bahuchet, Dr. Alan Goodall, Mike Kavanagh, Beryl Kendall, Doyle McKey, Dr. S. K. Sikes, Dr. G. Vattier-Bernard, and Mary Sue Waser for providing information from other parts of Africa. Dr. Carl Koford has kindly allowed me to make use of unpublished data from his own studies on *Colobus guereza* in Uganda. I thank officials of the Game Departments in Kenya, Tanzania, and Uganda for answering my queries. In addition I would like to thank Mr. George Dyke and Mr. Francis Weiss for giving me much valuable information on the London fur trade.

References

Allen, G. M., Lawrence, B. and Loveridge, A. (1936). Scientific results of an expedition to rain forest regions in eastern Africa. III. Mammals. *Bull. Mus. comp. Zool. Harvard Univ.* **79**, 29–126.

Allen, G. M. and Loveridge, A. (1933). Reports on the scientific results of an expedition to the southwestern highlands of Tanganyika Territory. II. Mammals. *Bull. Mus. Comp. Zool. Harvard Univ.* **75**, 45–140.

Allen, J. A. (1925). Primates collected by the American Museum Congo Expedition. *Bull. Amer. Mus. Natur. Hist.* **47**, 283–499.

Bachrach, M. (1946). "Fur: A Practical Treatise" (revised Ed.). Prentice-Hall, New York.

Bahuchet, S. (1972). Étude écologique d'un campement de pygmées Babinga (Région de la Lobaye, République Centrafricaine). *J. Agr. Trop. Bot. Appl.* **19**, 509–559.

Basilio, A. (1962). "La Vida Animal en la Guinea Española" (2nd Ed.) Instituto de Estudios Africanos, Madrid.

Bauchop, T. and Martucci, R. W. (1968). Ruminant-like digestion of the langur monkey. *Science,* **161**, 698–700.

Berhanu, L. (1975). Present status of primates in Ethiopia and their conservation. *Symp. 5th Congr. Int. Primatol. Soc. Nagoya, 1974*, pp. 525–528. Japan Science Press, Tokyo.

Blixen, K. (1937). "Out of Africa." Putnam, London.

Booth, A. H. (1956). The distribution of primates in the Gold Coast. *Jl W. Afr. Sci. Assoc.* **2**, 122–133.

Booth, A. H. (1960). "Small Mammals of West Africa." Longmans, Green, London.

Brass, E. (1925). "Aus dem Reiche der Pelze." Neuen Pelzwaren-Zeitung und Kürschner-Zeitung, Berlin.

Brown, L. H. and Urban, E. K. (1970). Bird and mammal observations from the forests of southwest Ethiopia. *Walia* **2**, 13–40.

Buechner, H. K. and Dawkins, H. C. (1961). Vegetation changes induced by elephants and fire in Murchison Falls National Park, Uganda. *Ecology* **42**, 752–766.

Butler, H. (1966). Some notes on the distribution of primates in the Sudan. *Folia Primatol.* **4**, 416–423.

Dandelot, P. (1971). *In* "The Mammals of Africa: an Identification Manual" (J. Meester and H. W. Setzer, eds.), Part 3. Smithsonian Inst. Press, Washington, D.C.

Dandelot, P. and Prévost, J. (1972). Contribution a l'étude des primates d'Ethiopie (simiens). *Mammalia* **36**, 607–633.

Dunbar, R. I. M. and Dunbar, E. P. (1974). Ecology and population dynamics of *Colobus guereza* in Ethiopia. *Folia Primatol.* **21**, 188–208.

Dunbar, R. I. M. and Dunbar, E. P. (1975). Guereza monkeys: Will they become extinct in Ethiopia? *Walia* **6**, 14–15.

Gautier, J. P. and Gautier-Hion, A. (1969). Associations polyspécifiques chez les Cerco-pithecidae du Gabon. *Terre Vie* **23**, 164–201.

Guest, N. J. and Leedal, G. P. (1954). Notes on the fauna of Kilimanjaro. *Tanganyika Notes Rec.* **36**, 43–49.

Haddow, A. J. (1952). Studies on *Cercopithecus ascanius schmidti. Proc. zool. Soc. Lond.* **122**, 337–373.

Haddow, A. J., Smithburn, K. C., Mahaffy, A. F., and Bugher, J. C. (1947). Monkeys in relation to yellow fever in Bwamba County, Uganda. *Trans. Roy. Soc. Trop. Med. Hyg.* **40**, 677–700.

Happold, D. C. D. (1972). Mammals in Nigeria, *Oryx* **11**, 469.

Harrisson, B. (ed.) (1971). "Conservation of Nonhuman Primates in 1970." S. Karger, Basel.

Heller, E. (1913). New races of ungulates and primates from equatorial Africa. *Smithson. Misc. Collect.* **61** (17), 12pp.

Hopkins, B. (1970). The Olokemeji Forest Reserve. IV. Check lists. *Nigerian Fld* **35**, 123–144.

Horwich, R. H. and Manski, D. (1975). Maternal care and infant transfer in two species of *Colobus* monkeys. *Primates* **16**, 49–73.

Hrdy, S. B. (1974). Male-male competition and infanticide among the langurs (*Presbytis Entellus*) of Abu, Rajasthan. *Folia Primatol.* **22**, 19–58.

Jeffrey, S. M. (1970). Ghana's forest wildlife in danger. *Oryx* **10**, 240–243.

Jeffrey, S. M. (1975). Ghana's new forest national park. *Oryx* **13**, 34–36.

Johnston, H. H. (1886). "The Kilima-njaro Expedition." Kegan, Paul, Trench, London.

King, F. W. (1974). International trade and endangered species. *In* "International Zoo Yearbook" (N. Duplaix-Hall, ed.), Vol. XIV, pp. 2–13. Zoological Society of London, London.

Kingdon, J. (1971). "East African Mammals: An Atlas of Evolution in Africa," Vol. I. Academic Press, London.

Kingston, B. (1967). "Working Plan for the Kibale and Itwara Central Forest Reserves." Govt. of Uganda Forest Dept., Entebbe. (Cyclostyled.)

Kingston, T. J. (1971). Notes on the black and white colobus monkey in Kenya. *E. Afr. Wildl. J.* **9**, 172–175.

Koford, C. B. (1972). "Guereza Monkeys of Kibale Forest, Uganda." Unpublished manuscript.

Langdale-Brown, I., Osmaston, H. A., and Wilson, J. G. (1964). "The Vegetation of Uganda and its Bearing on Land-Use." Govt. of Uganda, Entebbe.

Laws, R. M., Parker, I. S. C., and Johnstone, R. C. B. (1970). Elephants and habitats in North Bunyoro, Uganda. *E. Afr. Wildl. J.* **8**, 163–180.

Leskes, A. and Acheson, N. H. (1971). Social organization of a free-ranging troop of black and white colobus monkeys (*Colobus abyssinicus*). *Proc. Int. Congr. Primatol. 3rd.* **3**, 22–31.

Lind, E. M. and Morrison, M. E. S. (1974). "East African Vegetation." Longman, London.

Lowes, R. H. G. (1970). Destruction in Sierra Leone. *Oryx* **10**, 309–310.

Marler, P. (1969). Colobus guereza: territoriality and group composition. *Science*, **163**, 93–95.

Marsden, W. and Wright, T., (eds.) (1908). "The Travels of Marco Polo." Everyman's Library ed., J. M. Dent, London.

Mittermeier, R. A. (1973). Colobus monkeys and the tourist trade. *Oryx* **12**, 113–117.

Mohnot, S. M. (1971). Some aspects of social changes and infant-killing in the Hanuman langur, *Presbytis entellus* (Primates: Cercopithecidae) in Western India. *Mammalia* **35**, 175–198.

Napier, J. R. and Napier, P. H. (1967). "A Handbook of Living Primates: Morphology, Ecology and Behaviour of Nonhuman Primates." Academic Press, London.

Neumann, O. (1896). Geographische Verbreitung der Colobusaffen in Ost-Afrika und deren Lebensweise. *Sber. Ges. Natur. Freunde Berl.*, 151–156.

Oates, J. F. (1974). "The Ecology and Behaviour of the Black-and-White Colobus Monkey (*Colobus guereza* Rüppell) in East Africa." Unpublished Ph.D. Thesis, University of London, London.

Oates, J. F. (1977). The guereza and its food. *In* "Primate Ecology" (T. H. Clutton-Brock, ed.). Academic Press, London, (in press).

Omar, A. and de Vos, A. (1970). The annual reproductive cycle of an African monkey (*Cercopithecus mitis kolbi* Neumann). *E. Afr. Agr. Forest. J.* **35**, 323–330.

Osmaston, H. A. (1959). "Working Plan for the Kibale and Itwara Central Forest Reserves." Uganda Protectorate Forest Dept., Entebbe. (Cyclostyled.)

Poland, H. (1892). "Fur-Bearing Animals in Nature and in Commerce." Gurney and Jackson, London.

Powell-Cotton, P. H. G. (1904). "In Unknown Africa." Hurst and Blackett, London.

Rahm, U. (1970). Ecology, zoogeography, and systematics of some African forest monkeys. *In* "Old World Monkeys: Evolution, Systematics and Behavior" (J. R. Napier and P. H. Napier, eds.), pp. 589–626. Academic Press, New York.

Rahm, U. and Christiaensen, A. R. (1960). Notes sur *Colobus polycomos cordieri* (Rahm) du Congo Belge. *Rev. Zool. Botan. Afr.* **61**, 215–220.

Richards, P. W. (1952). "The Tropical Rain Forest: An Ecological Study." Syndics of the Cambridge University Press, London.

Robinson, P. T. (1971). Wildlife trends in Liberia and Sierra Leone. *Oryx* **11**, 117–122.

Rosevear, D. R. (1953). "Checklist and Atlas of Nigerian Mammals." Government Printer, Lagos.

Rudran, R. (1973). Adult male replacement in one-male troops of purple-faced langurs (*Presbytis senex senex*) and its effect on population structure. *Folia Primatol.* **19**, 166–192.

Rüppell, E. (1835). "Neue Wirbelthiere zu der Fauna von Abyssinien gehörig," pp. 1–16. Siegmund Schmerber, Frankfurt am Main.

Sabater Pí, J. (1973). Contribution to the ecology of *Colobus polykomos satanas* (Waterhouse, 1838) of Rio Muni, Republic of Equatorial Guinea. *Folia Primatol.* **19**, 193–207.

Sanderson, I. T. (1940). The mammals of the North Cameroons forest area. *Trans. zool. Soc. Lond.* **24**, 623–725.

Sanderson, I. T. (1957). "The Monkey Kingdom: An Introduction to the Primates." Chanticleer Press, New York.

Schenkel, R. and Schenkel-Hulliger, L. (1967). On the sociology of free-ranging colobus (*Colobus guereza caudatus* Thomas 1885). *In* "Neue Ergebnisse der Primatologie" (D. Starck, R. Schneider, and H.-J. Kuhn, eds.), pp. 185–194. Gustav Fischer, Stuttgart.

Schwarz, E. (1929). On the local races and distribution of the black and white colobus monkeys. *Proc. Zool. Soc. London.*, 585–598.

Sikes, S. K. (1964). A game survey of the Yankari Reserve of Northern Nigeria. *Nigerian Fld.* **29**, 54–82.

Spinage, C. A. (1973). A review of ivory exploitation and elephant population trends in Africa. *E. Afr. Wildl. J.* **11**, 281–289.

Struhsaker, T. T. (1972). Rain-forest conservation in Africa. *Primates* **13**, 103–109.

Struhsaker, T. T. (1974). Of monkeys and men. *Anim. Kingdom.* **77**, 25–30.

Struhsaker, T. T. (1975). "The Red Colobus Monkey." University of Chicago Press, Chicago, Illinois.

Struhsaker, T. T. and Oates, J. F. (1975). Comparison of the behavior and ecology of red colobus and black-and-white colobus monkeys in Uganda: a summary. *In* "Socioecology and Psychology of Primates" (R. H. Tuttle, ed.), pp. 103–124. Mouton, The Hague.

Sugiyama, Y. (1965). On the social change of Hanuman langurs (*Presbytis entellus*) in their natural condition. *Primates* **6**, 381–418.

Swynnerton, G. H. and Hayman. R. W. (1951). A checklist of the land mammals of the Tanganyika Territory and the Zanzibar Protectorate. *Jl E. Afr. Natur. Hist. Soc.* **20**, 274–392.

Synnott, T. J. (1968). "Working Plan for the Mt Elgon Central Forest Reserve." Govt. of Uganda Forest Dept. Entebbe. (Cyclostyled).

Tappen, N. C. (1960). Problems of distribution and adaptation of the African monkeys. *Curr. Anthropol.* **1**, 91−120.

Thorington, R. W., Jr., and Groves, C. P. (1970). An annotated classification of the Cercopithecoidea. *In* "Old World Monkeys: Evolution, Systematics and Behavior" (J. R. Napier and P. H. Napier, eds.), pp. 629−647. Academic Press, New York.

Ullrich, W. (1961) Zur Biologie und Soziologie der Colobusaffen (*Colobus guereza caudatus* Thomas 1885). *Zool. Gart., Lpz.* **25**, 305−368.

Verschuren, J. (1975). Wildlife in Zaïre. *Oryx* **13**, 25−33.

Wilcox, R. T. (1951). "The Mode in Furs." Scribner, New York.

Wilkinson, A. F. (1974). Areas to preserve in Sierra Leone. *Oryx* **12**, 596−597.

Wing, L. D. and Buss, I. O. (1970). "Elephants and Forests." *Wildl. Monogr., Chestertown* no. 19.

14

Presbytis entellus, The Hanuman Langur

JOHN R. OPPENHEIMER

I. Taxonomy and Distribution

A. The Species

Presbytis is one of five genera, of the subfamily Colobinae, family Cercopithecidae, that occurs in Asia. The genus is comprised of 14 species and 84 subspecies (Napier and Napier, 1967). The species with the widest distribution is *P. entellus* Dufresne, 1797 (Figs. 1 and 2). Its English name in India is the "common langur" or "hanuman monkey," and in Sri Lanka (Ceylon) is the "grey

469

Fig. 1. Adult female *P. entellus* sitting in a tree in a typical langur position. Photo taken about midway between Aurangabad and Ellora in Maharashtra (see also Figs. 4. 7, and 10).

wanderoo" (Pocock, 1939) or "gray langur" (Ripley, 1967). "Langur" and "hanuman" are the Hindi names for this monkey, but there are many other vernacular names, as there are fourteen major languages and approximately 250 dialects within India alone.

The geographic range of *P. entellus* (Fig. 3) lies between 70° and 95°E longitude and between 33° and 7°N latitude, covering most of India, Nepal, Sikkim, and Sri Lanka, and extending slightly into Pakistan, the southern edge of Tibet (Napier and Napier, 1967; Pocock, 1939) and into Bangladesh (Husain, 1974). There have also been unconfirmed reports of langurs in Afganistan (Vogel, personal communication), but, the monkeys are absent from the Sind and Punjab regions of northwestern India and Pakistan (Blanford, 1888). Though the species was reported not to occur naturally east of the Hooghly River in West Bengal (Blyth, cited in Jerdon, 1874; Hutton, cited in Blanford, 1888), its presence has been confirmed well to the east in Assam (Gupta, 1974; Spillet, 1966). The geographical range of *P. entellus* overlaps those of *P. geei* in Assam along the Manas River, of *P. pileatus* in Assam and Bangladesh, of *P. johni* in the Western Ghats from Coorg to Cape Comorin, and of *P. senex* in Sri Lanka (Prater, 1965).

Fig. 2. Adult and subadult males of an all-male group during a midday rest period at the edge of a river. Note tail curled toward the head on the moving male in the lower left corner of the photo. Photo taken at Corbett National Park, in Uttar Pradesh, to the north-east of New Delhi (see also Figs. 5, 6, 8, and 9).

B.   The Subspecies

Presbytis entellus has been divided by taxonomists into fifteen subspecies, which are listed in Table I and whose locations can be seen in Fig. 3. These subspecies have been differentiated on the basis of pelage colors and pattern, pelage length, and on slight differences in body size and skull measurements (Pocock, 1939).

In the absence of physical barriers, such as rivers or mountain ranges, the size of the geographical range of each subspecies is probably associated with the geographical range of the forest type(s) within which each lives, and the related climatic and edaphic factors. Within the geographical range of *P. entellus* in India and Nepal there are sixteen climatic forest types (Sagreiya, 1967). *Presbytis e. ajax* and *P. e. achilles* live in Himalayan Moist Temperate and Subalpine Forests, and Alpine Scrub. The fur of these subspecies is much longer than in the others at lower altitudes to the south (Pocock, 1939). *Presbytis e. schistaceus* inhabits the Subtropical Pine, Tropical Dry, and Tropical Moist Deciduous Forests in the foot-hills of the Himalayas and on the Terai. To the south of the Ganges, *P. e. entellus* has a geographical range that stretches 1900 km from east to west (Fig. 3) and

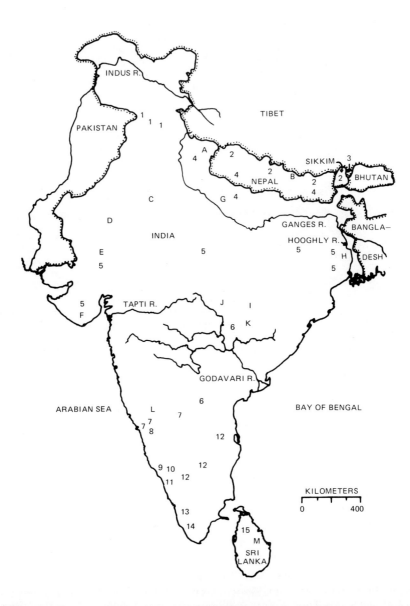

Fig. 3. The geographical range of *P. entellus* with international boundaries of *H. sapiens* indicated by solid lines accompanied by dots, and coast line and rivers shown by solid lines alone. The numbers indicate locations where subspecies have been captured (see Table I), and the letters indicate the locations of recent field studies (see Table II).

TABLE I

Subspecies of *Presbytis entellus* Dufresne, 1797, with an Indication of Location (See Fig. 3) and Altitudinal Range[a]

---

1. *P. e. ajax* Pocock, 1928.   Ranges from 1829 m at Chamba to 2987 m at Kulu. Has been observed above the snow line
2. *P. e. achilles* Pocock, 1928.   Ranges from 1630 to 3658 m in Sikkim and Nepal.
3. *P. e. lania* Elliot, 1909.   Described on basis of one incomplete specimen from Chumbi, Tibet (Elliot, 1913)
4. *P. e. schistaceus* Hodgson, 1840.   Ranges from 335 m at Ramnagar to 2332 m at Khati, and possibly up to 2743 m
5. *P. e. entellus* Dufresne, 1797.   Ranges from 61 m at Midnapore, West Bengal, and Palanpur, Gujarat, to 305 m at Hazaribagh, Bihar
6. *P. e. anchises* Blyth, 1844.   Ranges 305 m at Nimar to 610 m in the Nallamalai Hills
7. *P. e. achates* Pocock, 1928.   Ranges from 457 m at Bellary to 610 m at Dharwar
8. *P. e. iulus* Pocock, 1928.   Known from two specimens taken at Jog, Gersoppa Falls at 396 m
9. *P. e. aeneas* Pocock, 1928.   Ranges from 76 m at Makut to 610 m at Wottekolle
10. *P. e. elissa* Pocock, 1928.   Known from one troop at Nagarhole
11. *P. e. dussumieri* Geoffroy, 1843.   Known from one specimen obtained on Malabar Coast
12. *P. e. priam* Blyth, 1844.   Ranges from 259 to 488 m in Dharmapuri, Shevaroy, and Palkonda Hills, and possibly to 1829 m in Nilgiri Hills
13. *P. e. priamellus* Pocock, 1928.   Known from one specimen obtained in Cochin
14. *P. e. hypoleucos* Blyth, 1841.   Known only from one specimen obtained in Travancore
15. *P. e. thersites* Blyth, 1847.   Sri Lanka (Ceylon).

---

[a] From Napier and Napier, 1967; Pocock, 1939.

which consists, for the most part, of Dry Tropical Deciduous Forest. This subspecies also exists in the Tropical Thorn Forest in Parts of Gujarat and Rajasthan, and in the Tropical Moist Deciduous Forest on the eastern and southeastern edges of its ranges. The large number of subspecies found along the southwestern edge of the range (Fig. 3) reflects the shift from Tropical Moist Deciduous Forest along the coast to the Tropical Semi-Evergreen and Tropical Wet Evergreen Forests at higher elevations, as in the Nilgiri Hills. The fact that I mention forest types for India is only to indicate a general uniformity of past habitat type. Except in the hilly or mountainous areas with steep elevations and in a few protected low land areas, India has been pretty well stripped of natural forest, and this process is accelerating in Nepal.

It should be mentioned that some taxonomic subspecies may not be valid, and that future work might possibly uncover new subspecies. McCann (1933)

Fig. 4. Adult female with infant grooms adult male, while on the ground. Note the darker pelage color on the back, haunch, and proximal two-thirds of tail on the male.

has observed that there can be a great deal of variation in color pattern even within a single troop (Fig. 4). Thus, subspecies based on single specimens from adjoining areas, as in southwestern India (Table I and Fig. 3), are subject to question. Vogel (in preparation) has revised the taxonomy of the species and has divided it into four subspecies.

Behavioral differences occur among the local populations of *P. entellus*. The species has been divided into northern and southern forms on the basis of tail posture. The northern form carries the tail curled toward the head (Fig. 2), whereas the southern form carries the tail curled to the rear. The dividing line between the two forms is said to be the Tapti and Godavari Rivers (Jay, 1965; Yoshiba, 1968); however, the taxonomic subspecies *P. e. anchises* has a geographical range that seems to fall on both sides of the behavioral dividing line (Fig. 3). Differences have also been recorded between the vocal repertoires of two allopatric subspecies: *P. e. schistaceus* in the forest of the Kumaon Hills (A in Fig. 3) and *P. e. entellus* in the thorn forest of Rajasthan (C in Fig. 3) (Vogel, 1973). Additional data on differences in vocal repertoire of subspecies have also been obtained for *P. e. achilles* in Nepal (B in Fig. 3). These differences may be caused by differences in habitat type and population structure at the study sites (Bishop, 1975). Differences in the frequency of intertroop agonistic encounters among the different populations have also been noted (Vogel, 1975, 1976).

Information on social behavior and intratroop interactions can be found in Jay (1962, 1963a,b, 1965).

II. Activity, Habitat, and Diet

A number of field studies on *P. entellus* have been done in the past 15 years at various locations in India, Nepal, and Sri Lanka (Table II, Fig. 3). Some of these studies have included information on the activity patterns, habitat, and diet of the monkeys.

A. Activity Patterns

The two most prominent diurnal activities of the hanuman langur are eating and resting (Oppenheimer, 1973; Starin, 1973; Yoshiba, 1967). At Dharwar, eating occurs during 30 to 60% of the daylight hours and along with resting accounts for 80% of the diurnal activity (Yoshiba, 1967). Around Simla, 40% of the daylight hours are spent eating (Sugiyama, 1976). During the year at Singur (Fig. 3,H, Table II), the monkeys spend about 30% of their time eating and about 38% of their time resting. Moving, playing among juveniles (Figs. 5 and 6), nursing of young (Fig. 7), and social grooming (Fig. 4) are the next most important activities within the langur troop. The adult male spends much of this time in the canopy or on some other high object watching for, and ready to give warning of, approaching danger (Oppenheimer, 1973, 1977, in preparation).

The amount of time spent in each activity varies with the season. At Singur, more time is spent resting in the monsoon months, June through August (45%),

TABLE II

Location of Recent Field Studies Done on *Presbytis entellus* (See Fig. 3)

A. Bhimtal, Kumanon Hills, Uttar Pradesh, India (Vogel, 1971, 1973)
B. Melemchigaon (2591 m), Helambu Valley, Sindu District, Nepal (Bishop, 1975)
C. Sariska Wildlife Sanctuary, Rajasthan, India (Vogel, 1971, 1973)
D. Jodhpur (241 m), Rajasthan, India (Mohnot, 1971a,b, 1974)
E. Mt. Abu (1165 m), Rajasthan, India (Hrdy, 1974)
F. Gir Wild Life Sanctuary (226–649 m), Gujarat, India (Kurup, 1970; Rahaman, 1973; Spillett, 1968; Starin, 1973)
G. Kaukori (122 m), Uttar Pradesh, India (Jay, 1962, 1963a, 1965; Yoshiba, 1968)
H. Singur (7 m), West Bengal, India (Oppenheimer, 1973, 1976)
I. Raipur, Madhya Pradesh, India (Sugiyama, 1964; Yoshiba, 1968)
J. Ramtek (500 m), Maharashtra, India (Tiwari and Mukherjee, 1973)
K. Orcha (762 m), Madhya Pradesh, India (Jay, 1963a, 1965)
L. Dharwar (550–760 m), Mysore, India (Sugiyama, 1964, 1965a,b, 1966, 1967; Sugiyama and Pathasarathy, 1969; Sugiyama *et al.*, 1965; Yoshiba, 1967, 1968)
M. Polonnaruwa, Sri Lanka (Hladik and Hladik, 1972; Ripley, 1967, 1970)
N. Simla (1800–2200 m), Tara Devi (1500–1800 m), Hatto Peak (2700–3200 m), Himachal-Pradesh, India (Sugiyama, 1976)

Fig. 5. Play wrestling between two juveniles of an all-male group at river's edge.

than in the winter months, December through February (25%); less time is spent eating in the monsoons (25%), than during the winter (36%) (Oppenheimer, 1973, 1977). At the Gir Forest in the hot dry month of April almost half the monkeys' daylight activity is spent eating and the other half resting; during the monsoon month of July, they spent less time eating than resting (Starin, 1973).

In addition to the seasonal shifts, there are diurnal changes in activity. Eating occurs primarily in the early morning and late afternoon, whereas resting occurs mainly during the middle of the day (Jay, 1965; Mohnot, 1971a, 1974; Oppenheimer, 1973; Ripley, 1970; Starin, 1973).

Moving or travel from place to place within the home range occurred on the average during 9% of the daylight hours at Singur, although this changed with

Fig. 6. Play copulation between two juvenile males of an all-male group at river's edge.

Fig. 7. Older infant nursing from mother.

the season. During the hot dry months of March to May about 11% of the day-light hours were devoted to moving, whereas during the cool months of September to November only 8% were (Oppenheimer, in preparation). At the Gir, during the dry month of April, more than half of the moving took place on the ground; but in the wet month of July, when the foliage was lush, almost all the moving took place in the trees. Although the same number of major moves occurred in both months, they tended to be shorter in July (Starin, 1973). Bisexual troops (see Section III, B) moved about 0.36 km/day in the forest of Dharwar (Sugiyama *et al.*, 1965), from 0.50 to 1.40 km/day in the very dry environment of Jodhpur (Mohnot, 1974), and as much as 3.50 km/day at Mt. Abu (Hrdy, 1974). All-male groups (see Section III, B) moved from 2.0 to 3.0 km/day at Jodhpur (Mohnot, 1974), and from less than 0.01 to 14.50 km/day at Dharwar, where the group could travel 1 km in 15 minutes (Sugiyama *et al.*, 1965).

Although langurs are arboreal monkeys (Fig. 1), they do spend time on the ground (Figs. 2 and 4—7), which varies with the location, the habitat, and the season. At Singur, the langurs were on the ground only 5% of the daylight hours (Oppenheimer, 1973, in preparation), whereas at Kaukori, where trees were rare, they were on the ground 70—80% of the time (Jay, 1965). At Orcha they spent 30—50% of their time on the ground (Jay, 1965), and, similarly, at Mt. Abu 30—70%, depending on the feeding conditions and on the presence or absence of people and dogs (Hrdy, 1974). At Dharwar they spent between 20—40% of their time on the ground (Yoshiba, 1968). At the Gir, one troop that lived in riverine forest spent only 10% of its time on the ground, whereas another troop living in a more open habitat was on the ground 30—40% of the time, much of which was devoted to searching for and eating worms and pupae (Rahaman, 1973). The seasonal variation in time spent on the ground ranged from 12% during the cool dry winter to less than 1% during the monsoons at Singur (Oppenheimer, 1973, in preparation), and from 11% during the dry hot month of April to less than 1% during the wet month of July at the Gir (Starin, 1973). During the winter months at Singur, much of the time on the ground was spent raiding crops in the fields, since many of the trees had lost their leaves; during the dry hot months the monkeys rested on the ground and drank water from the ponds (Oppenheimer, 1973, in preparation).

B. Habitat and Diet

Langurs have a sacculated stomach specialized for digestion of leaves (Ayer, 1948), and are able to exist for long periods without drinking water. Because of this, they are able to live in very dry areas, where other species would be unable to survive (Jay, 1965). The hanuman monkey occurs in forests as well as in agricultural areas and in or near the towns and villages of man (Jerdon, 1874).

In the following discussion I have presented the studies done in a particular area under the subspecies that has been reported to predominate in that area by Pocock (1939; see Table I and Fig. 3). These subheadings are being used to give geographical cohesiveness or distinction to the various studies done. The actual subspecies involved, such as *P. e. schistaceus* at Kaukori, is open to question, as the population may look most similar to *P. e. entellus* (Sugiyama, personal communication). However, this similarity may be because of similarity of climatic conditions rather than close genetic affiliation.

1. *P. e. ajax*

A total of twelve troops was studied for 5 months from August 1972 through January 1973 in three areas in and around Simla (Fig. 3, Number 1 on right, 1500—3200 m elevation) in Himachal-Pradesh (Sugiyama, 1976). The region is hilly with slopes of 15 to 20° and is covered with moist temperate oak-coniferous forest. Heavy snowfalls occur during the winter months and the snow may remain on the ground for several weeks at a time. Annual rainfall ranges between 160 and 180 cm. Temperatures range between $-3°$ and $30°C$ of 36 species of plants eaten (see Table IV), 50% provided leaves, 22% buds, 25% flowers, 56% fruit, and 14% miscellaneous items like shoots, bark, and pith. Fruits, berries, beans, nuts, and seeds were eaten more frequently as a major food item than were leaves. Probably because of high altitude and low temperatures, the species composition of the diet of these langurs had little overlap with that of langurs in the lower, warmer regions to the south.

2. *P. e. achilles*

One troop of langurs has been studied for a year at Melemchigaon (B in Fig. 3), in Nepal (Bishop, 1975, personal communication). It moved between 2438 and 3048 m elevation (the upper elevation being near the tree line). In addition to snow, which sometimes reached a depth of 30 cm for brief periods during the winter months, over 200 cm of rain fell during the monsoons from June to September. The temperatures ranged from a minimum of $-2.8°$ to $22.2°C$. Within the troop's range were an oak (*Quercus semecarpifolia*) forest, a mixed broadleaf forest, and some fields near a Sherpa village. Oak leaves and leaves from shrubs were eaten throughout the year, but seasonal items, such as acorns, wheat, barley shoots, potatoes, mushrooms, and *Ilex* fruits, made up the bulk of the diet. Soil was eaten in the winter and spring months. One troop was seen at 4084 m, and was stated by local residents to go as high as 4267 m during the month of August. This was above the tree line (3570 m) and probably at the upper limit of the Dry Alpine Scrub Zone. During the rest of the year this troop lived at lower altitudes, possibly as low down as 2744 m.

3. *P. e. schistaceus*

A large troop of langurs was studied at Kaukori (G in Fig. 3; 122 m elevation), near Lucknow, on the Gangetic Plain (Jay, 1962, 1965). Most of the land was devoted to agriculture, and only 1% was occupied by forest. The annual rainfall ranged between 76 and 127 cm, with 51 to 102 cm falling during the monsoon months alone. Temperatures during the hot summer ranged between 38° and 48°C; wind and dust storms occurred during this period. Agricultural crops provided 90% of the monkey's food, and the forest provided the remainder.

4. *P. e. entellus*

A population of 900 langurs has been studied around Jodhpur (D in Fig. 3; 241 m elevation) on the eastern edge of the Indian Desert (Mohnot, 1971a,b, 1974). The area is naturally covered with open scrub, xerophytic vegetation. On the average, 30 cm of rain fall annually (range 9–53 cm), most of which occurs during July, August, and September. Temperatures average at the minimum

9.2°C and at the maximum 42.2°C, with a range of 2.0° to 46.8°C. The langurs drank water most frequently during the hot months. They primarily lived near water sources, and agricultural and horticultural areas. Leaves made up most of the diet, but buds, flowers, fruit, roots, stems, and sap were also eaten (Table III). Including fruit trees, vegetables in the gardens, and grasses, 90 species of plants of over 36 families were eaten (Table IV). In addition, the people fed the monkeys a large number of different food items.

The Gir Wildlife Sanctuary, located in part of the Gir Forest (Fig. 3,F; 226–648 m elevation), is primarily known for its remnant population of *Panthera leo*, the Asiatic lion (Spillett, 1968). The forest itself is an isolated patch of dry mixed deciduous forest, 1230 km² in area, surrounded by thorny scrub (Rahaman, 1973; Sagreiya, 1967; Spillett, 1968; Starin, 1973). Along the rivers, within the forest, is a dense riverine vegetation. The mixed deciduous portion of the forest has open and dense areas dominated by teak (Rahaman, 1973; Starin, 1973). The annual rainfall, primarily occurring during the monsoon months, averages 89 cm, with as little as 41 cm falling during 1966 (Spillett, 1968). The minimum temperature during the winter is 7.2°C, and the maximum temperature during the summer is 43.3°C. The langurs drank water during the hot summer months (Starin, 1973), as well as during the monsoons (Rahaman, 1973). Their diet consisted of young leaves (flush), mature leaves, flowers, fruit, seeds, stems, pith, bark, roots, and sap (Table III), as well as some soil invertebrates (Rahaman, 1973; Starin, 1973). During the 3 months of these two studies, 55 species of plants from 26 families were eaten (Table IV).

Two troops of langurs were studied in villages around Singur (Fig. 3,H; 7 m elevation) in Hooghly District, West Bengal (Oppenheimer, 1973, 1977, in preparation; Graves and Oppenheimer, 1975). Although, previously the area was probably covered with dense forest, it is today devoted to clear-cut agricultural fields, and villages with orchards, bamboo groves, kitchen gardens, and a variety of fruit trees surrounding the houses and ponds. Here the langurs dwell primarily within the villages. The average annual rainfall for the area is about 145 cm (Indian Meteorol. Dept., 1962); 211 cm fell during 1971 at the study site. During 1971 the average minimum temperature was 22.8°C and the average maximum was 29.7°C (range 12.2°–38.9°C). Drinking from ponds occurred during the hot dry months. The diet (Table III) consisted primarily of leaves (78%) and fruit (66%) from 68 species and 32 families of plants (Table IV). During the dry cool months, when many of the trees lost their leaves, field crops, such as potatoes, okra, cucumbers, and pumpkins were an important addition to the diet. Ants and soil were occasionally eaten.

5. *P. e. anchises*

Three troops were studied at Orcha (K in Fig. 3; 762 m elevation) in the Abujhmar Hills (Jay, 1965). The area is covered with deciduous forest, except

TABLE III

Species Composition of Plant Diet in Terms of Part(s) of Plant Eaten

Location and source	Leaves			Buds	Flowers	Fruit	Seeds	Sap	Bark	Miscellaneous	Total number of species
	Young	Old	Total		Percentage of species						
Jodhpur Mohnot (in prep.)	8.3	72.6	81.0	7.1	38.1	42.9	2.4	2.4	—	9.5	84
Gir Forest Rahaman (1973)	—	—	85.4	—	14.6	26.8	—	—	2.4	—	41
Starin (1973)	33.3	61.1	75.0	—	5.6	41.7	13.9	5.6	5.6	16.7	36
Singur Oppenheimer (in prep.)	—	—	77.9	1.5	23.5	66.2	—	1.5	—	—	68
Dharwar Yoshiba (1967)	—	—	94.6	—	21.6	32.4	2.7	—	—	2.7	37
Polonnaruwa Ripley (1970)	66.7	19.3	71.9	8.8	24.6	28.1	10.5	1.8	—	—	57
Hladik and Hladik (1972)	67.4	46.5	83.7	—	41.9	76.7	—	—	—	—	43
Mean	43.9	49.9	81.4	2.5	24.3	45.0	4.2	1.6	1.1	4.1	200

TABLE IV

Number of Species of Each Plant Family Eaten, and Indication of Plant Parts Eaten, by *Presbytis entellus* at Six Different Locations[a]

Family	Simla, Him'—Pra' (1)[b]	Gir Forest, Gujarat (2)[b]	Jodhpur, Rajasthan (3)[b]	Singar, West Bengal (4)[b]	Dharwar, Mysore (5)[b]	Polonnaruwa, Sri Lanka (6)[b]	Total number species (7)[b]
Acanthaceae			4:Fr,Lv,Sd				4
Aceraceae	2:Lv						2
Alangiaceae		1:Lv				1:Fr,Fl,Lv,Bd	1
Amaranthaceae			3:Lv				3
Amaryllidaceae			1:Bulb				1
Ampelidaceae	2:Fr,Lv	1:Lv,Sm	1:Fr,Fl,Lv				2
Anacardiaceae		1:Bk		3:Fr,Fl,Lv,Sp		2:Fr,Fs	4
Annonaceae		1:Fr		3:Fr,Fl,Lv		1:Fs	5
Apocynaceae		4:Fr,Lv,Sd	2:Fr,Fl,Lv	3:Fr,Fl,Lv	2:Fr,Lv	1:Fr,Fl,Fs	7
Araliaceae	1:Fr,Lv						1
Asclepiadaceae			2:Fl,Lv,Sp	1:Lv			3
Bambuseae				1:Lv	1:St		2
Berberidaceae	1:Lv						1
Bixaceae					1:Lv		1
Bombacaceae				1:Fr,Fl,Lv,Bd	2:Lv		2
Boraginaceae			2:Fr,Lv				2
Burseraceae		2:Fr,Lv					2
Capparidaceae		1:Lv	2:Fr,Fl,Lv			2:Fs,Lv	4
Carifoliaceae	2:Fr						2
Caricaceae				1:Fr,Lv			1

Family	1	2	3	4	5	6	Total
Celastraceae	1:Fl,Lv		1:Fl,Lv	1:Fl,Lv		2:Fr,Fl,Fs,Lv	5
Chenopodiaceae			2:Fl,Lv,Sm,Rt	1:Lv			2
Combretaceae	3:Fr,Fl,Lv,Sp,Fs		1:Fl,Lv	2:Fr,Lv	3:Fl,Lv	1:Fs	9
Commelinaceae			1:Lv				1
Compositae	1:Fl,Lv		1:Fl,Lv	1:Fl,Lv			2
Coniferae	2:Fr						2
Convolvulaceae	1:Fr,Lv	2:Lv	2:Lv		2:Lv		7
Cornaceae	2:Fr						2
Cruciferae			5:Fr,Fl,Lv,Sm,Rt	2:Fl,Lv			5
Cucurbitaceae			1:Fr,Lv	5:Fr,Lv	1:Fr	1:Fr	7
Dilleniaceae	1:Lv			1:Fr,Lv			1
Dioscoreaceae		1:Lv					1
Ebenaceae		1:Fr,Lv,Sd			1:Fr,Lv	4:Fr,Fl,Fs,Lv	5
Euphorbiaceae		2:Fr,Fs,Lv	1:Fl,Lv	1:Lv		5:Fr,Fl,Fs,Lv,Sd	8
Fagaceae	2:Fr,Bk,Pi	1:Fr					2
Gramineae	3:Fr,Bd,Bk,Pi,St		6:Lv,Sm			1:Fl,Bd	10
Guttiferae						2:Fr,Sd	1
Lauraceae			1:Lv	1:Lv			3
Leguminosae	2:Fr,Fl,Lv	13:Fr,Fl,Fs,Lv,Sp,Bk	14:Fr,Fl,Lv,Sp,Plant	6:Fr,Fl,Lv	10:Fr,Fl,Lv,Sd	14:Fr,Fl,Fs,Lv,Sd	48
Liliaceae	1:Fr,Fl,Lv,Bd					1:Fr,Fl,Fs,Lv,Bd	1
Loganiaceae							1
Loranthaceae					1:Lv		1
Malpighiaceae							1
Malvaceae		2:Fs,Lv	4:Fr,Fl,Lv,Sd	2:Fr,Lv			7
Melastomaceae						1:	1
Meliaceae		2:Lv	1:Fr,Lv	1:Lv	1:Lv	3:Fr,Fs,Lv,Sp	4
Menispermaceae			1:Lv				1
Moraceae	1:Fr	3:Fr,Fs,Lv,Sd,Sm,Pi	3:Fr,Fl,Lv	5:Fr,Lv	2:Fr,Lv	7:Fr,Fl,Fs,Lv,Bd,Sd	12

TABLE IV (continued)

Number of Species of Each Plant Family Eaten, and Indication of Plant Parts Eaten, by *Presbytis entellus* at Six Different Locations[a]

Family	Simla, Him'—Pra' (1)[b]	Gir Forest, Gujarat (2)[b]	Jodhpur, Rajasthan (3)[b]	Singar, West Bengal (4)[b]	Dharwar, Mysore (5)[b]	Polonnaruwa, Sri Lanka (6)[b]	Total number species (7)[b]
Moringaceae				1:Fr,Fl,Lv			1
Musaceae				1:Fr,Fl,Lv			1
Myrtaceae		2:Fr,Fl,Lv,Sd	3:Fr,Fl,Fs	3:Fr,Fl,Lv	1:Fl,Lv	1:Fr,Fl,Fs,Lv	6
Nyctaginaceae			1:Fr,Fl,Lv		1:Lv		2
Orchidaceae		1:Rt					1
Palmae				3:Fr,Fl			3
Polygonaceae	3:Fl,Bd		1:Lv				4
Portulacaceae			1:Lv				1
Rhamnaceae		1:Fl,Lv	2:Fr,Lv	1:Fr,Fl,Lv	5:Fr,Fl,Lv	1:Fl	8
Rosaceae	5:Fr,Fl,Lv, Bd,Bk,Pi						5
Rubiaceae		2:Fl,Lv	2:Fr,Fl,Fs,Lv	1:Lv	1:Lv	4:Fr,Fl,Fs,Lv	8
Rutaceae		1:Fr,Lv	2:Fl,Lv	6:Fr,Fl,Lv		2:Fs	7
Salvadoraceae							2
Sapindaceae	1:Fr	2:Fr,Fs,Sd,Sm		1:Fr,Fl,Lv		4:Fr,Fl,Fs,Bd,Sd	7
Sapotaceae				2:Fr,Lv		2:Fr,Fl,Fs,Bd	4
Simarubaceae			2:Fl,Fs,Lv,Bd				2
Solanaceae			5:Fr,Fl,Lv,Bd	4:Fr,Lv			6
Sterculiaceae		1:Bk				2:Fr,Fl,Fs,Lv	3
Tiliaceae		2:Fr,Fs,Lv,Sm	1:Fr,Lv	2:Fr,Lv	1:Lv	2:Fr,Fl,Fs,Lv,Sd	7
Trapaceae			1:Fr				1
Ulmaceae		1:Lv				1:Fr,Fl,Fs,Lv, Bd,Sd	1

	(1)	(2)	(3)	(4)	(5)	Total
Umbelliferae		2:Fl,Lv,Sm				2
Verbenaceae	1:insects			2:Fr,Fl,Lv	2:Fr,Fl,Fs,Lv	4
Zygophyllaceae		1:Lv				1
Miscellaneous	3:Fr,Lv,Bd	6:	2:Fr		2:Fr,Fl,Fs	10
Total No. of						
Species	55 + 1	90	69	37	71	299 +
Families	27	36 +	32	17	27	72 +

[a] Key to abbreviations: Bd, bud; Bk, bark; Fl, flower; Fr, fruit; Fs, flush (young leaves); Pi, pith; Rt, root; Sd, seed; Sm, stem; Sp, sap; St, shoot; Su, stipule; Lv, Leaf.

[b] Key to numbers in parentheses: (1) 5-month study by Sugiyama (1976); (2) separate studies by Rahaman (1973) and Starin (1973) which covered the months of April, July, and August; (3) over 4 years of study (Mohnot, 1974, in prep.); (4) 1 year study (Oppenheimer, 1976); (5) Yoshiba (1967); and (6) on studies by Ripley (1970) and Hladik and Hladik (1972). (7) Includes 17 species recorded by McCann (1933) during March and April at Abu Hills, Rajasthan.

for 3% devoted to agriculture. The annual rainfall amounted to 203 cm and the temperatures during the summer ranged between 32° and 48°C. The diet consisted of leaves, fruit, and sprouts, which came almost entirely from the forest.

6. *P. e. achates*

A large number of langurs was studied at Dharwar (I in Fig. 3; 550–760 m elevation) in the State of Mysore (Sugiyama, 1964; Yoshiba, 1967). The region is partly covered with dry deciduous and teak forests, and partly with grassland and cultivated fields. The rainfall averages about 90 cm, most of which falls during the months of May to October; the temperatures range between 13° and 40°C. Wind and dust storms occur. The diet consisted primarily of leaves (95%) and fruit (66%), but also included buds, flowers, nuts, and bark from trees, shrubs, and grasses of 37 species and 17 families (Tables III and IV). Crops may contribute up to 20% of the langur's food. Insects are also eaten: termites, insect larvae in leaf galls, and caterpillars (Yoshiba, 1967). Many of the troops lacked access to water during the dry season, and thus were dependent upon the moisture content of the vegetation they ate (Sugiyama, 1974) and on dew (Sugiyama, personal communication).

7. *P. e. thersites*

Four troops have been observed at Polonnaruwa (M in Fig. 3), an archeological site 5.1 km² in area, in Sri Lanka (Hladik and Hladik, 1972; Ripley, 1967, 1970). The shrub layer had been removed from half of the forest to enable archeological work to be done. Some parts of the forest were as high as 15 m, whereas other parts were as low as 5 m. The annual rainfall ranged between 127 and 191 cm and the annual mean temperature was about 27°C. Although two man-made ponds were located at the study site, drinking of water was rarely observed. Plant material, primarily leaves, fruit, and flowers (Table III), from 71 species and 27 families (Table IV), and clay from termite nests, were eaten.

8. *P. entellus*

The species, as a whole, thus inhabits a wide range of habitat types and environmental conditions. It lives in habitats that range from hot dry desert and cold alpine scrub to dense wet tropical evergreen forests. It ranges in altitude from 7 to 4267 m, and a single troop in the Himalayas may shift from 610 m to as much as 1523 m in altitude during the year (Bishop, 1975, personal communication). The environmental temperatures range from −3° to +48°C, and the annual precipitation (rain and snow) from 9 to well over 200 cm.

The diet of the langurs is quite varied, despite its being composed almost entirely of plants. All parts are eaten from the leaves to the roots and from the bark to the pith and sap. Leaves are eaten on the average from 81% of the species, fruit from 43%, and flowers from 24% (Table III). During the year langurs eat between 37 and 90 different plant species of 17 to 36 different families; more species of the family Leguminosae are eaten than any other family (Table IV). In all, more than 299 species from over 72 families have been recorded to occur in the diet of the common langur (Table IV). A list of 37 species of trees, shrubs, and herbs that were eaten at two or more of the study sites is given in Table V. Some of these species were eaten frequently, whereas others were eaten only occasionally. Nonetheless, this list gives a representative cross section of the plant species eaten by langurs. Crop species, such as spinach, cauliflower, and eggplant, are important when natural vegetation is seasonally restricted or when the area is primarily devoted to agriculture. Insects and annelids are occasionally eaten (Oppenheimer, in preparation; Rahaman, 1973; Yoshiba, 1967), as is earth (Bishop, pers. com.; McCann, 1933; Oppenheimer, in preparation; Ripley, 1970). Although water is usually obtained from the vegetation and from dew, it may be obtained from ponds and other water bodies, if available during the hot dry time of the year in otherwise moist climates (Fig. 8); langurs have also been observed to drink from ponds even during the monsoons in dry climates.

III.   Population Density, Structure, and Dynamics

A.   Density and Home Range Size

Because of the greater availability of food in forest environments, primate populations are denser there than in grassland environments (Crook, 1970).

Population densities of *P. entellus* range from 2.6 to 134.6 individuals per km^2 (Table VI). The lower densities (2.6–16.6/km^2) are most typical of populations living in open grassland or agricultural fields where the langur troops have large home ranges (1.5–13.0 km^2) (Jay, 1965), which may be separated by uninhabited regions. The higher densities (57.9–134.6/km^2) are most typical of forest-dwelling populations where the troops have small, sometimes overlapping, home ranges (0.05–0.20 km^2). This contrast is well illustrated by the data from the grassland and forest at Dharwar (Table VI). Intermediate densities (4.6–50.0/km^2) appear to be typical of populations living in close association with towns and villages of man (*Homo sapiens*) where the inhabitable environment is limited because of high population densities of *H. sapiens* (913/km^2 at Singur; Graves and Oppenheimer, 1975), concentrated and limited food resources, and/or low rainfall. The langur troops living in this type of area have intermediate sized home ranges of 0.38 to 0.95 km^2.

The population living in the forest at Orcha, however, is an exception to the above scheme. There the population density and home range size are the same as that proposed above as typical for a population dwelling in grass or cropland.

TABLE V

Species of Plants that occurred in the Diet of *Presbytis entellus* in Two or More Locations, and Indication of Part Eaten[a]

Family and species identification[b,c]	Gir Forest, Gujarat	Abu Hills, Rajasthan	Jodhpur, Rajasthan	Singur, West Bengal	Dharwar, Mysore	Polonnaruwa, Sri Lanka
Alangiaceae						
1. *Alangium salvifolium*, T	Lv	—	—	—	—	Fr,Fl,Fs,Lv
Anacardiaceae						
2. *Lannea coromandilica*, T	Bk	—	—	—	—	Fr,Fs
3. *Mangifera indica*, T	—	—	Fr,Fl,Fs	Fr,Fl,Lv	—	Fr,Fs
Annonaceae						
4. *Annona squamosa*, S	Fr	—	—	Fr,Lv	—	—
Apocynaceae						
5. *Carissa carandas*, S	Fr,Lv	—	Fr,Fl	Fr,Fl,Lv	—	—
6. *Tabernaemontana coronaria*, S	—	—	—	Lv	Fr,Lv	—
7. *Wrightia tinctoria*, S	Fr,Lv,Sd	Fl,Lv,St	Fl,Lv	—	—	—
Bombacaceae						
8. *Bombax malabaricum*, T	—	Fr,Fl,Sd	—	Fr,Fl,Lv, Bd	Lv	—
Capparidaceae						
9. *Capparis decidua*, S	Lv	—	Fr,Fl	—	—	—
Chenopodiaceae						
10. *Spinacia oleracea*, H	—	—	Plant	Lv	—	—
Combretaceae						
11. *Anogueissus latifolia*, T	—	Fl,Lv,St	—	—	Lv	—
12. *Terminalia belerica*, T	Fl,Lv,Sp	Fl,St	—	—	—	—
Cruciferae						
13. *Brassica oleracea*, H						
a. var. *botrylis*	—	—	Fl,Lv,	Fl	—	—
b. var. *capitata*	—	—	Fl,Lv	Lv	—	—

Euphorbiaceae						
14. *Phyllanthus emblica*, T	Fr,Lv	—	—	Lv	Fr,Lv	—
Leguminosae						
15. *Acacia arabica*, T	Fl,Lv	Lv	—	Fr,Lv	—	—
16. *Acacia catechu*, T	Lv,Bk,Sp	—	—	—	Lv	—
17. *Bauhinia racemosa*, S	Lv	—	Lv	—	Fr,Fl,Lv	Fs
18. *Butea monosperma*, S	—	—	Lv	—	Fl,Lv	—
19. *Cassia fistula*, T	—	—	—	—	Fr,Lv	Fr,Fl,Fs,Lv
20. *Tamarindus indica*, T	Fr,Fs,Lv	—	Fr,Lv	Fr,Fl,Lv	Fr,Lv	Fr,Fl,Fs,Lv
Malvaceae						
21. *Hibiscus esculentus*, H	—	—	Fr,Fl	Fr,Lv	—	—
Meliaceae						
22. *Azadirachta indica*, T	Lv	Fl,Fs	Fr,Fs	Fr,Lv	Lv	Fs
Moraceae						
23. *Ficus bengalensis*, T	Fr,Fs,Lv,Pi	Fr,St,Su	Fr,Fs	Lv	Fr,Lv	Fr,Fs
24. *Ficus glomerata*, T	Fr,Sd	St	—	—	—	—
25. *Ficus religiosa*, T	Fr,Fs,Lv,Sm	—	Fr,Fs	Fr,Lv	—	Fr,Fs,Lv
26. *Streblus asper*, S	—	—	Fr,Fs	Fr,Lv	Lv	Fr,Fl,Fs,Bd,Sd
Myrtaceae						
27. *Eugenia jambolana*, T	Fr,Fl,Lv,Sd	—	Fr,Fl,Fs	Fr,Lv	—	Fr,Fl,Fs,Lv
Rhamnaceae						
28. *Ziziphus mauritiana*, T	—	—	Fr,Lv	Fr,Fl,Lv	—	Fl
Rutaceae						
29. *Feronia limonia*, T	—	—	Fr,Fl,Lv	—	—	Fs
30. *Glycomis pentaphylla*, S	—	—	—	Fr,Lv	—	Fs
Sapindaceae						
31. *Sapindus emarginata*, S	Fr,Fs,Sd	—	—	—	—	Fr,Fs

Table V (continued)
Species of Plants that occurred in the Diet of *Presbytis entellus* in Two or More Locations, and Indication of Part Eaten[a]

Family and species identification[b,c]	Gir Forest, Gujarat	Abu Hills, Rajasthan	Jodhpur, Rajasthan	Singur, West Bengal	Dharwar, Mysore	Polonna-ruwa, Sri Lanka
Solanaceae						
32. *Capsicum annuum*, H	—	—	Fr,Lv	Fr	—	—
33. *Lycopersicon esculentum*, H	—	—	Fr,Fl,Fs	Fr	—	—
34. *Solanum melogena*, H	—	—	Fr,Fs	Fr	—	—
Tiliaceae						
35. *Grewia villosa*, S	Fr,Fs,Lv,Sm	—	Fr,Lv	—	—	—
Ulmaceae						
36. *Holoptelea intergrifolia*, T	Lv	—	—	—	—	Fr,Fl,Fs, Lv,Bd,Sd
Verbenaceae						
37. *Tectona grandis*, T	Insects	—	—	—	Lv	—

[a]See footnotes, Table IV.
[b]T, tree; S, shrub or small trees; H, herb.
[c]Common names of some species according to number: 3, mango; 4, custard apple; 5, karanda; 10, spinach; 13a, cauliflower; 13b, cabbage; 20, tamarind; 21, okra or lady's finger; 22, neem; 23, banyan; 25, pipal; 27, jambul or black plum; 29, wood apple; 32, chili; 33, tomato; 34, eggplant or brinjal; 36, Indian elm; 37, teak.

Fig. 8. Adult males drinking water at edge of river during dry season.

It has been suggested (Ripley, 1970) that the Orcha forest is suboptimal because of its high rainfall, which is reported to be 203 cm per year (Jay, 1965; Yoshiba, 1968). However, it is the carrying capacity of a habitat, rather than simply rainfall, which ultimately must account for population size and density. A seasonally dry deciduous forest consisting of only a few species might support a smaller population than an agricultural-horticultural area with multiple annual cropping regardless of total rainfall. Sugiyama (1976) found that home range size increased as the amount of food available decreased.

The population density at Singur also appears low in relation to its habitat and the troop home range size. An adjusted population density for Singur would be about $20/km^2$, if the unused agricultural land were omitted from the calculation. Within the suitable village habitat the home ranges were contiguous (Oppenheimer, in preparation).

Within the home range there is a smaller area, which is sometimes split up, where the troop spends most of its time—the core area. Core areas usually include the important sleeping and food trees. At the Gir the food trees were evenly scattered throughout the home range so that the core area was undefinable (Rahaman, 1973). At Dharwar the core areas for seven forest troops were about $0.09 km^2$ (range $0.05-0.16$), approximately one-half the size of the home ranges (Sugiyama *et al.*, 1965). Within the village habitat of Singur the core areas were $0.12 km^2$, about one-quarter the home range size (Oppenheimer,

TABLE VI

Population Density, Structure, and Home Range Size

Location[a]	Habitat	Density/km²	Size of bisexual troop		Size of all-male group[b]	Home range (km²)		Adult sex ratio in troops	One male: multimale: all male
			One male[b]	Multimale[b]		Bisexual troop[b]	All-male group[b]		
B. Melemchigaon	Forest and cropland	15.2	—	32	—	2.17	—	1:1.3	0:1:0
D. Jodhpur	Desert scrub and cultivation	18.1	35.4 (29–76)	(81)	15.7 (4–28)	0.95 (0.6–1.3)	11.25 (10.5–12.0)	1:4.3	20:0:11
E. Mt. Abu	In or near town	50.0+	20.8? (15–40)	22	6?	0.38	—	1:5.7	7:1:1?
F. Gir	Forest	127.0	20.1 (16–35)	33.3 (18–48)	2 (1–3)	0.20	—	1:5.3	3:8:2
G. Kaukori	Cropland	2.6	—	37.0 (20–54)	3	6.50	—	1:3.2	0:2:1
H. Singur	Village	ca. 4.6	12.8 (10–14)	—	?	0.43	—	1:5.6	2:0:1?
I. Raipur	?	—	26.8 (17–44)	32.1 (12–61)	6.7 (1–14)	—	—	1:5.7	9:7:3
K. Orcha	Forest	6.2	—	18.6 (10–28)	1	4.09 (1.3–6.5)	—	1:1.4	0:3:1
L. Dharwar	Forest	85.3–134.6	15.5 (9–23)	16.1 (11–24)	11.8 (2–32)	0.17–0.19 (0.10–0.32)	16.62	1:4.2	19:8:3
	Grassland	16.1	14.7 (9–21)	19.5 (18–21)	8.7 (7–10)	1.50	16.62	1:5.6	9:2:3
M. Polonnaruwa	Forest	48.3–67.6	—	30 (5–60)	—	? (0.05–0.19)	—	—	0:4:?
N. Simla	Forest	24.6	23 (19–98)	50.1 (19–98)	1.8? (1–5)	1.90 (0.66–4.91)	—	1:3.7	1:7:9
Overall Range		2.6–134.6	9–76	10–98	1–32	0.05–7.68	10.5–16.6		

[a] See listing in Table II.

in preparation). In the very dry area around Jodhpur the average core area was 0.26 km² (range 0.14—0.42), which was also about one-quarter of the average home range size (Mohnot, 1974). Even with the very small home ranges at Polonnaruwa, smaller than most of the core areas mentioned above, there were parts that were rarely used (Ripley, 1967). The core areas at Kaukori and Orcha were larger than most home ranges found in other studies, and ranged between approximately 0.8 and 2.0 km², about one-fifth to one-third home range size. In addition, the location of core areas within the home ranges at both Kaukori and Orcha changed with the season (Jay, 1965). The size of the core area relative to the home range depends on the distribution and density of food and sleeping trees and upon the amount of food produced. Thus if the food trees are concentrated at a few places within the home range, the core area (the area of maximum use) may be relatively quite small.

Although few data are now available, it appears the home ranges of all-male groups are large regardless of habitat type (Table VI), and are much larger than those of bisexual troops, with which in part they may overlap (Jay, 1965; Mohnot, 1974; Sugiyama et al., 1965). They have several sleeping sites and their core areas are dispersed (Mohnot, 1974).

B. Structure of Population

Three types of population structure occur in the hanuman langur. The first consists of a population broken up into bisexual troops in which each troop has a number of adult males, or at least one adult male and a number of subadults.* Subadult males are 4—6 years of age and adult males are above 6 or 7 years (Jay, 1965; McCann, 1933; Sugiyama, 1964). In such troops the sex ratio among adults may be one male to two females (Table VI). The second type of population is broken up into bisexual troops with only one adult male (subadults are absent), and one or more all-male groups. In these bisexual troops there may be, on the average, one adult male to six adult females (Table VI), though in the total population (including the all-male groups) the adult sex ratio may be as close as 1 male to 1.8 females (Mohnot, 1974). The third type of population is a mixture of single-male and multimale bisexual troops, and all-male groups.

It is thought that a population consisting of multimale troops is one which is located in a more favorable environment, whereas a population consisting of single-male troops is one living in a less favorable environment (Aldrich-Blake, 1970; Sugiyama, 1965b).

The type of structure present in the population to be independent of habitat type, amount of rainfall, population density, troop home range size, and even

---

*Vogel (1977) has suggested that troops with two or more adult males should be considered separately from troops with one adult male and one or more subadult males.

average troop size (Table VI). For instance, in forest habitats with similar rainfall and population density, the Gir has 2.7 times as many multi as single-male troops, and Dharwar has 2.4 times as many single as multimale troops (Table VI). A similar contrast occurs in the open habitats of Kaukori and Dharwar, which have similar annual rainfalls (89 and 100 cm) and low popula-tion densities. At Dharwar, the forest and grassland areas have very different population densities, but both areas have more single than multimale troops. It may be more accurate to state, though more difficult to document, that the structure of a population is determined by the population size relative to the carrying capacity of the environment, rather than by absolute density per unit area.

The number of sleeping sites available, and the amount of water and food present during the dry season are probably the main factors in determining suitability or carrying capacity of an environment (see discussion in Crook, 1970). Single-male troops are probably selected for in environments where climatic factors lead to a seasonal scarcity of food (Ripley, 1970). A single male can fertilize all the adult females in the troop and thus the additional males are expendable (Aldrich-Blake, 1970).

However, genetic factors probably also play a role in determining the type of population structure present at a particular location. In the forest at Polonnaruwa, *P. e. thersites* had multimale troops with an average size of 30 (Ripley, 1967, 1970), whereas *P. senex senex* (the purple-faced langur) had single-male troops with an average size of 8.4 (Rudran, 1973). *Presbytis e. thersites* had home range sizes two to five times larger than, and a population density one-quarter to one-third (Table VI) as large as, *P. s. senex* (home range: 0.009–0.077 km^2; density: 215/km^2) (Rudran, 1973).

C.   Structure and Size of Population Subunits

The size of bisexual troops ranges from 5 to over 120 (Jay, 1965), though the mean troop sizes for different populations lie between 12.8 and 37.0 (Table VI). Larger troop size may be correlated with more time spent on the ground, particularly in agricultural areas where resources are widely dispersed (Jay, 1965), and where the troops are exposed to harassment from man and dogs (Ripley, 1970). Also larger groupings may be temporary aggregations of two or more bisexual troops (see below). Multimale bisexual troops are slightly larger than single-male troops, but this difference can be accounted for by the greater number of males, rather than any difference in the number of adult females. The composition of bisexual troops from seven locations is given in Table VII. Adult males on the average make up 2.3–19.4% of the troop and adult females 31.0–59.3%.

All-male groups on the average range in size from 2 to 17.6 males (Table VII), although single males and one group of 59 males (S. Kawamura, cited in Sugiyama *et al.*, 1965) have also been observed. These groups primarily consist of adult and subadult males, but may contain juveniles only 1 year of age (Sugiyama, 1964; Yoshiba, 1968).

Bisexual troops have been noted to aggregate, as well as to break up into subunits. In the dry area around Nagpur and Raipur, troops came together for as much as an hour near water sources. This may account for the large troop size of 120 (cited above), as it was observed near a water reservoir; other troops observed in the area were as small as 13 (Jay, 1963a, 1965). At Polonnaruwa, bisexual troops divided into subgroups during the day to engage in various activities (Ripley, 1970).

D.   Population Dynamics

It has been generalized that a single-male troop of primates could, over time, become an age-graded multimale troop due to birth and maturation of additional males in the troop. The single-male troop represents the structure of the founding populations and the structure at high population densities, and the multimale troop, at low population densities. In addition, the multimale troop would become unstable as the population density increased and eventually division would lead to new single-male troops (Eisenberg *et al.*, 1972). However, as mentioned in Section III,B, density per unit area is a poor indicator of the type of population structure present at a particular location.

An early hypothesis (McCann, 1933) about the dynamics of population structure of *P. entellus* was that during the nonbreeding season the troops were multimale, but that with the approach of the breeding season one dominant male would expel the other males. This would result in formation of single-male troops and all-male groups. At the end of the breeding season the langurs would again form multimale troops.

However, most recent studies have shown that the overall population structure present in a particular locality tends to be stable throughout the year and from year to year (Jay, 1965; Mohnot, 1971a,b, 1974; Sugiyama, 1967; Yoshiba, 1968). At Jodhpur, a 5-year study has shown that, with one exception, all troops were and remained single-male troops (Mohnot, 1974). The one exception was a troop with 26 males which, 6 months after the study began, was transformed into a single-male troop. The alpha male expelled the other 25 males and was, in turn, displaced by the beta male, who became the troop's adult male.

The process whereby the troops at Jodhpur remained single-male troops was an internal one. As the infants grew older and became juveniles, tension between

TABLE VII

Composition of Bisexual Troops and of All-Male Groups (Means and Ranges)

Location and source[a]	No. of troops counted	Adult Male[b]	Adult Female[b]	Subadult Male[b]	Subadult Female[b]	Juvenile Male[b]	Juvenile Female[b]	Infant Male	Infant Female	Mean no. of individuals[b]
					Part A. Bisexual troops					
D. Jodhpur	6	1.0	20.0	—	6.1	2.6	4.9	4.5	4.5	43.6
F. Gir Forest Kurup (1970)	1	1	10	3	—	4		—	—	18
Rahaman (1973)	9	2.2 (1–6)	13.9 (3–27)	1.2 (1–4)	4.9 (1–19)		4.1 (1–9)		4.0 (2–6)	30.3 (16–48)
Starin (1973)	2	2.0 (1–4)	8.0 (4–10)	1.5 (0–3)	1.8 (0–5)		9.0 (6–13)		3.5 (0–5)	25.8 (22–31)
G. Kaukori	1	6	19	2	3	5		9	5	54
H. Singur	2	1.0	5.0 (4–6)	—	—	—	4.3 (4–6)		2.5 (0–4)	12.8 (10–15)
I. Raipur	16	2.8 (1–8)	15.7 (5–30)	Included in adults			4.6 (1–11)		6.1 (1–16)	29.1 (12–61)
K. Orcha	3	3.6 (2–6)	6.0 (4–9)	0.3 (0–1)	1.3 (1–2)	0.7 (0–2)	2.3 (0–5)	3.0 (1–4)	1.3 (1–2)	18.6 (10–28)
L. Dharwar Sugiyama (1964)	38	1.7 (1–7)	8.0 (4–14)	Included in adults			5.3 (0–12)			15.1 (9–24)
Sugiyama et al. (1965)	7	1.0	8.9 (6–13)	0.3 (0–2)	—		4.9 (1–10)			15.0 (10–24)

Part B. All-male groups

Location[a]	n										
D. Jodhpur	11	9.6	—	3.0	—	5.0	—	—	—	—	17.6 (4–28)
F. Gir Forest Rahaman (1973)	2	1.0	—	1.0 (0–2)	—	—	—	—	—	—	2.0 (1–3)
I. Raipur	3	6.1 (1–12)	—	Included in adults	—	0.6 (0–2)	—	—	—	—	6.7 (1–14)
L. Dharwar Sugiyama (1964)	6	6.7 (2–15)	—	3.9 (0–10)	—	1.4 (0–7)	—	—	—	—	11.8 (2–32)
Sugiyama et al. (1965)	2	6.0 (5–7)	—	2.5 (1–4)	—	1.0 (0–2)	—	—	—	—	9.5 (6–13)

[a]See locations in Table II and Fig. 3.
[b]Numbers in parenthesis indicate range.

them and the troop's adult male gradually increased. The adult male separated
the juveniles from their mothers and from the troop with aggressive displays
and chases. The actual expulsion from the troop lasted a number of days during
which the juveniles' attempts to return to their mothers and the troop were
repulsed by the adult male. At this point, the juveniles joined an all-male group
(Mohnot, 1974).

At Dharwar the process is slightly different as the adult males tolerated the
presence of their male offspring in the troop. However, due to successful attacks
from outside by all-male groups, the adult male of the bisexual troop was displaced on the average once every 4.5–5 years (Sugiyama, 1965b). The new
dominant male expelled the young males that grew up in the troop along with
the other members of the all-male group (Sugiyama, 1964, 1965a,b). At Dharwar
(Sugiyama, 1965b), Jodhpur (Mohnot, 1971b, 1974), and Mt. Abu (Hrdy, 1974),
infants of both sexes were in some cases killed by the new adult male during
reorganization of the troop (see Section III,E). In one case, Sugiyama (1967)
observed that three males from an all-male group displaced the adult male of
a troop and all three males remained in the troop for a month. At that time, the
breeding season commenced and one of the males expelled the other two. This
gives some indication of the basis for McCann's (1933) hypothesis.

Attacks on a troop may also be made by members of other bisexual troops.
At Dharwar, the adult male of one troop entered a troop without an adult male,
killed the infants, and copulated with the females (Sugiyama, 1966). At Mt. Abu,
a weak adult male was able to take over his troop again with the aid of five other
males, possibly from an all-male group; all six then attacked a neighboring troop
(Hrdy, 1974). At Polonnaruwa, where troops are multimale and all-male groups
are lacking, the adults of both sexes of one troop may attack another troop, and
may even pass through the home range of a third troop in order to do so
(Ripley, 1967).

In some cases, attacks from the outside lead to troop division. A non-troop
male fought with the male of a single-male troop and took some of the six adult
females away with him (Sugiyama, 1964). At Jodhpur the alpha male of an all-
male group attacked a troop and joined it. After a couple of months he left,
accompanied by twelve females and four young. He died 6 months later and
the females and young rejoined their original troop (Mohnot, 1974).

The ideal pattern of population dynamics in *P. entellus* has been hypothesized
to be as follows. In situations with low density relative to carrying capacity
the population will be divided into multimale age-graded troops. As the relative
density increases, the number of attacks on troops (from other troops or all-male
groups) will increase, and the size of the all-male groups will grow. Single-
male troops will become the prevailing pattern of population structure. The
increased number of attacks on the single-male troops will lead to high turnover

in troop males and to increased infant mortality and social disruption. This in turn, may lead to a reduction in the density relative to the carrying capacity and a gradual reappearance of multimale troops in a relatively peaceful social environment (Rudran, 1973; Sugiyama, 1965b).

The all-male group provides a substitute social environment within which the young males can grow up, and a place where a dominance order can be established among the adult males (Sugiyama, 1964). All-male groups tend to be peaceful and inactive for certain periods (Mohnot, 1974; Sugiyama, 1964). The young males engage in playful wrestling (Fig. 5) and chasing (Sugiyama, 1964), and perfect their sexual techniques (Fig. 6). The adult and subadult males engage in short, but intense, bouts of agonistic behavior until the dominance order is established or reestablished. Severe wounds may sometimes be incurred (Fig. 9). It is usually the dominant male of the all-male group that fights with and replaces the adult male of the bisexual troop (McCann, 1933).

Fig. 9. Agonistic interaction between two males of an all-male group. Male at upper right has lost part of left cheek, but has just forcefully rejoined group at the end of its rest period (Figs. 2, 5, 6, and 8), and reasserted his dominance. Observed at "Crocodile Pond," Corbett National Park, by Charles Southwick, Bikas Pal, Elizabeth Oppenheimer, and the author, on March 28, 1971.

E. Reproduction and Mortality

The troop has its foundation in its adult females and their progeny (Fig. 10) (Sugiyama, 1965b). The females reach sexual maturity at about 3.5 to 4 years (Jay, 1965), and it is these females that hold the troop in its home range; new leader males adapt themselves to the home range of the troop they have taken over (Sugiyama, 1965b). One suggested stimulus for attacks on bisexual troops is the desire of non-troop males to copulate with estrous females (Sugiyama, 1965b). Adult females come into estrus every 30 days, and estrus lasts 5–7 days. The estrous female signals her receptivity by giving head shakes, dropping her tail on the ground, and presenting to a male (Jay, 1965). If the male ignores her, she may hit him, pull his fur, or even bite him. Gestation lasts about 160 to 185 days (Mohnot, 1974). The female resumes sexual activity again at weaning, when the infant is about 10 to 12 months old (Jay, 1965). Thus females give

Fig. 10. The troop's foundation: mother and offspring embracing and greeting.

birth about once every 2 years; however, estrus can be induced by removal of the infant prior to weaning (Sugiyama, 1965b).

The inducement of estrus is probably the explanation for infant killing by an adult male just after he has taken over a troop (Hrdy, 1974; Sugiyama, 1965b). The mother abandons her infant after it has become severely wounded or after she has become worn out from protecting her infant from the adult male. On the average, females come into estrus within 8 days after the loss of their infants, and give birth about 6 to 7 months later. Thus in populations where troops are attacked frequently and the troop male holds his position for 4 years or less, there is a strong selection for males who kill the infants of their predecessors, as this increases the likelihood that their own progeny will reach reproductive age. It has also been suggested that removal of the infants allows the new adult male to establish a strong social bond with the females (Sugiyama, 1965b).

Although births occur throughout the year (Jay, 1965; Mohnot, 1974; Oppenheimer, 1976a, in preparation; Sugiyama *et al.*, 1965), concentrations of births have been observed to occur during the dry season. In north-central India, births tend to occur in April and May (Jay, 1965), in the west between March and May (Mohnot, 1974), and in the south between December and March (Sugiyama *et al.*, 1965). At Dharwar it was noted that there was a slight shift from year to year: November and December in 1961 and, a year later, December 1962 and January 1963 (Sugiyama *et al.*, 1965). In populations where male attacks are frequent and are accompanied by infant killing, the birth season may shift dramatically or may disappear (Yoshiba, 1968).

Reproduction and mortality have been best studied at Jodhpur (Mohnot, 1974). Over a 5-year period the population increased from 836 (17.4/km^2) to 905 (18.8/km^2), with an average net increase of 1.9% per year. There were 186 births. Twenty-two of the infants died within 4.5 months of birth. Sixteen adults died due to electrocution, road accidents, old age, or unknown causes. Fifty individuals, mostly females, disappeared. In June 1971, one troop of 82 individuals was reduced to 11 due to a mass mortality thought to be caused by famine (Mohnot, 1974); however, the circumstances strongly suggest poisoning by algae toxins in the drinking water (see Olson, 1951).

IV.   Relationship with Other Animals

A.   Birds and Reptiles

The interactions between langurs and birds range from playful attacks on the birds, to symbiosis, to attacks on langurs by birds. On three occasions, langurs have been observed near great horned owls, *Bubo bubo* : once they ignored each other, once a juvenile displaced the owl, and once two juveniles ran up to and jumped over the owl and ran off when it turned to look at them (Starin, 1973). Langurs and birds, particularly peafowl (*Pavo cristatus*), associate with each other on the ground or in the trees and respond to each other's alarm calls or movements (Jay, 1965; Rahaman, 1973; Starin, 1973). Sometimes langurs and peafowl even slept 'in the same trees, although once when the langurs entered a tree quickly the peafowl moved away (Starin, 1973). Jungle crows (*Corvus macrorhynchos*) sometimes flock in to feed on insects disturbed by passage of the monkeys (Rahaman, 1973). Twice hawks, probably *Accipiter badius*, have been observed to attack the monkeys: once it tried to grab an infant on the ground, but was deflected by the adult male of the troop (Rahaman, 1973); on the other occasion, it flew around a juvenile langur, too large to be prey, for 15 minutes, landed at its side and pecked it in the neck thrice, watched it for another 15 minutes some 40 m away, and then left (Starin, 1973).

Interactions with reptiles have been seen less often. On two occasions, langurs and the monitor lizard (*Varanus* sp.) have fed peacefully close together. On one occasion, a python (*Python molurus*) caught and swallowed an adult female langur (Rahaman, 1973).

B.   Mammals

1.   Lagomorphs and Rodents

In the Gir Forest, hares and squirrels are met with daily, and are usually ignored (Starin, 1973). However, langurs may be attracted by alarm calls of squirrels (*Funambulus* sp.), and occasionally a juvenile may try to catch one (Rahaman, 1973).

2.   Ungulates

Chital deer (*Axis axis*) are often seen close to or beneath trees the langurs are in. The two species respond to each others alarm calls (Rahaman, 1973), and by combining the good eyesight of the langurs with the good olfactory and auditory senses of the chital, both species benefit from the increased likelihood that predators will be detected sooner than if they were separate (Schaller, 1967, p. 64−65; Starin, 1973). The chital also eat fruit and vegetation that the langurs drop to the ground, and in one instance a chital nibbled the leaves off a branch held by a langur (Schaller, 1967, p. 62). This source of food is particularly important for the chital during the dry summer months, when less grass is available (Rahaman, 1973; Starin, 1973). The barasingh, *Cervus duvauceli*, also responds to the alarm calls of langurs (Schaller, 1967), and other ungulates like the nilgai (*Boselaphus tragocamelus*), the four-horned antelope (*Tetracerus quadricornis*), the sambar (*Cervus unicolor*), and domestic cattle have been observed to feed near the langurs (Starin, 1973).

3. Carnivores

The smaller carnivores, such as the mongoose (*Herpestes* sp.), may peacefully feed nearby the langurs (Rahaman, 1973), but the presence of larger carnivores causes a great deal of alarm among the langurs. Within the geographical range of *P. entellus*, lions are only present in the Gir Forest. Even there they may only rarely eat a langur, most likely one that was sick or previously killed by a leopard, *Panthera pardus* (Rahaman, 1973; Starin, 1973). Leopards are said to occur throughout India (Prater, 1965), and do include langurs in their diet (Schaller, 1967). At the Gir one leopard was able to approach the tree in which langurs were situated, and the langurs took notice of him and gave alarm calls only after a villager saw him and chased him off with shouts and rocks (Starin, 1973). The tiger, *Panthera tigris*, also once occurred throughout the geographical range of *P. entellus*, but now it is primarily restricted to souvenir stores in Calcutta, Bombay, and New Delhi, and a small number of national parks. In Kanha National Park, langurs are commonly eaten by tigers, but more often are eaten by leopards (Schaller, 1967, pp. 281 and 311).

Today, of the carnivores, the most frequent harasser and occasional killer of langurs may be the pariah dog, *Canis familiaris*, who lives in or near the villages of man in India (Oppenheimer and Oppenheimer, 1976) and in the forests of Sri Lanka. At Polonnaruwa dogs often chase the langurs and occasionally catch an infant (Ripley, 1970). At the Gir two dogs caught and ate a subadult male that was on the ground (Rahaman, 1973). Most of the dog—langur interactions probably occur when the langurs are raiding crops in the field or when, for some other reason, they are on the ground. In Singur, where the monkeys were on the ground only 5% of the daylight hours, they were attacked by dogs on the average of once every 93 hours of ground time; however, this low rate probably indicates that the langurs only came down when dogs were absent from the area (Oppenheimer, 1976b). On the Coromandel coast an adult male langur was caught while eating in a field and was killed by the dogs (A. Minchin, cited in Pocock, 1939, p. 101). On occasion, single dogs may be chased off by the troop's adult male (Oppenheimer, in preparation; Sugiyama, 1964). At Jodhpur where the dogs also often chase and bark at the monkeys, they have been observed eating and resting with the langurs, and being groomed by langurs (Mohnot, 1974).

4. Other Primates

The hanuman langur appears to get along quite well with other species of monkeys. At Polonnaruwa they live in the same forest with the purple-faced langur, *P. senex*, and the bonnet macaque, *Macaca radiata*. Little is said about their interactions, but the two species of langurs are so well adapted that one gives its morning spacing calls one-half hour prior to sunrise and the other after sunrise (Ripley, 1967, 1970). At Dharwar troops of bonnet macaques and hanuman langurs may intermix and travel together throughout the day (Sugiyama, 1967). At Kaukori rhesus macaques, *Macaca mulatta*, sometimes used the same food and water sources and slept in the same trees as the langur troop, and, on occasion, troops of both species intermingled on the ground or in the trees (Jay, 1965). At Jodhpur (Mohnot, 1974), Kaukori, and Halwapura (Jay, 1965) outcast adult rhesus monkeys lived within langur troops. The male and female rhesus that lived in the Kaukori troop were, because of the more pronounced aggressiveness of rhesus, dominant to the langurs and were able to settle disputes among the langurs (Jay, 1965).

The relationship with *H. sapiens* is more complex. The hanuman langur is considered sacred by the Hindu religion, and with this protection those langurs that lose their fear of man and live near human habitation may cause, at times, considerable destruction to crops and property (Hrdy, 1974; Jerdon, 1874; McCann, 1928; Pocock, 1939). The langurs also receive protection from the Buddhist religion, which forbids killing of animals. Thus in Nepal the langurs are protected by both religions, though if they enter the agricultural fields they may be shot (Bishop, personal communication). The importance of this religious protection is illustrated by two points. First, there are non-Hindu tribal peoples in India that hunt and eat *P. entellus*: the Kathkaris of Konkan, who hunt with bow and arrow (McCann, 1928); the Jumsare tribe (Crump, cited in Pocock, 1939, p. 100); the Mulcers of Travancore, who consider the flesh of *P. johnii* to be of higher medicinal value (Pocock, 1939, p. 108); and the Santhals. Second, village-dwelling langurs are absent in Moslem Bangladesh (personal observation), though Husain (1974) indicates they may be found in villages with Hindu populations. *P. pileatus*, along with rhesus, occurs in the Madhupur Forest just south of Mymensingh (personal observation). Rhesus monkeys can also be found in Hindu sections of the old city in Dacca and in some Hindu villages in the countryside (personal observation; W. Akanda, personal communication). Rhesus monkeys are also considered sacred by the Hindus (K. K. Tiwari, personal communication), probably because the sacredness of hanuman has been generalized to all monkeys.

The reason why the hanuman langur is sacred to the Hindus can only be explained in part here. Animal worship existed among the peoples of the Indus Valley civilization prior to the Indo-European invasion about 2000 BC and prior to the development of Hinduism (Basham, 1971). All of the Hindu gods and their various forms are associated with a particular animal—donkey, elephant, mouse, and swan, etc. There are two long epic works that are of great importance in Hindu literature and culture. The earliest epic is the

"Ramayana" (Griffith, 1915), which may date back to 1000 BC. Very briefly, Rama, who is really the Hindu god Vishnu in the shape of a human being, is sent to the earth to defeat Ravana of Lanka (Sri Lanka?). Ravana kidnaps Rama's wife, Sita, and the bulk of the epic relates to the search for Sita and her eventual recovery. A large number of animals take part in the search and recovery, but Hanuman, one of the monkey subleaders and son of the Vedic wind god, Vayu, is the most outstanding and most important. The second and larger epic (which runs to twelve thick volumes) is the "Mahabharata" (Ray, 1884). It is a story about the adventures of the sons of the royal Bharat family from whom the Hindu Indians have figuratively descended, and because of this the country of India is also called Bharat. In this second epic the "Ramayana" is retold briefly as one of the episodes and, of course, Hanuman appears; however, Hanuman appears in other episodes as well, and by some rather interesting process, turns out to be the brother of Bhima, one of the Bharat sons. Though I am tempted to quote a number of passages, I feel it is best to refer those of you who are interested to the translations cited above.

An indication of the importance of the worship of Hanuman is the inclusion in Chinese mythology of a story associating a Hanuman-like monkey with the introduction of Buddhism into China about 650 A.D. The story involves Buddha's confrontation with and subjugation of "Monkey," though "Monkey" is eventually given the rank of "Buddha Victorious in Strife" (Christie, 1968, pp. 123–130; Wu Ch'êng-ên, 1961, p. 349).

Whether Hanuman of the Indian epics was a *P. entellus* is questionable, but he did have a long tail, which he used along with fire to destroy Ravana's city (rhesus have short tails). This description of Hanuman, the wind god's son, does fit that of hanuman, the langur, who was probably abundant in the forests of northern India, where the early Hindus dwelled.

The kind of damage done by the langurs today is more limited in scope than that of their namesake, though it can bring financial disaster to the small farmer who is already at the mercy of floods, droughts, population pressure, and inflation. In the quest for food, the langurs may easily eliminate a field of vegetables or reduce the amount of fruit produced by an orchard or grove. They trample the plants in kitchen gardens or break tiles on the roofs of houses in their passage. They grab food from shops or remove objects, such as a mirror, from someone's open windowed bedroom (Hrdy, 1974; Jerdon, 1874; Oppenheimer, in preparation; Pocock, 1939). They may, where they are purposely fed, become so brave that they will swipe the food right out of a person's hand, even though it was supposed to be that person's lunch (Oppenheimer, personal observation).

The human response to all this is to shout or throw rocks, frequently just on sight of the monkeys, to station guards in the fields, and, occasionally, to call in police, tribals, or someone else, to shoot at least one or two of the offending horde (Bishop, personal communication; Blyth, cited in Jerdon, 1874; Crump, cited in Pocock, 1939, p. 100; Mohnot, 1974; Oppenheimer, 1976b). During observations between February 1971 and February 1972 at Singur, each troop was harassed on the average once every 2 hours during the diurnal period (based on over 4600 hours of observation). The harassment consisted primarily of rock throwing (45%) and shouting (41%), and was carried out mostly by young boys (71%). More than 50% of this harassment was motivated by human play, 33% in order to protect the fields, gardens, and orchards, and 11% to protect the houses (Oppenheimer, 1976b). In one village in Howrah District, West Bengal, people took up drums and other noise makers in the hopes of driving the monkeys away (Oppenheimer, personal observation). It is reported that another village caught all its langurs, loaded them into a railway car, and had them released at stations farther down the line (McCann, 1928). In some cases, people take advantage of the destructive abilities of the monkeys by luring them onto a disliked neighbor's property with the use of scattered grain (McCann, 1928).

However, all the above direct relations between man and the hanuman langur are, in a sense, trivial. It is rather the unconscious activities of man that have the most profound effect upon the survival of the hanuman langur. The most important of these is the destruction of forests in order to obtain firewood and land for agriculture. This even occurs in the towns and villages, where stands or groves of trees and bamboo may be cut down to provide space for growing more vegetables (P. Nundi, personal communication). It is this behavior of man that forces the langur into increased contact with *H. sapiens* and to become more dependent upon the food of *H. sapiens* in order to survive. It also brings about increased contact between hanuman and man which may result in increased transmission of diseases between these two species to the possible detriment of both (Graves and Oppenheimer, 1975). Thus the survival of the hanuman langur gradually becomes a case of whether it can successfully compete for food with the descendents of the Bharat family and whether the cultural traditions of a society can withstand the pressures of logarithmic growth.

Langurs, ideally, control their population size by shifting the structure of their populations from multimale to single-male troops and by going through a period of social disruption caused by "marauding" all-male groups (Section III,D). Social disruption also occurs in human populations: assault, rape, terrorism, and limited and unlimited warfare, declared or otherwise. Despite this, the population of *H. sapiens* in the world has continued to increase, and within the geographical range of *P. entellus* it will double in the next 27−32 years at the present rates of growth (Ehrlich and Ehrlich, 1972).

It appears that man is less able biologically to handle the problems that face him, and it may become an academic question as to whether he can ensure survival of the langur by creation of protected zones. Most national parks within the range of *P. entellus* have resident populations of *H. sapiens* and their cattle. The Gir Wildlife Sanctuary, which is located in a national forest, and which holds probably the last remaining population of the Asiatic lion, is being stripped of foliage to feed the cattle, and of trees to feed man's fires. India, which was once covered almost entirely with forest, now has forest on only 17% of its land area (Sagreiya, 1967), and much of this is scrubby second growth or monocultural stands.

Would preservation of a small number of local populations or demes of *P. entellus* mean that the species was being preserved? Probably not. It would only mean that some fraction of the gene pool, which in total represents the species, was being maintained over the short term. At a minimum, studies are needed to establish the extent of genetic variability present in demes of *P. entellus*, and to work out how large a deme must be so that this variability can be maintained over the long term.

V. Summary

The common or hanuman langur, *Presbytis entellus*, is the species with the largest geographical range of the genus; it is found throughout most of India, Nepal, Sikkim, and Sri Lanka, and in the southernmost tip of Tibet in the Himalayas. Together, its 15 subspecies inhabit many types of environments from the desert fringe to tropical wet forest to alpine scrub, from almost sea level to over 4000 m in altitude. It is an arboreal species that can adapt to living on the ground when and where trees are absent. Most of the daylight hours are devoted to resting, and to eating primarily the leaves (81%), fruit (43%), and flowers (24%) of over 299 plant species from more than 72 different families. Man's crops may be eaten when the trees seasonally lose their leaves, or where the forest has been replaced by agriculture. Water may be drunk during the driest time of the year. Population densities range from less than 3/km² in grass- and cropland to more than 130/km² in forest; troop home range size ranges from 13.00 km² in cropland to 0.05 km² in forest. The core areas are usually one-half to one-quarter the size of the home range. The population structure ranges from multimale bisexual troops, in areas where the population size may be below carrying capacity of the environment, to single-male bisexual troops accompanied by all-male groups, in areas where the environment is at least subject to seasonal stress. On the average, bisexual troops range in size from 13 to 37 individuals and all-male groups from 2 to 18, with an overall adult sex ratio of about one male to two females. Males reach adulthood after 6 or 7 years, and females after 4 years of age. Young males may be forced to leave the troop and their mothers because of harassment from the troop's adult male, or from a new adult male that has displaced the previous male. New adult males, who were usually the dominant males of all-male groups, are known to kill infants sired by their predecessors, which in turn brings the females into estrus within 2 weeks. Such a system increases the likelihood that some of the new adult male's progeny may reach maturity and may also help to slow population growth and reduce population size. Mature females come in estrus once every 30 days. Young are born after approximately a 6 month gestation and are weaned at about 10 to 12 months of age, at which time the female's estrous cycle would normally resume.

Langurs have mutually beneficial associations with other species, such as the peafowl and the chital deer, but they may also be subject to predation by pythons, tigers, and leopards, and may be harassed and occasionally killed by pariah dogs. Langurs live peacefully with other species of monkeys: the purple-faced langur, and the bonnet and rhesus macaques. However, because of human destruction of their natural abode, the forest, they have been forced to come down to the ground and into close contact with man. This situation is tolerated by the Hindu and Buddhist populations of man, but traditional beliefs are being strained and may break down under the continued pressure of increasing human population density. Today, even with religious protection, langurs are killed when they successfully compete with man for his food or when they cause damage to the abodes of man by breaking or displacing a few roof tiles. Although the langurs are only occasionally killed by Hindus and Buddhists, they are subjected to a great deal of harassment, primarily rock throwing and shouting. Some langurs are killed for food by tribal peoples. The survival of the hanuman langur is very closely tied to how well man can control his own population increase, and his own tendency to destroy natural environments.

Acknowledgments

I wish to thank N. Bishop, S. Mohnot, R. P. Mukherjee, D. Starin, Y. Sugiyama, K. K. Tiwari, and C. Vogel for sending me their material prior to publication, and the Haraders, D. Parrack, and R. Thorington for sending material that was unavailable to me in Bangladesh. The Bangladesh Agriculture Research Institute, the Bangla Academy, and the Dacca Museum graciously extended to me the use of their library facilities. N. Bishop, P. J. Dolhinow, C. M. and A. Hladik, D. Starin, Y. Sugiyama, K. K. Tiwari, and C. Vogel kindly read the first draft of this paper, and made helpful comments and criticisms.

This paper was written while the author was supported by the U.S. Public Health Service through Research Grant No. 5 RO7 A110048-15 from The National Institutes of Health to The Johns Hopkins University International Center for Medical Research.

References

Aldrich-Blake, F. P. G. (1970). Problems of social structure in forest monkeys. *In* "Social Behaviour in Birds and Mammals" (J. H. Crook, ed.), pp. 79—99. Academic Press, New York.

Ayer, A. A. (1948). "The Anatomy of *Semnopithecus entellus.*" Indian Publ. House Ltd., Madras.

Basham, A. L. (1971). "The Wonder that was India." Fontana Books, Rupa and Co., Calcutta.

Bishop, N. H. (1975). Vocal behavior of adult male langurs (*Presbytis entellus*) in a high altitude environment. Paper presented at the Amer. Assoc. Phys. Anthropol., Denver, Colorado, April, 1975.

Blanford, W. W. (1888). "The Fauna of British India, including Ceylon and Burma: Mammalia." Taylor and Francis, London.

Christie, A. (1968). "Chinese Mythology." Paul Hamlyn, London.

Crook, J. H. (1970). The socio-ecology of primates. *In* "Social Behaviour in Birds and Mammals" (J. H. Crook, ed.), pp. 103—166. Academic Press, New York.

Ehrlich, P. R. and Ehrlich, A. H. (1972). "Population, Resources, Environment: Issues in Human Ecology," 2nd ed. W. H. Freeman & Co., San Francisco, California.

Eisenberg, J. F., Muckenhirn, N. A., and Rudran, R. (1972). The relation between ecology and social structure in primates. *Science* **176** (4037), 863—874.

Elliot, D. G. (1913). "A Review of the Primates", Vol. III. American Museum of Natural History, New York.

Graves, I. L. and Oppenheimer, J. R. (1975). Human viruses in animals in West Bengal: An ecological analysis. *Hum. Ecol.* **3**, 105—130.

Griffith, R. T. H. (Transl.). (1915). "The Ramayan of Valmiki." E. J. Lazarus & Co., Benares.

Gupta, K. K. (1974). Fauna of Orang Wild Life Reserve. *Rhino-J. Kaziranga Wild Life Soc.* **2**, 20—30.

Hladik, C. M. and Hladik, A. (1972). Disponibilites alimentaires et domaines vitaux des primates a Ceylan. *Terre Vie* **26**, 149—215.

Hrdy, S. B. (1974). Male-male competition and infanticide among the langurs (*Presbytis entellus*) of Abu, Rajasthan. *Folia Primatol.* **22**, 19—58.

Husain, K. Z. (1974). "An Introduction to the Wildlife of Bangladesh." F. Ahmed, Motijheel, Dacca.

Indian Meteorological Department. (1962). "Memoirs of the Indian Meteorological Department," Vol. 31, Part III. Government of India Press, Simla.

Jay, P. C. (1962). Aspects of maternal behavior among langurs. *Ann. N. Y. Acad. Sci.* **102**(2), 468—476.

Jay, P. (1963a). The Indian langur monkey (*Presbytis entellus*). *In* "Primate Social Behavior" (C. H. Southwick, ed.), pp. 114—123. Van Nostrand, Princeton, New Jersey.

Jay, P. (1963b). Mother-infant relations in langurs. *In* "Maternal Behavior in Mammals" (H. L. Rheingold, ed.), pp. 282—304. Wiley, New York.

Jay, P. (1965). The common langur of north India. *In* "Primate Behavior: Field Studies of Monkeys and Apes" (I. DeVore, ed.), pp. 197—249. Holt, Rinehart and Winston, New York.

Jerdon, T. C. (1874). "The Mammals of India: A Natural History of All the Animals Known to Inhabit Continental India." John Wheldon, London.

Kurup, G. U. (1970). Field observations on habits of Indian langur, *Presbytis entellus* (Duffresne) in Gir Forest, Gujarat. *Rec. Zool. Surv. India* **62**(1—2), 5—9.

McCann, C. (1928). Notes on the common Indian langur (*Pithecus entellus*). *J. Bombay Natur. Hist. Soc.* **33**, 192—194.

McCann, C. (1933). Observations on some of the Indian langurs. *J. Bombay Natur. Hist. Soc.* **36**(3), 618—628.

Mohnot, S. M. (1971a). Ecology and behavior of the hanuman langur, *Presbytis entellus* (Primates, Cercopithecidae) invading fields, gardens, and orchards around Jodhpur, Western India. *Trop. Ecol.* **12**(2), 237—249.

Mohnot, S. M. (1971b). Some aspects of social changes and infant-killing in the hanuman langur, *Presbytis entellus* (Primates: Cercopithecidae), in Western India. *Mammalia* **35**, 175—198.

Mohnot, S. M. (1974). "Ecology and behaviour of the common Indian langur, *Presbytis entellus* Dufresne." Ph.D. Thesis (Zoology), 299 pp. University of Jodpur, India.

Napier, J. R. and Napier, P. H. (1967). "A Handbook of Living Primates." Academic Press, New York.

Olson, T. A. (1951). Toxic plankton. *Proc. Inserv. Training Course Water Works Problems*, pp. 86—95. Univ. of Michigan, Ann Arbor, Michigan.

Oppenheimer, J. R. (1973). Effects of environmental factors on the activity of village dwelling langurs (Primates) in West Bengal. *Proc. Indian Sci. Congr.* **60** (Part IV), 157 (Abstr.).

Oppenheimer, J. R. (1976a). *Presbytis entellus*: Birth in a free-ranging primate troop. *Primates* **17** (4), in press.

Oppenheimer, J. R. (1976b). Human—langur (*Presbytis entellus*) interactions in two Bengali villages. *Ann. Meet. Anim. Behav. Soc. June, 1976*, Boulder, Colorado.

Oppenheimer, J. R. (1977). Aspects of the diet of the hanuman langur, *Presbytis entellus. In* "Proceedings of the 6th Congress of the International Primatology Society," (D. Chivers, ed.) vol. I. Academic Press, London.

Oppenheimer, E. C. and Oppenheimer, J. R. (1976). Certain behavioural features in the pariah dog (*Canis familiaris*) in West Bengal. *Appl. Anim. Ethol.*, **2**, 81—92.

Pocock, R. I. (1939). "The Fauna of British India, Including Ceylon and Burma; Mammalia," Vol. I: Primates and Carnivora. Taylor and Francis, London.

Prater, S. H. (1965). "The Book of Indian Animals." Bombay Natural History Society, Bombay.

Rahaman, H. (1973). The langurs of the Gir Sanctuary (Gujarat)—A preliminary survey. *J. Bombay Natur. Hist. Soc.* **70**, 295—314.

Ray, P. C., (transl.). (1884). "The Mahabharata of Krishna-Dwaipayana Vyasa." Vana Parva, 2nd ed. Bharata Press, Calcutta.

Ripley, S. (1967). Intertroop encounters among Ceylon gray langurs (*Presbytis entellus*). *In* "Social Communication among Primates" (S. A. Altmann, ed.), pp. 237—254. Univ. of Chicago Press, Chicago, Illinois.

Ripley, S. (1970). Leaves and leaf-monkeys: the social organization of foraging in gray langurs, *Presbytis entellus thersites. In* "Old World Monkeys" (J. R. Napier and P. H. Napier, eds.), pp. 481–509. Academic Press, New York.

Rudran, R. (1973). Adult male replacement in one-male troops of purple-faced langurs (*Presbytis senex senex*) and its effect on population structure. *Folia Primatol.* **19**, 166–192.

Sagreiya, K. P. (1967). "India—The Land and People: Forests and Forestry." National Book Trust, India, New Delhi.

Schaller, G. B. (1967). "The Deer and the Tiger: A Study of Wildlife in India." Univ. of Chicago Press. Chicago, Illinois.

Spillet, J. J. (1966). The Kaziranga Wildlife Sanctuary, Assam. *J. Bombay Natur. Hist. Soc.* **63**, 494–533.

Spillett, J. J. (1968). A report on Wild life surveys in south and west India, November–December 1966. *J. Bombay Natur. Hist. Soc.* **65**, 1–46.

Starin, E. D. (1973). A preliminary study of the Gir Forest langur. B. A. Thesis. Friends World College, Huntington, New York.

Sugiyama, Y. (1964). Group composition, population density and some sociological observations of hanuman langurs (*Presbytis entellus*). *Primates* **5**, 7–38.

Sugiyama, Y. (1965a). Behavioral development and social structure in two troops of hanuman langurs (*Presbytis entellus*). *Primates* **6**, 213–247.

Sugiyama, Y. (1965b). On the social change of hanuman langurs (*Presbytis entellus*) in their natural condition. *Primates* **6**, 381–418.

Sugiyama, Y. (1966). An artificial social change in a hanuman langur troop (*Presbytis entellus*). *Primates* **7**, 41–72.

Sugiyama, Y. (1967). Social organization of hanuman langurs. *In* "Social Communication among Primates" (S. A. Altmann, ed.), pp. 221–236. Univ. of Chicago Press, Chicago, Illinois.

Sugiyama, Y. (1976). Characteristics of the ecology of the Himalayan langurs. *J. Human Evol.* **5**, 249–277.

Sugiyama, Y. and Pathasarathy, M. D. (1969). A brief account of the life of hanuman langurs. *Proc. Nat. Inst. Sci. Ind.* **35B**, 306–319.

Sugiyama, Y., Yoshiba, K., and Pathasarathy, M. D. (1965). Home range, mating season, male group and inter-troop relations in hanuman langurs (*Presbytis entellus*). *Primates* **6**, 73–106.

Tiwari, K. K. and Mukherjee, R. P. (1973). Studies on social behaviour in the common langur (*Presbytis entellus*) around Ramtek near Nagpur. *Proc. Indian Sci. Congr. 60* (Part 4), 157. (Abstr.)

Vogel, C. (1971). Behavioral differences of *Presbytis entellus* in two different habitats. *Proc. 3rd. Int. Congr. Primatol., Zürich, 1970*, Vol. 3, pp. 41–47. Karger, Basel.

Vogel, C. (1973). Acoustical communication among free-ranging common Indian langurs (*Presbytis entellus*) in two different habitats of north India. *Amer. J. Phys. Anthropol.* **38**, 469–480.

Vogel, C. (1975). Intergroup relations of *Presbytis entellus* in the Kumaon Hills and in Rajasthan (North India). *In* "Contemporary Primatology" (S. Kondo, M. Kawai, and A. Ehara, eds.), 5th Int. Congr. Primatol., Nagoya, 1974, pp. 450–458. Karger, Basel.

Vogel, C. (1976). "Oekologie, Lebensweise und Sozialverhalten der grauen Languren in Verschiedenen Biotopen Indiens. Fortschritte der Verhaltensforschung, *Beihefte zur Z. Tierpsychol.* **17**, Verlag Paul Parey, Berlin.

Vogel, C. (1977). Ecology and sociology of *Presbytis entellus. In* "Use of Nonhuman Primates in Biomedical Research" (M. R. N. Prasad and T. C. Anand Kumar, eds.). Indian National Science Academy, Delhi.

Wu Ch'êng-ên. (1961). "Monkey," (translated by A. Waley). Penguin Books Ltd., Harmondsworth, Middlesex.

Yoshiba, K. (1967). An ecological study of hanuman langurs. *Primates* **8**, 127–154.

Yoshiba, K. (1968). Local and intertroop variability in ecology and social behavior of common Indian langurs. *In* "Primates: Studies in Adaptation and Variability" (P. C. Jay, ed.), pp. 217–242. Holt, Rinehart and Winston, New York.

15

The Douc Langur: A Time for Conservation

I. Introduction

Douc langurs (*Pygathrix nemaeus nemaeus*, Linnaeus, 1771) were located and observed on Mt. Sontra near DaNang, Vietnam for a 10-week period, June to August, 1974 (see Table I). The primary purpose of this research was to survey

513

TABLE I

Locality Records for *Pygathrix nemaeus*

Locality	Elevation (m)	Province	Location	Museum or Reference[a]
Trang Bom	100	Bien Hoa	10°57'N, 107°00'E	USNM
Ban Me Thout	⟨400–500⟩	Darlac	12°40'N, 108°03'E	FMNH
Djiring	1000	Lam Dong	11°35'N, 108°04'E	Thomas, 1928
Nha Trang	50	Khanh Hoa	12°15'N, 109°11'E	Bonhote, 1907
Dinh Quan	150	Long Khanh	11°12'N, 107°21'E	USNM
Mt. Sontra	696	Quang Nam	16°07'N, 108°18'E	USNM
Nui Ba Dinh	986	Tay Ninh	11°22'N, 106°11'E	Morice, 1875
Col des Nuages	⟨300–400⟩	Thua Thien	16°12'N, 108°08'E	USNM
Dalat	1500	Tuyen Duc	12°03'N, 108°26'E	USNM

[a]USNM, United States National Museum, Smithsonian Institution, Washington, D.C., FMNH, Field Museum of Natural History, Chicago, Illinois.

a comparatively "safe" area where doucs had been reported. Baseline data was collected on habitat utilization, group size and composition, sex ratio, and aspects of social behavior. The project was designed to provide data as a basis for early conservation action. Suggestions were made concerning the placement of six reserves throughout Vietnam. Other nonhuman primates observed or heard on Sontra were the crab-eating macaque [*Macaca fascicularis validus* (Elliot, 1909)] and the light-cheeked gibbon [*Hylobates concolor* (Harlan, 1826)].

II. Distribution and Status

The douc is distributed in Laos, Vietnam and on the Island of Hainan (Simons, 1966). The report of the douc on Hainan comes from a report by Meyer (1892) from the Dresden Museum who examined a male specimen of this genus that had been sent to the museum from the island. This is the only known report from the island. The geographical limits of the genus are poorly known so that the estimate of 8°–23°N,100°–111°E probably gives the outside limits since it includes all of Laos, Vietnam, Cambodia, Hainan, and part of Thailand (Napier and Napier, 1967, p. 293).

As is well known, much of Southeast Asia and especially Vietnam has been subjected to extensive open and guerilla warfare for more than three decades. In fact, Vietnam has a history of warfare dating back more than 2000 years. Because of this warfare, human populations have been uprooted and moved numbers of times being displaced from hamlet to hamlet then ultimately to the cities for safety. Without attention, traditional food sources have gone unattended resulting in a heavy subsistence hunting of local fauna, including the douc langur.

Warfare has also degraded the natural habitat of this species since large amounts of sophisticated munitions as well as mines, booby traps, and live ammunition have been distributed throughout the forest. Bombs 500—700 pounds and their resulting craters scar the landscape and leave 30—foot deep breeding grounds for mosquitos. Literally, millions of craters exist; as many as 2,600,000 were estimated for 1968 alone (Orions and Pfeiffer, 1970, p. 552).

Nowhere else and at no other time have defoliants been utilized on such a massive scale. War on the plants of Vietnam and secondarily the animals began in late 1961 (Westing, 1971, p. 56). By 1969 one acre in six in South Vietnam had been sprayed. Most forests sprayed and especially those in areas surrounding Saigon were treated with Agent Orange (a 1:1 mixture of 2,4-D and 2,4,5-T), some with Agent White (a 4:1 mixture of 2,4-D and picloram), and some with Agent Blue (dimethylarsenic acid). By 1970 one-quarter of Vietnam's mangrove association had been sprayed and killed (Odum *et al.*, 1974). At the present time, no regeneration of these areas has been reported and, in fact, there seems little chance of this occurring since the mangroves are being reduced to charcoal in the face of severe fuel shortages throughout the country. It has been suspected that these plants do not regenerate since tens of thousands of acres sprayed years ago still show no sign of growth (Westing, 1971, p. 61).

Defoliants such as Orange, Blue, and White were utilized less frequently in the highlands except around military bases and in special areas of heavy military activity. However, burning of areas which had previously been defoliated or cleared in some alternate way occurred in a number of areas. Areas of Mt. Sontra had been cleared of heavy tropical foliage only around installations such as Air Force headquarters at the base of the mountain and near small guard stations partway up the mountain. Halfway up the mountain was a large radar installation (Panama). At the top was an abandoned Navy radar base.

III. Physical Description

Doucs are readily distinguishable from the smaller less arboreal macaque (*Macaca fascicularis*) due in part to their larger size and brilliant coloring. They are a comparatively large, long-tailed monkey. The body of *Pygathrix nemaeus nemaeus* is gray with annulated hairs of gray, black, and white. The face is naked and yellow-orange (10YR8/4)* and appears to vary according to the amount of sun to which the animal is exposed. I found that all zoo animals that were exposed to direct sunlight had dark yellow faces (7.5YR6/8) whereas those that were always kept inside or in indirect sunlight generally had quite light faces (10YR8/4). This tanning ability probably varies among individuals as well as population. The face of *P. n. nigripes* is black.

The eyes appear almond-shaped and are set at a slightly inclined angle, the lateral aspect of the eye being slightly higher than the medial. The eye shape and their location imparts an Asiatic expression to the douc's faces. The eye color is dark brown (7.5YR3/2) and there is a black line that completely encircles the rim of the eye.

The face is ringed with long white hairs which are much longer in adult males than in females or juveniles. I have observed that the lengthening of the beard commences at puberty, which appears at approximately 5 years of age in douc males. Characteristically, both sexes have vibrissae along the supraorbital torus. The crown of the head is gray to black with a band of black short hairs immediately superior to the supraorbital torus. Chestnut (2.5YR4/8) bands ring the face and neck separated by a white band around the neck.

The upper limbs are gray as is the body to the wrist, where white cuffs often proceed up toward the elbow. In males these cuffs are much wider than in females and in *Pygathrix n. nigripes* these cuffs are completely lacking. Hands and feet in both subspecies are black.

A prominent white triangular rump patch surrounds the long (589–770 mm) tail of the same color. Adult and juvenile males could be distinguished from females since a circular white spot was found on either side and superior to the corners of the rump patch (see Fig. 1). This feature develops between 8 and 18 months of age (Sample of eight males at San Diego, Cologne, and Basle). The same dimorphic coloration was found to occur in all male *Pygathrix n. nigripes* examined by this investigator (Lippold, museum notes).

Lower limbs are black in color from the hip to the knee in both subspecies of *Pygathrix*. The knee to the ankle in *P. n. nemaeus* is maroon to chestnut in color while in *P. n. nigripes* this area is black. The inguinal region is white in both sexes and subspecies.

Total lengths of specimens examined by VanPennen (1969, p. 104) and this investigator ranged from 1015 to 1350 mm, tail 589–770 mm, hindfoot 175–200, ear 33–40 mm, and greatest length of skull 107–122 mm.

Two features emerge from the measurements and indexes presented in Table II. First, in both species males are larger than females. Second, *P. n. nemaeus* appears slightly larger than *P. n. nigripes* in all measurements and

---

*Numbers in parentheses refer to a standardized designation of color based on hue, value, and chroma (Munsell color chart, 1954).

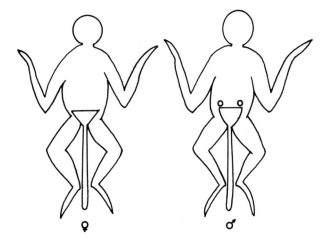

Fig. 1. Dimorphism in douc langur, posterior aspect.

TABLE II

Average Skull Indexes for *Pygathrix nemeaus* [a]

	P. n. nemaeus (6/3)	*P. n. nigripes* (8/3)
Brain case length		
Male	88.3	86.2
Female	82.7	84.0
Facial length		
Male	59.0	58.9
Female	57.4	56.3
Palate index		
Male	108.8	117.2
Female	108.4	114.1
Skull breadth		
Male	93.7	90.3
Female	97.3	90.4
Interorbital index		
Male	30.2	26.1
Female	24.7	21.6
Face braincase index		
Male	50.6	49.6
Female	50.4	45.6

[a] After Groves, 1970, p. 583.

indexes except that of palate index. Caution must be urged in these interpretations due to the small sample size. However, a generalization that cannot be avoided is the lack of substantial size differences between the sexes of both subspecies. This lack of sexual dimorphism in size is consistent with the general lack of size dimorphism in arboreal species.

IV. Description of Study Area

Sontra, the highest mountain of a complex on a peninsula forming the eastern boundary of the Bay of Tourane, was chosen for this preliminary study (see Fig. 2). Doucs have been recorded there by VanPeenen (1969, p. 71) and by Gochfeld (1974). The peninsula, approximately 10 km north of DaNang, is surrounded on the north and east by the South China Sea. Sontra, the highest peak rises from sea level to 696 m in less than 3 km and is situated at 16°07′N and 108°18′E. Other peaks on the peninsula reach heights of 647, 621, and 384 m.

Man's activities have left their mark on Sontra. Except for some inaccessible northern slopes, the area shows signs of moderate to heavy disturbance. An all-weather road has been built to service military installations along the high ridgeline. Extensive burning as a method of vegetation reduction has taken

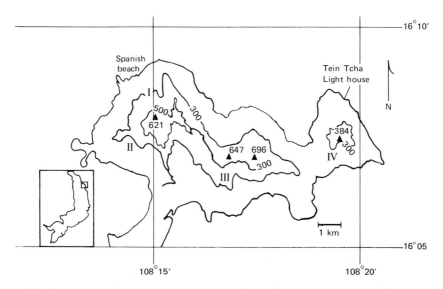

Fig. 2. Map of Mt. Sontra. Insert shows location of peninsula on northeast coast of South Vietnam. Study areas noted by Roman numerals (I, II, III, and IV).

place at the south base and on some of the lower slopes. Woodcutters continue to fell tall trees for the manufacture of charcoal (a native heat source and means of cooking) even though the area had been expressly off limits as a military installation. Hunting was also outlawed but animals were continually killed even though the area was constantly patrolled. After being in the area for a time the people confided to me that a douc had been killed on Sontra the day before I arrived.

A. Climate

Sontra lies within the monsoon belt of Southeast Asia but demonstrates a different record of precipitation and winds from more southerly areas. Records from Tien Tcha lighthouse on the northeastern end of the island were available (Fig. 3) (VanPeenen *et al.*, 1971, p. 128). The light is at the base of the mountain complex and, because of its location, probably records less precipitation than actually falls on the cloud-covered higher slopes. Tien Tcha records an average of 254 cm of rainfall per year. Rainfall is heaviest September through December but sizable amounts of precipitation occur in all months. Often it is the case that clouds cover the higher slopes (above 300 m) and deposit rain while the base of the mountain remains completely dry. According to VanPeenen *et al.*

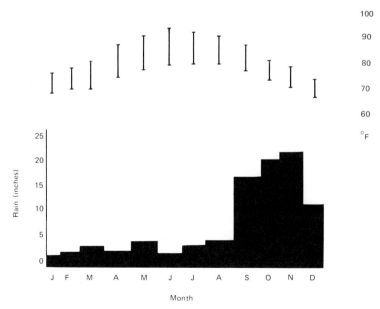

Fig. 3. Climatic data from Tien Tcha Lighthouse over a 31-year period (averaged). (After VanPeenen *et al.*, 1971, p. 129.)

(1971, p. 128), there are 41 streams on the mountain. Of these, three flow continuously beginning above the 450 m level and collect as many as eight tributaries.

B.   Soils and Forest Type

Rocky stream beds and boulders of volcanic origin are found all over the peninsula. On one of the southern slopes lateritic soils are quite common. There is fairly thin humus under the rapidly decaying leaf cover and evidence of extensive leaching is everywhere (VanPeenen, *et al.*, 1971).

Many trees are deciduous and some had lost their leaves while others had not. A large number of plant species were identified but no single tree species was considered dominant since most species appeared to occur in equal numbers. Much of the forest was secondary and only in undisturbed places was climax rain forest found.

As already mentioned there were only a few areas where relatively undisturbed rain forest occurred. These areas are on the northern slopes between 300 and 650 m. In these areas trees attained a height of 40 m or more and young trees formed a secondary canopy. Vines and epiphytes are common but undergrowth is not dense. Visibility was good (up to 10 m) and movement through the forest was comparatively easy. At this altitude the rain forest is maintained by frequent precipitation and fog often hangs over this area spilling down into the valleys as far as 350 m.

Secondary moist forest predominates above 300 m where tall trees have been removed by man's activities. No unbroken upper canopy exists but there are occasional tall trees remaining. Most trees are around 15 m tall and form a closed canopy. Undergrowth is dense with vines, shrubs, thorny palms, and bamboo clusters occurring close together producing an impassable tangle. It was impossible to move through this forest type with backpacks or equipment and passage had to be negotiated with a machette. In fact, cover was so heavy that the forest was almost dark by day. Plant species which characterize this area include both fig and oak-like trees; VanPeenen *et al.*. (1971) collected undergrowth plants which included Myrtaceae and Dioscoreaceae.

Dry forest is common and extensive at elevations from 0 to 300 m, especially on the south side of the peninsula facing DaNang. "Short shrubs and woody vines with waxy leaves are the dominant vegetation" (VanPeenen, *et al.*, 1971, p. 140). Short trees and dense undergrowth are common. Undergrowth is composed of Verbenaceae, Acanthaceae, Myrsinaceae, Rubiaceae, and Leguminosae. Razor-edged grasses were common.

Grasslands were found wherever extensive yearly burning took place. At the south base of Sontra many parklike meadows occurred. Above 200 m there were several green meadows and near the Tien Tcha lighthouse an extensive area had been cleared. Two acres of this clearing were covered with groves of bananas, pineapple, and bamboo.

V. Mammals

Tree shrews (*Tupaia glis modesta* Allen, 1906) were abundant in rain and secondary forest. Short-nosed fruit bats (*Cynoptercus brachyotis* Miller, 1898) were common in all habitats except rain forest. VanPeenen *et al.* (1971) found that almost all females that he trapped between March and May were pregnant or lactating. Three other families of bats were found on Sontra: Rousettus, Taphozous, and Rhinolophus.

Macaca fascicularis validus (Elliot, 1909), the crab-eating macaque, was collected by VanPeenen and also observed by this investigator. VanPeenen reports that macaques were commonly observed in groups of three to six individuals during the 1965–1966 period. However, more recently fewer animals (two to three) are seen and all are suspicious of humans and vehicles and flee when approached.

Macaques were seen on several occasions ranging from sea level to approximately 300 m; none were observed in the rain forest. Small groups of two to three individuals and single animals were most often observed. Informants, who had spent more time on the peninsula, reported seeing as many as ten animals at one time near the beach. It is believed that these animals and their relatives are the "rock apes" so often reported by military men who have served in Vietnam.

Pangolins (*Manis javanica* Desmarest, 1882) occur on the south slopes of Sontra. Yellow-handed squirrels (*Callosciurus flavimanus* (Geoffroy, 1831) were seen in secondary moist forest and rain forest at about 250 m. Long-nosed tree squirrels [*Dremomys rufigenis* (Bladford, 1878)] were found in all forested areas. Striped tree squirrel [*Tamiops rodolphei* (Milne-Edwards, 1867)] were observed in tall rain forest trees above 440 m elevation (VanPeenen *et al.*, 1971).

Five species of rats were reported by VanPeenen *et al.* (1971). The yellow-haired rat [*Rattus huang* (Bonhote, 1905)] was found in the rain forest. Kloss rat (*Rattus moi* Robinson and Kloss, 1922), while never abundant, was located in all types of forested areas. Norway rats [*Rattus norvegicus* (Berkenhout, 1769)] were found around piers and near the naval base only; they were not found on the mountain. Roof rat (*Rattus rattus molliculus* Robinson and Kloss, 1922) occurred in all habitats from the base to the summit of Sontra. The noisy rat (*Rattus sabanus revertens* Robinson and Kloss, 1922) was found on the ground in a variety of habitats above 100 m (VanPeenen *et al.*, 1971).

Porcupine (*Atherurus macrourus stevensi* Thomas, 1925) was located in grass-lands and dry forests. Ferret badgers (*Melogale personata* Geoffroy, 1831) were found in all habitats below 550 m while the small civet (*Viverricula indica thai* Kloss, 1929) was found only in secondary dry forest. Palm civets (*Paradoxurus hermaphroditus cochinensis* Schwartz, 1911) were found in all habitats and elevations. Mongooses (*Herpestes javanicus exillis* Gervais, 1841) were found in secondary forests or cleared areas. Muntjacs (*Muntaicus muntjak annamensis* Kloss, 1928) occur on Sontra. Informants reported these deer to be numerous especially in dry and wet second growth forests (VanPeenen *et al.*, 1971).

VI. Materials and Methods

The prime objective of this study was to establish the fact that doucs continue to exist in Vietnam. To this end, Mt. Sontra was chosen because doucs had been reported there as late as 1968—1969 (VanPeenen *et al.*, 1971; Gochfeld, 1974). Surveys were done in four study areas to establish tentative numbers, distribution, acitivity patterns, and habitat utilization of the species. Another objective was to gather behavioral information to compare with data collected at the San Diego Zoo for the last 4 years.

While 210:50 hours were spent in the field only 36:40 hours were in direct observation of doucs. Military operations hampered observations. Air operations as well as ground troops at times disturbed both the animals and investigator.

Possibly as a result of the high temperature and humidity douc activity peaks were found to be in the early part of the day from 5:30—7:30 hours and again in the later afternoon from 17:00 until immediately before dusk. By dusk the animals had moved to sleeping trees and did not come down until morning. Periods of activity coincided with times of least amount of heat and disturbance.

Field equipment consisted of a pair of 7 × 35 binoculars, stopwatch, compass, and camera. Photos were taken with a Nikon F equipped with a 135- and 400-mm lens. Tree markers were permanently placed on selected large trees in the study areas for future reference and location. Although the routine varied somewhat due to dependence on others for transportation, it usually started at 4:30 hours and ended after 18:00 hours. Since the area was a major infiltration route for enemy forces no night observations were attempted. Handwritten notes were taken and elaborated each evening. Valuable data was learned from talking with Vietnamese who had seen or eaten the monkeys. In preparation for these discussions, the observer learned some Vietnamese and pictures of zoo animals were shown for identification. It took only a short time to become known as the "monkey lady."

Observation conditions were difficult due to the heavy secondary forest coupled with high humidity and seasonal heat. Thick secondary forest made foot travel extremely difficult. Ground tracking was done by first locating areas where the monkeys had been feeding. These areas were recognizable since doucs drop huge amounts of fruits and leaves. Feeding trees were then plotted on the study map and then could be followed. Thickness of the secondary forest made prolonged observation impossible. However, some behavioral data was collected. A large amount of locomotor activity was observed; much of it in direct response to meeting the observer.

VII. Comparative Data

Up to the time of the present survey no field study had ever been attempted with the express intent of observing *Pygathrix nemaeus*, probably due in large part to the dangerous conditions. Some data especially concerning the distribution of *Pygathrix nemaeus* was found in the following sources: Allen, 1906; Bonhote, 1907; Elliot, 1909; Kloss, 1916; Robinson and Kloss, 1922; Thomas, 1925, 1928; Osgood, 1932; Pocock, 1941; VanPeenen, 1969, VanPeenen *et al.*, 1971 and Gochfeld, 1974. However, with the exception of VanPeenen and Gochfeld, this information was concerned with the location and physical description of the animals and contained little or no behavioral data. The remarks by VanPeenen and Gochfeld contain short but valuable behavioral descriptions.

VIII. Reaction to Observer

In all cases, both *Pygathrix* and *Macaca* would flee upon contact. There were several instances when groups were observed for a time while unaware of my presence. In no case were animals ever habituated to the observer. The behavioral patterns which characterize the reactions of doucs to my presence follow below.

Whenever the group detected the investigator they fled immediately. In some cases the adult male stayed behind and threatened the observer by brachiating back and forth, a behavior also noted by VanPeenen *et al.* (1971, p. 134–35). In other cases, along with this threatening behavior, a male would jump toward the observer and then quickly away in a display behavior. Both males and females vocalized with a "threat-warning bark" as soon as the observer was spotted or whenever danger was perceived.

In several instances doucs were observed without their knowledge. When the observer was discovered, juveniles rather than adults edged cautiously to within 10 m of the observer. A "threat-warning bark" from an adult would result in the return of the juveniles. Juvenile males were seen to display on several occasions in this context with a nervous, curious bravado before they finally fled. This behavior of quickly jumping from branch to branch combined with hand slapping has also been noted in zoo doucs (Lippold, observation notes).

At times doucs were as close as 15 m above the observer, quietly feeding and resting. Movement to the tree above the observer was noisy when the observer was not suspected. In fact, doucs sounded like elephants moving through a heavily wooded area because of their crashing and banging of branches. However, at other times, flight could be, similarly, extremely quiet and swift.

The animals were not always vigilant and when surprised they uttered a characteristic "threat bark" and warning calls. Panic diarrhea was common, much to the chagrin of the observer.

IX. Population Dynamics

Only one early estimate of group number was located and that can be credited to Osgood (1932) (see Table III). He states that groups of from 30 to 50 were seen traveling together in the rain forests of Col des Nuages not too far to the

TABLE III

Douc Langur Comparative Group Sizes and Composition

Source	Location	Group size	Socionomic: sex ratio
Osgood (1932)	Col des Nuages	30–50	
VanPeenen et al. (1971)	Mt. Sontra	3–5	Multimale group with infants and juveniles; 1:2 ratio
Gochfeld (1974)	Mt. Sontra	5	Single male, 3 females, infant; 1:3
		7	Multimale, more females than males 1:2
Present study (1974)	Mt. Sontra	8	Single male, multi-male groups,
		9	females outnumber males, sex ratio
		11	1:2.4, solitary or peripheral animals of either sex (information for three separate groups).

north of DaNang. The next reference comes from VanPeenen (1969, p. 105) where he notes "... [doucs were] observed in small groups almost always in the trees." Another later reference collected by VanPeenen *et al.*, (1971, p. 134−135) in 1965−1967 but reported in 1971 states "... A group of 3 langurs watched quietly from a tree about 60 m away...."

Gochfeld (1974, p. 135) observed langurs between May 1967 and May 1968 on frequent trips from DaNang to the top of Mt. Sontra. Two groups were observed feeding in the high canopy. One group consisted of five individuals. The composition of the group was as follows; one adult male, three adult females, and one infant (sex not specified). The second group consisted of five adults of which two were adult males, three adult females, one juvenile, and one infant (juvenile and infant, sex not specified).

Three groups along with several single adult animals were observed in this study. Group 1 consisted of eight members: one adult male, three adult females, one juvenile (male) of 3 or 4 years of age, and three infants. Of the infants, one male appeared to be approximately 1 year old while the other two, both females, appeared less than 6 months old. These young females were probably born at around the same time somewhere between February and April. All age estimates were made based on the author's observation of zoo animals of known age from San Diego, Cologne, Stuttgart, and Basle (see Table IV for age determinations).

Group 2 was composed of nine members: one adult male, four adult females, two juveniles (a female of around 3 years and a male of 2 years), and two infants (a male of approximately 1 year and a female of about 6 months).

Group 3 was the largest, with eleven members. There were two adult males, four adult females, three juveniles (two females nearly 3 years old and one male about 2 years), and two infants, both females. These females were more than 6 months but less than 1 year old.

Single animals of either sex were seen occasionally. Most single animals sighted were adults or nearing adulthood (approximately 4−5 years for females, 5−8 for males) but did not seem to be old. At least one lone female that was observed several times was somewhere between 4 and 6 years of age.

Native estimates of group size varied. One very knowledgeable individual remarked that he usually saw small groups of doucs (three to ten individuals) but on occasion had seen a group numbering sixty individuals. All efforts to locate this group met with failure. This large number seemed difficult to imagine but it differed from Osgood's 1932 estimate by only twenty individuals.

The sex ratio computed from adult members of all groups was 1 male to 2.4 females. In all groups, females outnumbered males. There was one multimale group in this study and also that of Gochfeld and there were solitary individuals of either sex.

The brief nature of this study made it impossible to ascertain whether these lone animals were between group associations or were true solitary animals. Zoo studies have provided some informative suggestions as to a phenomenon of individual and, hence, gene exchange between groups. In three of four captive colonies studied by this author (San Diego, Basle, and Stuttgart), a female of from 2 (Stuttgart) to 4 (San Diego and Basle) years of age was forcefully driven from the group by the adult male or males. It seems likely that if this phenomenon occurs in the wild it might provide a mechanism whereby females might join another group thereby maintaining and assuring heterozygosity and variation in the species.

Utilizing both Gochfeld's (1974) observations and those of this study, the number of mature animals was compared to immatures. Twenty-four animals were considered adult while sixteen were immature. Using these numbers, sixteen (or 20%) out of a total of sixty animals were immature. According to Southwick (personal communication), in order to have the population maintain itself at least 50% of the population must be immature. If this estimate is correct, the doucs of Sontra with only a 20% replacement rate are declining and not replacing themselves.

Several reasons for this low rate of replacement can be suggested. The doucs are hunted assiduously for their meat. Even though the mountain was declared off limits by the military, subsistence hunters shot a few animals every month. Soldiers that were stationed on the mountain also shot at animals and probably killed some from time to time. In an effort to curb target practice and hunting the military levied a high fine and imprisoned any soldier caught with a douc in their possession. However, to my knowledge no one was brought in or fined while I was there, although at least one douc was killed the day before my arrival.

No doubt some loss of young was as a result of falling, exploratory risk, and disease. Juveniles and older infants were found to be the boldest and most curious. There is no way to determine the effects of habitat destruction through burning and use of defoliants since no population numbers were known prior to these problems. Reports vary as to the use of defoliants on the mountain but in areas where defoliation has taken place a lowering in the birth rate of animals and a wholesale destruction and replacement of plant and animal communities has occurred (Odum *et al.*, 1974). An overall lowering of births plus a lessened survival rate are well known results of stress situations in zoo populations (Sedgwick, personal communication) and may be a factor here. Since the only estimations of group size prior to large-scale warfare are those of Osgood, in 1932, few true comparisons should be attempted. It is safe to say that the total population of doucs has decreased due to habitat distruction, hunting, and lowering of the birth rate. Added to these are natural causes such as disease, food shortages, and trauma.

Average group size for the Sontra area must be considered tentative since it is based on only three sources; VanPeenen, Gochfeld, and this study. It is also difficult to apply this data to the whole of Vietnam since Sontra is in many ways a special case. However, some suggestions can be attempted. Group size for doucs is smaller than the 30–50 suggested by Osgood in 1932. An average group probably consists of four to fifteen individuals. There are bisexual groups some with multiple adult males, while others have a single adult male. Adult females always outnumber males in the study groups on an average of about 2:1.

Single animals of both sexes were seen, however, it was difficult to determine whether this was a common or temporary condition brought on by the unstable conditions and the uncertainty of war. On a number of occasions, lone animals were encountered and no other animals could be located in their vicinity. Often these animals were young adults; females and males were found in almost equal numbers.

X. Birth Peaks

Age estimations of wild doucs were based on knowledge of size, coat color, and behaviors collected from 4 years of observation of zoo-born animals at San Diego, Cologne, Basle, and Stuttgart (Table IV). Based on field age estimates of month of birth of infants and juveniles, a birth peak most probably occurs between February and June.

Observation of more than one youngster of nearly the same age within a group was evidence that several births occurred at the same time. This concentration of births is also seen in zoo animals. For instance, females from established groups will often give birth in the same month, sometimes within the same week. Examples come from Cologne (Hick, 1973, p. 143) where Tanya gave birth on January 20, 1973 followed by Sonja on January 25, 1973. In 1974 a repeat of this pattern took place where Tanya gave birth on August 17 followed 7 days later on August 24 by Sonja (Hick, personal communication). This situation also occurred at San Diego where females gave birth March 13 and March 28 of the same year and then again October 16 and October 26, 2 years later.

Another factor which further complicates the situation is that particular females tend to give birth in nearly the same month year after year. In collections such as those at Cologne, Sonja and Tanya gave birth in January of 1973 then again in August of 1974, an interval of just 20 months. At Memphis 24 months separate April births of the same female. One female (Ginny) at San Diego produced a female infant June 17, 1973, which lived only 4 months, after losing several infants born in previous years (February, March, October, and August). Ginny then produced a male infant (Câu) 1 year later on June 24, 1974.

Birth season is most likely environmentally related but factors affecting it appear to be very complex and flexible. The highest amount of rainfall on Sontra

TABLE IV

Age Determination in Douc Langurs

Infant 1 0–8 months	Natal coat—infant face black with two light stripes beneath eyes. Head black with chestnut color band extending from forehead to ears. Body light chestnut with wide black stripe running from shoulders to rump patch. Arms, legs, and rump patch slightly darker chestnut. Hands and feet black, tail reddish gray. Behaviorally, closely associated with mother. Clinging to mother in first months, brief exploratory trips away from mother. Some solid food by end of first month.
Infant 2 8–18 months	Adult coat—infant face black, white cuffs and rump patch and tail. Circular spots above rump patch develop in male early in this stage (8–12 months). Face remains black interocular, around mouth, and on most of chin. Fully developed scrotum and testicles descend and change in color from brown to pink penis. Behaviorally more independent. Solid foods important, some nursing, and much nonnutritive suckling. Males spend time in rough and tumble play with age mates. Females spend time in grooming. Weaning generally begun by 12–13 months.
Infant 3 19–24 months	Adult coat—infant face gradually changes to color of adult. Face change occurs first around eyes and on cheeks, last place to change is around nose, mouth, and chin. Body size larger than infant 2 stage. Behaviorally, females spend more time in grooming and care of infants and are most often associated with adult females. Males take part in peer rough and tumble play. Weaning tantrums frequent in infants of both sexes; duration and treatment of infant tantrums dependent on mothers.
Juvenile 2–4, 5 years	Adult coat and face. Larger in size than infant 3. Body size and facial color are distinguishable from younger animals. Long face whiskers and beards still undeveloped in males.
Adult Females 4 years	Multiparous females have long nipples. Triangular rump patch but lack white circular spots above patch. Smaller than adult males and lack long facial whisker and beards. Red coloring in inguinal area during estrus and while pregnant.
Adult males 5 years	Larger body size than females. White circular spots over each corner of rump patch (developed by infant 2 stage). Penis bright red when erect. Red in inguinal area like female when female is in estrus. Behaviorally more aggressive when threatened. Protective of group and in some males paternalistic.

occurs from September through December with the largest amounts in October and November. Based upon the age of the youngsters observed, births were less frequent during this September to December period.

Weather data presents an interesting picture. As already mentioned the wettest months are September through December. At the same time the largest number and highest intensity of typhoons occur. Since fewer births seem to occur at this time of year it suggests a relationship between birth peaks and environmental factors.

A concentration of births appears to occur in advance of the peak of fruiting season. By the time that an infant is ready to eat solid food, and fruits are available. If birth peaks are environmentally related, as they appear to be, then it follows that when environmental factors change the fruiting season, birth peaks should be expected to be offset. This is exactly the situation that one might expect to occur in a zoo since food is provided and often constant humidity and temperature is maintained.

As is apparent in Fig. 4 the births which have been recorded at zoos have, in fact, spanned the calendar. However, successful births do have a higher incidence in months where seasonal fruiting is taking place in Vietnam.

It is not unique to suggest that birth peaks are an evolutionary phenomenon which reflects a species adaptation to a particular environment. Over many generations a species adapts to the ecological situation and a birth peak is developed at a time when conditions are most advantageous to the birth and rearing of young. Because of this phenomenon it seems quite likely that birth peaks would vary depending upon local conditions and that no sweeping statements should be made for all doucs in Vietnam. As Kurland (1973, p. 248) so succinctly states "... since storms and flooding might be the determinates of vegetative and fruiting seasonality then primate feeding strategies and population dynamics [i.e. birth peaks] may have evolved under regularly occurring density independent factors."

XI.   Interspecific Competition

Sontra seemed to be virtually free of natural predators of both doucs and macaques. No dogs, few wild carnivores such as civets, mongoose, badger, and no cats were in evidence. Some reptiles such as snakes were encountered but these seemed to pose little problem for the arboreal primates.

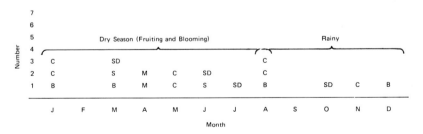

Fig. 4. Recorded birth months of captive living doucs. B, Basle; C, Cologne (Köln); S, Stuttgart; SD, San Diego; M, Memphis.

The major natural forces limiting population growth would probably have been availability of food, disease, and natural calamities such as falling. However, these probably have had little chance to structure the population recently since it has been under such heavy outside intervention from the human population. Monkeys are food sources and as mentioned earlier in the paper, guns which were designed by man to hunt men are turned on the monkeys. At this time, the only chance for survival of the doucs lay in their ability to move quickly and silently through the high tree tops.

In areas where trees had been cut for firewood and charcoal, the doucs are at a disadvantage since their only escape routes are on the ground and into the underbrush. In every case when they were encountered by the investigator in these situations they were amazingly fast and out of view in a matter of seconds.

Doucs confined themselves primarily to heavily forested areas of either primary of secondary growth. Only on a few occasions were they seen on the road and a few times they were seen to descend to the banana plantations at sea level.

XII. Feeding Behavior

Doucs ate a wide variety of leaves but concentrated upon the new growth. Certain fruits which were abundant were also consumed. Sometimes the group remained in a few fruit trees for as long as 1 hour, consuming fruits and leaves. At no time were any insects seen to be consumed. Reports from informants related that buds from deciduous trees were eaten when the tree was without leaves. Many leaves were only partly eaten and dropped to the floor of the forest. Some animals were seen to feed from the same branch as others but this was not a common occurrence and was seen only when food was closely spaced. It appeared that whenever possible doucs preferred to feed on a branch by themselves. Infants were seen to feed beside their mothers but no active food sharing was witnessed. Instances of active sharing have been reported by Kavanaugh (1972, pp. 406–704) at San Diego Zoo and by Gochfeld (1974, p. 167) on Mt. Sontra. This investigator has witnessed this mode of passive food sharing reported by these investigators. However, food sharing did not seem to be a primary strategy of food dispersal or distribution and happened only occasionally. In all cases it appeared to take place due to the nature and proximity of the food.

During the survey time the doucs food sources consisted of fig fruits and leaves and another fruit tree which has small 8 cm × 8 cm round bitter orange fruits. Fruits, flowers, and leaves of this tree were sought and relished. Natives from DaNang would also collect this orange-apple in large quantities. The availability of fruiting trees seemed to determine the movement of the doucs since they would move from one fruiting tree to another. Where these trees were close together, the animals were spaced close together; where the trees were farther apart, the animals were also far apart. No more than two to three animals were found in a single tree. It is likely that doucs shift areas depending upon the food supply and this may be the reason that doucs could not be located in one area 7 months after this survey was completed.

Doucs were seen to feed by using their hands to bring food to their mouths more often than bringing their mouth to food. Much food was tested and then dropped before an acceptable leaf or fruit was consumed. In cases where it could be tested, the fruits that were rejected were very ripe or ripe by human standards—never unripe or green. Leaves that were dropped were mature and tough, never new growth.

Food was at times gathered at one place, by nipping off a terminal bud, then carried by mouth to a more secure location. Feeding was at all times tension free and no dominance interactions could be seen during these times. Group dispersion tended to be wide during feeding, although this did depend somewhat upon the nature of the food source.

It is probable that most water comes from the foods consumed since the investigator never saw the animals drinking directly from a water source. On a few instances doucs did lick accumulated dew off of leaves in the early morning. In the zoo, water consumption is seen after monkey chow or compressed dry food pellets are eaten. In zoos where pellets are utilized in low quantities and vegetables and fruits are fed more often, little direct water consumption is seen.

XIII. Locomotor Behavior

Movement usually occurred when the observer was spotted and a threat bark issued. Animals of both sexes and all ages except infants uttered this bark. On several occasions doucs would sit watching me, fully aware of my presence. It was at these times that juvenile males would cautiously edge toward me, more interested than frightened. The only time that an adult female moved toward me was to retrieve her youngster. Sometimes adult males would allow youngsters to move as close as to within 5 m of me before issuing a threat bark. The threat bark always stopped the advance of the youngsters, although sometimes it took several barks before the youngsters would return to the group. It was only after several encounters with the same group that this behavior was observed. While the groups never really became habituated to me, they did learn to recognize me. I always wore the same type and color of clothing and made a special attempt not to wear military khaki. I also tried to be in the same place at the same time of day so that my behavior and movements might be predictable to the doucs.

It was impossible to discover which animals initiated movement or which animal or animals lead the group. However, in some cases, adult males stayed behind to threaten the observer.

Arboreal pathways through the forest appeared established and groups moved quadrapedially along these routes. Little branchiation or bipedal locomotion was seen. Horizontal jumps of incredible length were made on many occasions by all animals except infants. The jump would begin like a dive, with the arms over and above the head and the legs propelling the animal. Landings were always rear feet first. A jump which covered a distance of 5 to 6 m was not uncommon. Vertical movement from one story to another was seen on occasion. The investigator was continually impressed by the spectacular acrobatics and aerial ability of the doucs.

In most areas where trees overlapped, established pathways were utilized. The most favored of these were solid branches which, because of their continued use, had been cleared of leaves and smaller branches. In times of panic some poor choices were made and at least one juvenile fell a distance of 10 m before breaking his fall by catching a limb close to the ground. It is not difficult to see how bones could be broken and lives lost in wholesale panic.

When traveling through rain forest, animals moved in single file. In secondary forest arboreal pathways were not as established and movement seemed less organized. It appeared that each animal had to make its own decisions as to the most advantageous pathway for movement or escape.

XIV. Social Behavior

Due to the brief duration of this study as well as the extreme difficulties presented by wartime conditions, few behavioral observations could be made. Attempts were made to collect data on grooming, play, dominance, and sexual behavior.

Social grooming (allogrooming) occurred somewhat more often than solitary (auto) grooming. Allogrooming was especially common between adult females. Juvenile females were also seen to spend large amounts of time in grooming of adult females, infants, and peers. In almost no case were they seen to groom adult males.

Adult females groomed adult males far more often than males reciprocated this attention. An invitation to groom initiated by an adult male was always accepted quickly by an adult female. Social grooming occurred most frequently in the afternoon, before a nap or immediately before sleep. Most intense grooming of both a social and solitary type occurred before rest. Episodes of social grooming lasted anywhere from a few seconds to 1 hour. Solitary grooming usually lasted for a shorter period of time.

Douc langurs groom in much the same way as other primates. Both the hands and mouth are utilized, separately or together. Hair is searched for detritis of any sort and this is then removed with the fingers or mouth. Special areas for attentive grooming in infants were the anogenital region, the face, and the ears. In older animals the head, neck, face, and arms seemed to receive special attention.

XV.   Dominance Behavior

No complete dominance sequences were observed. However, a loose dominance organization appears to exist. For instance, not every adult male in a group received an equal amount of grooming. In any case, all adult males seemed dominant over all females, juveniles, and infants. There were cases where a number of females banded together and dominated the most dominant male. Dominance in juveniles appeared based on sex and size. Large juvenile males were dominant over smaller females.

Data from 4 years of observation at the San Diego Zoo (Lippold, observation notes) suggests that a hierarchy exists among adult females. Low ranking females with infants become more central than females without infants. At Basle and Stuttgart (Lippold, observation notes) a very old and perhaps sterile female became somewhat peripheral and then was driven out of the group by the dominant male. This situation, though, has not occurred at San Diego, and the old female has retained her high status even without a youngster.

At San Diego there are two adult males and the oldest is always dominant. The second less dominant male (Max, 5 years) was born at the zoo and has been socialized within this group. No competition for estrus females occurs. However, some peripheralization did occur when Max became sexually mature.

XVI.   Sexual Behavior

No sexual behavior was observed in the field. In the zoo, females experience an estrus cycle of 28 to 30 days in which they appear receptive approximately 10 days of this period. The cycle is repeated until a female is impregnated. Although reports vary (Hollihn, 1973, p. 185), this investigator and Hick (personal communication) found that animals of either sex invite sexual behavior. A female will characteristically assume a prone position, then look over her shoulder to the male. Males will intently stare at a female then look to a location where copulation will occur. The female then moves to this position and the male mounts her.

During estrus a female's inguinal area becomes red and continues in this condition throughout pregnancy. During the entire pregnancy females permit or invite copulation and copulations have been observed on the day of delivery. The estimated time of gestation varies between 165 (Hick, personal communication) and 180—190 (Lippold and Brockman, 1974, p. 7) days.

XVII. Consort Behavior

Although this behavior was not apparent from field observations, there is good zoo evidence that doucs form loose consort relationships. The male is protective of a pregnant female and he sleeps close to her all during pregnancy. After the infant is born, this behavior continues during the first year of the infant's life (Lippold, observation notes).

XVIII. Play Behavior

Since animals could not be seen for sustained periods of time, play was infrequently recorded. However, at times, both solitary and social play was observed and most play behavior occurred between infants and juveniles. In zoo situations, wrestling and chasing has also been observed between juveniles and adult males and between at least one adult female and a juvenile male (Lippold, observation notes).

Invitations for social play usually included a play face where the head is thrown back and the chin thrust forward with the mouth open and partly smiling. Other means to invite play included tail pulling, running past another, jumping up and down on a branch, and ear pull and run. In some cases, careful grooming sessions turned into social play.

Social play involved short periods of wrestling, chasing, and jumping through trees and a kind of tag (where one animal would chase another then touch him and run away). Social play most often occurred between peers. The three groups of doucs observed in this study were small so that all older infants (stages 2 and 3) and juveniles played together. Sometimes social play, such as wrestling and chasing, involved only a pair of animals.

Play characteristically occurred after eating and before rest. Late mornings, early afternoons, and then again before rest at night were the most popular playtimes. Adults were always tolerant of infants but juveniles might be threatened if their play became too rough. In a zoo situation a young male was able to wrestle and behave in a rough and tumble manner with the older adult male far longer than he could with the adult females. If the young male would invite any of the adult females to play, the three would threaten him as a group (Lippold, observation notes).

Solitary play involved jumping up and down on tree limbs, manipulation of inanimate objects such as leaves, twigs, and fruits, and displays (jumping and running from branch to branch while hand slapping). Play bouts lasted for a few minutes at a time and were followed by a rest period. Play would start then stop then restart and in some cases this sequence was repeated more than 15 times. Play was seen only when the group rested for periods longer than 30 minutes.

XIX. Conservation Measures

Reports vary as to the continued existence of *Pygathrix nemaeus nigripes*. This author did not observe *P. n. nigripes* in the field but did examine skeletal remains and skins of this subspecies at the Field Museum of Natural History in Chicago, Illinois and at the Smithsonian Institution in Washington, D.C.. The Smithsonian collection had two specimens that had been collected as recently as 1961 near Dalat, Vietnam. For more recent distribution, the staff from the zoo in Saigon reported that they believed that the black-footed douc may still exist to the southwest of Saigon around Tinh Bien (10°30′N, 105°E) near the Vietnam—Cambodian border. However, this area could not be checked by this observer because of the hostilities near there in 1974.

Pygathrix nemaeus nigripes collected in 1932 and now housed in the Field Museum came from the vast heavily forested plateau around Ban Me Thout (12°40′N,108°03′E). It is possible that doucs still exist in this area. The fact that they have not been reported recently from this area may just reflect the fact that zoologists did not collect from this area. It is, therefore, impossible to assess the status of *P. n. nigripes* without further exploration.

The continued existence of *P. n. nemaeus* has been confirmed by the present study. Doucs also still exist in Laos, since a survey of Bangkok animal dealers in the summer of 1974 by the author revealed that all doucs that were exported to world zoos came from southern Laos and not Vietnam.

Because the present report is a preliminary survey, it would be misleading to estimate total population or the density of animals per unit area. From all the evidence, doucs survive although their numbers are probably reduced from prewar populations. Man and technology have been the worst enemies of doucs. Habitat destruction through war and economic pursuits such as hardwood cutting and charcoal preparation have adversely effected the environment upon which the douc depends for existence. Subsistence hunting has more directly reduced their numbers. It is foolish to suggest that the species can naturally adapt to these unnatural calamities.

The new united government of Vietnam administered by Vietnamese for Vietnamese is now in a position to affect change. With the end of the war will come population stability and a more consistent production of food. More food will lessen the hunting pressure on the wild native animals.

Soon it should be possible to establish reserves where the beautiful native animals can live in their natural habitat fully protected. In the best interest of conserving the animals, many reserves rather than one should be established. In the north at least two reserves should be set up; the first near the Vietnam-China border around Lao Cai (22°30′N, 104°E). A second reserve should be near the border of Vietnam and Laos around Kim Cnong (18°15′N, 105°E) not far from the Laotian town of Nape where doucs were reported in the 1930's. Mt. Sontra (16°07′N, 108°18°E), the area covered in this survey, would also make a good reserve since so much is known of its ecosystem.

In the highlands, Ban Me Thout (12°40′N, 108°03′E), where doucs were collected in the 1930's would be a good central highlands reserve. Dalat (11°56′N, 108°25′E), an area which has been spared the ravages of war and an area where the black-footed douc may still survive, would make an ideal south-central reserve area. Finally, if some native forest exists around Tinh Bien (10°30′N, 105°E) this would be a good southwest reserve area near the Vietnam—Cambodian border. This plan would then establish a reserve system which would cover the entire nation setting aside six areas of diverse ecosystem type.

Important resources needed to ensure the survival of the douc langur include: forests with multiple stories and a fairly unbroken canopy, and a variety of food resources including trees which bear fruit throughout the year. Of course, the most important and probably the most difficult to ensure is the protection of the doucs from man.

Governmental restrictions which expressly forbid the hunting of endangered animals have been a part of Hanoi's policy since 1963. From that year according to Professor Dao Van Tien, Dean of the Faculty of Biology at the University of Hanoi, the primates (macaques, langurs, and gibbons) have been protected by a governmental decree (IPPL Newsletter, 1974). It is obvious that for the leaders of Vietnam, saving animals as well as people has been a major priority for some time.

XX. Summary

This chapter discusses the results of observation of three groups of douc langurs living on Mt. Sontra near DaNang, Vietnam. The animals are part of what is thought to be a somewhat diminished population from former times. No total estimate of doucs on Mt. Sontra or in Vietnam was attempted. Subsistence hunting and destruction of the habitat have been cited as responsible for the endangered status of this beautiful primate.

A major goal of the present study was to confirm the existence of doucs on

Mt. Sontra and to ascertain the population density and feasibility of longer behavioral studies. The doucs must be the focus of a longer more extensive field study so that the actual number and status of the population can be determined. At some future date this investigator would like to accomplish this goal.

Suggestions were made concerning areas where reserves might be established so that conservation of the doucs in their native habitat could be undertaken. The world would be diminished greatly by the loss of the animals of Vietnam. Now there is hope for the people and animals of Vietnam—the war is over and now is a time for conservation.

Acknowledgments

I would like to thank the following colleagues and friends for their help and encouragement: Dr. Charles Southwick, for his continued consultation through all the difficult times of preparation, Dr. David Horr and Dr. Peter Rodman, for their advise concerning field equipment and conditions, Dr. Patti Scollay, for bringing the San Diego Doucs to my attention and for critically reading this manuscript, Dr. Geoffery Bourne for asking me to write this chapter—it has served to further strengthen my commitment and dedication to the doucs and their survival, Dr. Richard Thorington of the Smithsonian Institution, and Dr. Phillip Herskovitz of the Field Museum of Natural History for allowing me to study the skins and skeletal materials of *Pygathrix nemaeus*.

I would like to acknowledge three grants from the San Diego State Foundation which have sustained the douc research. Most of all, deep appreciation and gratitude goes to my father Robert M. Lippold who has always supported my research with his heart and his pocketbook.

References

Allen, J. A. (1906). Mammals from the Island of Hainan, China. *Bull. Amer. Mus. Hist.* **22**, 463–490.

Anomymous (1974). *Int. Primate Protection League (News.)* **1** (2), 2–3.

Bonhote, J. L. (1907). On a collection of mammals made by Dr. Vassal in Annam. *Proc. Zool. Soc. London*, pp. 3–11.

Elliot, D. G. (1909). Description of apparently new species and subspecies of monkeys of the genera *Callicebus, Lagothrix, Papio, Cercopithecus, Erythrocebus* and *Presbytis*. *Ann. Mag. Nat. Hist.* **4**, 244–274.

Gochfeld, M. (1974). Douc langurs. *Nature (London)* **247**, 167.

Groves, C. P. (1970). The forgotten leaf-eaters and the phylogeny of the Colobinae. *In* "Old World Monkeys" (J. R. Napier and P. H. Napier, eds.), pp. 555–588. Academic Press, New York.

Hick, U. (1973). Wir sind umgezogen. *Z. Kolner Zoo.* **4** (16), 127–145.

Hollihn, U. (1973). Remarks on the breeding and maintenance of Colobus monkeys

(*Colobus guereza*) Proboscis monkeys (*Nasalis larvatus*) and *Douc langurs* (*Pygathrix nemaeus*) in Zoos. *Int. Zoo. Yearb.* **13**, 185—188.

Kavanaugh, M. (1972). Food sharing behaviour within a group of douc monkeys (*Pygathrix nemaeus nemaeus*). *Nature, (London)* **239**, 406—407.

Kloss, C. B. (1916). On a collection of mammals from Siam. *J. Natur. Hist. Soc. Siam.* **2**, 1—32.

Kurland, J. (1973). A natural history of Kra macaques (*Macaca fascicularis* Raffles, 1821) at the Kutai Reserve, Kalimantan Timur, Indonesia. *Primates* **14**, 245—263.

Lippold, L. K. and Brockman D. K. (1974). San Diego's douc langurs. *ZooNooz* **47** (3), pp 4—11.

Meyer, A. B. (1892). Communication to Zoological Society December 20, 1892. *Proc. Zool. Soc. London*, p. 665.

Morice, A. (1875). "Coup d'oeil sur la faune de la Cochinchine Francaise." H. Georg, Lyon.

Munsell soil color chart (1954). Munsell color company, Inc., Balt. Maryland.

Napier, J. R. and Napier, P. H. (eds.). (1967). "A Handbook of Living Primates" Academic Press, London.

Odum, H. T. *et al.* (1974). "The Effects of Herbicides in South Vietnam," Part B: Working Papers. National Academy of Sciences, Washington, D.C.

Orions, G. H. and Pfeiffer, E. W. (1970). Ecological effects of the war in Vietnam. *Science* **168**, 544—554.

Osgood, W. H. (1932). Mammals of the Kelley-Roosevelts and Delacour Asiatic Expeditions. *Field Mus. Natur. Hist. Zool. Ser.* **18**, 193—339.

Pocock, R. I. (1971). "The Fauna of British India, including Ceylon and Burma," Mammalia, Vol. 2. Taylor and Francis. London.

Robinson, H. C. and Kloss, C. B. (1922). New mammals from French Indochina and Siam. *Ann. Mag. Natur. Hist. Ser.* **9**, 87—99.

Simon, N. (ed.) (1966). "Red Data Book." Int. Union Conservation Nat., Geneva.

Thomas, O. (1925). The mammals obtained by Mr. Herbert Stevens on the Sladen-Goodman Expedition to Tonkin. *Proc. Zool. Soc. London*, pp. 495—506.

Thomas, O. (1928). The Delacour Exploration of French Indochina. *Mammal. Proc. Zool. Soc. London*. **3**, 831—841.

VanPeenen P. F. D. (1969). "Preliminary Identification Manual for Mammals of South Vietnam." United States National Museum, Smithsonian Institution, Washington, D.C.

VanPeenen P. F. D., Light, R. H., and Duncan, J. F. (1971). Observations on mammals of Mt. Sontra. S. Vietnam. *Mammalia* **35** (1), 126—143.

Westing, A. H. (1971). Ecocide in Indochina. *Nat. Hist. No.* **803**, pp. 56—61.

16

The Lesser Apes*

DAVID J. CHIVERS

I. Introduction

Preeminently qualified for arboreal habits, and displaying among the branches amazing activity, the Gibbons are not so awkward or embarrassed on a level surface as might be imagined. . . . It is, however, in the trees that they are seen to most advantage; there, free and unembarrassed, they appear almost

*Dedicated to the memory of C. R. Carpenter.

to fly from bough to bough, and assume in their gambols every imaginable attitude; hanging by their long arms, they swing themselves forward with admirable facility, seizing, in their rapid launch, the branch at which they aimed: they throw themselves from a higher to a lower perch with consummate address, and again ascend to the loftiest with bird-like rapidity. In all these movements their long arms are of utmost advantage. . . . M. Duvaucel states that the siamang is gregarious, and that each troop is conducted by a chief, whom the Malays believe to be invulnerable. They salute the rising and the setting of the sun with the most terrific cries, which may be heard at the distance of several miles. . . . Raffles states it to be bold and powerful, but docile and affectionate. Gentleness, intelligence and docility may be regarded, indeed, as characteristics of gibbons generally.

W. C. Linnaeus Martin (1841)

The lesser apes or gibbons (family Hylobatidae) are among the most appealing members of the order Primates. They are almost wholly arboreal and are distributed throughout the mainland and islands of Southeast Asia constituting the Sunda Shelf. They have a spectacular arm-swinging form of locomotion and loud complex calls of considerable purity, which capture the whole spirit, both joyful and melancholic, of the tropical rain forest.

The different species and subspecies are distinguished by coloration, markings, and calls, but all are characterized by their long arms, lack of a tail, upright posture, and dense, sometimes long, hair, and attractive markings around the face (which contribute to their appealing expressions). It is this attractiveness, along with their apelike intelligence and the clearance of large areas of forest, which place their future in such jeopardy.

In this chapter I intend to survey the variety of gibbons, describing briefly their appearance and distribution, summarizing the recent views of various authors, and commenting on the evolution of this primate family. The geologically unstable nature of the Sunda Shelf region has provided unique opportunities to study evolution in action—not just of a primate but of one of our closer relatives, an ape.

I shall then present data on their ecology and behavior, on their monogamous, frugivorous, and territorial tendencies. In the light of these observations, and with information on the numbers of gibbons in certain countries, it is possible to comment on their future prospects and to suggest steps that should be taken to ensure that adequate gene pools of each species or subspecies are conserved.

II. Classification

Some authorities place all but the siamang in the genus *Hylobates* (Elliott, 1913; Schultz, 1933; Simonetta, 1957), while others consider the gibbons (including the siamang) to be monogeneric (Martin, 1841; Forbes, 1897; Kloss, 1929). Groves (1972) recommends a three-way split at the subgeneric level:

(a) *Symphalangus*, the large black siamang of Sumatra and the Malay Peninsula
(b) *Nomascus*, the concolor gibbons of Laos, Vietnam, and southern China
(c) *Hylobates*, the majority of gibbons, including the hoolock of Assam, Bangladesh, and Burma, the well-known "lar" gibbons of Thailand, the Malay Peninsula, and Sumatra, the "grey" gibbons of Borneo and Java, and the curious Kloss gibbons of the Mentawai Islands (Fig. 1).

Fig. 1. Distribution of the lesser apes (from Groves, 1970; Chivers, 1974; Marshall and Marshall, 1975; Wilson and Wilson, 1976b).

TABLE I

Classification, Distribution, Coloration, and Diagnostic Calls of the Lesser Apes

Common name	Scientific name [a]	Weight (kg.) [a]	Distribution	Coloration [b]	Calls Male	Calls Female
Siamang	*syndactylus*	10.5	Malay Peninsula and Sumatra	M — black, throat sac gray or pink; infants, black	Scream	Bark-series, ca. 18 seconds
Concolor (crested or white-cheek)	*concolor*	5.7	Laos, Vietnam, Hainan, South China	SD — male, black ± white cheeks; female, golden, black patches; infants, whitish	Grunts, squeals, whistles	Rising notes and twitter, ca. 10 seconds
Hoolock (white-browed)	*hoolock*	6.7	Assam, Burma, Bangladesh	SD — male, black, white eyebrows; female, golden, white eyebrows; infants, whitish	Diphasic, accelerate, variable	Diphasic, variable, accelerate; lower than, alternate with male, 15 seconds
Kloss	*klossi*	5.8	Mentawai Islands (Indonesia)	M — black, all ages	Quiver-hoot, moan	Slow rise and fall, ± bubble, ca. 45 seconds
Pileated (capped)	*pileatus*		Southeast Thailand and Cambodia	SD — male, black; white hands, feet, head ring; female, silvery gray; black chest and cap; infants, grey	Abrupt notes, diphasic with trill after female call	Short rising notes, rich bubble, 18 seconds

Common name	Species		Distribution	Coloration[b]	Call type	Call description
Grey (Müller's)	*muelleri*		Borneo, except southwest	P — mouse-gray to brown, cap and chest dark (more so in female, pale face ring in male)	Single hoots	As *pileatus* but shorter. 10 or 15 seconds
Silvery (moloch)	*moloch*	5.9	Java	M — silvery gray, all ages; cap and chest darker	Simple hoots	Like *lar* at start, ends with short bubble, 14 seconds
Agile (dark-handed)	*agilis*	5.9	Malay Peninsula, Sumatra (most), southwest Borneo	P(AD) — light buff + gold, red or brown; reds + browns; brown or black, female, white eyebrows; male, white brows + cheeks	Diphasic hoots	Shorter, higher-pitched, rising notes, stable climax, 15 seconds
Lar (white-handed)	*lar*	5.5	Thailand Malay Peninsula North Sumatra	AD — black or light buff, white face-ring, hands + feet; P(AD) — dark brown; buff; brown(red)/buff	Simple and quiver hoots; Simple and quiver hoots	Longer, climax fluctuates, 18 seconds; Longer climax, fluctuates, 21 seconds; Longer climax fluctuates, 14/17 seconds

[a] From Schultz (1973).
[b] M, monochromatic; SD, sexually dichromatic; AD, asexually dichromatic; P, polychromatic.

There are many confused and conflicting reports in the literature and insufficient field observations to identify precisely the various species and subspecies. While some are clearly good biological species (Mayr, 1963), others are probably not yet reproductively isolated from their neighbors. On the basis of their extensive travels through Southeast Asia, Marshall and Marshall (1975) provide strong support for Groves' new assessment. They recognize nine species of hylobatid of varying separation in terms of geography, pelage color and markings, and calling behavior. These two studies provide the basis for the classification presented here (Table I).

A. Siamang

First reported from Bencoolen by Miller (1778), the Sumatran siamang was not brought to the attention of zoologists for another 40 years (Raffles, 1821). The Malayan siamang was first described by Thomas (1908) and it has since been shown to differ slightly from its Sumatran counterpart in various features particularly its smaller body size (Kitahara-Frisch, 1971). Although fossil siamang have been found in Java (Hooijer, 1960), it is today restricted mostly to the mountain regions of Sumatra and Peninsular Malaysia (Wilson and Wilson, 1976a; Chivers, 1971, 1974).

This black ape is about twice the weight of all other gibbons with a relatively broader chest, shorter legs, longer arms, a higher face, a large throat sac (inflated while calling and yawning), and webbing between toes II and III (Schultz, 1933) (Figs. 2a, 3a—c). MacKinnon (1976) suggests the siamang to be at the upper limit of size for exploiting the terminal branch frugivorous niche, for which hylobatids seem to be adapted through their suspensory behavior. Certainly it is more sedentary than the other gibbons. It is also much noisier: resonating "booms" (mostly by the male) alternate with series of "barks (and booms)" from the female, shattering "screams" by the male, and streams of loud "bark-chatters" from both sexes (Chivers, 1974).

B. Concolor

Also known as the crested, white-cheeked, black or Indo-Chinese gibbon, it occurs east of the Mekong river in Laos and Vietnam, with five subspecies in succession from *gabriellae* in the hotter, moister south, through *siki, leucogenys*, and *lu* to *concolor* in the cooler drier north, and *hainanus* on the island of Hainan (Groves, 1972). This subgenus has been known since Harlan's description in 1826; it formerly extended much farther north into China, and may still occur across the Red river and on the mainland north of Hainan (van Gulik, 1967).

Infants are whitish, slowly turning gray and then black, except for white cheeks. Adult males remain this color, but females become golden except for a black patch on the crown (Fig. 3e). Males have a very small throat sac, but the

characteristic calls are very high pitched and pure in tone, lacking the low reson-
ance of siamang calls. In addition to the white cheek patches, the peaked
arrangement of hairs on the head, and the shape of the nose, make the concolor's face very distinctive from those of other hylobatids (Fig. 2b).

C.   Hoolock

This species was also described by Harlan (1834). It also is known as the
white-browed gibbon, and resembles concolor in its former eastward extension
into China and the sequence of color change from infancy to the sexually dis-
chromatic adults (Fig. 3f). It occurs at the northwest limits of gibbon distribution,
with a western race between the Brahmaputra and Chindwin rivers, and an
eastern race between the Chindwin and Salween rivers (Groves, 1967).

The adults have a divided white eyebrow band (Fig. 2c); their calls are unusual in that there is no striking sexual dimorphism, and the pattern of short
"who-hah" and great calls (rapid accelerating series of alternate low and high
notes) is very variable (McCann, 1933; Marshall and Marshall, 1975; Marler and
Tenaza, 1976).

D.   Kloss

The last hylobatid species to be discovered was originally described as a
dwarf siamang since individuals of all ages and both sexes are black, with some
degree of webbing between the toes (Miller, 1903; Schultz, 1973). Kloss (1929)
placed it intermediate between *Symphalangus* and *Hylobates*, and Groves (1972)
and Tenaza (1975b) have now shown it to be more closely related to species
of the subgenus *Hylobates*. Locally it is referred to as the bilou (Figs. 2e and 3d).

It is found only on the four islands of the Mentawai group off West Sumatra.
Males chorus before dawn; females chorus 2 hours or more later with beautiful
long calls of pure notes ascending and descending slowly, with or without a long
trill in the middle (Tenaza, 1976).

E.   Pileated

This was one of the last species of gibbon to be described in the nineteenth
century, by Gray (1861). Occurring in southeast Thailand and Cambodia, it is
also sexually dichromatic (Fig. 2d). The adult male is black with white eyebrows,
hands, and feet, and a whitish ring round the crown; the adult female is silvery
gray with a black triangular cap and chest patch narrowing down to the pubis.
She also has a white band above the eyes which apparently disappears with age
(Fig. 3g). The infants are gray with the black spreading faster in maturing males
than females to give the adult pattern; this is the reverse of the concolor where
the juvenile has the color and markings of the adult male.

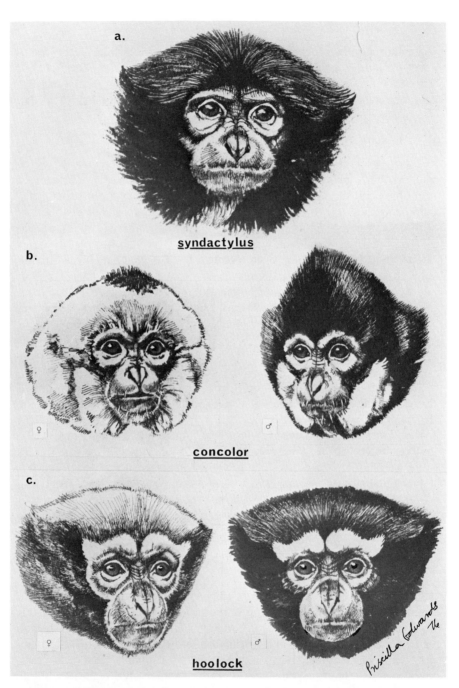

a.

<u>syndactylus</u>

b.

♀   ♂

<u>concolor</u>

c.

♀   ♂

<u>hoolock</u>

Fig. 2. Facial characteristics of the lesser apes (drawings by Priscilla Edwards). (a) *syndactylus*; (b) *concolor*, male and female; (c) *hoolock*, male and female.

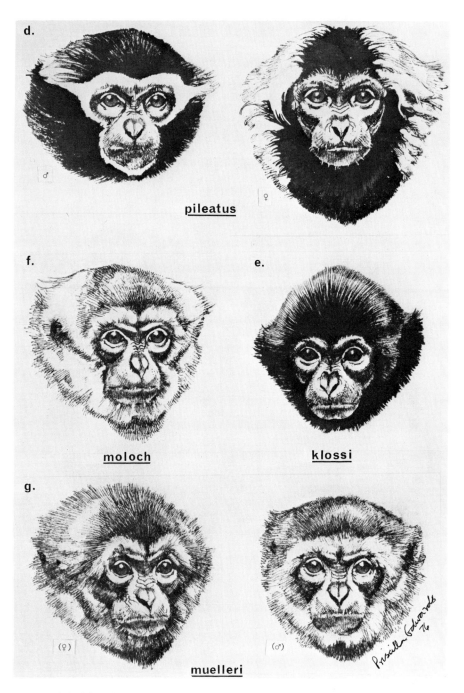

Fig. 2(d)—(g). (d) *pileatus*, maie and female; (e) *klossi*; (f) *moloch* (Javan); (g) *muelleri* (Borneo), male and female (probably not distinctive).

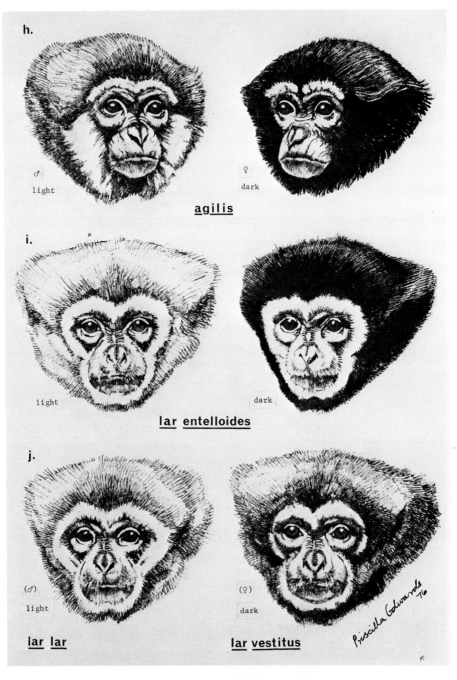

Fig. 2(h)–(j). (h) *agilis* (Malayan), dark to pale, white eyebrows (joined or separate), males with white on cheeks; (i) *lar entelloides* (Thai), dark phase almost black, pale phase creamy; (j) *lar lar* (Malayan) and *lar vestitus* (Sumatran), color varies from dark brown to buff.

Fig. 3. The lesser apes.
(a) *syndactylus*, subadult male in Malaya; (b) male and female calling at Twycross Zoo, England; (c) male and female calling in Malaya; (d) *klossi*, young male on Siberut, Mentawai Islands (courtesy of Rich Tenaza).

The female "great call" starts with a few notes of rising pitch, which give way to a rapid series of notes forming a rich bubbling call. As in most other species, the male call is less dramatic and he is mostly silent during the female call, coming in with a short series of diphasic gasping notes ending in a soft trill.

Fig. 3(e)—(g). (e) *concolor*, male and female grooming at Clères, Normandy France (courtesy of Marcel Hladik); (f) *hoolock*, adult male (left) in Assam (courtesy of Ron Tilson); female (right) in Calcutta Zoo (courtesy of R. P. Mukherjee); (g) *pileatus*, subadult female (left) (courtesy of Jeremy Raemaekers) and male (right), brachiating at Twycross Zoo.

Fig. 3(h)—(m). (h) *muelleri*, male and female at London Zoo; (i) *moloch*, female in Ujung Kulon Reserve, Java (courtesy of Walter Angst); (j) *agilis*, young male in private collection on Penang (courtesy of Paul Gittins); (k) *lar lar*, dark phase in Kuala Lumpur Zoo; (l) *lar entelloides*, adult male on Ko Klet Kaeo, Gulf of Siam (courtesy of Joe Marshall). (m) *lar vestitus*, adult female in the Gunung Loeser Reserve, Sumatra (courtesy of Herman Rijksen).

F. Grey

This name is used here to cover the rather variable (in appearance) group of gibbons inhabiting all but the southwest part of Borneo; it includes the subspecies *muelleri* and *abbotti* (Groves, 1972) and *funereus*, but not *albibarbis* (Groves, 1971). Color varies from mouse-gray to brown, with black on the cap and chest, more marked in *muelleri* than in the paler *abbotti* (Figs. 2g and 3h). Groves points out, however, that patterns of geographical variation tend to be swamped by individual variation within local populations. Male calls are single as in the lar gibbon, and female calls are somewhat shorter than those of the pileated gibbon but otherwise are almost identical.

First mentioned by Martin (1841), the gibbons of Borneo were confused with concolor in the nineteenth century because of an error in the type locality (see Groves, 1972), and, more recently, with the moloch of Java (Fooden, 1969), apparently because of superficial similarities in coloration.

G. Silvery

The silvery or grey gibbon of Java (*moloch*) is now mainly restricted to the west half of the island. The young are creamy in color; males and females are the same silvery gray color, often darker on the belly and with the trace of a black cap (Groves, 1972) encircled by a pale ring and with a white facial rim (Marshall and Marshall, 1975; Figs. 2f and 3i). Calls recorded by the Marshalls appear to show structural resemblances to those of *lar* and *agilis*, in particular, but the female great call tends to be shorter with a weak climax.

Finally we come to the two better-known "species," which are almost entirely allopatric and which are widely distributed through Thailand, Malaya, Sumatra, and the southwest of Borneo.

H. Agile

Also known as the dark-handed gibbon (Fig. 3j), it occurs (1) across the Malay Peninsula between the Mudah and Perak rivers on the west, and north of the Kelantan river on the east, extending north into Yala and Narathiwat provinces of Thailand, (2) over all but the north of Sumatra (probably up to a line running from Medan south to Lake Toba and then west to Singkil), and (3) in Kalimantan, south of the Kapuas and west of the Barito rivers (the *albibarbis* of Groves, 1972) (Chivers, 1971, 1974; Wilson and Wilson, 1976a; Marshall and Marshall, 1975; Gittins, personal communication).

In the Malay Peninsula, individuals are mostly dark in color, ranging from reddish to dark brown; it is not true, however, to say that the pale phase is rare. At least 25% of gibbons in two localities in West Malaysia are buff (Gittins, personal communication). The inappropriateness of trying to apply dichromatism to this species is more clearly emphasized by an analysis of 72 clear sightings of *agilis* in Sumatra (Wilson and Wilson (personal communication); here the animals are clearly lighter in color than in Malaysia, and the mixture of colors is complex (see tabulation below):

black	24%	red	13%	light golden-buff	29%
brown	7%	red + brown	1%	buff + red	13%
		red, buff, brown	1%	buff + brown	13%

They were all black in the eastern lowlands and swamp forests, where siamang are absent. Marshall and Marshall (1975) report that agile gibbons in Borneo are polychromatic between extremes of gray, brown, and buff as in *muelleri*, with a darker crown and chest and pale face rim reminiscent of *pileatus*.

Gittins (personal communication) points out that the white face ring commonly ascribed to the "lar" group, is composed of two parts in Malayan *agilis*, probably under separate genetic control. All young animals have a complete broad ring of white around the face. In adult females, however, there is only white above the eyes (usually thin, either separate or joined, bands); in males there is always some degree of whitening of the cheeks, often very marked, in addition to the white above the eyes (Figs. 2h and 3j). Groves (1972) makes similar observations, and confirms (personal communication) a strong tendency for sex differences in facial markings among Sumatran *agilis*, a feature not confirmed in the field (Carolyn Wilson, personal communication).

The population in Kalimantan is also polychromatic; basically brown, individuals vary between gray, brown, and buff, and are difficult to distinguish from *muelleri* (Marshall and Marshall, 1975). This last point leads Groves (personal communication) to query that these are two specifically distinct populations.

There is more uniformity between the calls of the three populations. The female "great call" is similar to that of *lar*, but is shorter by about 5 seconds with fewer, shorter note and a higher pitch with no changes in frequency at the climax. The male calls, by contrast, are very distinctive. They are diphasic since noise is emitted on both inspiration and expiration (Chivers, 1974); females also emit these calls on occasion (Gittins, personal communication). Marler and Tenaza (1976) suggest that this interpopulation divergence in male songs only has arisen through contact between them in response to selection against hybridization.

I. Lar

Also known as the white-handed gibbon, again we have three populations: (1) over most of Thailand (and the extreme east and southeast of Burma), extending down into Peninsular Malaysia on the west, divided by Groves (1972) into a northern race, *carpenteri*, and a southern race, *entilloides*—a division not supported by Marshall (personal communication); (2) most of Peninsular Malaysia south of the Perak and Kelantan rivers (*lar*); and (3) Aceh, the northern province of Sumatra, northwest of Lake Toba (*vestitus* rather than *albimanus*) (Chivers, 1971, 1974; Marshall *et al.*, 1972; Marshall and Marshall, 1975; Wilson and Wilson, 1976a).

The Thai variety is clearly asexually dichromatic, with almost black animals in the dark phase and light buff animals in the pale phase (Fig. 2i and 3l). Historically this has influenced gibbon taxonomy to the extent that asexual dichromatism has been imposed on most species of the subgenus *Hylobates* (Schultz, 1933; Carpenter, 1940; Fooden, 1969; Groves, 1970, 1972). Data collected recently in Malaysia (Chivers, 1974) and in Sumatra (Wilson and Wilson, 1976a; MacKinnon, personal communication) support the view advanced above for *agilis*, that in Sumatra and Malaya (as well as in Borneo) gibbons are much more variable in color.

In Malaysia the extremes of color variation are less than in Thailand—from dark brown to buff—and some individuals are a mixture (Figs. 2j and 3k) out of 37 individuals clearly viewed, 48% were classed as dark phase, 22% as pale phase, and 30% were not easily assignable to either category (Chivers, 1974). In Sumatra, Wilson and Wilson (1976a) clearly observed 22 animals: 5% brown, 27% red, and 68% golden. Rijksen (personal communication) has observed the full range of colors (Figs. 2j and 3n). In view of Gittins' observations on white markings on the face of *agilis*, it would be wise to investigate this phenomenon more closely in *lar*, since observations of complete face rings in Thailand may not be applicable further south; several females in Malaysia have narrower and less complete face rings than their mates.

The female great call usually lasts 20 seconds or more; it is longer and more complicated than that of *agilis* (see melograms in Chivers, 1976b, and sonograms in Marler and Tenaza, 1976, and in Marshall and Marshall, 1976). Hoots by males are more quavering than those of *agilis* and lack the inspiratory gasp so characteristic of the latter. As in most other gibbons, the male is quiet during the female great call, entering with a coda when it ends and continuing to hoot until her next call.

In summary, some general remarks can be made about coloration, calling, chromosomes, and sympatry.

Fooden (1969) has suggested that coloration follows three patterns among gibbons: (1) the siamang, grey, silvery, and Kloss gibbons are monochromatic, (2) the concolor, hoolock, and pileated gibbons are sexually dichromatic, with males dark and females light in color, and (3) lar and agile gibbons are asexually dichromatic, with either males or females in the dark or pale phase (Table I). We have seen how recent observations indicate the need for a fourth category, of polychromatism, to which *muelleri* and the Malayan and Sumatran races of *lar* and *agilis* should be moved. Groves (1972; personal communication) is convinced, however, from his study of museum material that *lar* and *agilis* are dichromatic. He claims that Malayan *lar* is very variable in the dark phase and the pale phase is rare; conversely, the Sumatran *agilis* is rare in the dark phase, and very variable in the pale phase. Further study will hopefully resolve this dispute between museum and field workers. What is especially intriguing is that irrespective of phylogeny, those gibbons inhabiting the northern mixed deciduous forests, in contrast to the tropical evergreen rain forests of the Sunda Shelf (see below), exhibit similar sexual dichromatism.

Typically group calls, lasting about 15 minutes, are composed of a duet between male and female, in which the female has the most distinctive and longest call, followed immediately by a coda from the male; during the 2–4 minutes between these "great calls" there is often much hooting, mostly by the male. Female *klossi* have the most distinctive call of all, but they do not duet with their mate; in contrast, adult *hoolock* duet together albeit with calls that are poorly differentiated. The *syndactylus* and *concolor* vary even more from this pattern, mainly because of the quality of sounds emitted, but there are close similarities between (1) the females of *pileatus* and *muelleri*, (2) the females of *lar*, *agilis*, and *moloch*, (3) the males of *lar* and *moloch*, and (4) the males of *agilis* and *muelleri*, and, to a lesser extent, *pileatus*.

The concolor gibbon has 52 chromosomes (diploid no.) with three acrocentric pairs of which one bears satellites; the siamang has 50 chromosomes with one acrocentric pair with satellites; all other gibbons have only 44 chromosomes, no acrocentic pairs, but one metacentric (see Groves, 1972).

The only extensive sympatry occurs between the siamang and lar or agile gibbons in the mountains of Malaya and Sumatra, respectively (Chivers, 1974; Wilson and Wilson, 1976a). The only other documented case is between *lar* and *pileatus* over about 100 km² of the Khao Yai National Park in southeast Thailand (Marshall et al., 1972). However, Brockelman (personal communication) has recently found hybrid individuals. Mixed groups of lar and agile gibbons have been found in the upper reaches of the Mudah river in Peninsular Malaysia (Gittins, personal communication); the situation has to be investigated more fully, since it may be the result of human disturbance. In many cases, however, areas of possible overlap between adjacent "species" have still to be visited.

III. Evolution

At this stage it should be helpful to hypothesize on events leading to the evolution of these nine species of gibbon (1) to clarify the diversity of the lesser apes as outlined above, and (2) to appreciate the significance of the extinction of any one of these species.

Gibbonlike fossils have been found in Miocene deposits in Africa and Europe, but nothing has been found during the Pliocene, even in Asia. The similarity of Miocene fossils to modern forms may result from parallelism rather than direct phyletic relationships (Simons and Fleagle, 1973). Molecular evidence from studies of myoglobin, however, support a phylogenetic separation between gibbons and other hominoids of 18 million years or so (Romero-Herrera *et al.*, 1973).

It is the frequent volcanic activity and the effects of glaciations, such as changes in sea level and vegetation, during the Pleistocene, which have made the Sunda Shelf such an exciting region for the study of plant and animal evolution. The different populations have evolved (and are still evolving) in isolation from each other during periods of high sea level or restricted vegetation. Studies of changes in populations of other animals, especially other primates, provide clues to help our interpretation of events in gibbon evolution (Medway, 1970, 1972a).

First we must summarize current understanding of the sequence of geological, climatic, floral, and faunal events during the last few million years (Table II; Haile, 1971; van Heekeren, 1972; Medway, 1972a; Ashton, 1972; Verstappen, 1975; Whitmore, 1975). The Sunda Shelf seems to have been formed toward the end of the Miocene, with the uplifting of land in the South China Sea to form islands which subsequently became joined to the Asian mainland, e.g., islands first appeared at the site of West and Central Java and the enlarged land surface became joined to the mainland at the end of the Pliocene (van Heekeren, 1972). By this time all major existing plant genera had migrated into the region, dispersing rapidly through the main hill systems as semi-evergreen dipterocarp forest (Ashton, 1972).

The Early Pleistocene was moister, with corresponding evolution of mixed dipterocarp forests. It was also a time of considerable volcanic activity, which produced further changes in land surfaces; these eruptions have continued intermittently to the present day. The spread of large mammals into Sundaland occurred at this time; their affinities with the Villafranchian fauna of India and Burma results in this first wave being referred to as Siva-Malayan. This migration route was possibly blocked subsequently by the growth of tropical rain forest, and later faunas originated from south China—Sino-Malayan (Medway, 1972b; see also, de Terra, 1943).

The Middle Pleistocene (1 million—250,000 years B.P.) contained the driest period, coinciding with major glaciations in temperate regions and the largest depression of sea level. It was presumably the time when savanna flora and fauna (including man) spread over large areas of the exposed Sunda Shelf, reducing and isolating tracts of tropical forest. A major pluvial followed, causing an extensive spread of tropical forests and the advent of true forest animals into Sundaland, including all the primate genera.

Marked climatic fluctuations continued into the Upper Pleistocene with alternate depressions and regressions of the sea by 200 feet or more causing continual changes in land area, with floral zones moving several hundred feet up and down mountains. The periods of isolation to which most areas were subjected produced new habitats and vegetation and corresponding adaptations of the fauna. It is during these last 200,000 years that the present distribution and diversity of gibbons must have been produced. With the knowledge that siamang and a smaller gibbon were present together on Java nearly one million years ago, we can now proceed to establish a time scale against the model of hylobatid evolution.*

The first stage is to consider a population of ancestral hylobatids distributed throughout much of the south Asian mainland and extending out onto the Sunda Shelf less than 3 million years ago (Fig. 4a). Then we must evisage the origins of the three subgenera from this probably large, black ancestor, with the more conservative siamang being isolated out on the southern edge of the Sunda Shelf, the curious concolor on the mainland to the northeast, and the predominant *Hylobates* in between (Fig. 4b). The last mentioned group then spread out into Sundaland, and, in time, when the savana spread and the sea level rose it became separated into several populations. (Fig. 4c).

The mainland form gave rise to the hoolock, and possibly the Kloss once it was isolated out on the west, while the island form out on the eastern edge of the Sunda Shelf gave rise to the monochromatic ancestor of the pileated, grey, moloch, lar, and agile gibbons. When the sea level lowered again the eastern population spread out west and south through the expanding forest, the ancestors of *lar/agilis* and *moloch*, respectively. It also seems likely that individuals crossed the rivers draining into the South China Sea to go northwest and found the *pileatus* population, the ancestral *hoolock*, having moved further northwest back into mainland Asia (Fig. 4d–e).

This brings us to the Upper Pleistocene and to the final events occurring during the last glaciation, which had two peaks with two major regressions of the sea (and lowland forest?)

The *lar/agilis* ancestor spread north, back onto the Asian mainland; the siamang contracted slowly northward into the mountains of Sumatra; the moloch became isolated in Java (Fig. 4f). These changes were accentuated in

*New datings indicate that the Djet beds in Java may date from nearly 2 m years ago.

TABLE II

Sundaland during the Pleistocene: Background for Hylobatid Evolution[a,b]

Era	Years before present	Climate glaciations	Sea level ± present (feet)	Fauna			
				Man	Gibbons	Other primates	Other mammals
Miocene				Islands emerge from the sea at the site of Java			
Pliocene	12,000,000			Volcanic activity uplifts Sunda Shelf along three main axes			
	3,000,000	Donnau	−600	Villafranchian Upper Siwalik Siva-Malayan			Stegodon Archidiskodon Merycopotamus Hippopotamus
Pleistocene lower	1,000,000	Gunz		Spread of savanna fauna Homo erectus		Meganthropus	Unicornis Megacyon Buboisea Leptobos
	750,000			Djetis Sino-Malayan	syndactylus ?moloch		
middle	400,000	Mindel	−400 −100	Spread of forest fauna Trinil	syndactylus ?moloch	Pongo Presbytis	Hyaena, Manis, Caprolagus, Elephas, Cervus, Bos, Sus, Tapirus, Axis, Rhinoceros, Panthera, Hippopotamus, Felis, Viverra
						Macaca	

				Homo sapiens			Panthera, Cervus,
250,000	Riss	−250					Stegodon, Axis,
100,000		−100	Ngandong	H. s. soloensis			Elephas, Bibos,
							Rhinoceros, Sus,
							Hippopotamus
	upper						
40,000	Wurm I	−220	Wadjak	H. s. sapiens			Manis, Tapirus,
	Wurm II	−220	Niah				Sus, Bos, Cervus,
							Muntiacus, Mustela,
10,000		+50			moloch	Tupaia	Rattus, Ratufa,
8,000						Macaca	Sundasciurus
						Presbytis	
present	Holocene	0					

[a] Adapted from Van Heekeren (1972) and Medway (1972a).

[b] It is becoming conventional to regard the Miocene as ending 5M years B.P. and the Pliocene at 2M years B.P. Furthermore, new datings place the peak of the Djetis at 1.5M years B.P. and the Trinil at 1M years B.P., which have the effect of extending the earlier phases of hylobatid differentiation.

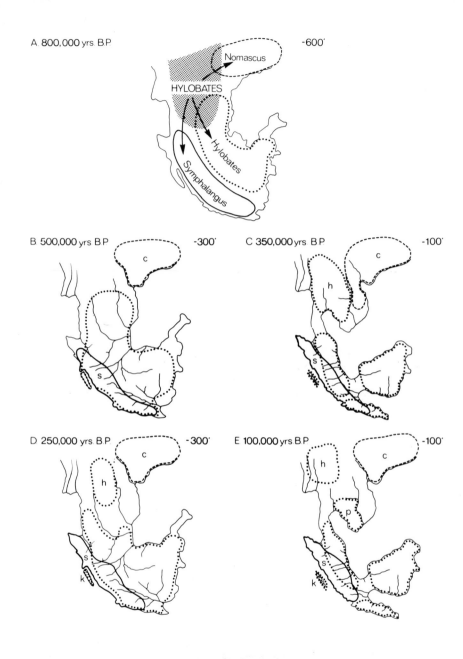

A 800,000 yrs B.P.　　　　　　　　　　　-600′

Nomascus

HYLOBATES

Hylobates

Symphalangus

B 500,000 yrs B.P.　　　　　-300′

c

s

C 350,000 yrs B.P.　　　　　-100′

c

h

s

D 250,000 yrs B.P.　　　　　-300′

c

h

s

k

E 100,000 yrs B.P.　　　　　-100′

c

h

p

s

k

Fig. 4. A model for the evolution of the lesser apes.

(a) about 800,000 years ago, lowest sea level, trivergence of *Hylobates* across the Sunda Shelf—ancestral concolor to east, ancestral siamang to south, ancestral "lar" in center; (b) about 500,000 years ago, spread of central population at a time of very low sea level; (c) about 350,000 years ago, breakup of central population by savanna and rise in sea level; (d) about 250,000 years ago, *hoolock* evolves in isolation to northwest; (e) about 100,000 years ago, rise in sea level, second breakup of central population.

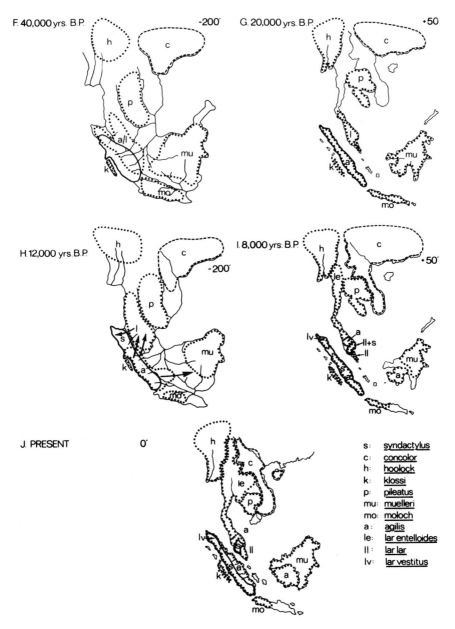

F. 40,000 yrs. B.P. -200'

G. 20,000 yrs. B.P. +50

H. 12,000 yrs. B.P. -200'

I. 8,000 yrs. B.P. +50'

J. PRESENT 0'

s: syndactylus
c: concolor
h: hoolock
k: klossi
p: pileatus
mu: muelleri
mo: moloch
a: agilis
le: lar entelloides
ll: lar lar
lv: lar vestitus

(f) about 40,000 years ago, *pileatus* evolves in isolation prior to lowering of sea level; (g) about 20,000 years ago, high sea level isolated *lar, agilis, moloch,* and *muelleri* from each other; (h) about 12,000 years ago, low sea level permits movement of *agilis* and *syndactylus* into Malaya, and, simultaneously or later, *lar* into Sumatra and *agilis* into Borneo; (i) about 8,000 years ago, rise in sea level brings about final evolution in isolation to achieve (j) present distribution. (±figure indicates deviation of sea level from present, in feet; distribution of subgenera indicated as follows: *Symphalangus,* solid line; *Nomascus,* broken line, *Hylobates,* dotted line.)

the interstadial leading to the differentiation of *agilis* in Sumatra and *lar* on the mainland of the Malay Peninsula, from where it spread slowly northward between *hoolock* and *pileatus* (Fig. 4g). When the sea level lowered for the last time, the siamang spread across into the Malay Peninsula, along with *agilis*, which also migrated east into Borneo, and some *lar* moved across into the north of Sumatra (Fig. 4h). The isolation and adaptations were completed when the sea rose again (Fig. 4i), to give rise to the present distribution and diversity (Fig. 4j).

This scheme differs from Groves (1972) in postulating (1) an early separation of most of the species now exhibiting sexual dichromatism from the monochromatic ancestor of the "lar" group, and (2) the origin of this "lar" group out on the east of the Sunda Shelf, rather than on mainland Asia. MacKinnon (personal communication) points out that the various populations would have been migrating back and forth between the mainland and "islands" as the climate and vegetation changed, with the edges of the Sunda Shelf, especially the east, being the main distribution centers.

This scheme does provide a solution to Kitahara-Frisch's (1971) dilemma on finding that the morphological variation in the three hylobatids common to Malaya and Sumatra is not correlated; the siamang is larger in Sumatra, the lar gibbon is larger in Malaya, and the agile is thk same size in both countries. The model described herein conforms to the rule that the "mainland" population is larger than the "island" form. Presumably the *agilis* population in Malaya is too small and recent for differences to have emerged.

We have seen how the population in the vicinity of present-day Borneo could have evolved into populations of increasing color variability: (1) the relatively monochromatic *moloch* and *muelleri*, (2) the more polychromatic *lar* and *agilis* in Malaya and Sumatra, until we find (3) the asexually dichromatic populations of *lar* in Thailand.

Perhaps the ancestral *concolor* was originally north of the Red river, from where it slowly shifted south between the South China sea and the Mekong river. It is clear from the Chinese literature that there were gibbons over much of China during the first millenium A.D. (van Gulik, 1967); what is not clear is their identity. Van Gulik claims that they were *agilis*, but he would seem to be in error. While the eastern specimens are clearly *concolor* (as expected) the majority of drawings show a complete face ring; one of the best, from the west, is clearly of *hoolock* (Groves, 1972), points out that this is the only drawing of the series to be signed by the artist, the best sign of authenticity). Groves suggests that the drawings with complete face rings might reflect artistic license; they would certainly have known of *lar* and *agilis* from travelers going over the mountains and rivers into Thailand or overseas to the south.

It is most likely that it was concolor which was distributed over much of China, as far north as the Yellow river. It is less likely that *hoolock* was able to

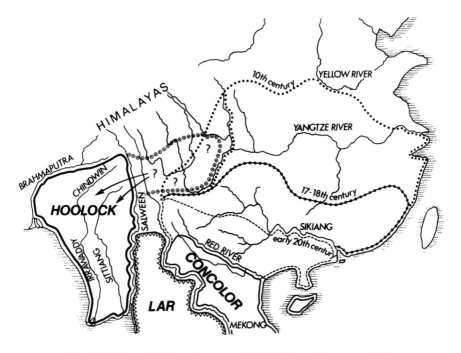

Fig. 5. Gibbons in China (from van Gulik, 1967 and Groves, 1972).

cross the river and mountain barriers into the province of Yunnan, and even less likely, if the scheme of evolution present here is anywhere near the truth, that *lar* or *agilis* ever occurred in China (Fig. 5). Nevertheless, these three possibilities remain unresolved. Tiwari (personal communication) points out, however, that much of the Indian fauna has originated apparently from the vicinity of southern China—maybe the hoolock crossed in the opposite direction! Furthermore, since this migration of animals into India continued until the Garo-Rajmahal gap closed about 40,000 years ago with the development of the Ganges and Brahmaputra river systems, it is unlikely that the hoolock gibbon reached Assam until after that date (otherwise gibbons would now be distributed over at least part of the Indian subcontinent).

What we can be sure of, with relevance to conservation, is that the geographical range of the lesser apes has contracted southward in historic times into the evergreen rain forest rather than the deciduous monsoon forests. MacKinnon (1974) postulates a similar phenomenon for orangutans, whose ancestors we know to have been even more widely distributed over South and East Asia during the Pliocene and Pleistocene, and whose range and numbers have been far more drastically reduced.

It is also difficult to know how important a factor human civilization has been in reducing the spread and numbers of these Asian apes, and how much more suitable for gibbons is evergreen rather than deciduous forest. We must remember that waves of human populations have been moving southward through Sundaland throughout the last million years, the critical period in the evolution of modern gibbons—first *Homo erectus*, and then increasingly advanced forms of *Homo sapiens* (de Terra, 1943; van Heekeren, 1972), each exerting greater effects on the vegetation.

The leaf-monkeys, *Presbytis* sp., are relatively sedentary in their habits and specialized in their feeding. Hence we find parallels in their evolution for (1) the spread of siamang and agile gibbons into Malaya from Sumatra, (2) the widespread distribution of *lar* and *agilis*, (3) the complex events in Borneo, and (4) special events in the Mentawai Islands (see Medway, 1970). In Thailand and Malaysia we find *obscura* (which occurs nowhere else), the superspecies *aygula-melalophos* (occurring everywhere else), and *cristata* only in mangrove swamps on the west coast. Elswhere, *cristata* has a widespread distribution in coastal and riverine habits, showing a preference for disturbed and riparian habitats, often far inland (Wendell Wilson, personal communication). Borneo contains two species not found elsewhere—*frontata* and *rubicunda*—the former closely paralleling the grey gibbon in its distribution (Medway, 1970).

The macaques, *Macaca* sp., by contrast, are represented in the region by only four species, each being more numerous and widely distributed than most of the other primate species; the exception is the curious *pagensis* of the Mentawai Islands (Tenaza, 1975a; Wilson and Wilson, 1976a). Their terrestrial habits make them more mobile and their omnivorous habits enable them to thrive in disturbed forest and around human habitation. Furthermore, their distribution conforms more with a pattern of evolution out from the Asian mainland (centered on Burma and Thailand), rather than one having several foci out on the Sunda Shelf (Medway, 1970). He points out that the stump-tail, *arctoides*, clearly evolved outside Sundaland, since it is adapted to deciduous forests. The pig-tail, *nemestrina*, has recently spread south through the Malay Peninsula, across to Sumatra and into Borneo via Bangka and Belitung, but not as far as Java. The smaller long-tail, *fascicularis*, represents the earliest stage of macaque evolution with fossils dating from the Middle Pleistocene; while it extends east only as far as the Mekong, it is widely distributed throughout Sundaland—on Java, Bali, the Lesser Sunda Islands and the Philippines, including Palawan (Medway, 1970).

After a detailed study of the distribution and morphology of Asian colobine monkeys, Brandon-Jones (personal communication) has concluded that their present distribution can only be explained satisfactorily by postulating two successive climatic deteriorations during the last glacial period. In the light of his findings and a brief survey of gibbon distribution and interrelationships he suggests tentatively that the first deterioration would have made all parts uninhabitable for gibbons except for Southeast China, Vietnam, and the Mentawai Islands (he does not believe that gibbons had reached Borneo by this time). In the following period of climatic remission, *klossi* crossed into Sumatra and differentiated into the relatively monochromatic progenitor of the "lar" group. The second deterioration was less severe than the first, but relict populations of the "moloch" group survived only in northwest Sumatra, northeast Borneo, and west Java. During the following amelioration of conditions, populations served as the foci for the recent dispersion of gibbons.

The most important population, now *lar vestitus*, expanded southward through Sumatra and across into Borneo, as *agilis*, which in turn gave rise to *lar* upon reaching the Malay Peninsula. The northern forms, *pileatus* and hoolock, may represent the end products of this radiation. Alternatively, they could be relicts of the interstadial dispersion of gibbons. This question, he believes, along with the problem posed by the present endemism of the siamang to Sumatra and the Malay Peninsula, can only be resolved by further taxonomic and paleontological research. The other relict populations in Borneo and Java evolved into *muelleri* and *moloch*, respectively, with *klossi* surviving on the Mentawai Islands and concolor in eastern Asia.

This model contradicts some aspects of the one presented here, in particular, in the extremes of climate proposed, in discounting sea level changes as critical factors, in the recent nature of gibbon radiation proposed, and in the stress placed on "rafting" as a means of population spread in primates. Nevertheless, it provides a fresh look at a subject previously dominated by archaeological and anthropological tradition, and confirms the need for a reconsideration of the rate and antiquity of the current hylobatid radiation. The botanists firmly believe, however, that the antiquity of the evergreen rain forests (reflected in the great diversity of tree species) is coupled with its persistance throughout Sundaland, at least since the regression of savanna in the Upper Pleistocene (Ashton, 1972; Whitmore, 1975). Yet, while they discount the presence of substantial corridors of savanna through Sundaland in the last 100,000 years, they acknowledge that there has been some change in the distribution of particular forest types.

Further evidence might narrow such differences of opinion. We cannot hope for greater accuracy or precision until the distribution of vegetation types and changes in sea level and climate have been documented more fully. Pollen analysis and geological techniques have an important role to play here. In the meantime, we can only speculate with the few pieces of the jigsaw already available; we can also only wonder why the gibbons of the northern deciduous forests are dichromatic, why the eastern gibbons tend toward gray coloration, why the older forms are black and the newer ones more colorful, and whether, as in many birds and mammals (Medway, personal communication), "island" forms are darker in color than "mainland" forms.

It should now be clearer that we are dealing not with just one ape species but with at least nine unique populations. Hence, the problems for conservation are magnified. Each has particular attributes that require understanding and preservation, and yet throughout Southeast Asia they act as indicators of the "health" of the tropical rain forest, so important for man's survival through its potential wealth of animals and plants and its properties of soil and water retention. After summarizing what we know about their ecology and behavior, we must examine the current status of the various gibbons and try to assess the nature and extent of the threat that man now imposes on their survival.

IV. Socioecology

Detailed studies of ecology and behavior have been conducted only on the lar gibbon in Thailand and Malaysia, the siamang and agile gibbon in Malaysia, and the Kloss gibbon in the Mentawai Islands (Table III). Incidental observations on all these species, and on hoolock, moloch, and grey gibbons are also available—almost nothing is known of the concolor gibbon in the wild.

Following Carpenter's (1940) pioneering study of the lar gibbon in Thailand, no further interest was shown in free-ranging lesser apes until Kawamura (1958, 1961) returned to Thailand in 1958 and Ellefson (1968, 1974) went to Malaysia in 1964 to study the same species. In the 1968, my wife and I went to Malaysia to study the siamang for 2 years (Chivers, 1971–1974); observations have been continued to this day (Chivers *et al.*, 1975) and extended to the lar gibbon (MacKinnon, 1976; Raemaekers, 1976) and to the agile gibbon (Gittins, 1976). During studies of a lar gibbon population introduced to the island of Ko Klet Kaeo, Berkson *et al.* (1971) initiated a comparative study of the sympatric lar and pileated gibbons in Thailand's Khao Yai National Park; this is now being developed by Brockelman and Marshall. Since 1970, Tenaza (1975a), followed by Tilson (1976), have paid special attention to the Kloss gibbon during their primate studies on the Mentawai Islands; at present Whitten is concentrating on their feeding ecology and behavior.

Apart from these field studies, all we know about the naturalistic behavior of gibbons comes from observations made during studies of orangutans MacKinnon, 1974; Rodman, 1973, 1976; Rijksen, 1976; Galdikas-Brindamour, unpublished data), during studies of bird migration (McClure, 1964), and during a census of Indonesian primates (Wilson and Wilson, 1975, 1976a,b). All studies are detailed by Baldwin and Teleki (1974).

As predicted by Ellefson (1974), when he wrote his thesis in 1967, we find that all species of lesser ape studied so far share the same major features of social organization, feeding, and ranging behavior. They all live in monogamous family groups (adult pair and dependent offspring) in small, stable areas of tro-

TABLE III

Field Studies of the Lesser Apes

Species	Country	Full field study	Incidental observations or brief study/survey
syndactylus	Malaysia	Chivers (1974), Koyama (1971); MacKinnon (1976); Raemaekers (1976).	Carpenter (1938); McClure (1964); Kawabe (1970); Ellefson (1974); Fleagle (1976)
	Sumatra	Brotoisworo (in prep.)	MacKinnon (1974); Wilson and Wilson (1976a); Rijksen (1976)
hoolock	Assam		Tilson (1976); Zoological Survey of India (K. K. Tiwari)
	Burma		McCann (1933)
klossi	Mentawai Islands	Tenaza (1975a) Tilson (1976); Whitten (in prog.)	
pileatus	Thailand	Brockelman (in prog.)	Marshall *et al.* (1972)
muelleri	Borneo	Leighton (in prog.)	MacKinnon (1974); Rodman (1973); Wilson, and Wilson (1975)
moloch	Java		Angst (per. comm.); Brotoisworo (per. comm.); Marshall and Marshall (1976)
agilis	Borneo		Galdikas-Brindamour (in prog.);
	Sumatra		Marshall and Marshall (1976) Wilson and Wilson (1976a,b); Gittins (in prog.)
	Malaysia	Gittins (in preparation)	Ellefson (1974); Chivers (1974, 1976b)
lar	Thailand	Carpenter (1940); Berkson *et al.* (1971); Brokelman *et al.* (1973, 1974)	Kawamura (1958, 1961); Marshall *et al.* (1972); Brockelman (in prog.)
	Malaysia	Ellefson (1974); MacKinnon (1976); Raemaekers (1976).	McClure (1964); Chivers (1974, 1976b)
	Sumatra	Wangsadinata (in prep.)	MacKinnon (1974, 1976); Wilson and Wilson (1976a,b); Rijksen (1976).

pical rain forest from which conspecifics are excluded by calling and chasing behavior (territories), and within which they are mainly frugivorous. Marshall and Marshall (1976) refer to the predawn calls by the males of some species; these precede "the show of force" occurring in all species, in which each gibbon pair daily advertises its territory by loud calls (dominated by the female) accompanied by gymnastics.

There are, however, greater variations between species than predicted by Ellefson. The siamang group lives in a territory, which, at 20 hectares, in less than one-half the size of a lar gibbon territory (Ellefson, 1974) and nearly twice the size of a Kloss gibbon territory (Tenaza, 1975a; Tilson, 1976). In spending at least half the daily feeding time eating leaves, the siamang is rather less frugivorous than the lar gibbon, which spends about 80% (Carpenter, 1940), 70% (Ellefson, 1974), or 60% (Raemaekers, 1976) of feeding time on fruit. MacKinnon (1976) has further shown that the smaller lar gibbon feeds for half as long as the siamang, and travels for twice as long each day. The siamang is also less intense in its territorial behavior, more cohesive in its group daily behavior, and the adults are more tolerant of the maturing offspring than the lar (Chivers, 1972, 1974; cf. Ellefson, 1968, 1974).

All species appear to be almost exclusively arboreal, which has special significance for their conservation. Through their suspensory behavior they appear to be adapted for exploiting the terminal branch niche (Ellefson, 1974; Grand, 1972; Andrews and Groves, 1976), and, in particular, the products of fig trees —new leaves, flowers, and fruit (MacKinnon, personal communication; Chivers, 1974). The short cycles of, and lack of synchrony within or between, species of fig (*Ficus*) make it one of the few resources that are usually continuously available in an small tract of forest. Fig trees are more numerous, however, in forest that has been disturbed to some extent (Corner, 1952).

Thus, selective logging, if carefully controlled, can sometimes produce an habitat enriched for frugivorous animals such as primates (Chivers, 1974; Wilson and Wilson, 1975). Since such disturbance is so difficult to control to the required level, however, it is wholly impractical to advocate a policy of selective logging in the conservation of gibbons. Brockelman (personal communication) stresses that any logging is detrimental to gibbons because (1) sleeping and rest trees are removed, (2) increased light penetration alters the microclimate and character of the forest, (3) many food trees are shade tolerant, and (4) many figs require shade conditions for germination and early growth. Furthermore, the significance of particular food tree species, such as figs, to gibbons needs fuller investigation since exceptions are already coming to light (Gittins, personal communication). It may be the great diversity of trees species in the diets of lesser apes that is the significant factor in their success (Raemaekers, personal communication)—another topic for detailed study, requiring special attention to altitudinal differences.

The tropical rain forests of the Far East, extending through the Malay archipelago from Sumatra in the west to New Guinea in the east, are broken into western and eastern blocks by monsoon forests in the seasonally dry central part of the archipelago—west Philippines, Sulawesi, the Moluccas, and Lesser Sunda Islands (Whitmore, 1975). These two blocks of evergreen rain forest correspond to the Sunda and Sahul Shelves; the eastern block extends into northern Australia and out into the Pacific with increasing floristic poverty, and the western block persists in regions with ever-wet climates—parts of the Kera Isthmus, Southeast Thailand and Cambodia, Southwest Ceylon, and the Western Ghats of India.

Thus, the western evergreen rain forests of Malaya, Sumatra, Borneo, and West Java coincide with the center of gibbon distribution; Brockelman (1975) suggests that they can carry twice the density of gibbons as the deciduous monsoon forests of Assam, Burma, northern Thailand, Laos, Cambodia, and Vietnam. The latter are composed of two main formations: tropical semi-evergreen and tropical moist deciduous forests, which differ according to the extent and intensity of the dry season (Whitmore, 1975).

Wyatt-Smith (1953) recognized five climatic climax forest formations in the Malay Peninsula. The forest canopy up to about 4000 feet above sea level is dominated by trees of the family Dipterocarpaceae; within this range there is altitudinal zonation into lowland, hill, and upper dipterocarp forests, with the main transition zones at about 1000 and 2500 feet, respectively (Fig. 6a, b). The first two belong to the tropical lowland evergreen rain forest formation of Whitmore (1975), and the last, along with the oak-laurel forest up to 4500 feet, belongs to the world-wide formation of lower montane forest. Above 4500 feet above sea level (6000 feet in Sumatra and Java) the upper montane forests are characterized as montane ericaceous forests. With the exception of siamang, gibbons are rarely found in the upper montane zone (see below). Tree species diversity decreases markedly with altitude; lowland forests provide the richest habitats for arboreal mammals.

So as to gain a full perspective of the socioecology of the lesser apes, let us look closely at those of the Malay Peninsula, two species for which considerable information is now available.

The large black siamang lives in the forests alongside lar gibbons, two species of leaf-monkey (*Presbytis obscura* and *P. melalophos*), two species of macaque (*Macaca fascicularis* and *M. nemestrina*), and one prosimia—the slow loris *Nycticebus coucang*). The slow loris is nocturnal and the pig-tailed macaque is the only species commonly found on the ground; otherwise the overlap in habitat use and diet is quite considerable.

There is some seasonality in these forests (Whitmore, 1975) with a short dry season at the start of the year following the wettest of two monsoon seasons;

Fig. 6. Forests and siamang in Malaya. (a) View across the lowland and hill diptero-carp forests; (b) aerial view of the Kuala Lompat Post of the Krau Game Reserve and surrounding forest; (c) female siamang hangs eating fruit of *Aglaia*; (d) siamang family group rest in Tualang tree, *Koompassia excelsa*—from left to right: male, subadult male, juvenile male, infant male, female.

there is 1700–3600 mm of rain annually according to locality (it is wettest on the east coast and central part of the west coast, and driest in the center of the Penin-sula). The flora is very diverse; Poore (1968) has recorded 373 tree species of 139 genera in 23 hectares of forest across the Pahang river from the main siamang study area.

Siamang live in family groups of an adult male and female (supposedly paired for life) and up to three offspring, born at intervals of 2 to 3 years. For the first year of life the infant is cared for by the female, but in the second year it is the adult male who carries the infant about during the day and from whom the in-fant learns independence of movement. The juvenile stage, marked by the early stage of independence of the young from its parents, lasts from about $2\frac{1}{2}$ to

5—6 years of age. The adolescent or subadult siamang reaches social maturity after physical maturity by the age of 8 or 9 years.

His or her departure from the natal group is brought about by the motivation to mate, and by the increasing intolerance of the adult pair, especially the male, to the continued presence of this old offspring within the group. This intolerance is greatest during the 6 months every 2—3 years when the adult pair are mating. The young male advertises his availability for a mate by calling apart from the rest of the group; the sexually dimorphic calls of siamang provide clear indication of the status of the calling animal. It seems that a young female is then attracted to his part of the forest. Note that the permanency of adult pairings contrasts with the relatively long period of instability while young adults search for compatible mates (Chivers, 1974; Chivers et al., 1975). The adult pair currently under observation have been together for at least 15 years and must each be nearly 25 years old (if the oldest offspring in 1968 was their first, produced when they were about 10 years old); their oldest offspring had not found a suitable mate after 4 years (but this is probably an extreme case because of suboptimum habitat near the river).

The siamang group is very cohesive. The average separation of individuals during the day is about 8 m, and rarely are animals more than 30 m apart (Chivers, 1974). This cohesion is achieved by social perception rather than by overt communicative signals (Chivers, 1976a). In general, young siamang tend to follow the female closely without straying far from the male. Thus while the female usually leads group progressions, the adult male can influence the timing or direction of such movements.

The more individuals fight with each other, the more they groom. There are however, only one or two brief scraps each day in a siamang group, whereas pairs of siamang associate together for $1\frac{1}{2}$ hours/day, actually grooming each other for about 40% of this time. It is clear that social contact dominates the cleaning function of this behavior, since it is such a lengthy process involving mainly the older animals. The infant and juvenile are groomed as they settle for the night with female and male, respectively.

As siamang travel about their territory, with the juvenile following the female, the male in the center, and the subadult at the rear, they repeatedly use the same arboreal pathways. Eighty percent of the time is spent in the central 45% of the territory, where 75% of their food trees are situated (Chivers et al., 1975). At any one time they balance intensive use of a part of this core area with frequent sorties to the periphery. The latter behavior appears to combine elements of food searching and boundary patrol.

On the average, siamang spend 50% of feeding time eating young leaves and 40% eating fruit—the balance of 10% being accounted for by flowers, buds, and insects. In some seasons leaf consumption may rise to over 60% and fruit-eating may decline to less than 30% (Chivers, 1974). Of feeding time 30% involves the leaf shoots, unripe fruit (technically flowers), and the ripe fruit of species of *Ficus*. Day ranges appear to be orientated about the locations of trees with ripening fruit or other nutritious foods (the taste of these fruits is unpleasantly sour to man).

It appears that territories of 15—30 hectares provide a suffiient long-term supply of food for the whole group. In 14 months the siamang fed in only 5% of the trees in their territory; this has increased by several percent in subsequent observation periods. Siamang appear to maintain the exclusiveness of their home range by loud group calls which are given on about 30% of days over the year and which carry for several kilometers. These distances signals are only reinforced by border disputes on about 15% of this time. The result is stable, often well-spaced territories; the intermediate buffer zones (a concept still to be verified) seem to allow for increased ranging in times of food scarcity and possibly to facilitate range acquisition by the offspring of the resident group. While the limits of the core area and territory have changed little over the 6 years in our study area, the outer limits of ranging (home range) are more variable with time.

The group is usually active for about $10\frac{1}{2}$ hours each day, of which just over half the time is spent feeding; they travel nearly 1 km about their small territory of 25 hectares. The pattern of behavior outlined in the Appendix varies according to the stage of the annual climatic cycle, the longer floral cycles, and the 2- or 3-year reproductive cycle.

The most obvious differences between the two species are the faster and more independent activity of group members of the lar gibbon family group. In particular, the common way in which feeding and traveling behavior are combined, best described as "foraging," is seen infrequently in siamang. Lar gibbons feed less than siamang and travel individually over a broad front, eating a little here and there and converging to feed together in the same tree a few times each day. They also engage in more social behavior than siamang, both within and between groups; groups call on 80—90% of days, call twice a day, and meet neighbors more often, with the female playing a more active role in such conflicts. The adults are more aggressive to their older offspring. By contrast, life in a siamang family is much slower, more coordinated, and more tolerant.

Observations of lar gibbons in the Krau Game Reserve (MacKinnon, 1976; Raemaekers, 1976) are showing them to be less frugivorous than described by Carpenter and Ellefson. Nevertheless, although similar in social structure and diet, there are important differences in feeding and ranging behavior, which enable these closely related species to coexist together in the same habitat. At 50—60 hectares, the territories of lar gibbons are more than twice the size of those of their larger relative. Where both are consuming the same foods, the smaller species ranges more widely, eating small quantities of food in each of the many trees they visit. The siamang group feed together for longer in fewer, more localized food sources.

Thus, while there may be a similar biomass of the two species in the forest as a whole, siamang are more localized in their feeding and ranging at 2 kg/hectare, whereas the lar gibbons are dispersed more evenly through the forest at 0.5 kg/hectare (relating body weights to home range size, rather than to area of habitat available, as in Table V). Siamang feed three to four times more than they travel each day, whereas lar gibbons feed little more than they travel. These differences in feeding and ranging strategies seem mainly to be a function of body size (MacKinnon, 1976).

The concentration of feeding and travel activity in the morning, the timing of group calls, and the early settling for the night by both species, afford temporal separation from the monkey species, which feed most actively at dawn and dusk, sometimes in the same food trees, and have a long midday siesta (Chivers, 1973). Siamang escape the midday heat by moving into the lowest levels of the canopy. The only common trees in which monkeys feed, whose flowers and fruit are apparently neglected by the apes, are species of *Parkia* and *Dillenia*; otherwise the overlap in terms of tree species is extensive. Each species, however, usually consumes different parts of the tree, often at different stages of maturity (Curtin and Chivers, 1976).

Observations of the Kloss gibbon in the Mentawai Islands (Tenaza, 1975a) bring out the following points of similarity to, and difference from, mainland hylobatids: (1) groups of 3.4 individuals live in territories of only 7 hectares (14 hectares according to Tilson, 1976); (2) males and females do not sing together, but chorus separately with their neighbors—males mainly between 0330 and 0530 hours (before dawn) once every 2.5 days, with a secondary peak from 0800–0900 hours; females mainly between 0630 and 0830 (after sunrise) once every 5 days; (3) while males are relaxed while calling 150–500 apart, females display vigorously across the territorial boundary only 10–50 m apart. Tenaza claims, however, that it is the males that establish territories, although he saw only three instances of this.

As in other gibbons, he suggests that monogamy and territoriality are inseparable, with the proximal basis for monogamy being territorial hostility between adults of the same sex resulting in each territory containing only one adult of each sex. There are no large carnivores on the Mentawai Islands; since sea eagles and ospreys prey on fish, the only predators on primates are pythons and man (for more than 1000 years at least).

The small size of their territories, in which safe night sleeping places are scarce, compares with those of 4 to 6 hectares in the experimental colony on Ko Klet Kaeo Island in the Gulf of Siam (Berkson *et al.*, 1971; Brockelman *et al.*, 1973, 1974). The latter groups are provisioned, but until data on feeding behavior of Kloss gibbons is collected by Whitten during 1976, we cannot know whether their small territories reflect a sustained local abundance of food, a preference for a very varied diet with a high content of leaf matter, or a lack of competitors, especially primates.

It would seem unlikely that the larger territories in mainland gibbons are artifacts of disturbance (Tenaza, 1975a); rather it would seem that the isolated population of gibbons on Siberut are particularly crowded and threatened by man (Tilson, personal communication), especially as they are unusual among gibbons in having two calls only used in encounters with man (Tenaza, 1976). It will be very interesting to know details of their diet, since they appear to enjoy very reasonable reproductive success.

V.   Population Sizes

It is almost impossible to assess numbers of each species of gibbon with any real accuracy. There is enough disagreement about the actual areas of each country, let alone the continually changing areas of forest. It is also important (but very difficult from area figures usually available) to distinguish between different forest types, since some are unsuitable habitats for gibbons (see below). Nevertheless, I have tried to collect the relevant data for each country in which gibbons occur; with estimates of population density in these countries this permits calculations of potential population size.

It must be stressed that, because these calculations are of the numbers of animals that should be present, they are likely to be overestimates; this is particularly likely since density estimates are usually obtained from study areas which have been chosen for the abundance of hylobatids. We would be advised to apply the formula produced by the Wilsons (personal communication) who found in their survey of Sumatra that all species of gibbon occurred in only one-third of the total forest area. The final cautionary note is that these populations are being reduced rapidly; gibbons die every day as direct or indirect consequence of human activity.

On the assumption that most suitable habitats outside protected parks and reserves will have been cleared within 15—30 years, it is also possible to predict the size of each population. These estimates are more reliable since one has greater confidence in the minimum area of suitable habitat that will remain, and in the relatively high population density that should be sustained therein. Since some forest is likely to remain outside these sanctuaries, especially in highland areas, these estimates are pessimistic.

Given these serious limitations, these estimates are useful in that they confirm (1) that populations of most species are currently in a healthy state, and (2) that they will be drastically reduced for all species within a very short time unless current trends are quickly slowed, halted, or reversed. By clarifying the situation in this way, there will hopefully be a more positive management of wildlife resources in the tropical rain forests of Southeast Asia.

Gibbons do not usually inhabit swamp and montane forests, although *agilis* occurs at high densities in certain swamp forests in Riau province of Sumatra (Carolyn Wilson, personal communication). Most species are animals of lowland and hill forest; only the siamang shows preference for the higher forests, although it is rarely found in submontane forests. Medway (1972b) recorded *lar* up to 4700 feet above sea level on Gunung Benom in Malaysia and siamang near the summit at 6000 feet. Rijksen (personal communication) comments that he has seen neither hylobatid above 3300 feet, and the lar gibbon rarely over 2200 feet above sea level. Tilson (personal communication) recorded *hoolock* between 110 and 1800 feet in Assam, but Carter (1943) collected 76 specimens in Burma between 625 and 4200 feet above sea level.

Chivers (1974) and Wilson and Wilson (personal communication) have collected data (shown in the following tabulation) on altitudes of three species.

Heights above sea level (feet)

		Range	Mean	n
Siamang	Malaysia	50–4500	1655	39
	Sumatra	50–4000	1433	36
Lar	Malaysia	50–2500	930	45
	Sumatra	600–1000	790	9
Agile	Malaysia	50–4000		
	Sumatra	50–3120	977	50

Hence one must make allowance for these habitat preferences in calculating the area of distribution of each species.

Counts of group size (Table IV) from various parts of the Sunda Shelf confirm earlier assessments (Carpenter, 1940) that gibbon families are usually composed of three to four animals. It is more difficult, however, to assess their density in rain forest (Table V). Three different methods have been used: (1) a careful count of all animals in a small study area (chosen presumably for its high abundance of gibbons); (2) extrapolation from such counts to encounters over a wider area; and (3) extrapolation from counts of calling frequency in a known population to groups calling daily in larger areas of survey.

The larger the area surveyed, the lower the result tends to be; with siamang, for example, scores vary from 1 to 13 animals/km², with an average of about 6/km². Allowing for overestimates in study areas, the figure of 2½ siamang/ km² has been chosen as the basis for a conservative estimated of their overall

TABLE IV

Average Size of Gibbon Groups

Country	Species	Group Size $\overline{x}$	n	Author
Assam	*hoolock*	3.5	7	Tilson (personal communication)
		3.4	11[a]	Tilson (personal communication)
Thailand	*lar*	4.4	21	Carpenter (1940)
		3.0	9	Tilson (personal communication)
Malaysia	*lar*	3.3	28	Ellefson (1974)
		3.5	6	MacKinnon (1976)
		2.7	15	Chivers (1974)
		4.0	4[b]	Chivers (1974)
	agilis	4.5	6	Gittins (personal communication)
	syndactylus	3.0	6	MacKinnon (1976)
		3.6	18	Chivers (1974)
		3.8	13	Chivers (1974)
Sumatra	*lar*	4.4	5	Rijksen (1976)
		3.0	26[c]	Wilson and Wilson (personal communication)
	agilis	3.0	26[c]	Wilson and Wilson (personal communication)
	syndactylus	3.8	16	Wilson and Wilson (personal communication)
		3.3	6	MacKinnon (1976)
		4.1	7	Rijksen (1976)
	klossi	3.4	14	Tenaza (1975a)
		3.5	15	Tilson (1976)
Kalimantan	*muelleri*	4.0	10	Rodman (1976)
		3.5	8	MacKinnon (1976)

[a]Survey.
[b]Study areas only.
[c]Scores for *lar* and *agilis* combined.

numbers. A similar figure seems applicable for the rare moloch gibbon in Java (Angst, personal communication). Siamang may be very abundant in some parts of Sumatra (Riksen, 1976; MacKinnon, 1976).

Sumatran and, especially, Bornean hylobatids appear to occur at higher densities than their Malayan counterparts (Table V). This may be due to differences in vegetation, but, more likely, the decrease in species competing for fruit and similar resources may be the main factor in their occurrence at higher densities, as is suggested by Gittins (personal communication) for the agile gibbon in the Malay Peninsula. Although there is a great variety of primates in Borneo, there is only one hylobatid in any locality. Current studies should shed additional light to this problem.

Kloss gibbons, living in much smaller territories than their congeners, occur at the highest densities. The lowest densities appear to occur in the north—in Thailand—where Brockelman (1975) suspects that densities of at least one group (of four gibbons)/km² decline below that level in the mixed deciduous forests in the north of the country. The higher density estimated for *hoolock* (Tilson, personal communication) might indicate a better adaptation to this kind of habitat; similar adaptation would be predicted for *concolor*, for which no data are available.

Thus, estimated densities of gibbons vary approximately from two to four gibbons/km² in the north, through five or six in the Malay Peninsula, to eight or so in Sumatra (twenty in the Mentawai Islands), and to ten in Borneo.

The most accurate estimates of forest area are probably those of Brockelman (1975), based on his study of Earth Resources Technology Satellite (LANDSAT-1) images taken in 1972 and 1973 and on estimates of forest area by the Royal Thai Forest Department derived from overflights during 1961 and 1967. Good data are also available for West Malaysia from the Economic Planning Unit of the Prime Minister's Department, the Office of the Chief Game Warden and the Forest Department in Kuala Lumpur, and for East Malaysia from the Forests Department in Kuching, Sarawak (Table VI).

The least accurate are probably those for Indonesia, which accounts for about one-fifth of the $1\frac{1}{2}$ million km² of gibbon habitats. Those used here are based on Basjarudin (1973) with modifications suggested by Brotoisworo and the Wilsons (personal communication), who point out that these figures date from Dutch surveys in 1929. Although intensive logging has only started in the last 8 years, even if the 1929 figures were accurate they are certainly overestimates at present—one can only guess by how much. It is noteworthy, however, that the calculation that 55% of Sumatra is still forested is identical to that for Malaysia; the figure of 78% for the more isolated forests of Kalimantan is not very different from the 60–70% suggested by MacKinnon (personal communication).

If there is a danger of overestimating gibbon numbers for this reason, they may be underestimated through too conservative an estimate of population density; these two errors might cancel each other out. If they do not, the error would be of the order of 100%. Actual numbers are not as important as obtaining an estimate of the size of each gibbon population (Table VI). Whether the total number of gibbons is 2 million or 8 million, we can be confident that each species is currently numerous, with the exception of *moloch* (Table VII).

We can be equally confident that these numbers will decline drastically within 15–20 years if current trends continue to the point where gibbons survive only in the national parks and game reserves already in existence. These 65,000 + km² account for less than 4% of total gibbon habitat. There are few signs that

TABLE V

Density of Gibbons[a]

Species	Country	Location	Area surveyed (km²)	groups	No. of gibbons	Gibbons/sq. km²	Author
syndactylus	Malaysia	Survey	1130	568–967		1.9–3.3	Chivers (1974)
		Ulu Gombak	19.2	24	91	4.7	
		Ulu Sempam	43.2[b]	13	49	1.1	
			0.9[b]	4	16	17.8	
		Kuala Lompat	17.6	9	34	1.9	
			0.4[b]	1	5	12.5	
	Sumatra	Survey	4.0	6	18	4.5	MacKinnon (1976)
			8.2			9.2	Wilson and Wilson (personal communication)
		Ranun	3.0	6	20	6.6	MacKinnon (1976)
		W. Langkat	35.0	4	18	15[c]	
		Ketambe	1.5	6	24	16.0	Rijksen (1976)
hoolock	Assam	Survey	8.0	7	25	14.0	Tilson (personal communication)
klossi	Mentawai Islands	Siberut	2.5	12		30	Tenaza (1975)
				15	53	25 max.	Tilson (personal communication)
muelleri	Borneo	Ulu	5.0	8	28	10.5	MacKinnon (1976)
		Segama north	15.0	4	14	15[c]	
		Kutai	2.7	10	40	11.7	Rodman (1976)
			0.4	1	2	4.1	
		ITCI	1.1	8	29	22.2	Wilson and Wilson (1975)

			[a]				
moloch	Java	Ujong Kulon	20			2.6	Angst (personal communication)
agilis	Malaysia	Survey	175	199–275		4.6–6.3	Chivers (1974)
		Sg. Dal	4.0	17	76	19.0	Gittins (personal communication)
	Sumatra	Survey	9.8			6.1	Wilson and Wilson (personal communication)
lar	Malaysia	Survey	1244	1068–1634		3.4–5.3	Chivers (1974)
		Ulu Gombak	8.3	14	56	6.8	
		Ulu Sempam	17.0	6	24	1.4	
			0.9[b]	3	12	13.3	
		Kuala Lompat	4.9	6	24	4.9	
			0.4[b]	1	5	12.5	
			4.0	6	21	4.1	
	Sumatra	Ranun	3.0	4	14	7.0	MacKinnon (1976)
		W. Langkat	35.0	2	6	18[c]	
		Ketambe	1.5	4	18	12.0	Rijksen (1976)

[a]Number per km^2.
[b]Study area only.
[c]Compare encounter frequency with that from area where density more accurately known.

579

TABLE VI

Estimates of Forest Area and Population Size

Country[a]	Total area (km²)	Area of hill and lowland forest (km²)	% total area	Area of protected forest (km²)	Species	Total area of distribution (km²)	Protected area (km²)	No. of groups	No. of gibbons
Assam	121,900	47,900	39	235	*hoolock*	31,000	235		80,600
Burma	678,000				*hoolock*				
Laos	236,700				*concolor*				
Vietnam	334,300				*concolor*				
Cambodia[a]	181,000	26,320	15		*concolor*				
					pileatus				60,000
Thailand[a]	514,910	94,452	18		*pileatus*	13,500	2,632	10,000	40,000
					lar	75,000	17,724	15,000	60,000
					agilis	2,500	0	2,500	10,000
Peninsular Malaysia[b]	131,000	66,950	51	8,150	*agilis*	5,735	679	5,750	23,000
					lar	54,720	7,741	54,750	219,000
					syndactylus	13,150	4,841	8,900	34,000
Sumatra[d]	472,300	260,000	55	18,280	*syndactylus*	131,350	18,280	105,330	398,000
					lar	40,900	6,370	25,865	77,000
					agilis	205,120	11,910	129,718	388,000
Mentawai Islands	6,000	4,200	70	100	*klossi*	4,200	100	24,000	84,000
Java[d]	127,000	28,000	22	2,422	*moloch*	9,340	1,145		20,000
Borneo									
Kalimantan[d]	539,500	419,000	78	11,410	*agilis*	108,940	3,550	113,750	455,000
East Malaysia[c]	198,000	128,700	65	1,740	*muelleri*	547,700	9,600	511,400	2,190,800

Sources of information:

[a] Thailand and Cambodia (West of Mekong), Brockelman (1975; personal communication);
[b] West Malaysia, personal observation, Ratnam (personal communication), Sargent (personal communication);
[c] East Malaysia, Proud (personal communication). Scriven (personal communication),
[d] Indonesia, Wilson and Wilson (personal communication), Basjarudin (1973), MacKinnon (personal communication), Darsono (personal communication), Brotoisoworo (personal communication).

TABLE VII

Estimates and Predictions of Population Sizes for Each Gibbon Species[a]

Species	Country	Present population estimate[b]	Population prediction for 1980[c] [( ) percentage reduction]
syndactylus	West Malaysia	34,000	29,000 (15)
	Sumatra	133,000	110,000 (17)
		167,000	139,000 (16)
concolor	Laos		
	Vietnam	228,000	22,800 (90)
hoolock	Assam	80,000	1,000 (99)
	Burma	452,000	90,000 (80)
		532,000	91,000 (83)
klossi	Mentawai Islands	84,000	3,000 (96)
pileatus	Thailand	40,000	10,5000 (74)
	Cambodia	60,000	12,000 (80)
		100,000	22,500 (77)
muelleri	East Malaysia	584,000	23,000 (96)
	Kalimantan	1,241,000	73,000 (94)
		1,825,000	96,000 (95)
moloch	Java	20,000	6,900 (65)
agilis	West Malaysia	33,000	4,000 (88)
	Sumatra	256,000	71,000 (72)
	Kalimantan	455,000	21,000 (95)
		744,000	96,000 (87)
lar	Thailand	125,000	60,000 (52)
	West Malaysia	73,000	46,000 (37)
	Sumatra	52,000	38,000 (27)
		250,000	144,000 (42)
		3,950,000	621,200 (84)

[a]*Notes:* (1) Population estimates for *lar* and *agilis* in Sumatra have been reduced by 33%, and for *lar* in Malaysia and *syndactylus* in Sumatra by 67%. (2) It is assumed that about 30% of Laos and Cambodia remain forested, and that 33% of this area is stocked with 4 gibbons/km²; it is also assumed that only 10% will remain as suitable habitat in 15 years. (3) It is assumed that 50% of Burma remains forested, and that 33% of that area is stocked with 4 gibbons/km²; it is also assumed that only 20% of suitable habitat will remain in 15 years.

[b]Based on 4 gibbons/km² for *lar, pileatus, agilis, muelleri,* and *hoolock* (2 for *lar* in Thailand); 2.5 gibbons/km² for *syndactylus* and *moloch*; and 20 gibbons/km² for *klossi*.

[c]Based on 6 gibbons/km² for all species except *klossi* (30), *muelleri* (10), and *concolor, lar* (Thailand), and *hoolock* (4).

581

current trends are being slowed, let alone reversed, so we should be prepared for this 96% reduction of gibbon populations. Yet, calculations suggest that the total reduction will only be of the order of 84% (Table VII). While this may stem from too conservative an estimate of current population sizes, it may indicate the desired slowing of current trends.

It would seem that there are already sufficient areas to sustain viable populations of all species for which data are available, even of *moloch*, but, as will be discussed below, the existence of such areas is not enough; they must be well stocked and efficiently protected. The predictions for Burma and "Indo-China" are based on the assumption that similar actions are being taken as in Thailand, Malaysia, and Indonesia; if this assumption proves false the outlooks for *hoolock* and *concolor* are very gloomy, especially for the latter whose range has suffered severe upheavals for many years. The hoolock gibbon, currently little disturbed by technology, should flourish if hunting and cultivation are not excessive.

There are good chances of forest remaining outside protected areas in 15–20 years. Rodman points out, for example, that some of the large logging concessions in Kalimantan will never be clear-cut and will continue to sustain large populations of gibbons. There are indications that *syndactylus, pileatus,* and *lar* might not be affected as seriously as the other species during the next few years (Table VII). This would result either from current population estimates being too conservative or, more likely, from adequate reserves having being set aside in the right places for sustaining populations at the highest densities. The siamang, in particular, seems likely to continue to flourish in the mountains Sumatra and Malaya. While population decline will probably be drastic, large populations of *hoolock, muelleri,* and *agilis* should also survive.

This "optimism" is only justified if human attitudes and activities alter markedly over the next few years. We shall now proceed to assess the current situation in four main parts of gibbon distribution—Assam, Thailand, Malaysia, and Indonesia—and to consider what has to be achieved if these optimistic predictions are to be realized.

VI. Discussion

A. Current Situation

1. Assam

Today the hoolock gibbon is restricted mainly to the forest reserves of Assam, Burma, and Bangladesh. A recent reconnaissance by Tenaza (personal communication) indicates a greatly reduced range in Burma and Bangladesh already. In August 1972, Tilson (personal communication) surveyed South and Central Assam and found hoolocks in the United Khasi-Jaintia Hills, the Garo

Hills south of Goalparo, the Mikir Hills, and the hill forests that border Nagaland. Forest Department officials stated that they were still to be found in the
forests of Nagaland and Manipur, and in scattered populations near the Kaziranga Wildlife Reserve.

Most populations were found in mixed evergreen forests between altitudes
of 600 and 2000 feet, although they can be found at lower altitudes. Some of
these forests are secondary growth, but this does not seem to interfere with
breeding. There is no forest in Assam, however, where hoolocks are officially
protected. In the few forest reserves where they still occur, selective logging is
proceeding without consideration for their resource requirements.

In other areas, large tracts of primary forest are being clear-felled, which
further isolates remaining populations. In February 1973, Tilson submitted
a proposal to the Assam Forest Department suggesting that a forest reserve in
the Sibsagar district be officially designated as a hoolock gibbon reserve, but
no action has been initiated. Because of the prevalence of Hinduism in Assam,
these gibbons are not hunted, except in Nagaland where the indigenous people
use them as a source of food.

2.   Thailand

Brockelman (1975), with the help of Dr. Joe Marshall, officials of the Royal
Thai Forest Department, Mr. Peter Rand, and others, has recently conducted an
investigation into current population sizes (see above) and the future prospects
for lar and pileated gibbons in Thailand in view of present policies, practices,
and trends.

The future of gibbons in Thailand is threatened most by (1) exploitation of
forests for lumber and farming land, (2) hunting practices, and (3) the demands
for pets and research, both local and international. Although government
restrictions on legal exports means that this should not be a significant drain
on populations, smuggling of young animals still continues, which could
become critical as populations dwindle for other reasons. By contrast, legislation has probably had little or no effect on hunting—shooting the adults for food
and selling any surviving young as pets.

The main threat, however, comes from the irreversible destruction of forests
all over the country; concessions have been allocated in most forests and access
to remote areas is facilitated by the construction of new roads. Squatters following the loggers prevent any regeneration with their slash- and-burn farming
techniques. The protection of areas of forest from clearing and hunting will
be crucial to the survival of these wholly arboreal primates: this will require
a rapid change in attitudes through education and enforcement.

The northern mixed dry forests are a less suitable habitat than the southern
evergreen rain forests. In the fourteen southern provinces, however, forest is
mainly confined to the three mountain ranges. The most extensive forest extends

from the Thai-Burmese border into Phang-gna and Phuket provinces; the largest intact area of 5000 km² contains wildlife sanctuaries totaling 1564 km². The second largest area is of 4630 km², extending from northern Nakhon Si Thammarat to Satun province at the Malaysian border; it is severed by roads in about four places and contains a game reserve of 1288 km² in the south and a national park of 1056 km² in the north. Between these two ranges lie several forested areas, totaling about 1644 km², mostly at lower elevations, but they are being depleted. The forested mountains up to the Malaysian border, wherein agile gibbons are found, are also unprotected.

There are probably as many lar gibbons in this southern part of the country as in the west, north, and northeast combined. If the present reserves and parks are satisfactorily protected they should be large enough to ensure the survival of *H. lar* in Thailand; several are at least 1000 km² (those of 60 km² are probably too small). Most of these parks and sanctuaries, however, have been established only recently and full protection of the trees and the wildlife is not yet a reality. Increased confrontation between protectors and exploiters is bound to occur, and the outcome is still very much in doubt.

The situation with respect to the pileated gibbon in southeastern Thailand is less favorable. Brockelman suggests that suitable habitat within the largest parks and sanctuaries could support about 10,000 individuals, but elsewhere forests are being rapidly destroyed. An area of overlap between lar and pileated gibbons of at least 100 km² occurs in Khao Yai National Park, fortunately, the best protected area in the country (at least the central part). This is at the northwest limit of the range of the pileated gibbon. About two-thirds of the remaining habitat of the pileated lies in Cambodia, about which little is known except that satellite images show rapid exploitation in some parts.

We cannot be assured yet that all areas set aside to be saved in Thailand will be able to support gibbon populations, as exploitation is extremely difficult to control, and sufficient budgets and manpower for protection have not been committed. Conservation, however, is increasingly becoming a public issue.

3. Malaysia

The same problems apply here as in Thailand and Indonesia, although the emphasis differs. Hunting poses a less serious threat than in Thailand, and only now are export controls becoming effective. The advanced technological development of Malaysia, however, means that the greatest threat of forest clearance has reached an advanced stage. Ratnam (personal communication) points out that hydroelectric schemes pose an even greater threat than land development schemes. He believes that the lowland forest, amounting to less than one-half of the forest remaining in the Peninsula, is being cleared at the rate of 2500 km²/year.

Since the main National Parks and Reserves are in highland areas, this will mean that only 650 km² of lowland forest (possibly containing as many as 4000 gibbons) may remain in 10–15 years. The viability of this area is weakened since it is composed of several scattered small reserves. While the higher forest may remain more or less intact, we have already seen how a large part of it is unsuitable for gibbons. Vast tracts of forest still remain in the Main Range running the length of the west side of the Peninsula, and in the eastern Trengganu Highlands (from which siamang are absent). These two ranges are joined along the Pahang–Kelantan border by mountains including the highest in the country and the largest National Park, Taman Negara, of 4325 km². The proposed Endau-Rompin National Park of about 2600 km² further to the south does not contain such a rich flora and fauna. The only other two large reserves are the Krau Game Reserve of 530 km², covering the Gunong Benom massif in the center of the country, and the Grik Wildlife Reserve of 680 km² in the northwest.

The Malayan Nature Society have taken the initiative in showing how economic development can be linked with the conservation of natural resources. They have presented a scheme (Ng, 1974) for conserving large tracts of forest representative of each ecosystem for (1) the protection of soil and water, (2) the protection of stock, (3) the conservation of genes, (4) the more productive management of these ecosystems, (5) scientific research, especially for comparing exploited with undisturbed areas, (6) education, and (7) tourism.

The Forest Department reports that in 1973 there were 83,362 km² of primary and secondary forest in the Peninsula (Scriven, personal communication). Of these, about 66,950 km² are of types providing suitable habitats for gibbons (Ratnam, personal communication; Sargent, personal communication). The Forest Department hopes that a total of more than 50,000 km² will soon be gazetted as permanent forest estate; this would include protected and amenity forest as well as productive forest scientifically managed for the supply of timber and other forest produce.

The extension of protected forests, the more productive management of these forests, and the efficiency of the government departments concerned again make one optimistic that viable populations of all three lesser apes—siamang, lar, and agile gibbons—will continue to thrive in the Malay Peninsula, even if most other forest is cleared.

The prospects are equally encouraging for East Malaysia where similar conservation efforts are under way. About 60–70% of Sabah is still forested, although only about 25% is completely untouched. Three reserves have been established—the Kinabalu National Park of 805 km², Palau Gaya of 36 km², and the Sepilok Forest Reserve (with its orangutan sanctuary) of about 40 km² (Scriven, personal communication).

Similarly, about 60% of Sarawak is still forested according to their Lands and Surveys Department; about one-third of these forests are forest reserves or protected forests where hunting is forbidden and exploitation for timber may be modest (24,000 and 3000 km², respectively—Annual Report of the Forests Department, Sarawak, 1973). There are now four National Parks gazetted covering an area of 657 km², of which the largest is Gunung Mulu at 529 km², with a further 200 km² in six localities proposed (Proud, personal communication; Scriven, personal communication). The Forests Department are showing special interest in all primates, and several important projects have just been started. They are aimed at assessing current population sizes and distribution, establishing protected reserves in suitable areas, and setting up rehabilitation centers to try and return confiscated apes to the wild (Proud, personal communication). This has been tried with considerable success for orangutans at three locations in Indonesia; with gibbons there has been a single success by Monica Borner in Sumatra (Carolyn Wilson, personal communication) and several efforts in the Sepilok Reserve, Sabah (MacKinnon, personal communication).

4.   Indonesia

Virtually no forest or gibbons remain in East Java. Sanctuaries in Central and West Java, like the Ujung Kulon Reserve (360 km²) harbor gibbons at densities of about 2½ km². Angst (personal communication) reports that gibbons occur only in the lower eastern 20 km² of the reserve. The primary nature of the forest in West Java has been affected by lava flow following the eruption of Krakatau in 1885. Moloch gibbons are probably otherwise restricted to forest isolates east and south of Bandung (Brotoisworo, personal communication). It seems, therefore, that this gibbon species is represented by hundreds of animals (personal communications from Angst, Brotoisworo, Marshall, and Groves), rather than the thousands estimated from figures of available forest (Tables VI and VII).

In Kalimantan (Indonesian Borneo), as in Sumatra, vast logging concessions have been granted and much of the lowland forest will soon be at least selectively logged. There are, as yet, few reserves in Kalimantan. The Kutai Reserve (3060 km²) is the largest nature park in the island, but about half has already been selectively logged (Carolyn Wilson, personal communication); the Tanjong Puting Reserve is the only sanctuary for *agilis* in Borneo at the present time. Clear-felling for farm land depends apparently on the logging companies. Both Rodman and Wendell Wilson (personal communications) have found grey gibbons living at high densities in isolated tracts of selectively logged forest.

The largest Indonesian nature reserve is in Sumatra—the Gunung Loeser Reserve (4165 km²), which is continuous with the Langkat and Kloet Reserves. These 6500 km² cover a large part of the mountains of the province of Aceh, but, since hylobatids rarely occur over 3000 feet there, the area suitable for gibbons is considerably less (Rijksen, personal communication). The lesser apes are protected in Indonesia; only on the Mentawai Islands is hunting a problem. There is a pet trade in which siamang are more popular than the smaller gibbons (Wilsons, personal communications).

On the Mentawai Islands, the Teiteibatti Wildlife Reserve has been recently established covering 65 km² of central Siberut to protect the forest and its animals from intensive logging. Less than 5% of Siberut has been logged, probably more on Sipora and the Pagi Islands; about 70% of the four islands is primary forest, supporting 50–80,000 gibbons (Tilson, personal communication). The Reserve should soon be extended to 100 km². The unique primates on these islands merit the special attention that is now being given to them through an integrated WWF/Survival International project for wildlife conservation and the socioeconomic development of the endemic people (Whitten, personal communication); the concept of the project is itself unique and has a much wider application.

B.   Prospects for Conservation

It is to be hoped that integrated projects, aimed at helping man and animals simultaneously, will soon be initiated elsewhere. Indeed, this seems to be inherent in the concept of Biosphere Reserves now being promoted by UNESCO through their Man and Biosphere (MAB) program; this seeks to establish reserves on the sites of existing national parks, with the important proviso that the requirements of economics and conservation be satisfied equally. Reserves are divided into core and buffer areas; the core is left inviolate, while the buffer zone is essential to protect the core area from human activity, but it provides excellent opportunities for tourism, education, ecological research, and land management studies.

Taking this one stage further, we see a solution to the dilemma facing mankind. Man and forest animals can only coexist successfully if (1) large areas of forest are set aside and left untouched by man, (2) we accept that a major part of the land area is cleared of forest and devoted to human use, and (3) the intervening area is managed rigidly to minimize disturbance of the forest ecosystem while yielding some benefits to man.

In many countries 50% or more of the land area is being cleared of forest; further clearance should be prohibited. A satisfactory compromise might be reached if 20% can be preserved inviolate, and the remaining 30% managed carefully. Careful management means restricting logging to about five trees/ hectare, and exerting control to allow regeneration. While it has been shown that gibbons can survive in an habitat disturbed in this way, it must be stressed that the balance is so delicate that a policy of selective logging could so easily be abused to the extent that it is incompatible with the survival of gibbons and other animals. Nevertheless, it may be the only solution over a part of the habitat if the needs of man and animals are to be satisfied.

It is not just a matter of saving forests for gibbons and other animals. People are slow to realize that large tracts of forest are essential for man's survival: (1) forests are renewable and poorly understood resources, which would bring many benefits to man if controlled properly, (2) forests over watersheds are essential in stabilizing water flow into the lowlands and in controlling sediment flow, and (3) upland slopes are notoriously poor for sustained farming (from Brockelman, 1975). Brockelman argues that forest clearance because of expanding human populations is no longer justified in the long term for either ecological or economic reasons.

Thus, the following actions are necessary: (1) control over, and an eventual ban on, clear-felling of forests; (2) the establishment of large reserves throughout each country and their vigorous protection from any form of disturbance; (3) light and sustained use of all remaining forests under careful government control; (4) strong enforcement of export restrictions; and, perhaps of greatest importance (5) full education of the people about conservation and the best way to use the natural resources, including the importance of forests, in general, and about the general and academic interests of gibbons, in particular. Only with strong popular backing can firm government legislation and the efficient deployment of manpower halt the current drain on gibbon populations.

While the lesser apes are united by certain morphological and behavioral characteristics, each species has become adapted to distinctive differences in environment. Hence, requirements for conservation vary accordingly. *Moloch*, some races of *concolor*, and, possibly, *pileatus* are in danger of extinction. The position of *klossi* is very delicately balanced, and all races of *concolor* must be in jeopardy. The healthiest populations are *syndactylus*, *lar*, *agilis*, and *muelleri*; lack of hard data on *hoolock* must cause concern for its survival. Even these healthy populations, however, are threatened by current practices to the extent that they will be reduced to relict populations within 15 to 20 years.

Thus, action to establish large reserves in the correct localities and to ban exports are most urgently needed in Java, "Indo-China," and Thailand, but are no less important in the long term in Malaysia, Indonesia, Burma, and Assam. It is not possible or appropriate to be any more specific in such recommendations. The protection of areas from clearance and hunting will be crucial to the survival of these wholly arboreal primates; this will require a rapid change in attitudes by education and enforcement.

Thousands of gibbons are currently being made homeless during forest clearance. While some are temporarily providing an important source of protein for the local people, the majority are being wasted. In Indonesia, however, coordinated efforts are being made to capture gibbons in such areas and to translocate them to areas of low density (Carolyn Wilson, personal communications). This practice, increasingly successful with orangutans, is more difficult with territorial species such as gibbons. There is also the risk of introducing disease; such efforts should be restricted, therefore, to areas of low population density. Similarly, it is exceedingly costly and difficult to breed gibbons in captivity in significant numbers, because adults will not tolerate crowding in the way that the more sociable primates will (Berkson *et al.*, 1971; Brockelman *et al.*, 1973, 1974). Whatever element of management of wildlife resources is introduced, it can never cope with more than a small, albeit important, part of conservation. Hence, the conservation of large tracts of forest so as to maintain sufficient gene pools of these relatively large animals is of prime importance.

Studies in captivity can bring important benefits to human society, as well as yielding valuable information for conservation efforts. Such studies are justified in these terms, so long as it realized that their direct contribution to the conservation effort is likely to be small and that trade and breeding programs cannot be allowed to involve these species whose survival is seriously jeopardized. In the first instance, these studies should be conducted in the countries of origin; only in the cases of *lar, agilis, muelleri* and, possibly, *syndactylus* can export to countries with more funds and expertise in research techniques be justified.

So I conclude this effort to start synthesizing the wealth of information that has been accumulating on the lesser apes over the last decade, and to advocate steps that should be taken to ensure their survival. It is a formidable task that must be continued with great vigor over the next few years; we have only scratched the surface in our efforts to understand this important and fascinating group of hominoid primates. In the next 10 years we must make further progress in our understanding of gibbon biology, and we must work for a marked decline in the rate of forest clearance and destruction of wildlife, if our efforts are not to be in vain. There has to be a dramatic change in attitudes to the forest environment; this change has occurred among conservation organizations, but it has yet to permeate through to the local peoples. Action is needed now; it is almost too late.

VII. Summary

The lesser apes or gibbons are distributed throughout the deciduous monsoon and evergreen rain forests of the islands and mainland of Southeast Asia. They are characterized by their long arms and erect posture—adaptations to suspensory behavior, in general, and to brachiation, in particular—and by their loud and complex calls of considerable purity—apparently adaptations to monogamy and territoriality.

Nine species, of varying status, are currently recognized on the basis of differences in coloration and pelage markings, calls, and distribution. These fall into three groups, probably best distinguished at the subgeneric level: (1) the large monochromatic (black) siamang in the southwest; (2) the sexually dichromatic concolor in the northeast; and (3) the sexually dichromatic hoolock and pileated in the north, the monochromatic Kloss in the west, the polychromatic grey and silvery in the east (and south), and the polychromatic (and asexually dichromatic) lar and agile gibbons in the center. Calls are sexually dimorphic and based on a common pattern throughout.

The tripartite split into *Symphalangus*, *Nomascus*, and *Hylobates* is suggested to have occurred about one million years ago, and *Hylobates* is postulated to have broken up into distinct populations during subsequent changes in sea level and vegetation during the glacial periods in the last part of the Pleistocene; *hoolock* and *klossi* separated off first, then *pileatus*, and finally *moloch*, *muelleri*, *lar*, and *agilis*. Since each has special attributes, the problems of conservation are magnified.

Socioecological studies have revealed that all species are monogamous, territorial, and frugivorous (to varying extents). Group size varies between two and six, usually three or four, and solitary individuals are often observed. Territory size varies between about 10 hectares for *klossi* and 60 hectares for *lar*. Fruit consumption varies from between 40—50% for *sundactylus* and 50—70% for *lar*. Group life is highly cohesive, especially in *syndactylus*, with minimal stress among group members. Intergroup relations are frequent and noisy, especially in the smaller species, but overt aggression is minimized through the elaborate ritualized displays of calling and chasing.

Populations of all species except *moloch* and *concolor* are currently healthy, but those of *pileatus* and *klossi* are rather restricted. All species show a preference for lowland and hill forests, with only the siamang showing a tendency to exist in submontane forest. Lowest densities (two to four individuals/km²) occur in the northern deciduous monsoon forests; densities rise to about five or six in the evergreen rain forests of the Malay Peninsula, and to eight or ten gibbons/km² in Sumatra and Borneo respectively; *klossi* occur at the highest densities of twenty or more/km².

Estimates of present forest area and the areas of national parks and game reserves indicate that less than 4% of gibbon habitats are currently protected. The present levels of clear-felling for timber and farming, and of hunting for food and trade, will lead to an 80—90% destruction of gibbon populations within 15 to 20 years, with the extinction of *moloch*, and possibly of *klossi*, *pileatus*, and *concolor*, and with the survival of *hoolock*, *lar*, *agilis*, *muelleri*, and *syndactylus* in isolated forest relics.

Such a situation would be disastrous, not only for the tropical rain forest and its inhabitants, but also for man. Man's survival depends to no small extent on the persistence of large areas of forest to prevent water and soil from inundating the areas he inhabits. Furthermore, the forest soils, especially in hilly areas, will not sustain farming in the long term; man has hardly begun to realize the forest's full potential as a renewable resource. Hence, the following actions are needed urgently, with special priority in countries containing the more endangered species:

1. The drastic curtailment and eventual cessation of clear-felling
2. The speedy establishment of effectively protected forest reserves in which no disturbance is tolerated, preferably covering at least 20% of the present habitat of each gibbon species
3. The careful management of the remaining forest area, with very light timber extraction sustained in the long term
4. Strong enforcement of export restrictions, including a ban on the pet trade, reducing the movement of gibbons between countries to whatever minimum may be deemed essential by the government and scientific authorities concerned
5. The elimination of hunting with the provision of alternative sources of animal protein
6. The development of research programs to elucidate the faunal–floral relationships within the tropical forest ecosystems, and to determine an acceptable level of human activity in disturbed forests
7. Full education of the people about conservation, and the best way to use natural resources, so as to obtain their active involvement.

VIII. Appendix: A Day in the Life of the Malayan Siamang and Lar Gibbon

Day dawns soon after 0600 hours local time in these latitudes. At first light the siamang group stirs; the adult male, followed by the juvenile, moves in from his sleeping position in the terminal branches of one of the large emergent trees. The sub-adult, perhaps in a different tree, also stirs. The air is soon full of the sound of falling excreta as the group urinates and defaecates in concert; the female and infant eventually join in. If there is a fruiting tree nearby, the female may lead the group to feed immediately; otherwise they will rest further until after sunrise, especially on cloudy days.

After the first feed, which may include several trees and some traveling, the group often moves up into a high emergent tree between 0900 and 1100 hours to rest, scan the forest and, sometimes, to call in concert in response to a neighbouring group or some other stimulus. Although the male may start this calling, it is the female who directs the pattern of the chorus. After about 15 minutes the calling may be taken up elsewhere, and the group rests and grooms for about 30 minutes—sometimes all together in a huddle, sometimes in pairs—before the female continues to lead the group around their small territory.

Infrequently they meet a neighbouring group on the territorial boundary. In this event the female and young hide, and the males (including sub-adult) chase to and fro and sit and stare to each other for 15—60 minutes. The family group then rests before resuming the pattern of brief preiods of travel alternating with long bouts of feeding. In the early afternoon, in the heat of the day, activity decreases and individuals rest and groom for more than an hour. If satiated, or if heavy rain starts, the group will move up into a sleeping tree before 1600 hours (move than two hours before sunset). There may, however, be a final feeding bout before the siamang settle for the night—usually by 1700, sometimes after 1800 hours on very active days. (From Chivers, 1974.)

Compare the above with an abbreviated version of Ellefson's (1974) description of a day in the life of a group of white-handed gibbons in the south of the Malay Peninsula.

The gibbon group stirs about 20 minutes after it starts to get light. The buff infant male is with the buff female, but the black male, buff sub-adult male and black juvenile are in separate trees—all within 40 metres of each other; their coats are still damp from the rain of the previous evening, early morning clouds obscure the sun. They sit and look around, each perched on small branches in a hunched position with their long arms wrapped around their knees. The male shakes, swings down and away from the trunk, hangs, urinates, and defaecates—the others follow suit. They then move 50 metres to the large tree with berries in which they fed on the previous evening—by brachiating, climbing, bipedal running and jumping. On the way the juvenile squeals, intercepts the male and they embrace. The female is the first to reach the tree and start feeding, the sub-adult sits in a nearby tree. When he tries to enter 15 minutes later the male rushes towards him and shakes the branch; the sub-adult squeals and leaves the tree.

Soon the others leave the tree and forage on new leaves in small trees, while the sub-adult eats berries in peace. A neighbouring group starts calling; the adults look in that direction and then carry on feeding. About 15 minutes later, at 0700 hours, the female starts calling, the male joins in, and they move up into an emergent tree. The juvenile follows them and joins in, the sub-adult rests and then carries on foraging. The calling ends after 13 minutes, with the most frantic movements at the climax of each great call by the female. The group forage on towards their territorial boundary from their centrally situated sleeping trees—eating some new leaves here, shoots, berries or fruit there, and an occasional insect.

By 0800 hours they have reached their boundary, the pace slows, the male is more alert. He gives soft calls when he sees the neighbouring group; the two males sit watching each other. During the next 70 minutes this staring alternates with conflict-hooing acrobatic displays and one or two chases to and fro across the boundary. There is no physical contact, no apparent victory or gain. The other members of the group tire of the ceremony long before the males and moved back into their respective territories to rest and forage; the males eventually follow them. Around 1000 hours they feed together in a fruiting tree for 15 minutes; then they move on separately foraging to congregate 45 minutes later in a tree with new buds.

More foraging brings the group to a tree with new leaves, on which they feed. Around noon the female moves to the male and they rest and groom each other, the young play together nearby. After 10 minutes they resume foraging, drinking briefly from a hole in a tree trunk in the hand-dipping manner characteristic of gibbons. Soon after 1300 hours they reach a fruiting tree—all except for the sub-adult feed. The sub-adult enters after 10 minutes, but the male rushes over to him and grabs and kicks him; he struggles free squealing and leaves the tree to forage alone. After 30 minutes the male rests and the female approaches and grooms with him; the juvenile joins them. After only 5 minutes they resume foraging. Soon the sub-adult starts calling, he has lost contact with the group. After a time the adults call briefly and the sub-adult soon comes hurrying up.

Foraging and sporadic feeding continue until after 1500 hours, when the female climbs up into a tall tree with the infant and starts to settle for the night—lying on her back, grooming and playing with the infant. The others forage on for another 30 minutes until they too find sleeping places. They rest and groom themselves, shift position in their trees, drowse and eventually sleep. The sun is still bright and it will be light for another $2\frac{1}{2}$–3 hours. They have travelled $1\frac{1}{2}$ kilometres to a different sleeping place and have been active for nearly 10 hours, foraging for $4\frac{1}{2}$ hours, feeding for $2\frac{3}{4}$ hours, resting for about $\frac{1}{2}$ hour and engaging in social behaviour (territorial, play, grooming or calling) for $1\frac{3}{4}$ hours. They stay in this sleeping position for the next $14\frac{1}{2}$ hours.

Acknowledgments

I am very appreciative of financial support received from the Science Research Council, Commonwealth Foundation, Boise Fund, New York Zoological Society, Royal Anthropological Institute, the Royal Society (Government Grants-in-Aid), and the University of Cambridge during my studies in Malaysia in 1968–1970, 1972, and 1974; and of assistance received there from the Game Department, University of Malaya School of Biological Sciences, and the Institute for Medical Research. of Biological Sciences, and the Institute for Medical Research.
I thank Dr Geoffrey Bourne for the opportunity to contribute to this venture; my wife Sarah for her help in field work, data analysis, and preparation of this manuscript; Priscilla Edwards for her hard work in skillfully portraying the facial characteristics of each gibbon species; Raith Overhill for drawing some of the figures; the Visual Aids Unit, Department of Anatomy, for preparing photographs of all the figures for publication; and John Payne for his careful reading of the manuscript.
Most of all I thank those who have so generously contributed data, photographs,

ideas, and critical comments of earlier drafts of this manuscript: Dr. Walter Angst, Mr. Douglas Brandon-Jones, Dr. Warren Brockelman, Mr. Edy Brotoisworo, Mr. Chuck Darsono, Mr. Paul Gittins, Dr. Colin Groves, Dr. John MacKinnon, Dr. Joe Marshall, Lord Medway, Mr. Kenneth Proud, Mr. Jeremy Raemaekers, Mr. Louis Ratnam, Dr. Herman Rijksen, Dr. Peter Rodman, Mr. Ken Scriven, Mr. K. Sargent, Dr. Rich Tenaza, Dr. Ron Tilson, Dr. K. K. Tiwari, Dr. Tim Whitmore, Dr. Carolyn Wilson, and Dr. Wendell Wilson.

The broad scope of this chapter would not have been possible without their help. While I have tried to blend together the variety of opinions expressed on this complex subject so as to present an united front for gibbon conservation, it does not follow that all those listed above wish to be associated with all aspects of my treatment or final recommendations. I have leaned toward a frank, realistic, and optimistic approach, in the belief that this is the best basis on which to obtain the widest cooperation from the people in establishing reserves and enforcing laws.

References

Andrews, P. and Groves, C. P. (1976). Gibbons and brachiation. *In* "Gibbon and Sia-mang" (D. Rumbaugh, ed.), Vol. 4, pp. 167—218. S. Karger, Basel.

Ashton, P. S. (1972). The Quaternary geomorphological history of western Malesia and lowland forest phytogeography. *In* "The Quaternary Era in Malesia" (P. Ashton and M. Ashton, eds.), pp. 35—49. Dept. of Geography Misc. Ser. no. 13, University of Hull, England.

Basjarudin, H. (1973). Nature conservation and wildlife management in Indonesia. *In* "Coordinated Study of Lowland Forests of Indonesia" (K. Kartawinata and R. Atmawidjaja, eds.), pp. 89—115. BIOTROP and IPB, Bogor, Indonesia.

Baldwin, L. A. and Teleki, G. (1974). Field research on gibbons, siamangs, and orang-utans: an historical, geographical and bibliographical listing. *Primates* **15**, 365—376.

Berkson, G., Ross, B. A., and Jatinandana, S. (1971). The social behaviour of gibbons in relation to a conservation program. *In* "Primate Behavior: Developments in Field Laboratory Research" (L. A. Rosenblum, ed.), pp. 225—255. Academic Press, New York.

Brockelman, W. Y. (1975). Gibbon populations and their conservation in Thailand. *Natur. Hist. Bull. Siam Soc.* **26**, 133—157.

Brockelman, W. Y., Ross, B. A., and Pantuwatana, S. (1973). Social correlates of re-productive success in the gibbon colony on Ko Klet Kaeo, Thailand, *Amer. J. Phys. Anthropol.* **38**, 637—640.

Brockelman, W. Y., Ross, B. A., and Pantuwatana, S. (1974). Social interactions of adult gibbons (*Hylobates lar*) in an experimental colony. *In* "Gibbon and Siamang" (D. Rumbaugh, ed.), Vol. 3, pp. 137—156. S. Karger, Basel.

Carpenter, C. R. (1938). A Survey of wildlife conditions in Atjeh, North Sumatra. *Neth. Comm. Int. Nature Protection*, **12**, 1—33.

Carpenter, C. R. (1940). A field study in Siam of the behavior and social relations of the gibbon (*Hylobates lar*). *Comp. Psychol. Monogr.* **16**, 1–212.

Carter, T. P. (1943). The mammals of the Vernay-Hopwood Chindwin expedition, Northern Burma. *Bull. Amer. Mus. Natur. Hist.* **82**, 99–1.

Chivers, D. J. (1971). The Malayan siamang. *Malay Nat. J.* **24**, 78–86.

Chivers, D. J. (1972). The siamang and the gibbon in the Malay Peninsula. *In* "Gibbon and Siamang" (D. Rumbaugh, ed.), Vol. 1, p. 103–135. S. Karger, Basel.

Chivers, D. J. (1973). An introduction to the socio-ecology of Malayan forest primates. *In* "Comparative Ecology and Behaviour of Primates" (R. P. Michael and J. H. Crook, eds.), pp. 101–146. Academic Press, London.

Chivers, D. J. (1974). The Siamang in Malaya: a field study of a primate in tropical rain forest. *Contrib. Primatol.* **4**, 1–335.

Chivers, D. J. (1976a). Communication within and between family groups of siamang (*Symphalangus syndactylus*). *Anim. Behav.*, **57**, 116–135.

Chivers, D. J. (1976b). The gibbons of Peninsular Malaysia. *Malay. Nat. J.*, (in press).

Chivers, D. J., Raemaekers, J. J., and Aldrich-Blake, F. P. G. (1975). Long-term observations of siamang behaviour. *Folia Primatol.* **23**, 1–49.

Corner, E. J. H. (1952). "Wayside Trees of Malaya". Singapore.

Curtin, S. H. and Chivers, D. J. (1976). Leaf-eating primates of Peninsular Malaysia: the siamang and the dusky leaf-monkey. *In* "Ecology of Arboreal Folivores" (G. G. Montgomery, ed.), in press. Smithsonian Inst. Press, Washington, D.C.

de Terra, H. (1943). Pleistocene geology and early man in Java. *Trans. Amer. Phil. Soc.* **32** [N.S.], 437–464.

Ellefson, J. O. (1968). Territorial behaviour in the common white-handed gibbon, *Hylobates lar*. *In* "Primates: Studies of Adaptation and Variability" (P. Jay, ed.), pp. 180–199. Holt, Rinehart and Winston, New York.

Ellefson, J. O. (1974). A natural history of gibbons in the Malay Peninsula. *In* "Gibbon and Siamang" (D. Rumbaugh, ed.), Vol. 3, pp. 1–136. S. Karger, Basel.

Elliot, D. G. (1913). "A Review of the Primates." American Museum of Natural History, New York.

Fleagle, J. G. (1976). Locomotion and posture of the Malayan siamang and implications for hominoid evolution. *Folia primatol.* **26**, 245–269.

Fooden, J. (1969). Color-phase in gibbons. *Evolution* **23**, 627–644.

Forbes, H. O. (1897). "A Handbook to the Primates", Vol. II. Edward Lloyd Ltd., London.

Gittins, S. P. (1976). Ecology and Behavior of the gibbon *Hylobates agilis*. Ph.D. dissertation, University of Cambridge, England. (In preparation.)

Grand. T. I. (1972). A mechanical interpretation of terminal branch feeding. *J. Mammal.* **53**, 198–201.

Gray, J. E. (1861). List of Mammalia, tortoises and crocodiles collected by M. Mouhot in Camboja. *Proc. Zool. Soc. London*, pp. 135–140.

Groves, C. P. (1967). Geographic variation in the hoolock or white-browed gibbon (*Hylobates hoolock* HARLAN 1834). *Folia Primatol.* **7**, 276–283.

Groves, C. P. (1970). Taxonomic and individual variation in gibbons. *Symp. Zool. Soc. London* **26**, 127–134.

Groves, C. P. (1971). Geographic and individual variation in Bornean gibbons, with remarks on the systematics of the subgenus *Hylobates*. *Folia Primatol.* **14**, 139–153.

Groves, C. P. (1972). Systematics and phylogeny of gibbons. *In* "Gibbon and Siamang" (D. Rumbaugh, ed.), Vol. 1, pp. 1–89. S. Karger, Basel.

Haile, N. S. (1971). Quaternary shorelines in West Malaysia and adjacent parts of the Sunda Shelf. *Quaternaria* **15**, 333–343.

Harlan, R. (1826). Description of an hermaphrodite orangutan lately living in Philadelphia. *J. Acad. Natl. Sci. (Philadelphia)* **5**, 229–236.

Harlan, R. (1934). Description of a species of orang from the Northeastern province of British India, lately the Kingdom of Assam. *Trans. Amer. Philos. Soc. Philad.* (NS) **4**, 52–59.

Hooijer, D. A. (1960). Quaternary gibbons from the Malay Archipelago. *Zool. Verh. (Leiden)* **46**, 1–41.

Kawabe, M. (1970). A preliminary study of the wild siamang gibbon, *Hylobates syndactylus*, at Fraser's Hill, Malaysia. *Primates* **11**, 285–291.

Kawamura, S. (1958). The preliminary survey of the white-handed gibbon in Thailand. **1**, 157–158.

Kawamura, S. (1961). A pilot study on the social life of white-handed gibbons in northwestern Thailand. *Nature Life S.E. Asia* **1**, 159–169.

Kitahara-Frisch, J. (1971). Evolution of the siamang (*Symphalangus syndactylus*) in Southeast Asia during the Pleistocene. *Proc. 3rd. Int. Congr. Primatol. Zürich, 1970,* **1**, 67–73.

Kloss, C. B. (1929). Some remarks on the gibbons, with the description of a new subspecies. *Proc. Zool. Soc. London,* pp. 113–127.

Koyama, N. (1971). Observations on mating behaviour of wild siamang gibbons at Fraser's hill, Malaysia. *Primates* **12**, 183–189.

McCann, C. (1933). Notes on the colouration and habits of the white-browed gibbon or hoolock, *J. Bombay Natur. Hist. Soc.* **36**, 395–405.

MacKinnon, J. R. (1974). The behaviour and ecology of wild orang-utans (*Pongo pygmaeus*). *Anim. Behav.* **22**, 3–74.

McKinnon, J. R. (1976). A comparative ecology of the Asian apes. *Primates,* **18**, 4.

McClure, H. E. (1964). Some observations of primates in climax dipterocarp forest near Kuala Lumpur, Malaya. *Primates* **5**, 39–58.

Marler, P. and Tenaza, R. R. (1976). Signalling behavior of wild apes with special reference to vocalizations. *In* "How Animals Communicate" (T. A. Sebeok, ed.), in press. Indiana University Press, Bloomington, Indiana.

Marshall, J. T. and Marshall, E. R. (1975). Letter to the Editor. *Mammal. Chrom. Newsl.* **16**, 1, 30–33.

Marshall, J. T. and Marshall, E. R. (1976). Gibbons and their territorial songs. *Science* **193**, 235–237.

Marshall, J. T., Ross, B. A., and Chantharojvong, S. (1972). The species of gibbon in Thailand. *J. Mammal.* **53**, 479–486.

Martin, W. C. L. (1841). "Natural History of Man and Monkeys. A General Introduction to the Natural History of Mammiferous Animals with a Particular View of the Physical History of Man and the More Closely Allied Genera of the Order of Quadrumana or Monkeys," pp. 413–446. Wright, London.

Mayr, E. (1963). "Animal Species and their Evolution." Harvard Univ. Press, Cambridge, Massachusetts.
Medway, Lord (1970). The monkeys of Sundaland. In "Old World Monkeys" (J. R. and P. H. Napier, eds.), pp. 513–553. Academic, Press, London.
Medway, Lord (1972a). The Quaternary mammals of Malesia: a review. In "The Quaternary Era in Malesia" (P. Ashton and M. Ashton, eds.), pp. 63–83. Dept. of Geography Misc. Ser. no. 13. University of Hull, England.
Medway, Lord (1972b). The Gunong Benom expedition 1967, VI. The distribution and altitudinal zonation of birds and mammals on Gunong Benom. Bull. Brit. Mus. Natur. Hist. 22, 103–151.
Miller, C. (1778). Extract from several letters . . . giving some account of the interior part of Sumatra. Phil. Trans. 68, 161–179.
Miller, G. S. (1903). Seventy new Malayan mammals. Smithson. Misc. Collect. 45, 1–73.
Ng, F. S. P. (1974). Blueprint for conservation in Peninsular Malaysia. Malay. Nat. J. 27, 1–16.
Poore, M. E. D. (1968). Studies in Malaysian rain forest. I. The forest on Triassic sediments in Jengka Forest Reserve. J. Ecol. 56, 143–196.
Raemaekers (1976). Synecology of Malaysian apes. Ph.D. dissertation, University of Cambridge, England. (In preparation.)
Raffles, T. S. (1821). Descriptive catalogue of a zoological collection made on account of the Honourable East India Company in the island of Sumatra and its vicinity. Linn. Trans. 13, 239–274.
Rijksen, H. D. (1976). The social behavior of the Sumatran orangutan. In preparation.
Rodman, P. S. (1973). Synecology of Bornean primates: I. A test for interspecific interactions in spatial distribution of five species. Amer. J. Phys. Anthropol. 38, 655–659.
Rodman, P. S. (1976). Diets, densities and distributions of Bornean primates. In "Ecology of Arboreal Folivores" (G. G. Montgomery, ed.), in press. Smithsonian Inst. Press, Washington, D.C.
Romero-Herrera, A. E., Lehmann, H., Joysey, K. A., and Friday, A. E. (1973). Molecular evolution of myoglobin and the fossil record. Nature (London) 246, 389–395.
Schultz, A. H. (1933). Observations on the growth, classification and evolutionary specialisations of gibbons and siamangs. Hum. Biol. 5, 212–255; 385–428.
Schultz, A. H. (1973). The skeleton of the Hylobatidae and other observations on their morphology. In "Gibbon and Siamang" (D. Rumbaugh, ed.), Vol. 2, pp. 1–54. S. Karger, Basel.
Simonetta, A. (1957). Catalogo e simonimia annotata degli ominoidi fossili ed attuali (1758–1955). Atti Soc. Tosc. Sci. Nat. (Pisa) B64, 53–112.
Simons, E. L. and Pleagle, J. G. (1973). The history of extinct gibbonlike primates. In "Gibbons and Siamang" (D. Rumbaugh, ed.), Vol. 2, pp. 121–148. S. Karger, Basel.
Tenaza, R. R. (1975a). Territory and monogamy among Kloss' gibbons (Hylobates klossi) in Siberut Island, Indonesia. Folia Primatol. 24, 60–80.
Tenaza, R. R. (1975b). Functions and taxonomic implications of singing among Kloss' gibbons (Hylobates klossi) in Mentawai Islands. Amer. J. Phys. Anthropol. 42, 334.
Tenaza, R. R. (1976). Songs, choruses and countersinging of Kloss' gibbons (Hylobates klossi) in Siberut Island, Indonesia. Z. Tierpsychol., in press.

Thomas, O. (1908). On mammals from the Malay Peninsula and islands. *Ann. Mag. Natur. Hist.* **2**, 301—306.

Tilson, R. (1976). Human evoked predator alarm calls of Kloss' gibbon. In preparation.

van Gulik, R. H. (1967). "The Gibbon in China. An Essay in Chinese Animal Lore." E. J. Brill, Leiden.

van Heekeren, H. R. (1972). "The Stone Age of Indonesia." Martinus Nijhoff, The Hague.

Verstappen, H. T. (1975). On palaeoclimates and land form development in Malesia. *In* "Modern Quaternary Research in South-east Asia" (G-J. Bartstra and W. A. Casparie, eds.), pp. 3—36. Balkema, Rotterdam.

Whitmore, T. C. (1975). "Tropical Rain Forests of the Far East." Clarendon, Oxford.

Wilson, C. C. and Wilson, W. L. (1975). The influence of selective logging on primates and some other animals in East Kalimantan. *Folia Primatol.* **23**, 245—274.

Wilson, W. L. and Wilson, C. C. (1976a). Behavioural and morphological variation among primate populations in Sumatra. *Yearb. Phys. Anthropol.*, in press.

Wilson, W. L. and Wilson, C. C. (1976b). The primates of Sumatra. An introduction to their distribution, density and socioecology. In preparation.

Wyatt-Smith, J. (1953). Malayan forest types. *Malay. Nat. J.* **7**, 45—55.

17

The Conservation of Eastern Gorillas

ALAN G. GOODALL and COLIN P. GROVES

I. Distribution and Taxonomy

A. Distribution

In modern times—at the beginning of the present century, if not today—the gorilla has been found in the following discrete areas (see Fig. 1):

1. A large area extending from the Sanaga river (Cameroon) south to the mouth of the lower Zaïre river, and east nearly to the Oubangui river, wherever

599

Fig. 1. Distribution of eastern gorilla populations. See text for a description of each numbered area. (n.b. These data are not up to date and some populations may now be extinct.)

there is forest. The most northwesterly locality recorded is Edea (3.47°N, 10.10°E); the most northeasterly is Barundu, 22 miles northeast of Nola (3.40°N, 16.15°E); the most northerly is Touki near Bertoua (4.35°N, 13.30°E); the most southerly is Tshela in Mayombe district (5.05°S, 12.50°E). Politically, these localities fall within the borders of Cameroon, Equatorial Guinea, Gabon, Central African Republic, Congo, Angola (Cabinda enclave), and Zaire.

2. A small area on the upper Cross River, from Tinto (5.33°N, 9.35°E) northwest to the fringes of the Obudu plateau (6.38°N, 9.06°E); straddling the border between Cameroon and Nigeria.

3. An isolated patch of forest near Djabbir, Bondo district, Zaire (3.55°N, 23.53°E).

4. Lowlands of eastern Zaire, from the Ulindi river north approximately to the equator. The most northwesterly locality is Lubutu (0.40°S, 26.55°E); the most northerly is Kilimanensa (0.35°S, 28.30°E); the most easterly is Pinga (1.00°S, 28.40°E); the most southeasterly is Bibugwa (2.30°S, 28.20°E); the most southerly is Shabunda (2.45°S, 27.34°E).

5. The Itombwe mountains, Zaire, between Mwenga (3.00°S, 28.28°E) and Fizi (4.18°S, 28.56°E).

6. The mountains centered on Mt. Kahuzi, Zaire (2.10–2.25°S, 28.40–28.50°E).

7. The massif of Mt. Tshiaberimu, Zaire (0.05°N to 0.40°S, 29.00–29.20°E).

8. The six extinct volcanoes of the Virunga range, straddling the Zaire–Rwanda–Uganda border (1.20–1.30°S, 29.24–29.42°E).

9. The Kayonza or Impenetrable forest, Uganda (1.00°S, 29.40°E).

Since no gorilla has been reported from area (3) since 1908, it seems likely that this population is extinct. The status of gorillas in area (2) is unknown, but they still existed less than 20 years ago (March, 1957). Within area (1), gorillas no longer occur in the Mayombe district of Zaire, just north of the mouth of the Zaïre river (Verschuren, 1975). Elsewhere, gorillas would seem to occur over most or all of their original range, though often in reduced numbers.

B.   Taxonomy

The position adopted here is that gorilla belongs to the genus *Pan*. Tuttle (1967), among others, has offered strong anatomical grounds for considering the gorilla and chimpanzee congeneric; other features in common between the two African apes, differentiating them from the Orangutan and man alike, are the nature and hue of the hair pigmentation, the presence of a white pygal tuft in the young, the drastic reduction of the musculature of the thumb, the terrestrial adaptations of the foot (nearer to *Homo* than to *Pongo*), the form of the skull, including the lateral buttressing of the orbits, the low position of the braincase, and the common existence of a frontomaxillary suture in the orbit, the relatively homomorphic incisors, the semisectorial nature of P_3 with only a small metaconid, the form of the wrist, features of blood protein chemistry, and of the chromosomes, and perhaps ethological characteristics such as the greater sociability than *Pongo*. This is not to deny that there are great differences between the gorilla and the chimpanzee; morphologically, many of them can be regarded as consequences of the gorilla's large size and stocky build, specializations perhaps for a montane environment (Groves, 1971). However, there are also striking differences in temperament, vocalizations, and social organization, and it is these rather than the morphological differences which might argue for retention of the genus *Gorilla*. Unfortunately, there is no fossil material bearing on the question of the date of separation of the two species, in spite of an attempt by Pilbeam (1969) to allocate two species of Lower Miocene hominoids to the respective lineages.

From the time of its discovery, up until 1927, numerous species and subspecies of gorillas were described, over half of them by Professor Paul Matschie of Berlin. In 1929, Coolidge reduced them all to a single species with two subspecies: *Gorilla gorilla gorilla* and *G. g. beringei*. This classification remained sacrosanct until Vogel (1961) showed, in a study of the mandible, that eastern gorillas (the *beringei* of Coolidge) could be sharply divided into two geographically distinct types. Raising *beringei* to the rank of a full species, he recognized within it two subspecies: *G. b. beringei* from the Virunga Volcanoes and *G. b. graueri* from the eastern lowlands and nearby mountains [areas (4), (5), and (7) mentioned previously]. In view of the clear-cut nature of Vogel's conclusions, and of a paper by Haddow and Ross (1951) exposing certain shortcomings of the methodology exployed by Coolidge, it is most unfortunate that Schaller, in his influential book, deliberately chose to be conservative and perpetuate the old two-subspecies fallacy.

Groves's conclusions on gorilla taxonomy were published in preliminary form in 1967 and more completely in 1970. Briefly, he found that Coolidge was correct in assigning all gorillas to a single species, but Vogel was fully justified in recognizing *graueri* as a distinct subspecies [in the 1967 paper, Groves called this *Gorilla gorilla manyema*, dating from Rothschild in 1908; but Corbet (1967) considered this latter name to be a lapsus for a name *mayema* which had previously been applied to a western gorilla and so is a synonym of subspecies *gorilla*]. The three subspecies are characterized below, using the more recent allocation of the gorilla to the genus *Pan*.

1. *Pan gorilla gorilla.*   Western, lowland or coast gorilla. Gorillas from areas (1), (2), and (3) belong here. This subspecies has a broad face, small jaws and teeth, short palate, a single mental foramen placed under the premolars; vertebral border of scapula straight; humerus long; hallux short, divergent; fur short, gray with lighter "saddle" in adult males, extending to thighs and not sharply defined from body color; a "lip" at upper end of nasal septum.

2. *Pan gorilla graueri.*   Eastern lowland gorilla. Gorillas from areas (4), (5), and (7) are included here. This subspecies has a narrow face, larger jaws and teeth, and longer palate; mental foramen often multiple, and further forward; vertebral border of scapula straight; humerus long; hallux short, divergent; fur fairly short, black with whitish "saddle" in adult males restricted to back and sharply defined; no "lip" on nasal septum.

3. *Pan gorilla beringei.*   Gorillas from area (8) are the typical representative of this race. This subspecies has a low broad face; very big jaws and teeth and very long palate; mental foramen multiple, often under canine; vertebral border of scapula sinuous; humerus shortened; hallux long, parallel to other toes; fur

long and silky, black, with whitish sharply defined saddle restricted to back in adult males; no "lip" on nasal septum.

It is to be noted that, except in the shape of the face, *P. g. graueri* is intermediate between the other two races, such that no division into two species is possible.

Recently, Cousins (1974) has amplified these descriptions somewhat, pointing out that *P. g. gorilla* can be distinguished additionally by (1) the more pronounced, less hair-covered supraorbital ridge and (2) further differences in the nose, with side-splayed and padded nostrils. By contrast the two eastern races have narrower nostrils and a supraorbital ridge padded with hair. Cousins has drawn typical noses of all three races; that of *graueri* is illustrated as being even narrower than that of *beringei*, and the latter has nostrils that are more raised above the level of both the bridge and the upper lip; in the text, however, he describes them as "similar"—notwithstanding, it still might be possible to differentiate them on the average, though not in every single case. None of the photos or living examples of *beringei* seen has quite the narrow, flat nose of many *graueri*.

Cousins also points out that most of the "Mountain" gorillas exhibited recently and currently in zoos are in fact *graueri*. At present only one zoo, Cologne, possesses genuine *beringei*.

Some eastern populations are less easy to allocate to *beringei* and *graueri*. Although mostly similar to gorillas from the lowlands [area (4)] and Itombwe [area (5)], gorillas from area (7), Mt. Tshiaberimu, have a *beringei*-like foot structure. More difficult is the question of the gorillas from area (6), Mt. Kahuzi. Groves (1967, 1970) assigned them to *beringei* while noting the existence of some divergent features. Recently, however, Casimir (1975a), on the basis of further skull material, proposed instead to allocate this population to *graueri*. It is to be noted, in addition, that photographs of Kahuzi gorillas published in several popular magazines (see, for example, Grzimek, 1974) depict animals with relatively short fur (especially around the face, where the fur is long and shaggy in *beringei*) and *graueri*-like noses. However, in postcranial features—scapula, humerus, and hallux—Kahuzi gorillas are invariably like *beringei*. This sort of intermediacy is, of course, to be expected in subspecies whose ranges abut. A possible course of action might be to create a new subspecies for Kahuzi gorillas. It is a little ironical that this population, the very one which may just be deserving of such a status, is the one which has never been awarded a separate name! For the moment, it is better to leave it as *P. g. beringei*, "with a touch of *graueri*."

Nothing can be said about the Kayonza forest gorillas except that the few skulls available seem to resemble *graueri*; no postcranial material is known. In the matter of absolute size, much of what has been written is mythical. Cousins (1972) quotes and dissects a number of published claims for gorilla heights and weights, showing how frequently estimates or extravagantly measured records creep into the literature and become established. He also shows that captive gorillas become amazingly obese and quite untypical of gorillas as a whole. The heaviest authenticated wild male gorilla would seem to be one shot by Raven on Mt. Kahuzi, weighing 460 lbs (210 kg), whereas one zoo specimen was 776 lbs (350 kg)! On the basis of very few specimens, Groves (1970) calculated average heights and weights for adult males of the three races as shown in the following tabulation.

	Height (cm)	Weight (lb)	[kg]	Records
gorilla	166.6	307.3	[140]	37
graueri	175	360.3	[165]	4
beringei	172.5	342.9	[155]	6

Further measurements quoted by Cousins (1972) indicate that Kahuzi gorillas can be very large, like *graueri*, but that the average for *beringei* may be too high. He thinks the latter probably is very heavy, as indicated, but would average a little shorter in stature than *gorilla*.

Mountain gorillas have more fundamentally asymmetrical skulls than western gorillas, and invariably have the left side longer than the right whereas in western gorillas the right side is commonly longer (Groves and Humphrey, 1973). Skulls of *graueri* are more like western gorillas in this respect. It is suggested that, in contrast to the orangutan, some cerebral dominance may exist in the gorilla, showing up in asymmetrical behavior which, because of the modeling effect of the jaw musculature on the skull, can be detected skeletally in the case of chewing asymmetry.

Breakdown of the alveolar margins of the cheek teeth is another common feature of mountain gorilla skulls. This often advances to the extent that the whole root is exposed and the tooth is lost. Colyer (1936) first drew attention to this, pointing out that it is a frequent (in Groves's experience, almost universal) occurrence in mountain gorillas, occurs in low frequency in western gorillas, and not at all in eastern lowland gorillas. C. R. S. Pitman (personal communication) pointed out that it is never seen either in skulls from the Kayonza forest. Colyer ascribed it to the presence of bamboo in the diet, but this is not a satisfactory explanation since its occurrence does not, in fact, correlate with the availability of bamboo.

Apart from geographical differences, it would seem that family likeness within a gorilla group can be strong enough to affect morphology. In 1971, by courtesy of the National Geographic Society, Groves spent 2 weeks in Camp Visoke (now the Karisoke Research Station), the study center for mountain gorillas in Rwands, organized and run by Ms. Dian Fossey. On this occasion he was able to study a series of thirteen gorilla skulls, of all ages and sexes, which had been picked up in the study area. Five of these skulls (all accompanied by partially complete skeletons) are from a single troop, killed by poachers at Tsundura on the slopes of Mt. Karisimbi, and were collected by Goodall in October 1970. By comparison with other skulls, and with specimens previously studied elsewhere, the Tsundura specimens show some unusual features: the males (but not the females) are small in size, and one of them a peculiarly short palate; a cleft in the mandibular condyle, seen in over half of *beringei* jaws, occurs on only one side of one jaw in the series; only two of the five have the usual flaring jaw angles; two of them have the mental foramen placed unusually far back; three of them have a "rocking" jaw which Vogel (1961) found to be very unusual in *beringei*. Even more remarkable is the finding that the Tsundura troop do not have the usual short humeri of mountain gorillas. The limb indexes are shown in the following tabulation:

Index	Tsundura	No. of skulls	Other *beringei*
Brachial	78.0—82.3	5	79.2—88.5
Intermembral	117.0—112.6	4	110.2—117.5
Femur—humerus	114.6—123.2	4	110.8—117.9

The sample consists of two adult males, two adult females, and a juvenile female. The similarity to each other of the whole series, including adults, suggests a high level of inbreeding. Whether it applies to gorillas as a whole, or rather implies an increasing isolation for mountain gorilla troops in recent times, cannot be decided at the moment.

II. Ecology

A. Geographical Regions and Biotopes

Gorillas live in climatically, and hence floristically, very diverse regions. The nine areas are briefly described below.

1. The main West African area is equatorial rain forest, extending in altitude from sea level to about 2500 ft on the Cameroon plateau. The average daily maximum temperatures for the hottest month vary from 90°F on the coast at Kribi, to 96°F inland at Batouri and Moloundou, and 99°F at Ouesso in the Sangha river valley; average daily minima for the coldest month do not fall below 68°F at Kribi or 63°F at Ouesso, but fall to 56°F in the Batouri district. Rainfall is relatively low (59.7 inches) at Mayombe but rises north along the coast to 122.3 at Kribi, and falls again to the low 60's inland; probably more important is humidity, which at 0600 hours in the wettest months is universally

94—98%, but at noonday in the driest months can be down to 45% at Mayombe and 51% at Batouri; at Krìbi it never drops below 77%.

2. The Cross River zone lies at an altitude of between 2000 and 5000 feet and it would seem (Sanderson, 1935) that gorillas are most commonly found in the montane forest belt. Temperatures here seem comparable to those in the Cameroon plateau region but, in spite of a high annual rainfall of 134.4 inches (at Mamfe), humidity is low, never exceeding 89% at 0600 in the wettest month and dropping to 55% at 1200 hours in the driest month.

3. There is no data at all on the forest near Bondo, but it cannot be too different from area (1).

4. The eastern Zaire lowlands—the "Utu region" of Schaller (1963)—lie at a mean altitude of some 2,000 feet, comparable to the major part of the western lowlands. The forest type is tropical rain forest. The temperature, rainfall, and humidity figures very closely match those for an inland locality such as Batouri except that the average minimum temperature in the coldest month is only 63°F, adding up to a narrower temperature range than most in area (1).

5. The Itombwe massif rises in places to 10,000 feet, but a "mean altitude" at which gorillas are common would be some 2200—2600 m (7200—8500 ft) where, for example, the type series of *graueri* were collected. Beyond this, and the information (Schaller, 1963) that the vegetation is montane forest with bamboo above 8000 feet, we have no data.

6. Mt. Kahuzi, and nearby Mt. Biega, again montane forest, has been well described by Goodall (1974) and Casimir (1975b). Gorillas inhabit mainly the zone between 2100 and 2400 m (6900—7900 ft). The maximum temperature does not vary markedly with the month, averaging 17.9°C (64°F), nor does the minimum at a nearly constant 10.4°C (52°F). However, rainfall varies markedly with an annual average of about 1800 mm (72 inches), relative humidity varying from about 85% at 0600 in the wettest month to only 50% at 1500 hours (not at noon!) in the driest. Therefore, this region never has temperatures as high as in the previous areas listed, although in the coldest months it is not much colder than they are. It has a rainfall comparable to that of the Cameroon plateau, and has a narrower humidity range, lower in the wettest month than in previous areas listed, but comparable, or not much lower, in the driest.

7. Gorillas seem to inhabit much the same altitudes on Mt. Tshiaberimu as on Mt. Kahuzi; Schaller's camp was at 8600 ft and Lubero, another gorilla locality, is at 7380 ft. The average daily maximum temperature are around 68°F, with a minimum of about 52°F; rainfall is about 70 inches per annum and an average humidity figure is 71% (from Schaller's figures and from a geographical handbook). There would seem, therefore, to be little or no difference between this area, climatically, and Mt. Kahuzi.

8. In the Virunga Volcanoes, gorillas occur from about 8000 ft to the moun-

tain peaks at over 13,000 ft, but it is clear that their activities center around 9500 to 11,000 feet. Schaller (1963) emphasizes the striking difference between the three eastern volcanoes (Mts. Muhavura, Gahinga, and Sabinio) and the three western ones (Mts. Mikeno, Visoke, and Karisimbi). In the western group, the bamboo zone extends down the mountainside as far as 7300 feet and up to 9200 feet; in the eastern group, the bamboo zone extends only down to 8500 feet, but up to 10,000 feet. Above this, and up to 11,000 feet (where the alpine moorland zone begins) in the eastern group is a forest of tree-heath and *Hypericum*, but in the western group this is reduced to a zone between 11,000 and 11,400 feet, and between 9,200 and 11,000 feet is the remarkable *Hagenia* forest, much of which, together with open herb-covered slopes, is prime *beringei* habitat. Spinage (1972) assigns the *Hagenia* and bamboo zones to the temperate climatic zone; here the average maximum temperature is about 60°F, with a minimum of only 40°F (freezing at night can occur at ground level). The average rainfall is only 72 inches, while the relative humidity is fairly constantly from 67 to 76%. In between the two volcano groups lies a mountane "saddle" at 8000 to 8500 feet, mostly clothed in bamboo, secondary herbs and shrubs; here too, gorillas occur occasionally. In general, then, the Virunga habitat offers the coldest climate of any gorilla habitat; not the wettest nor the most humid by any means, but perhaps the most constantly damp, the most misty and sunless, the most marshy, and also the most open.

It is because of the openness of the tree canopy, even in the *Hagenia* forest itself, that ground vegetation flourishes, often growing up to 6–8 feet high and providing abundant forage for gorillas. This is especially so on the more open slopes of the volcanoes where *Peucedanum*, and *Carduus* grow in abundance, and on the forest edges, especially in the saddle area, where nettles and *Galium* grow together in profusion. Temperate species of *Galium aparinae* and *Urtica dioica* are commonly thus associated in nitrophilous woodland edge communities (Passarge, 1967).

B. Feeding Behavior

1. Dietetic Diversity

Schaller (1963) listed the food items he recorded in his seven study areas, estimating in some cases the relative frequency of usage. Because of the striking vegetational differences, the eastern (Kisoro) and the western (Kabara) regions of the Virunga Volcanoes had to be treated separately. Lists of gorilla food plants from two areas in the Mt. Kahuzi region, Nyakalonge, and Mbayo-Tshibinda, have recently been supplied by Goodall (1974) and for the Bukulumisa—Tshibinda area by Casimir (1975b). Goodall's list of 104 species (which he believes is still incomplete) contains more species than all of Schaller's lists combined. It also contains nearly twice as many as the 56 recorded by Casimir in the same region. In part, this can be attributed to Goodall's data on several groups in the Nyakalonge area (mainly primary forest regions) but is undoubtedly due to a far greater reliance on visual observations to supplement the restricted, and often misleading, data obtained from trail signs alone. For example, the Kahuzi gorillas often discarded the leaves of the woody vine *Urera hypselodendron* and ate the bark. However, by direct observations, Goodall was able to see that the leaves are also eaten in large quantities, though not from every vine—hence the misleading trail signs. Similarly, *Galium spurium* was also identified as being eaten though trail signs often tended to indicate it was not. Neither of these food items (nor several others—see Goodall, 1977c), appear on Casimir's lists. Other items which Casimir positively identified (from trail signs alone) as being eaten by gorillas are doubtful, e.g. *Pteridium* spp. (frequently eaten by *Cercopithecus mitis*).

Therefore, in view of the little time spent by Schaller in some of his study areas (particularly, Kayonza forest, Mt. Tshiaberimu, and the Utu lowlands and his limited opportunity for visual observation on feeding behavior), it would hardly be surprising if long-term studies in these areas revealed many more food plants of gorillas. Dr. J. Sabater-Pi kindly supplied, prior to publication, a list of food plants for the western gorillas in the Mt. Alen region (Equatorial Guinea). This list contains 91 items and so is almost as varied as the Mt. Kahuzi list.

In Table I we can see dietary differences between different gorilla populations, as calculated from lists supplied by the above authors. Despite the shortcomings of the lists for some areas, the following general statements on gorilla feeding habits seem to be valid:

1. When considering all gorilla populations combined, a very wide variety of different food items have already been recorded (over 200).

2. The range of variety, or dietetic diversity, varies between regions.

3. The gorillas of the Virunga Volcanoes, though their actual diet varies markedly from the eastern to the western cluster (Schaller, 1963) (Goodall, 1971), all differ from the other gorillas in the following ways: their range of food items is quite narrow; half their dietary items are herbaceous; they eat flowers, and more root and bulb items than is seen in other regions; and they eat far more stems and far fewer leaves.

4. The four "intermediate montane" groups (Kayonza, Tshiaberimu, Kahuzi, and Itombwe), although the lists are not all complete, nonetheless do seem to have dietary tendencies in common in which they differ from both Virunga and true lowland gorillas; they eat many vines and comparatively fewer herbs; they consume much bark; and they eat more leaves (especially in Kahuzi) than in most other regions.

Due to the relatively small number of items listed for all groups, except

TABLE I

Composition of Diet in some Populations of Eastern Gorillas

	Visoke	Kabara	Kisoro	Kayonza	Tshiaberimu	Kahuzi	Itombwe	Utu	Alen
Total plants	26	29	27	27	19	118	25	17	93
Percentage consumed of									
Herbs	46	38	44	15	26	23	20	29	19
Vines	8	17	18	37	11	30	24	0	19
Shrubs	8	10	11	14	16	15	8	0	15
Trees	23	24	15	14	26	19	24	41	44
Grass/sedge	8	7	11	4	11	5	8	0	1
Ferns	4	3	0	11	0	4	12	12	0
Epipyhtes	4	0	0	0	0	4	0	0	1
Percentage Plant parts consumed									
Leaves	18	16	29	28	18	54	13	20	17
Bark/cortex	11	18	13	35	36	28	50	5	12
Stem	27	22	29	7	27	2	13	20	2
Pith/medulla	18	10	13	17	5	6	17	25	17
Root/bulb	4	14	8	0	0	1	0	10	3
Flowers	8	4	0	0	0	1	0	0	1
Fruit	4	6	8	10	10	4	4	15	38
Shoot/bud	2	2	3	3	5	1	4	5	11
Whole plant	7	6	0	0	0	3	0	0	?

Kahuzi, no other differences between the various areas can safely be assumed, although the apparent heavy utilization of stems by the Tshiaberimu gorillas— a similarity to Virunga—may be real and could relate to their flaring jaw angles, in which they resemble Virunga gorillas.

5. Lowland gorillas, both eastern and western, seem to have some common dietary preferences: much of their food material is derived from trees; fruit is prominent in their diet, especially in Mt. Alen, and to a lesser extent shoots and bulbs; they eat little bark, especially in Mt. Alen.

However, the meager data for Utu (i.e., the eastern lowlands) does not permit more than this sketchy generalization.

Differences in food items between the various gorilla populations cannot be entirely explained by the presence or absence of the particular items in the different regions. Schaller has already pointed out that while various plants were seen to be eaten by gorillas in one region, several items, although present, were not eaten by the gorillas in other regions. He suggested that these were "cultural differences" in feeding habits. Goodall (1974) found similar differences between the Kahuzi gorillas and other populations. However, he also recorded differential importance of some food items, e.g., *Galium*, a major food item in Virunga, is only a minor food item in the diet of the Kahuzi gorillas. Similar differences were seen with *Vernonia* shrubs and various epiphytic ferns. Evidence was also found to indicate that such differences in feeding habits existed between neighboring groups living in adjoining areas of the Kahuzi region and even extended to the level of individual differences within a single gorilla group.

2. Selectivity of Food Items

Attempts to interpret the selection of food items in terms of energy and protein values (Goodall, 1974), and protein quality in terms of various amino acids, together with various mineral contents (Casimir, 1975b), showed no definite correlations with any of these factors in respect to plant parts eaten as opposed to those rejected. Goodall showed that the Kahuzi gorillas could obtain ample supplies of energy and protein from their strictly vegetarian diet. This diet could also supply all of their water requirements throughout the year (although several animals were seen to lick "dew" from leaves, and trail signs on several days indicated that animals had been drinking from a small stream). However, he found that many items with low gross energy and protein values (especially barks) were also eaten in large quantities. Since the digestibility of such high fiber content foods is low in the nonruminant gorilla, he came to the conclusion that they were actually eaten for their beneficial physical action in the gorilla's gut. This seems to be confirmed by observations of the dung of the Kahuzi gorillas. When they are feeding on a variety of food items their dung consists of a number of discrete lobes which are usually attached to each other, rather like "popper beads," in a long chain. It is the long fibers of barks, such as *Urera hypselodendron*, which join up the lobes. When the gorillas are feeding only on bamboo shoots and the bases of *Cyperus* leaves their dung is a shapeless, watery mass.

Other factors which could possibly influence selectivity were noted by Goodall also found that most of the Kahuzi foods tasted extremely bitter but he shape, size, and texture, requiring varying degrees of preparation necessary for their ingestion. Thus choice was not always confined to the most tender, the most succulent, or the easiest item to ingest. Nothing is known of the influence of such factors as taste or odor of plant parts. Schaller (1963) pointed out that most of the gorilla foods have a bitter taste (to man). He saw no incident in which a gorilla raised an object to the nose and definitely seemed to smell it. Goodall also found that most of the Kahuzi foods tasted extremely bitter but he suggested that the sense of smell may be much more important to gorillas than Schaller indicates—both in its social functions and possibly in the detection of young bamboo shoots below the grounds. He saw many incidences of gorillas smelling strange objects, especially clothing or other items belonging to him— even his boots and his head! Often the gorillas first touched strange objects with their fingers and then sniffed these.

The role of "plant secondary compounds" in influencing the selectivity of food items by herbivores has recently been emphasized by Freeland and Janzen (1974). They point out that the production of such compounds by the plants is an important process which has evolved as part of a "defense" mechanism against being eaten by herbivores—a point often overlooked by zoologists. According to these authors, a generalized herbivore, in order to avoid poisoning or physiological damage by plant secondary compounds, should

1. Treat new foods with extreme caution.
2. Be able to learn quickly to eat or reject particular foods and need only to ingest minute quantities to do so.
3. Have the capacity to seek out plants containing highly specific classes of nutrients.
4. Have to ingest a number of different staple foods over a short period of time, and simultaneously indulge in a continuous food sampling program
5. Preferentially feed on the foods with which they are familiar and to continue to feed on them for as long as is possible.
6. Prefer to feed on foods that contain only minor amounts of toxic plant secondary compounds
7. Have searching strategies and a body size that neither maximizes the number of types of foods that are potentially available nor maximizes the total—but rather compromise between these two functions.

They also point out that the ability to detoxify different plant secondary compounds varies between individuals and is under genetic control.

Thus, in order to try and interpret the differences in diet between different gorilla populations, we have to start from the position of the individual animal (Goodall, 1977c). Under such influences and constraints as local abundance and availability of potential food items (Levins and MacArthur, 1969) and its own ability to detect and/or detoxify various plant secondary compounds (Freeland and Janzen, 1974) each animal chooses its own diet. Many animals have been shown to be able to select a diet of adequate nutritional value and quality (Harriman, 1973). Other secondary factors, such as the beneficial action of long bark fibers, like *Urera*, in the gut may also influence the choice of some food items (Goodall, 1974). Thus the final range of items eaten by the individual can be seen to be a compromise between a multitude of such influential factors.

Even further influences on selectivity are possible within the social nexus of the gorilla group. It has been shown that the choice of individual animals can be influenced by the choice of others (Itani, 1958). Many animals, especially during early stages of ontogeny, can learn which items are food from conspecifics, particularly in social mammals (Kawamura, 1959). Therefore, depending upon its own ability to deal with any plant secondary compounds which may be present, the young herbivore can then incorporate such items into its own diet. One could also expect selection to favor some degree of individual variation in feeding habits because of its role in reducing competition between group members for various limited resources. Thus the "optimal diet" of individuals may be polymorphic.

It is then clear that the food repertoire of a social group is the result of the interaction of the choice of individual members. This will be perpetuated by group tradition yet allowing for the incorporation of "new" food items by individual "initiators."' In these terms it is now easy to explain the cultural differences of Schaller, for the emphasis of each of the component influences— both genetic and environmental—will vary not only from region to region but within regions. Such behavior has obvious advantages to long-lived, social animals inhabiting changing habitats, for it not only enables them to exploit short-term abundances but to take advantage of long-term changes—a beautiful example of evolutionary adaptability on the part of what Ardrey (1961) unfortunately calls a "sorry paradox of incongruities on the road to extinction"!

Thus while further research on all gorilla populations will undoubtedly reveal many more food items, the general pattern is now clearly as follows:

1. A "staple" component of the diet consists of four or five species as exemplified in the Virungas by such species as *Galium simense* (whole plant), *Carduus afromontanus* (leaves, stem, and flowers), *Peucedanum linderii* (peeled stem and root pith), *Urtica massaica* (leaves), *Laportea alatipes* (leaves), and various epiphytic ferns *Polypodium* sp. (whole plant). In the Tshibinda-Kahuzi region the "staples" are *Urera hypselodendron* (leaves and bark), *Basella alba* (leaves),

Taccaza floribunda (leaves and bark), and *Lactuca* (many species) (leaves). In the Mt. Alen region "staples" are *Afromomum* (many sp.) (fruits, shoots, and pith) and *Musanga cecropioides* (fruits, leaves, buds, and shoots).

2. "Specially preferred" food items often with a restricted and localized distribution and requiring extra energy for their collection and/or preparation include: e.g., several semiparasitic species of *Loranthus* (leaves, bark, and flowers) in the Kahuzi region, together with *Galineria coffeioides* (pith), *Coffea* sp. (leaves), and many species of *Ficus* (bark, especially cambium layer). In the Virungas, similar items would be *Vernonia adolphi-frederici* (pith, flowers, and bark), *Lobelia wollastonii* (pith), and possibly a species of bracket fungus (*Ganoderma applanatum*) which Fossey and Harcourt (1977) report as being the sole food item over which they saw "squabbles."

3. "Seasonal items" are plants or plant parts which are eaten in large amounts, sometimes as staples, when seasonally abundant, e.g., in Kahuzi, *Arundinaria alpina* (shoots), *Myrianthus holstii* (fruits), *Syzygium guinense* (fruits); in the Virunga region, *Arundinaria alpina* (shoots) where present, *Pygeum africanum* (fruits), and *Rubus runsorrensis* (fruits).

4. "Occasional" food items are sometimes chosen. In Virunga, such species include *Hagenia abyssinica* (bark) and *Hypericum lanceolatum* (bark and rotten wood); in Kahuzi a much wider range includes such species as *Piper capense* (leaves, cortex, and stem), *Hagenia abyssinica* (bark and pith), and *Hypoestes* sp. (leaves).

5. "Very rarely" eaten items (or perhaps newly sampled) is the final category. Again many species in Kahuzi include *Brillantaisia* (several spp.) (leaves), *Sericostachys scandens* (leaves), and *Eucalyptus* sp. (bark); in Virunga, *Senecio alticola* and *S. erici-rosenii* (pith and rotten wood) together with such strange items as *Senecio trichoptyrygius* (dead leaves and pith of dead stem), and *Lobelia gibberoa* (pith of dead stem) are chosen.

Such a pattern of feeding behavior clearly illustrates the predictions of Freeland and Janzen mentioned above. Any differences between the gorillas of different regions could have implications for conservation plans, for example, are the gorillas of some regions "better fed" than those in other regions? Are the Virunga gorillas—with their apparently narrow dietetic diversity—any worse, or better, off than the Kahuzi gorillas who have a very wide dietetic diversity?

Before such questions are discussed in relation to the general ecology of particular populations, two further anomalies regarding the diet of free-living gorillas can quickly be elucidated here. First, while data from some areas now indicate that some gorillas ingest some amounts of animal matter (Fossey, 1974; Sabater-Pi, 1966, personal communication), Goodall (1974) have shown that the Kahuzi gorillas, at least, can obtain ample energy and protein from an entirely phytophagous diet. It seems more likely therefore that the small amounts of animal matter ingested are eaten for their Vitamin B_{12} Value and not for their protein content alone. Second, although Goodall saw free-living gorillas in Both Virunga and Kahuzi drinking, his analyses of the water content of many gorilla food items from the Kahuzi region showed that this would generally be sufficient to maintain the gorillas in water balance. However, it would not necessarily preclude any animals from drinking free-standing water, especially in the dry season.

C.   Ranging Behavior

1.   General Foraging Behavior

Although the gorillas of the Virunga region climb cautiously to collect such food items as epiphytic ferns and various fruits and some flowers, they are mainly terrestial feeders. In contrast to this, the gorillas of Goodall's main study group in the Tshibinda—Kahuzi region were seen to feed in every conceivable position from sitting on the ground to hanging in the trees suspended by one foot and one hand, partly upsidedown, and anything up to forty meters above the ground. A marked vertical distribution of feeding habits was usually observed within the group (especially in secondary forest) with the largest silverback feeding on the ground and many of the group feeding at varying heights above him. However, when feeding on the leaves of semiparasitic *Loranthus* sp., or the fruits of *Myrianthus holstii* and *Syzygium guinense*, all animals were seen to climb confidently, rapidly, and expertly. Even the largest silverback, who must have weighed at least 200 kg (440 lb), was frequently seen to climb over 20 m high, even among swaying, hanging vines. A slightly smaller adult male was an excellent climber and was invariably found feeding above ground.

It was quite obvious that the main causes of such differences in climbing behavior between the two gorilla populations were to be found in the differences in vegetational types present in the two regions and thus the distribution of available forage for the gorillas. In the more open vegetation of the Virunga region succulent herbs and vines grow in profusion at or near ground level. The shrub layer is limited or even absent in places and so too are trees from many of the steeper volcanic slopes. Woody vines are small and few in number with very restricted distribution.

In contrast to this, vegetation types of an extremely wide variety of physiognomy were found in the Kahuzi region: open swamps, herbaceous meadows, old tangled secondary regenerating forest with many different layers interconnected by innumerable vines, and, finally, mature primary montane forest with high canopies and sparse ground cover. Thus, the forest world of the Kahuzi gorillas offers a much more three-dimensional habitat and therefore a wider variety of resources for exploitation by generalized quadrupedal climbers.

2. Range Utilization

The overall home range sizes of two gorilla groups in the Kahuzi region, about 30 km², was found to be far larger than any recorded in the Virunga region [Goodall (1974) and data from Schaller (1963) and Fossey (1974)]. Casimir's study group in the Bukulumisa—Tshibinda area of the eastern Kahuzi region also had a similarly large home range (but this included incorrect data—see Goodall, 1977a,c). On the basis of very meager data, Jones and Sabater Pi (1971) quote home range sizes for the western gorillas of two study areas as "averaging from 5.6 to 6.75 km²."

Regarding home range utilization by eastern gorillas Schaller wrote "except for bamboo, gorilla habitats throughout the range show no conspicuous seasonal differences in the abundance of forage . . . . The only generalization regarding movement which can safely be made are that gorillas travel continuously within the boundaries of their home range and that they appear to arrive in a certain section of their home range at irregular intervals." While this appears to be largely true of the central Virunga gorillas (with possible exceptions when *Pygeum africanum* is in fruit), totally different patterns of home range utilization were shown by Goodall's main study group in Tshibinda—Kahuzi.

The study area in this region was a mosaic of eight vegetation types, namely, four primary forest types and four secondary forest types—with many forms of cultivation along the entire eastern boundary (Goodall, 1974).

Over 68% of the main study group's home range area consisted of secondary vegetation types and the remainder of primary. Some of the vegetation types, especially Bamboo and *Cyperus* swamp, were very localized in their distribution about the home range while, in contrast, the open secondary forest types were widespread, especially in the eastern section.

A detailed analysis of the daily routes taken by the group throughout a 7-month period (which included parts of two rainy seasons and the intervening dry season) showed that considerable changes occurred in the utilization of different parts of the home range. The group (1) covered different lengths of day journeys in different months; (2) visited some areas at different times of the year; and (3) visited some areas more than others.

Their patterns of home range utilization can be summarized as follows. The day journeys during the wet season months of April, May, and early June were not extensive ($\bar{x}$ = 596 m) and were mainly restricted to localized areas of open secondary regenerating forests. However, with the advent of the dry season in mid-June, and the corresponding decrease in both the abundance and regeneration of food plants (especially vines) in the exposed secondary forest areas, the main group were observed to significantly increase the length of the day journeys, the total number of visits, and revisits to various quadrats. This behavior was also correlated with the seasonal appearance of the fruits of *Myrianthus holstii* which were eaten by the study animals in vast amounts during this time. The mean day journey distance for July/August was 1240 m./day.

As the fruiting season of this species came to an end in early September, the new vegetative shoots of bamboo (*Arundinaria alpina*) were appearing above the ground in increasing numbers in the localized stands of bamboo forest. The behavior of the gorillas at this time showed significant increases in day journey length ($\bar{x}$ = 1519 m./day), and increases in the revisiting of localized areas of bamboo. Large amounts of the fruits of *Syzygium guinense* and the basal parts of *Cyperus latifolius* were eaten near the end of October from the localized concentrations of these species which occurred in or at the sides of the swamps and near the bamboo forest. Day journey-length decreased significantly during October ($\bar{x}$ = 1180 m./day). While overall monthly range size decreased during September, and again in October, the number of revisits to bamboo areas increased.

The nesting habits of the gorillas were not found to influence their migrations, for the gorillas generally nested wherever they happened to be feeding near nightfall. They used whatever vegetation was available in the immediate vicinity to construct nests unless they were in either *Cyperus* swamps or pure stands of bamboo, in which case they would travel to the mixed forest interface to nest. Thus there were no permanent, or even regularly used, nest-site locations, nor were any particular types of vegetation used exclusively for the construction of individual nests.

By November the open secondary forest areas had an abundance of new growth after the late rains and almost all the traces of heavy feeding by the main study group earlier in the year had disappeared. According to Adrien Deschryver (personal communication), Conservator of the Kahuzi—Biega National Park, the main study group moved back to these areas by December, thus repeating the annual cycle which, according to the most experienced pygmy tracker, has been the same pattern during the 15 years or more he has been hunting gorillas in this area.

D. Nesting

Schaller (1963) was the first to record differences in night nesting patterns between different populations; since then his observations have been extended by Jones and Sabater Pi (1971), Goodall (1974), and Casimir (1975a). Of nests seen by Schaller in the Kabara region, 97% were on the ground while the figure given by Goodall for the nearby Visoke gorillas is just under 90%. These figures contrast strongly with only 78% in Nyakalonge-Kahuzi, 54% in the Kayonza forest, 46% in Tshibinda—Kahuzi, 45% at Kisoro, 33% on Mt. Tshiaberimu, and only 22% in the Utu lowlands. The reasons for such differences are undoubtedly related to the vertical distribution of suitable nesting material. The nest height data from the two Kahuzi regions were significantly different (p = < 0.001 chi-square, in

Goodall, 1974). The many strata in the mixed secondary/primary forest of Tshibinda—Kahuzi offered more suitable opportunities for tree nesting than the mature primary forest regions near Nyakalonge, where on several occasions the whole group slept on the ground without making nests at all. On each occasion, suitable nesting material was only a very short distance away.

A further character varying in an interesting way is the presence of dung in the night nest. At Kabara, 73% of nests contained dung which had been slept on by the animal during the night (the dry fibrous dung of gorillas does not stick to the fur and is, therefore, not likely to provide a focus for infection); at Kisoro this occurred in 82% of the nests, giving a high figure for the Virunga Volcanoes as a whole Schaller (1963). This contrast with figures of 14.2% for Tshibinda—Kahuzi, 2.4% for Nyakalonge—Kahuzi (Goodall, 1974), about 10% for Mt. Tshiaberimu, Kayonza, and Utu (Schaller, 1963), and 8.7% for Bukulumisa—Kahuzi (Casimir, 1975a). Jones and Sabater Pi found 30.7% of nests not soiled with dung, 26.1% had dung present outside the nests, and 43.2% were soiled with dung (they do not specify whether this was slept on or not).

Unlike the differences in nest heights, these differences cannot be so easily explained by differences in local habitat—although differences in diet could well affect the feces and, therefore, make them more, or less, likely to be slept on. Again there were significant differences found between the neighboring gorilla populations in the Nyakalonge and Tshibinda areas of Kahuzi (Goodall, 1974). It was quite clear that many of the Nyakalonge gorillas defecated well outside their nests, especially the silverbacks. Individuals in Tshibinda—Kahuzi were seen to deliberately place their rump outside the nest rim before defecating. It is, therefore, suggested that such differences in defecating habits, like those in feeding habits, are cultural differences. Similarly, these differences exist not only between populations living in widely separate regions but also between groups living in the same region and even between individual members of the same group. As to the reason for the apparently unsanitary habit of the Virunga gorillas in consistently sleeping on their dung, explanations involving amount of dung or frequency of defecation are unconvincing. The Kahuzi gorillas defecated much and often but mainly carefully outside their nests. Goodall (1974) suggests that the dung of the Virunga gorillas is deliberately used to line their nests and thus insulating the occupant against the subzero ground temperatures at this altitude. This is completely unnecessary at lower altitudes and has, therefore, not been included into the cultural repertoire of these populations.

The remarkable claims of Jones and Sabater Pi (1971) regarding the orientation or gorilla nests in order to receive maximum benefit of the sun at dawn was not substantiated by considerable investigation into the nesting habits of both the Visoke and Kahuzi gorillas by Goodall (1974).

Jones and Sabater-Pi's claims appear to be based solely on finding more nests on either flat or eastern and southerly facing slopes as opposed to those where the terrain was exposed to the north and west. However, they have not discussed the distribution of suitable nesting material in relation to ground slope, ground slope itself, the positioning of individual nests with regard to the rising sun, or, finally, the time of rising of the gorillas in relation to the possible time the rising sun illuminated the individual nests. Perhaps more surprising, Goodall also found no general correlation between the location of gorilla nests and the availability of suitable shelter from the elements, though again individual gorillas differed.

E. Group Size and Composition

Differences in social organization between gorillas in different habitats have long been noticed by observers. Blancou (1951) stated that, in West Africa, coast-living gorillas live in small family groups consisting of a male, one or two females, and several juveniles, while those of the interior live in much larger groups: a typical group would contain one male, four adult females, and eight to ten juveniles or subadults. Schaller (1963) quotes records of group sizes from the literature and gives his own observations, and tabulates (in his Tables 19–20) available data on group composition; data of Kawai and Mizuhara (1959) and Fossey (1972) are in quite good agreement with his results.

To the nearest integer, the mean for group size for the data both of Schaller (1963), based on ten counts, and of Fossey (1972), based on six counts, and excluding the rather peculiar group VIII, is 17 animals at Kabara and Visoke; 7 at Kisoro (Schaller, 16 counts); 9 in the Kayonza forest, (Schaller, 13 counts); 9 on Mt. Tshiaberịmu (3 counts); 11 on Mt. Kahuzi [Schaller, 1 count; Casimir (1975), 1 count; Goodall (1974), 6 counts]; 13 on the Itombwe mountains (Schaller, 1 count); 12 in the Utu lowlands (Schaller quoting Cordier, 9 counts); 7 on Mt. Alen, Rio Muni (Jones and Sabater Pi, 1971, 8 counts); 6 in the Abuminzok–Aninzok region, Rio Muni (Jones and Sabater Pi, 1971, 5 counts); and about 8 in other West African records (Schaller, 6 counts). Thus, the western volcanoes tend to have larger groups than elsewhere, the intermediate mountains follow, and the West African lowlands have much smaller groups. Oddly, the eastern volcanoes, Kayonza and Utu, for all of which there seems to be good counts, do not fit into the picture: Utu has larger groups than in West Africa, and Kisoro and Kayonza have comparatively small ones. Precisely what all this may relate to is unclear. It should be pointed out, however, that all counts have a wide range with maxima of twenty or thereabouts in all areas for which numerous counts are available, except West Africa where the maximum is twelve and minima of two or three except at Kabara, where the minimum is five. There are difficulties in making consistent accurate nest counts in some forest areas since some tree nests can easily be overlooked or, as seen in Kahuzi, some animals may not make nests at all. Any real and consistent differences will only become clear when sufficiently accurate surveys are performed in many different regions.

Group composition also varies from place to place. Here the data are sparser than for actual sizes of groups: Schaller records composition analyses for six West African groups, six each from Kayonza and Kisoro, and ten from Kabara (plus three from Utu, one for Itombwe; too little to be meaningful). Of other authors, only Fossey (1972) and Goodall (1974) break down group membership (for six groups on Mt. Visoke, in the western volcanoes like Kabara, and six groups in the Kahuzi region, respectively). It would seem (Table IIa) that the number of silverback males varies from 1 or 1.5 per group in the lowland and intermediate areas (and Kisoro) to around 2 per group in the western volcanoes. As a percentage of the group, this means that the latter have fewer silverbacks than elsewhere. The number of blackback males is, however, greater relatively, as well as absolutely, in the western volcanoes, although West Africa also has a high percentage. Relative to the number of silverbacks, the number of blackbacks is much greater in the former, but relative to the number of females West Africa again has as high a tally as Kabara and Visoke. The significance of this is that males tend to leave their groups and wander alone, sometimes with a juvenile in tow. From Schaller's data, only one male per group would seem to be permanently associated, such that Itani and Suzuki (1967) refer to the gorilla group as fundamentally a one-male group. The "extra'" males may associate with a given group for a long time or may spend most of their lives alone, traveling with one group or another for only a few days at a time. Since the silverbacks form the bulk of the solitaries at Kabara and Visoke, we may hypothesize that either (1) a wandering silverback has less tendency to associate with groups in the western volcanoes than elsewhere, or (2) more blackbacks go wandering at Kisoro and Kayonza; (3) age-specific male mortality could also have some bearing on this: Schaller (p. 100) infers that sexual maturity could be a time of higher mortality in males; and (4) there is the possibility that males in different gorilla demes acquire silverbacks at different ages so that, for example, there are more blackback males at Kabara simply because the silvery color develops later than at Kisoro. However, the total group sex ratio still varies, so the last two explanations cannot be the entire story.

Dian Fossey's (1972) group VIII, consisting of two silverbacks, three blackbacks, and an aged female, who later died, leaving an all-male group, is so far quite without parallel in published gorilla studies.

It must also be observed that in the East African populations the ratio of adult females to young (infants plus juveniles) is quite constantly 1 : 1.2–1.5 (see Table II). Since this is a criterion of reproductive success, we will return to these figures later. Unfortunately, as yet there is little known of actual group dynamics over any considerable period of time, with the possible exception of the Visoke population.

TABLE II

Troop Compositions in Different Areas

(a) Prior to 1970[a]

	Western Africa	Kayonza	Kisoro	Kabara	Visoke
Absolute number per troop					
Silverback male	1.5	1.2	1.0	1.7	2.0
Blackback male	0.8	0.3	0.3	1.5	2.3
Adult female	3.0	1.8	2.3	6.2	5.8
Juvenile	0.7	?1.5	0.8	2.9	3.3
Infant	1.5	1.3	2.3	4.6	3.2
Percentage composition					
Silverback male	20.0	18.9	14.6	10.0	12.0
Blackback male	11.1	5.4	4.9	8.9	14.0
Adult female	42.2	29.7	34.1	36.7	35.0
Juvenile	4.4	24.3	12.2	17.2	20.0
Infant	22.2	21.6	34.1	27.2	19.0
Infant + juvenile	26.6	45.9	46.3	44.4	39.0
Troop composition ratios					
Blackbacks per silverback	0.5	0.25	0.3	0.9	1.15
Silverbacks per female	0.5	0.65	0.4	0.3	0.3
Blackbacks per female	0.3	0.2	0.1	0.3	0.4
Adult males per female	0.8	0.8	0.6	0.5	0.75
Juveniles per female	0.2	0.8	0.35	0.5	0.6
Infants per female	0.5	0.7	1.0	0.75	0.6
Infant + juvenile per female	0.7	1.5	1.35	1.25	1.2

(b) in 1971/1972 [b]

	1971				1972			
	Ngezi	Muside	Kabgende	Total Sabinio + saddle	Muhavura	Sabinio	Karisimbi	Kahuzi
No. Juveniles	2	0	2	4	1	4	2	14
No. infants	3	1	4	8	2	1	2	13
Total no infants + juveniles	5	1	6	12	3	5	4	27
No blackbacked adults	14	7	8	29	6	14	7	35
No. adult females if female: BB male ratio is								
1:0.1 (like Kisoro)	12	6	7	25	5	12	6	32
1:0.4 (like Visoke)	10	5	6	21	4	10	5	25
Juveniles per female								
Low	0.2	0.0	0.3	0.15	0.2	0.3	0.3	0.46
Juveniles								
High	0.2	0.0	0.3	0.15	0.3	0.4	0.4	0.56
Infants per female								
Low	0.25	0.2	0.6	0.3	0.4	0.1	0.3	0.41
High	0.3	0.2	0.7	0.4	0.5	0.1	0.4	0.52
Infants plus juveniles								
female, low	0.45	0.2	0.9	0.5	0.6	0.4	0.6	0.84
Female, high	0.5	0.2	1.0	0.6	0.8	0.5	0.8	1.08

[a] Source: Schaller (1963); Fossey (1972).

[b] Source: Harcourt and Groom (1972); Groom (1973); Goodall (1974).

[c] Plus two unidentified: one was an adult of unknown sex.

F.   Intergroup Relations

Fossey (1972) shows that the hoot-series, commonly followed by chest-beating, function in intergroup communication. This occurs more specifically between the silverbacks of two groups, or between a group male and a lone male. On Visoke if the comminucants are close, the chest beats are more liable to be replaced by ground-thumping or branch-breaking (vocal plus visual display) or by vegetation tearing or a sideways run (visual display alone). This lays to rest once and for all, one would hope, the old superstitions of the gorilla chest-beating in rage when about to attack, etc.

According to the general view, the group is the highest level of gorilla social organization. However, careful reading of Schaller's book (pp. 124—129), and inspection of his maps, suggest the possible existence of local communities combining several groups. A line drawn between Kabara and Rukumi approximately separates the ranges of groups IV, VI, and VIII from those of groups I, II, III, V, VII, and XI; indeed, only group VI crossed this boundary from time to time. Group IX, whose range seemed to be to the further west, only occasionally entered the study area, and so might be thought of as representing a third local community. All the intergroup interactions reported are between groups of the same community. Although group VI, as stated above, did range into the area occupied by the other putative community, no approaches with any group in the latter area were recorded.

Lone males seemed, from the evidence, to associate preferentially with a particular group, but in any case always within one or other community. Schaller describes the arrivals and departures of "the Lone Stranger," who would at times associate with groups IV or VI. If we include the solitaries whose wandering areas are described, the western community would have at least 68 members; the eastern one, would have 77 members.

This community organization, if substantiated by future research, would link the social organization of the gorilla to that of the chimpanzee. Nishida (1968) describes these local communities (he calls them unit groups) in the chimpanzees of the Mahali mountains, and compares them with other regions. The communities may number as many as 80 animals, and are generally of the order of 60 or so. Interunit group relations are characterized by avoidance, with some "subordinate" to others. Their ranges are largely separate, but overlap. The main differences from gorillas lies in the unstable nature and constant regrouping of the component subgroups. When two groups of a community of gorillas meet they may associate and even nest together for a night, but on separating it is widely believed that they always rearrange themselves into their original groups; there is almost no exchange between the different groups. However, there is evidence of several possible cases of females transferring from one group to another in Schaller's book, and in Kahuzi Goodall found widely varying nest counts in some areas which were indicative of some group transfers—even if only temporary in nature.

III. Conservation

A. Population Size and Recruitment Rates

Unfortunately we have little accurate information to say exactly how endangered the eastern gorilla is in terms of absolute numbers. The only general survey and census so far conducted (Emlen and Schaller, 1960; Schaller, 1963) estimated the entire population size to be between 5,000 and 15,000. The wide range of their estimates reflects the many difficulties in conducting accurate, short-term census work on forest-living primates. Although even the lower estimate may not be considered alarming by some for such a long-lived animal, two further facts must be considered: (a) The actual distribution of eastern gorillas within their total range of 35,000 square miles (90,650 km²) is most uneven. Schaller and Emlen found that gorillas were concentrated in some sixty more or less isolated pockets of forest which varied in size from 10 to 200 square miles. In addition, the degree of isolation of these pockets varied greatly, some being as little as 2 miles apart while others were over 30. (b) The sizes of some of these "pockets" and their gorilla populations are being reduced, drastically in some regions. In others, their floristic composition is being altered.

Most eastern gorillas live in Zaire—a most fortunate locality in many respects for in this country there are already seven National Parks. Three of these incorporate territory occupied by gorillas: the Virunga Park (8000 km²), Kahuzi-Biega (600 km²), and Maiko (10,000 km²), the latter being a new park covering part of the great Forest of Maniema in the Utu lowlands where the bulk of the *P.g. graueri* population lives. No recent reliable information is available about the status of these animals. According to Deschryver (personal communication) gorillas were being killed by local people in the Walikali region because of their alleged raids on local crops. It is not known how extensive are such killings.

The Itombwe mountains are not included in any reserve. Verschuren (1975) reports that the high peaks are intact but there is much erosion due to overgrazing by domestic stock. There is an obvious need to place some kind of conservation order on the area, but since the surrounding human population is dense, it will need strict and constant surveillance.

The Kahuzi-Biega Park has recently been the site of several reports [Goodall (1973) and Grzimek (1974)] and scientific investigations (Casimir and Butenandt, 1973; Casimir, 1975a,b; Goodall 1974, 1975). The number of gorillas in the park has been estimated by the conservator, Adrien Deschryver, as 200–500. From extrapolation of population densities in the Nyakalonge and Tshibinda areas and consideration of the relative proportions of primary and secondary forest within the park, Goodall (1975) favors a lower figure of some 150 animals. This would give a population density of some two to three animals per km². It must be stressed here that no census has yet been conducted in this region, though attempts are being made to get these conducted (Goodall, 1977b).

The area subjected to more detailed study, the Virunga Volcanoes, was estimated to have a population of between 400 and 500 (Schaller, 1963). About 200 of these were in the vicinity of Schaller's Kabara study area (an area of 30 square miles covering the southern and eastern flank of Mt. Mikeno and extending to the foot of Mt. Karisimbi and halfway to Mt. Visoke). He hypothesized at least 100 more for other areas in the western section. For the whole eastern area of the volcanoes, the Mts. Sabinio, Mgahinga, and Muhavura region, he estimated only 40 to 50 gorillas and 50 to 75 in the 8-mile long saddle between the two sectors of the volcanoes.

Most of these areas have been surveyed in more recent years by various workers from Dian Fossey's camp on Mt. Visoke. Based on these surveys, Fossey (in Goodwin and Holloway, 1974) estimates only 275 for the total area of the volcanoes. Harcourt and Groom (1972) found only 25 in the "saddle" and Groom (1973) reports 45 on Mts. Sabinio and Muhavura, stating that they are still declining. The implication is that Schaller's estimate for this region was too low, especially seeing that, since Schaller's day, gorillas have entirely disappeared from Mt. Mgahinga, and, as a resident species, from the entire Ugandan portion of the volcanoes. In Rwandese territory only on Mt. Visoke is there a really flourishing population, with over seventy animals known to exist. However, many of these animals spend large portions of their time over in the Zaire section of the park. It is thus clear that, with its own population of almost 200 animals, the Zaire section contains by far the greatest proportion of the entire Virunga stock. Overall the figures give a population density of some seven animals per km².

As yet no further surveys have been done in any other area inhabited by eastern gorillas. Schaller, in May 1959, estimated the population of Mt. Tshiaberimu (incorporated within the Virunga National Park) as only 30–40, and noted that the number was probably decreasing due to hunting and habitat destruction. It is thus questionable whether any still remain. The Kayonza population, which he estimated to be between 120 and 180 animals, is still unstudied and perhaps relatively undisturbed due to the inhospitable nature of the terrain.

At this point we can turn to Table IIb, which attempts to assess recruitment rates by calculation of adult femals: juvenile and infant ratios. Schaller (1963) finds that a female produces an infant on an average of every 4 years. Mortality is 40–50% in the first 6 years of life (i.e., approximately during infant and juvenile stages). The infant:juvenile ratio is nearly 2:1 (except in Tshibinda–Kahuzi where it is almost 1:1), the boundary between these two stages being about 3 years of age. These data give us a baseline. While it is relatively easy to assign animals to various age classes during field operations, with the exceptions of obvious cases such as silverback males and lactating females, sex determination is much more difficult.

The observations of Harcourt and Groom (1972) and Groom (1973) distinguished juveniles from infants, but only rarely were able to differentiate adult females from blackback males. Therefore, for the purposes of calculating female: young ratio, a way had to be found of assessing what proportion of Harcourt and Groom's "other adults" are likely to be females. The blackback to female ratio at Kisoro (Table IIa) is only 0.1:1; on Visoke, from Fossey's (1972) data, the ratio is 0.4:1. Therefore, an approximate upper and lower figure for the number of females likely to be included in the "other adults" is calculated. Following this, a low and a high ratio of juveniles and infants to adult females is calculated, together with a total young to female ratio.

Every one of the total young per female ratios for Harcourt and Groom's (1972) and Groom's (1973) samples are below those for Kisoro, Kabara, and Visoke, as recorded previously by Schaller and Fossey. This includes the sample for the southeast slopes of Mt. Karisimbi, where Groom (1973) considers that the population is vigorous and healthy. However, it must be noted that, unlike the Sabinio, Muhavura, and "saddle" counts, this does not represent the whole population for an area and so may not be truly comparable. The figures of 1.35, 1.25, and 1.2 young per female of Schaller and Fossey suggest either that the birth interval is below 4 years, or that preadult mortality is well under 50%. The comparable figures of Harcourt and Groom, e.g., 0.5 or 0.6 in 1971 declining to 0.4–0.5 (low and high, respectively) in 1972 for Mt. Sabinio, indicates a birth interval much above 4 years, or a preadult mortality above 50% or both, and a declining population [in absolute figures, the Sabinio population appeared to have declined even in 1 year; Harcourt and Groom (1972) estimated 31 to 42 animals the previous year while Groom (1973) counted a definite 32 in 1972 census].

The number of juveniles in the samples is, as Schaller noted in his figures, of the order of one-half (or slightly more) of the number of infants in most of the 1971–1972 census figures. The exception is Mt. Sabinio where, in 1972, fewer infants than juveniles existed, further evidence for a decline in the birth rate, or at least in surviving young. Indeed, throughout the census figures, particularly in the firmer (more reliable) 1972 figure (Groom, 1973), one notes that the juvenile to female ratio is less critically low than the infant to female ratio. There is simply no avoiding the conclusion that, as of 1972, the recruitment rate of gorillas in the eastern Virunga Volcanoes was unnaturally low and declining.

When the data for the Kahuzi region are compared it can be seen that the infant and juvenile to female ratio is almost 1:1. The data as presented hide some variation which was found within the Kahuzi region itself. The female to juvenile and infant ratio was lower in the primary forest regions near Nyakalonge (1:0.8), while for one group in the Tshibinda—Kahuzi area it was 1:1.2. According to Adrien Deschryver (personal communication), this latter group numbered only six when he first contacted it in 1966. Prior to this, there had been considerable illegal hunting of gorillas for meat by local pygmoid people. After 1966, regular patrols were made by armed guards in this area of the park and the gorilla population grew rapidly. It is possible that immigration has also served to swell the numbers of this and other groups in this particular area. Thus, at least in the eastern section of the Kahuzi—Biega region, the recruitment rate is healthy.

B. Threats to Gorilla Populations

While some populations of eastern gorillas may be increasing, such as those in well protected areas like the Kahuzi—Biega National Park, others are undoubtedly declining. This decline is the direct result of three interrelated factors: (a) predation and exploitation, (b) disturbance, and (c) habitat destruction. While overall population densities of eastern gorillas are by far the highest in the Virunga Volcanoes region, this population is declining, especially in eastern areas where it is occurring at an alarming rate.

Death of gorillas caused by other animal predators, such as the leopard (Schaller, 1963), have probably been minimal in the past and are likely to be even less so in the future, since the leopard itself has virtually been eliminated from many areas by poachers. However, gorilla mortality caused by local human populations appears to be occasionally severe in some regions, although many killings go undetected or unreported. Mention has already been made of the group of five gorillas which had been stoned to death that Goodall found near Tsundura in Virunga. Unlike the Utu gorillas which were killed as "pests" or the Kahuzi gorillas which were, in the past, killed for meat, the Tsundura gorillas (and others later—see Groom, 1973) were apparently killed in order to obtain parts of their bodies for "witchcraft" purposes. There also remains the possibility that any young present in groups which are thus annihilated may be captured and sold to unscrupulous dealers. In at least one other incident two young males were captured in 1969 and donated to the Cologne zoo by the Rwandese Government then in power, (Fossey, personal communication). The fate of their mothers and fellow groups members is unknown, but it is highly probable that they were killed.

Gorillas in many regions, both outside and inside protected areas, are subjected to disturbance of varying degrees by human visitors to their forest habitats. Such people come for many, often illegal, purposes.

1. Wood Gathering

This is probably by far the most common source of incursion into gorilla habitat, especially in areas where other sources of cooking fuel are scarce or expensive. The visits of large numbers of people over long periods of time can easily devastate extensive areas. In this way the boundaries of existing parks and reserves are quickly eroded. Miniature sawmills have been found in the Virunga Volcanoes. Since *Hagenia* is a very slow growing tree, heavy exploitation of this species would seriously affect the forest. In the south of the Kahuzi-Biega park, several hectares of forest on Mt. Shamulamba, bordering the Lushandja swamp, were destroyed by one of several illegal charcoal-producing industries.

2. Poachers

Their quarry varies from area to area and, in the case of some pygmoid tribes, can include virtually anything from squirrels to monkeys and even gorillas themselves. Often bushbuck (*Tragelaphus scriptus*) and Red Duiker (*Cephalophus nigrifrons*) are the main quarry, especially in the Virunga region. Goodall once came upon an encampment of some twelve to fifteen poachers near Mt. Karisimbi, who were about to start a large coordinated hunt. The antelopes are often hunted with spears, bows, and arrows and are usually chased to earth by dogs wearing "cowbells.". Snares are extensively set in many areas, and in some cases visited only occasionally. One bushbuck was caught in two such snares and eventually escaped, after the wires had amputated the two trapped feet. In Ruhengeri, such venison was sometimes sold openly from door to door. An elephant was wounded and chased out of the Virunga park by poachers in 1970. It died among the shambas where it was left to rot, apparently having been killed for the hairs on its tail—which are sold as "magic charm" bracelets.

Although no gorilla deaths were apparently directly caused by poacher's snares in Virunga, one young infant was found strangled in such a wire noose by Casimir in the Kahuzi region, (personal communication). Old pitfall traps, presumably for the giant forest hog (*Hylochoerus meinertzhageni*), were common in Nyakalonge—Kahuzi. Fortunately, the gorillas appeared to be able to circumnavigate these traps with ease. However, even if not the direct quarry themselves, gorillas are bound to be harassed or even accidentally killed by poaching on such vast scales. Worse still, as in Kahuzi in the past and in Virunga today, they are specifically hunted in some areas.

3. Honey collecting

Many incursions are made into forest areas to collect honey, a much sought after source of food, from the underground nests of some bee species, from naturally occurring tree nests, or from hives specially prepared from hollow logs. Apart from the disturbance to forest animals, this is fairly harmless in some cases and is almost essential to some pygmoid tribes. It is, however, illegal within park areas. Much greater damage is done in some areas when whole trees are deliberately set on fire in order to "smoke out" the bees. Many dead *Hagenia abyssinica,* their trunks black and charred by such treatment, can be seen in the Virunga region.

4. Smuggling

In the eastern area of the volcanoes, Groom (1973) gives the staggering figure of one group, of ten to twenty smugglers, passing every half hour along the route over the Sabinio—Mgahinga saddle between Uganda and Rwanda, and only slightly less traffic along the Mgahinga—Muhavura route. The groups make a noise as they go, to frighten away large, wild animals. They, like the poachers, sometimes camp overnight in the park. Groom describes the Sabinio—Mgahinga saddle as having been converted, by smugglers, herdsmen, and other trespassers into a strip of meadow 30 yards across, and notes that such a large open space would form a barrier to gorillas. The absence of gorillas from Mgahinga would certainly be, at least in part, a consequence of this situation.

5. Cattle Grazing

In dry periods, especially, cows are driven into the forest areas where they eat not only any available herbage but also the leaves of many shrubs which their herders cut down. Very quickly, large areas of forest are either eaten, slashed, or trampled underfoot. Such intrusion varies considerably from area to area, but takes place even within recognized areas of National Parks—nowhere more so than in the Rwandese section of the volcanoes. It is likely that anything up to a thousand cows were illegally in this park in recent years. Their herders had apparently divided up the park into so many "ranges," one for each herder! The long-term effects of such habitat destruction is as yet unclear since little is known about the regeneration of forest areas which have been so destroyed. However, in Kahuzi, Goodall found that much of the areas previously grazed by cattle was completely overgrown with *Veronia ruwenzoriensis* shrubs, a species which, although eaten by bushbuck and chimpanzees, was not eaten by gorillas. The extent of disturbance to gorillas by cattle and their noisy herders is considerable and could easily influence normal migrations.

6. Tourism

Other occasional visitors, such as tourists, cause minor disturbances often through inconsiderate and careless behavior which inevitably provokes gorilla charges and could result in severe injury to either party. While tourism must be encouraged in some areas of the National Parks, it must be done carefully. One day in Kahuzi, a party of no fewer than *sixteen* tourists were taken to "see" a gorilla group!

7. Cash Cropping and Habitat Destruction

Any healthy gorilla population could withstand even the losses caused by the above sources, but an additional, far more dangerous, threat to their continued existence comes from habitat destruction, not only on the localized scale within forests by cattle, but on a much larger scale involving the complete devastation of vast forest areas. There are several reasons for such practices, mainly economic and agricultural. In Rwanda, for example, some 8,000–10,000 hectares of their section of the Virunga Park was cleared of its unique *Hagenia* woodland in 1969 to provide shambas (small farm plots) to grow vegetables and *Pyrethrum*—the country's largest single source of foreign exchange. This action of the Rwandese government then in power, under the auspices of a develop- ment project organized by Fonds Europeen de Development (F.E.D.), was condemned by Dr. Van der Goes van Naters of I.U.C.N. in a letter addressed to the Commission of European Communities in June 1970 (personal communi- cation). He pointed out that no ecological or geological surveys had been conducted *before* the devastation of the forest. Spinage (1972) has argued forcibly the disastrous consequences of the removal of such a large natural "sponge" from this important catchment area in terms of increased rain water run off and erosion. Such consequences have been well known to scientists for many years. However, advice was either not requested or ignored in favor of the economic gains from the *Pyrethrum* crops. Since artificial pyrethrin- like insecticides have now been manufactured, and undoubtedly will soon be available in commercial quantities, it is doubtful whether Rwanda will con- tinue to benefit from her newly cultivated supplies.

8. Agriculture and Vegetational Changes

In many regions inhabited by eastern gorillas the type of agricultural prac- tice known as "slash and burn" is in widespread use. Literally, the vegetation is cut from large tracts of forest and burned—the resulting ash enriching the soil. Since the productivity of such areas decreases rapidly within a few years, they are often then used for grazing or abandoned. If left for sufficiently long periods, the resulting secondary regenerating forest may eventually be replaced by primary forest virtually indistinguishable from the original. However, if the human population density is high, such areas may be subjected to frequent, repeated burnings which can result in an even greater deterioration of the productivity and even cause a deflected climax community such as *Pteridium* ferns (Richards, 1964).

Areas of such ferns were extensive along the central western borders of the Kahuzi–Biega Park near Nyakalonge. Undoubtedly such areas would serve to isolate the "pockets" of existing gorilla populations even more or eliminate them completely when they grew too small to be viable. However, such slash-and-burn practices had slightly different results in the eastern regions around Mbayo, Tshibinda, and Mt. Bukulumisa, perhaps due to richer soils or less repeated clearing. In 1972, much of these areas were covered by secondary regenerating forest types, from herbaceous "meadows" to mature stands characterized by such trees as *Neobutonia macrocalyx* and *Dombeya goetzenii*. In even older areas, mixtures of primary and secondary forest species were evident. The forests further west were less disturbed and much primary forest still remained. The utilization of such varied forest types by the Kahuzi gorillas and their importance to gorilla ecology has already been described. Thus it can be seen that varying degrees of habitat destruction and different agricultural practices can have very different consequences in different regions.

C.   The Future of Gorillas

In some ways it is unfortunate that, because of the relative ease of observing gorillas in the Virunga Volcanoes region, this population has been the most studied. Its behavior and ecology have been assumed by some to be typical of the mountain gorilla. The comparative research conducted in the Virunga and Kahuzi regions (Goodall, 1974) outlined above clearly shows that gorillas, like many other primates (Crook, 1970), exhibit different patterns of behavior and ecology in different habitats. The Kahuzi research also shows that differences even exist within gorilla populations inhabiting particular regions, especially where there is considerable biotope variability. The fact that eastern gorillas (and no doubt western ones) are found in such a wide variety of forest biotopes is a tribute to the great plasticity of their behavior and ecology. Such plasticity has also enabled them to take advantage of the vast increases in available food resources in the secondary regenerating forest areas recently created by man. Schaller (1963) was the first to point out the role of rotational agricultural activity in creating the secondary forest conditions which are favored by gorillas.

This has been clearly shown to be true for the Kahuzi gorillas, particularly, during the rainy seasons. Therefore, Goodall (1974, 1977a) suggests that this population is living in forests more similar to those in which gorillas have evolved than are the Virunga population, who in fact are living in an extremely specialized (and perhaps relatively new) type of habitat. Thus the view expressed here is that the plight of the Virunga gorillas, like their behavior and ecology, must not be taken as typical of eastern gorillas as a whole. While some of the dangers threatening them with extinction, as outlined above, are similar if not the same as those threatening other gorilla populations, the solutions to the conservation problems they pose vary from region to region depending upon local conditions.

There is no doubt that some of these dangers, especially those in the Virunga Volcanoes, could be removed *simply by the implementation of the statutes and regulations of the internationally recognized National Park*. The recent new government in Rwanda has shown great concern over its conservation responsibilities. Aided by funds from the World Wildlife Fund, guard patrols have been increased in their section of the volcanoes. These were greatly facilitated by the gift of a Peugeot mini-bus from the manufacturers. If these patrols can be at least as effective as those in the Zaire section then there is at least some hope for the future. What is really needed is an international agreement between Zaire, Rwanda, and Uganda regarding effective patrols throughout the entire chain of the volcanoes. However, at best such measures will only save the existing population and enable it to live in a habitat which has been much reduced in both size and quality, especially in Uganda and Rwanda.

It is thus clear that the biggest single threat to the continued existence of eastern gorillas comes from the uncontrolled destruction of their habitat by man. Hope is on the way, certainly in Zaire where President Mbuto Seso Seko recently announced that soon   12−15% of this vast country will be set aside for nature reserves or National Parks. Since it is certain that the future of eastern gorillas depends mainly upon those responsible for conservation in Zaire, two things are now of importance (1) the location and distribution of these areas to be designated, in relation to the distribution of eastern gorillas, and (2) the type of management ultimately decided upon.

Successful conservation of any species demands full knowledge and consideration of the organism's natural ecology. In the case of eastern gorillas, it is now clear that this must take account of the findings of field studies in various regions. These have shown different patterns of feeding, nesting, and ranging behavior, which have been, in each case, well adapted to the local conditions. Therefore, given such a variety of gorilla behavior and ecology, it is this *variety* we must seek to conserve. This can be achieved by the designation of suitably varied areas for gorilla reserves and their subsequent proper management. Faced with the many demands upon land which rapidly increasing populations create, even the most ardent conservationist has to realize that not *all* eastern gorillas can be saved. However, if the present patterns of land usage continues, only very few will survive. Thus some order of priority has to be given to ensure that as many eastern gorilla populations survive as is possible. It is suggested here that primary consideration be given to ensuring the existing *variety* of populations and habitats in the *largest areas* of forest possible, a few large areas being ecologically more suitable than many small ones. Any gorillas living in small isolated pockets which are not designated as reserves, and are threatened with extinction, could be captured and sold to such zoos and research institutes as can prove they have suitable facilities and a good breeding record, and the resulting money used to finance the management of designated reserve areas. There also remains the possibility of the relocation of some animals from one area to another. Obviously great care would have to be taken to ensure compatible biotopes and the presence of known food types. The recent film made by Survival Anglia T.V., of Goodall's main study group in Tshibinda—Kahuzi, showed that "strange" infants may be accepted by a gorilla group, although in this particular case inadequate precautions had been taken to ensure its survival.

The areas which are designated for gorilla reserves, because of their variety, may require different management techniques to ensure the successful survival of their gorilla population. While this may entail the controlled neglect of some areas if may also require the artificial continuation, or even creation, of secondary/primary forest mosaics in other areas. Any timber which is felled as a result of such management techniques should be sold to the local population for their much needed fuel supply.

The following recommendations are made:

1. The formulation of realistic land use management plans and gorilla conservation plans to cover the whole variety of gorilla habitats, including the construction of management plans of any specially designated gorilla reserves or national parks. That the most effective conservation may require "interference" from man, should be recognized.

2. The immediate shelving of any plans to extend the existing *Pyrethrum* projects or to create cattle stations in the Virunga Volcanoes region, so that more facts can be collected about the future prospects of both gorillas and their habitat before both are irretrievably lost with obvious disastrous consequences on the local climate and geology.

3. The creation of a "gorilla research headquarters" perhaps at I.R.S.A.C. headquarters near Bukavu or within the auspices of I.N.C.N. (Zaire) itself. Field stations could be established at Kabara, Tshiaberimu, Itombwe, and other suitable areas. Each station should be manned by resident and/or visiting scientists of diverse fields of interest so the maximum amount of information regarding the management of gorillas and their habitat can be collected. Local personnel should be encouraged and trained wherever possible to ensure continuity of data recording and a national interest in the work.

4. A detailed survey should be undertaken of the distribution and status of gorillas throughout the whole of their range together with associated data on vegetation, local land usage, and any threats to existing habitats.

5. The findings of all such research should be made available as soon as possible, both to National and International Conservation bodies, so that the management of these resources can take place along well informed lines.

6. Various management techniques such as habitat creation, relocation of gorillas, and encouragement of tourist visits (especially in the Virunga Volcanoes) should be investigated on small-scale pilot schemes. After *thorough* field trials they could be incorporated into local management plans wherever suitable.

7. All statutes and regulations governing National Parks should be *strictly* enforced. Where existing regulations may forbid any interference implied by some management techniques, they should be reexamined by a team of experts and only altered if on balance such interference will be beneficial to the gorillas.

Although appreciation of wildlife and wilderness areas is spreading rapidly in Africa, as in the industrial world, it is still the case that esthetic and scientific considerations take second place to economic conditions in most countries. This is not to say that scientific spin-offs from conservation projects have not brought enormous and valued international prestige to the country concerned, but generally the initial impetus will have been economic, e.g., the tourist trade, more rarely soil conservation, or, sometimes, simply that the area in question was unsuitable for any other type of use.

So it is that while an African country, especially a very poor one like Rwanda, may express and feel high regard for the worldwide interest in, and appreciation of, its national parks and their fauna, it is the addition of an economic motivation that will provide the real impetus to strengthen the parks' security. The very fact that the concerned government office in Rwanda refers to Tourism in its title focuses attention on this. After all, the proper administration of the Parc des Volcans is going to involve heavy punitive measures against Rwandese citizens who, for the most part, consider their actions quite innocent and in the cause of pursuing their legitimate livelihood. The government can only justify its actions, to itself and especially to its citizens, if it is convinced that it is acting on behalf of its own long-term interests, and that preserving gorillas will, in the long run, be for the benefit of its people. An "international resource," the gorillas may well be, but it is hard to be entirely without sympathy for a destitute country which, while holding this resource in trust for future generations and for the international community, seeks at the same time to benefit economically from it and, crudely, to get a rake-off for itself (See Goodall, 1977c).

This is to say that conservationists should not themselves be blind to local factors in preserving wild areas and their priceless cargo of animal species. An example is provided by Spinage (1972), who stresses the importance of a forest cover on the Virunga Volcanoes as a catchment area; grazing and trampling by cattle seriously reduces soil permeability, cultivation accelerates erosion, and alternation of long dry spells with uncontrollable flash floods results. As the Virunga Volcanoes comprise above 10% of Rwanda's rainfall catchment area, and account for well above 10% of its total precipitation, the case for leaving them under undisturbed forest is unassailable. This is precisely the sort of national economic consideration which carries weight in conservation decisions.

A final point must be made. Rwanda has a population of over 4 million people at a density of 600 per square mile, and this is growing at 3% per annum. Land pressure is intense; the population has spilled over the border into both Uganda and Zaire. Experience in India, Indonesia, and elsewhere shows that under such conditions the niceties of soil conservation are ignored, and people against their own better judgement will open up the steepest, most unsuitable slopes for cultivation only to abandon them within a few years when the top soil has all been lost. Park administration, international finance, scientific concern, all may be in top gear, but this is the ultimate creeping threat, and it is out of our hands.

IV. Summary

The gorilla is listed as "vulnerable" in the Red Data Book of 1972. No information is available on the numbers of gorillas in West Africa, but in East Africa one subspecies, the mountain gorilla, is given a red page and stated to be "endangered." This chapter described the current status of the mountain gorilla and commented also on the likely status of the neighboring eastern populations of the eastern lowland subspecies, after providing a background of basic taxonomic, ecological, and behavioral information on all gorillas, particularly the two eastern subspecies. Data from recent field studies of eastern gorillas, in the Mt. Visoke area of the Virunga Volcanoes region and the Nyakalonge and Mbayo—Tshibinda areas of the Mt. Kahuzi region by Alan Goodall, was discussed in relation to differences in feeding and general ecology of these and other populations. Finally, their implications for the present status of eastern gorillas and its future conservation were discussed and positive measures to ensure their future survival were given.

Acknowledgments

One of us, A. G. G., would like to take this opportunity to express most sincere thanks to the following: The Governments and Park Officials of the Republics of Rwanda and Zaire, the Director-General and staff of I.N.C.N., and I.R.S.A.C.—Zaire (particularly zi, Dr. Ntika Nkumu, Dr. J. Verschuren, Dr. P. Kunkel,), the Science Research Council, the World Wildlife Fund, the Fauna Preservation Society, Liverpool University, the University of Gent, the late Dr. L. S. B. Leakey, Miss Dian Fossey and her staff at the camp on Mt. Visoke, Professor A. J. Cain, Professor R. A. Hinde, Citoyen J. Kalamo, Patrice Wazi-Wazi, all members of "Mission Scientifique Belge des Volcans de l'Afrique Centrale: 1971–1972," especially Mlle. Christine Marius-Weyns, Mrs. Irene Leppington, Mr. E. Djoleto-Nattey, Mrs. L. Turtle, Dr. T. Lawrence, Dr. R. G. Pearson, Dr. G. A. Parker, Dr. J. Bishop, Dr. S. Bradley, Dr. R. White, Dr. F. and Mme. E. Dondeyne, Dr. H. J. and Frau H. Schlichte, and, finally, my wife Margaret.

References

Ardrey, R. (1961). *"African Genesis."* Collins, London.

Blancou, L. (1951). Notes sur les mammifères de l'Equateur africain français, *Mammalia (Paris)* **15**, 143–156.

Casimir, M. J. (1975a). Some data on the systematic position of the Eastern gorilla population of the Mt. Kahuzi region (Zaire). *Z. Morphol. Anthropol.* **66**, 188–201.

Casimir, M. J. (1975b). Feeding ecology and nutrition of an eastern Gorilla group in the Mt. Kahuzi region (Republique du Zaire). *Folia Primatol* **24**, 81–136.

Casimir, M. J. and Butendandt, E. (1973). Migration and core area shifting in relation to some ecological factors in a Mountain Gorilla group in the Mt. Kahuzi region. *Z. Tierpsychol.* **33**, 514–522.

Colyer, J. F. (1936). *"Variations and Disease of the Teeth of Animals."* Dental Board of the U.K., London.

Coolidge, H. J. (1929). A revision of the genus *Gorilla. Mem. Mus. Comp. Zool. Harvard* **50**, 291–381.

Coolidge, H. J. (1936). Zoological results of the George Vanderbilt expedition of 1934. Part IV. Notes on four gorillas from the Sangha river region. *Proc. Acad. Nat. Sci. (U.S.)* **88**, 479–501.

Corbet, G. B. (1967). The nomenclature of the eastern lowland gorilla, *Nature (London)* **215**, 1171–2.

Cousins, D. (1972). Body measurements and weights of wild and captive gorillas, *Zool. Garten. N.F. (Leipzig)* **41**, 261–277.

Cousins, D. (1974). Classification of captive gorillas, *Int. Zoo Yearb.* **14**, 155–9.

Crook, J. H. (1970). *"Social Behaviour in Birds and Mammals,"* Academic Press, London.

Emlen, J. T. and Schaller, G. B. (1960). Distribution and status of the Mountain gorilla *(Gorilla gorilla beringei)* 1959. *Zoologica* **45**(1), 41–52.

Fossey, D. (1972). Vocalizations of the mountain gorilla *(Gorilla gorilla beringei). Anim. Behav.* **20**, 36–53.

Fossey, D. (1974). Observations on the home range of one group of mountain gorillas *(Gorilla gorilla beringei). Anim. Behav.* **22**, 568–581.

Fossey, D. and Harcourt, S. (1977). Feeding and ranging behavior of mountain gorillas *(Gorilla gorilla beringei)* in the Virunga Volcanoes region. *In* "Primate Ecology:

Studies of Feeding and Ranging Behavior in Lemurs, Monkeys, and Apes," (T. H. Clutton-Brock, ed.). Academic Press, London.

Freeland, W. J. and Janzen, D. H. (1974). Strategies in herbivory by mammals: the role of plant secondary compounds. *Amer. Natur.* **108**, 269—289.

Goodall, A. G. (1971). Preliminary report on the behaviour and ecology of the mountain gorilla (*Gorilla gorilla beringei*) in the Mt. Visoke study area. (Virunga Volcanoes). University of Liverpool, Liverpool, unpublished manuscript.

Goodall, A. G. (1973). "A bizarre expedition for contact with mountain gorillas." *Africana*, May.

Goodall, A. G. (1974). "Studies on the ecology of the mountain gorilla (*G. gorilla beringei*) of the Mt. Kahuzi-Biega region (Zaire) and comparisons with the mountain gorillas of the Virunga Volcanoes. Liverpool University, Liverpool, unpublished Ph.D. thesis.

Goodall, A. G. (1975). "Mountain Gorillas in Danger." *In* "Wildlife '75." (N. Sitwell, ed.). London Editions, Ltd., London.

Goodall, A. G. (1977a.) "Aspects of the feeding behaviour of a mountain gorilla group (*Gorilla gorilla beringei*) in the Tshibinda—Kahuzi region (Zaire)." *In* "Primate Ecology: Studies of feeding and ranging behaviour in Lemurs, Monkeys and Apes." (T. H. Clutton-Brock, ed). Academic Press, London.

Goodall, A. G. (1977b). "On habitat and home range in Eastern gorillas in relation to conservation." *In* "Recent Advances in Primatology," Vol. 2 Academic Press, London.

Goodall, A. G. (1977c). "The Ecology of some eastern gorillas". (In preparation).

Goodwin, H. A. and Holloway, C. W. (1972; revised 1974). "*Red Data Book*," IUCN, Morges.

Groom, A. F. G. (1973). Squeezing out the last mountain gorillas. *Oryx* **12**, 207—215.

Groves, C. P. (1967). Ecology and taxonomy of the gorilla. *Nature (London)* **213**, 890—3.

Groves, C. P. (1970). Population systematics of the gorilla, *J. Zool. (London)* **161**, 287—300.

Groves, C. P. (1971). Distribution and place of origin of the gorilla. *Man* **6** [N.S.], 44—51.

Groves, C. P. and Humphrey, N. (1973). Asymmetry in gorilla skulls: evidence of lateralized brain function? *Nature* **244**, 534.

Grzimek, B. (1974). "Schreiend raste der Gorilla auf mich zu. . . .!" *Tier* **14**(7), 4—9 and 46.

Haddow, A. J. and Ross, R. W. (1951). A critical review of Coolidge's measurements of gorilla skulls. *Proc. Zool. Soc. London* **121**, 43—54.

Harcourt, A. H. and Groom, A. F. G. (1972). Gorilla census. *Oryx* **11**, 355—363.

Harriman, A. E. (1973). *Amer. Natur.* **90**, 97—106.

Itani, J. (1958). On the acquisition and propagation of a new food habit in the natural group of the Japanese monkey at Takasaki-Yama. *Primates* **1**, 84—98.

Itani, J. and Suzuki, A. (1967). The social unit of chimpanzees, *Primates* **8**, 355—381.

Jones, C. and Sabater Pi, J. (1971). Comparative ecology of *Gorilla gorilla* (Savage and Wyman) and *Pan troglodytes* (Blumenbach) in Rio Muni, West Africa. *Bibliogr. Primatol. No. 13*. S. Karger, Basel.

Kawai, M. and Mizuhara, H (1959). An ecological study on the wild mountain gorilla

(*Gorilla gorilla beringei*) report of the Japan Monkey Center second gorilla expedition 1959. *Primates* **2**, 1—42.

Kawamura, S. (1959). The process of sub-culture propagation among Japanese macaques *Primates* **2**, 43—60.

Levins, R. and MacArthur, R. (1969). An hypothesis to explain the incidence of monophagy. *Ecology* **50**, 910—911.

March, E. W. (1957). Gorillas of eastern Nigeria. *Oryx* **4**, 30—4.

Merfield, F. G. (1957). "*Gorillas Were My Neighbours.*" Book of the Month, London. (Published in USA as "Gorilla Hunter").

Nishida, T. (1968). The social group of wild chimpanzees in the Mahali Mountains. *Primates* **9**, 167—224.

Passarge, H. (1967). Uber Saumgesellschaften in nordostdeutschen Flachland. *Fedes Rerpertorium.* **74**, 145—158.

Pilbeam, D. R. (1969). Tertiary Pongidae of East Africa: evolutionary relationships and taxonomy. *Peabody Mus. Bull. No. 31.* Yale Univ Press, New Haven, Connecticut.

Richards, P. W. (1964). "The Tropical Rain Forest." Cambridge Univ. Press, Cambridge.

Sabater Pi, J. (1966). Rapport preliminaire sur l'alimentation dans la nature des gorilla du Rio Muni (Oest Africain). *Mammalia* **30**, 235—240.

Sanderson, I. T. (1940). The mammals of the North Cameroons forest area. *Trans. Zool. Soc. Lond.* **24**, 623—725.

Schaller, G. B. (1963). "The Mountain Gorilla." Chicago Univ. Press, Chicago Illinois.

Spinage, C. A. (1972). The ecology and problems of the Volcano National Park, Rwanda. *Biol. Conserv.* **4**, 194—204.

Tuttle, R. H. (1967). Knuckle-walking and the evolution of Hominoid hands. *Amer. J. Phys. Anthropol.* **26**, 171—206.

Verschuren, J. (1975). Wildlife in Zaire, *Oryx* **13**, 25—33.

Vogel, C. (1961). Zur systematische Untergliederung der Gattung *Gorilla* anhand von Untersuchungen der Mandibel. *Z. Säugetierk*, **26**, 65—76.

Subject Index

A

Activity patterns
  of aye-aye, 43–44
  of galagine, 8–11
  of gibbon, 591–593
  of hanuman langur, 475–478
  of red ouakari, 187–189
Age
  determination of
    in douc langur, 527
    in red ouakari, 204
  of sexual maturity
    in hanuman langur, 500
    in red ouakari, 211–212
Aggression
  in aye-aye, 51–52
  in hanuman langur, 498–499
    intertroop, 498
  in red ouakari, 176
    dominance and, 221–222
    interspecific, 224, 225
Agriculture, gorilla and, 629–630
Algeria, Barbary macaque in, 269
  hunting, internal use, and export of, 274
Allogrooming
  in douc langur, 532–533
  in red ouakari, 212–215
Allomaternal behavior, see Aunting
Alopecia, in red ouakari, 195, 197
Amazonia
  definition of, 119–120
  future of conservation in, 157–159
  habitat destruction in, 124–125
  habitat types in, 120–124
  hunting in, 125–127

live capture in, 128–129
  primates of, 129–130
    Callithricidae, 130–138
American Association of Zoological Parks
    and Aquariums (AAZPA), 76
Anatomy, of aye-aye, 40–41
Assam, status of gibbon in, 582–583
Aunting, in lion tamarin, 84, 86

B

Baldness, in red ouakari, 195, 197
Behavioral research, use of baboon in, 405
Biogeography, of lion-tailed macaque,
    302–303
Birth(s), see also Breeding
  in captivity
    among Barbary macaques, 279–280
    among red ouakaris, 176–177
Birth periodicity
  in douc langur, 527–529
  in galagine, 24–25
  in hanuman langur, 501
  in red ouakari, 212
Blood, of baboon, 406–407
Body size
  of baboon, 399
  of douc langur, 516, 518
  of gorilla, 603–604
Brazilian Academy of Sciences, 81
Brazilian Forestry Development Institute
    (IBDF), 78, 80–81
  failure to protect lion tamarin in wild,
    87, 88
Brazilian Nature Conservation Foundation
    (FBCN), 78

639

Taxonomic Index

Tremarctos ornatus, 106, 107, 108
Trichomonas, 175, 232
Tropidurus torquatus, 74
Trypanosoma cruzi, 233
Tupaia, 559
  glis modesta, 521
Turdus rufiventris rufiventris, 74
Tyrannus melancholicus, 107
Tyfo alba, 32

U

Ulmus, 113
Unicornis, 558
Urera hypselodendron, 608, 611, 612
Uropsalis lyra, 106, 107
Urostigma, 74
Urtica, 113
  dioica, 607
  massaica, 612

V

Vandeleuria oleracea rubida, 248
Varanus, 502

Vateria indica, 308
Vepris bilocularis, 296, 298
Vernonia, 610
  adolphi-fredrici, 613
  ruwenzoriensis, 628
Viburnum acuminatum, 296, 298
Vicugna vicugna, 108
Viscum ramosissimum, 296, 299
Vitrus rotundafolia, 182, 183, 191
Viverra, 558
Viverricula indica thai, 522

W

Wrightia tinctoria, 488

X

Xipholena atropurpurea, 90

Z

Ziziphus mauritiana, 489